*CAMBRIDGE MONOGRAPHS*
*ON MECHANICS AND APPLIED MATHEMATICS*

GENERAL EDITORS
G. K. BATCHELOR, Ph.D., F.R.S.
*Professor of Applied Mathematics at the University of Cambridge*

J. W. MILES, Ph.D.
*Professor of Applied Mechanics and Geophysics, University of California, La Jolla*

# THE EARTH'S VARIABLE ROTATION: GEOPHYSICAL CAUSES AND CONSEQUENCES

# THE EARTH'S VARIABLE ROTATION: GEOPHYSICAL CAUSES AND CONSEQUENCES

KURT LAMBECK
*Professor of Geophysics*
*The Australian National University*

CAMBRIDGE UNIVERSITY PRESS
CAMBRIDGE
LONDON NEW YORK NEW ROCHELLE
MELBOURNE SYDNEY

CAMBRIDGE UNIVERSITY PRESS
Cambridge, New York, Melbourne, Madrid, Cape Town, Singapore, São Paulo

Cambridge University Press
The Edinburgh Building, Cambridge CB2 2RU, UK

Published in the United States of America by Cambridge University Press, New York

www.cambridge.org
Information on this title: www.cambridge.org/9780521227698



First published 1980
This digitally printed first paperback version 2005

*A catalogue record for this publication is available from the British Library*

*Library of Congress Cataloguing in Publication data*

Lambeck, Kurt, 1941–
The earth's variable rotation.

(Cambridge monographs on mechanics and applied mathematics)
Bibliography: p.
Includes index.
1. Earth – Rotation. 2. Geophysics. I. Title.
QB633.L35 525'.35 79–7653

ISBN-13 978-0-521-22769-8 hardback
ISBN-10 0-521-22769-0 hardback

ISBN-13 978-0-521-67330-3 paperback
ISBN-10 0-521-67330-5 paperback

To Meg, Fiona and Alexis

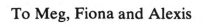

# CONTENTS

*Preface* ix

**1 A perspective** 1

**2 Some physical properties of the Earth** 6
2.1 Elastic deformation 6
2.2 Anelastic deformation 13
2.3 The core 21
2.4 Gravity field 25

**3 Rotational dynamics** 30
3.1 Rigid body rotation 30
3.2 Non-rigid body rotation: general formulation 33
3.3 Rotation of spheroid with fluid core 37
3.4 Rotational deformation 39
3.5 Damped linear motion 44

**4 Excitation functions** 47
4.1 Alternative formulations 47
4.2 Surface loading 50
4.3 Schematic excitation functions 52

**5 The astronomical evidence** 62
5.1 Observations of the Earth's rotation 62
5.2 Length-of-day 73
5.3 Polar motion 85
5.4 New observation techniques 104

**6 Tides** 107
6.1 Earth tides 109
6.2 Ocean and atmospheric tides 119
6.3 Satellite observations of tidal parameters 132
6.4 Tidal perturbations in rotation 138

**7 Seasonal variations** 146
7.1 Atmospheric pressure 147
7.2 Groundwater 159
7.3 Oceans 163
7.4 Winds 166
7.5 Astronomical and geophysical comparisons 179
7.6 Non-seasonal meteorological exictations in l.o.d. 190

**8 The Chandler wobble** 195
8.1 The Chandler period of the solid Earth 196
8.2 Dissipation in the solid Earth 203
8.3 The pole tide 211
8.4 Excitation of the Chandler wobble 220

**9 Decade fluctuations** 245
9.1 Core-mantle coupling 246
9.2 Oceanic and atmospheric contributions 268
9.3 Earthquakes 282
9.4 Conclusions 283

**10 Tidal dissipation** 286
10.1 Introduction 286
10.2 The problem 290
10.3 Astronomical evidence 299
10.4 Ocean tide dissipation 319
10.5 Discussion 334

**11 Paleorotation** 342
11.1 Polar wander 342
11.2 Tidal accelerations in the geological past 352
11.3 Paleontological clocks 360
11.4 Paleorotation and the lunar orbit 388

*Bibliography* 401

*Author index* 437

*Subject index* 445

# PREFACE

The Earth's rotation has occupied the interest of astronomers, mathematicians and geophysicists for at least the last 200 yr. This continued involvement, in what must have initially been thought of as a straightforward problem, is a consequence of the multitude of factors that perturb the rotational motion from what it would be if the Earth were wholly rigid. Forces and deformations in the atmosphere, oceans, crust, mantle and core all perturb the rotation to varying degrees from the idealized rigid body motion, and a complete discussion requires one to delve into many aspects of the Earth and planetary sciences. It is undoubtedly this interdisciplinary aspect that has drawn astronomers, oceanographers, meteorologists and solid Earth physicists to the subject.

Geophysical studies of the Earth's rotation have their roots in the works of Lord Kelvin, Sir George Howard Darwin and Sir Harold Jeffreys amongst others. Munk & MacDonald, in 1960, thoroughly reviewed the subject in their monograph *The Rotation of the Earth; a geophysical discussion.* Their work has dominated the subject ever since and it is unusual to find any aspect of the problem that they have not touched upon. Yet since 1960, and probably because of their very considerable effort at clarifying the subject, much new information, both of an observational and of a geophysical nature, has become available. Some of this is collected in symposia proceedings, in particular those edited by Marsden & Cameron (1966) and by Mansinha, Smylie & Beck (1970). Short reviews of recent results have been given by Rochester (1970, 1973) and Lambeck (1978*b*). Important developments over the last 15 yr include the following.

(1) Precise length-of-day data have become available owing to recent improvements in both universal time and in atomic time. These data have led to high-frequency information in the length-of-

day spectrum that was previously only suspected. They have also led to an improved understanding of year-to-year fluctuations in the seasonal terms and opened up the possibility of using the astronomical data as constraints on the atmospheric circulation.

(2) The pole positions from 1900 to 1970 have been re-evaluated and more reliable Chandler wobble parameters can now be estimated. Precise data have also become available from the analysis of satellite orbit perturbations.

(3) Geophysical knowledge of the Earth's interior has undergone a very considerable revision since 1960, leading to an improved understanding of the geophysical excitation functions that perturb the rotation. Considerably more information has become available on ocean and atmospheric excitation functions.

(4) An important recent literature exists on the interpretation of the ancient and medieval eclipse data, providing more reliable estimates of the secular tidal acceleration of the Earth's spin and of the Moon's orbital motion. The tidal dissipation question has also been re-analysed.

(5) New evidence on the Earth's acceleration over the geological past is available from various sources, and this has further consequences on the past evolution of the Earth–Moon system. This is perhaps the one area of progress not foreshadowed in Munk & MacDonald's monograph. Progress in paleomagnetism has also led to further insight into the question of polar wander.

Developments in space science and technology have spurred new interest in the subject of the Earth's rotation. Precise tracking of satellites for gravitational studies, laser ranging to the Moon for studying the lunar motion, long-baseline interferometry observations for deciphering extra-galactic radio sources, and the precise manoeuvring of interplanetary flights all require an equally precise tracking of the motions of the tracking stations and of the Earth's rotation axis. At the same time, these new techniques permit the rotational motions to be measured with a precision and resolution that will ultimately yield major improvements over conventional astronomical observations. The impact of these new methods on the geophysical discussion has been small until now, but in the next few years numerous new excitation functions can be expected to rise out of the measurement noise. It is hoped that the present discussion will point the way to the interpretation of these new data.

The progress since about 1960 forms the rationale for the present work. As in Munk & MacDonald's monograph, the emphasis is on the geophysical discussion, on the evaluation of the geophysical excitation functions driving the variable rotation so as to explain the astronomical record, and on the use of the latter for obtaining further geophysical insight into the Earth.

Discussions and correspondence with many colleagues over the last few years have contributed to this work and helpful comments on parts of this manuscript have been received from D. L. Anderson, D. E. Cartwright, J. R. Cleary, R. Hide, W. M. Kaula, J. L. Le Mouël, R. T. Merrill, R. J. O'Connell, C. T. Scrutton, F. D. Stacey, F. R. Stephenson and T. Yukutake. I am particularly grateful to F. D. Stacey who commented extensively on the entire manuscript. Perhaps I have not always followed their advice but it is a pleasure to record my gratitude to all of them. The research leading to this book has been supported at various times by the Institut de Physique du Globe of the Université de Paris VI; by the Département des Sciences de la Terre of the Université de Paris VII; by the Institut National d'Astronomie et de Géophysique; by the Délégation Générale à la Recherche Scientifique et Technique, the Centre National d'Etudes Spatiales; and by the Research School of Earth Sciences of the Australian National University. Even if they were not always aware of what I was doing, I do thank them now for their support.

Canberra — Kurt Lambeck
November 1978

ONE

# A PERSPECTIVE

A discussion on the Earth's rotation is conveniently separated into three parts: (i) precession and nutation, (ii) polar motion and (iii) changes in length-of-day (l.o.d.). Precession and nutation describes the rotational motion of the Earth in space and is a consequence of the lunar and solar gravitational attraction on the Earth's equatorial bulge. Polar motion, or wobble, is the motion of the rotation axis with respect to the Earth's crust. Changes in the l.o.d. are a measure of a variable speed of rotation about the instantaneous pole. We are primarily concerned here with the last two components of the motion.

The standard treatment of precession and nutation for a rigid Earth is that by Woolard (1953), but a more comprehensive treatment is by Kinoshita (1977). Observational evidence is discussed by Federov (1963). Further discussions are found in the symposium proceedings edited by Federov, Smith & Bender (1977). The main discrepancies between the observed and theoretical nutations are consequences of the presence of the liquid core. The problem of the precession and nutation of a shell with a liquid-filled spheroidal cavity continues to draw the attention of mathematicians and geophysicists (see, for example, Roberts & Stewartson 1965; Busse 1968; Toomre 1966, 1974). It is touched upon briefly in chapter 3.

Perturbations in the rotation from the rigid body state are caused by motions and deformations of the Earth by a variety of forces. Chapter 2 discusses some general aspects of the deformations of the solid part of the Earth. The theory of the rotation of the deformable Earth is outlined in chapter 3, while chapter 4 consists of some general observations on the form and characteristics of the geophysical excitation functions that drive the variable rotation. The source of information for the variable rotation are threefold, depending on the time scale considered. The primary source is the record of about 150 yr of observations collected by positional astronomers since the introduction of telescopes (chapter 5). Historical records, back to 1000 BC and earlier, contain observations of eclipses, occultations, conjunctions and other configurations of the celestial bodies. These are valuable for studying secular changes in the l.o.d. The deciphering of these records requires a background in Classical, Islamic and Oriental languages, literature and history in addition to a knowledge of positional astronomy (chapter 10). Evidence of the motion of the pole of rotation during the geological past is contained in the paleomagnetic records, while evidence for a changing l.o.d. is found in the fossil records of certain invertebrates. These data are discussed in chapter 11.

Figure 1.1 illustrates the polar motion amplitude spectrum based on astronomical observations taken since the beginning of this century. The spectrum is characterized by two main peaks, one centred at 12 months, the other at about 14 months. The former is a forced wobble driven by seasonal redistribution of mass within and between the atmosphere, oceans, and ground and surface waters (chapter 7). The amplitude of this oscillation is about 0″.10, or about 3 m on the Earth's surface. The second peak is the famous Chandler wobble, also referred to as the free Eulerian precession. It represents a free oscillation of the Earth and is associated with three main problems: (i) Can its period be quantitatively explained? (ii) How is it maintained against dissipation since any free oscillation in physics will eventually be damped? (iii) Where is the rotational energy dissipated? The search for the answers to these questions involves excursions into the physics of the core, mantle, oceans and atmosphere (chapter 8). Seventy years of polar motion data have not been sufficient to resolve in a satisfactory manner the existence of oscillations on a decade time scale (chapter 9). More certain seems to be the observation that the pole is drifting in a direction towards Greenland at a rate of about 0″.002–0″.003 $yr^{-1}$, and this is possibly related to an exchange of mass between the world's oceans and the ice caps, and to post-glacial rebound. These drifts would have been much larger at the time when melting was first initiated. On the geological time scale, paleomagnetic evidence suggests that the rotational pole has wandered over the Earth's surface relative to the continents. Whether it is the pole that has moved or the

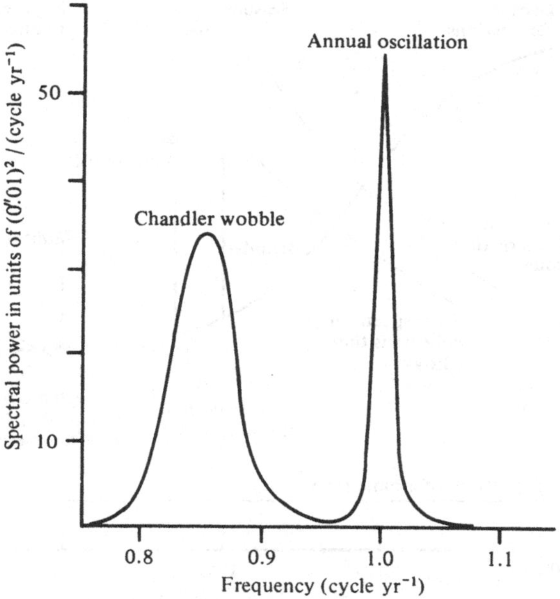


Figure 1.1. Wobble amplitude spectrum based on astronomical observations taken since the beginning of the twentieth century.

continents, or both, remains indeterminate, but the evidence is that there is no compelling reason to invoke polar wander (chapter 11).

Figure 1.2 illustrates the amplitude spectrum of the l.o.d. changes. The telescope observations and the historical records indicate a pronounced secular change: the l.o.d. is increasing at a rate of about 0.001–0.002 s every century. This is mainly a consequence of the work done by the Moon in raising the ocean tides, resulting in a transfer of angular momentum from the Earth's spin to the lunar motion and in a gradual increase in the Earth–Moon distance (chapter 10). Despite the smallness of this acceleration, its consequences – when integrated over geological time – are impressive: if this mechanism operated throughout the past, the Moon would have been very close to the Earth about $1.5\times10^9$ yr ago and the l.o.d. would have been only about 5 h. Due to the exchange of mass between ice caps and oceans and the readjustment of the mantle due to this redistribution fluctuations on a time scale of $10^4$ yr must be expected at times of glaciation and

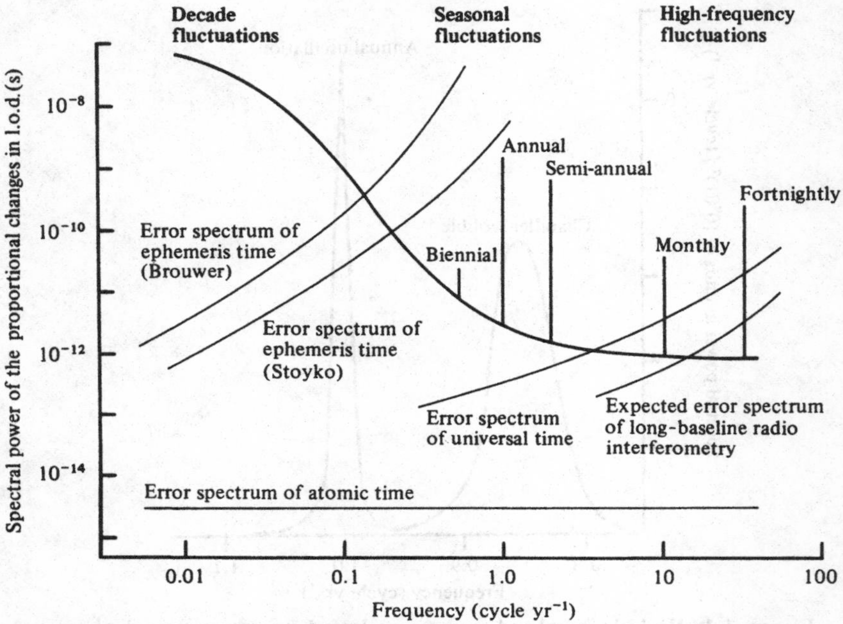


Figure 1.2. Spectrum of l.o.d. changes based on astronomical observations taken since the early nineteenth century. The observational error spectra are discussed in chapter 5.

deglaciation. Decade fluctuations are clearly evident in the l.o.d. records; changes in the l.o.d. of some 4–5 ms within 10–30 yr have occurred on several occasions since the record became reliable in the middle of the nineteenth century. Only the core is sufficiently mobile and has sufficient inertia to explain these fluctuations, the core motions being almost certainly transmitted to the mantle by electromagnetic forces (chapter 9). The decade fluctuations also exhibit trends that are similar to changes in a variety of global climatic indicators and lead to the interesting speculation that climatic changes on the decade time scale may be associated with internal processes. Seasonal changes in l.o.d. including annual, semi-annual and biennial periods, are almost entirely a consequence of a variable strength of the zonal wind circulation (chapter 7). This circulation also appears to be responsible for much of the higher frequency fluctuations in the l.o.d. Periodic tidal deformations (chapter 6) cause fluctuations near 27 d and 14 d. The precise determination of these tidal perturbations may ultimately shed some light on the anelastic behaviour of the Earth's mantle at frequencies beyond the seismic band (chapter 3).

# TWO

# SOME PHYSICAL PROPERTIES OF THE EARTH

Many aspects of the Earth's irregular rotation are consequences of deformations of the solid part of the Earth, and the quantitative evaluation of the excitation functions requires a knowledge of the planet's physical properties. The relevant properties include its shape and gravity, the variation of density and elastic parameters with depth, a measure of anelasticity and viscosity, and electrical conductivity. Information on these properties comes from a variety of sources, including seismology, geodesy and magnetism. These are discussed in most geophysics textbooks (see, for example, Stacey 1977; Kaula *et al.* 1980), and we discuss here only those aspects that are relevant to the subject of the Earth's rotation. Magnetic and electromagnetic data are discussed in chapter 9.

## 2.1 Elastic deformation

### 2.1.1 *Equations of motion*

The theory of the elastic deformations of the Earth is a classic subject in geophysics and is discussed in numerous books and papers (see, for example, Backus 1967; Takeuchi 1967; Jobert 1973*a*). The relevant equations describing the deformations in an element of volume are:

(i) A relation equating the rate of change of linear momentum of an element of volume with applied body forces **F** and surface forces, or stress, **T**

$$\rho\frac{\mathrm{d}\mathbf{v}}{\mathrm{d}t}=\mathbf{F}+\nabla\cdot\mathbf{T},$$

where **v** is the velocity of the element and $\rho$ its density.

(ii) Relations describing the deformation of the element. These consist of a relation between displacements $d_i$ and strain $e_{ij}$,

$$e_{ij} = \tfrac{1}{2}(\partial d_i/\partial x_j + \partial d_j/\partial x_i), \qquad i, j = 1, 2, 3,$$

and a relation between stress and strain. We are mainly concerned with a linear law or Hookean elasticity

$$T_{ij} = \lambda\,\Delta\delta_{ij} + 2\mu e_{ij}.$$

Here $\lambda$ and $\mu$ are the Lamé constants; $\lambda = K - \tfrac{4}{3}\mu$, where $K$ is the incompressibility or bulk modulus, and $\mu$ the rigidity. $\Delta = \sum_k e_{kk}$ is the cubic dilatation.

(iii) An equation of continuity

$$\partial\rho/\partial t + \nabla \cdot (\rho \mathbf{v}) = 0.$$

The initial state of the Earth can usually be considered as one of hydrostatic equilibrium, and in most problems one is concerned with small perturbations from this state. Furthermore, for these studies the Earth can be considered to be spherically symmetrical. Then the linearized form of the perturbation equations of conservation of linear momentum is, for small deformations,

$$\rho\partial^2\mathbf{d}/\partial t^2 = \nabla \cdot \mathbf{T} - \nabla(\rho g\mathbf{d} \cdot \mathbf{e}_r) - \rho\nabla U + g\nabla \cdot (\rho\mathbf{d})\mathbf{e}_r. \quad (2.1.1)$$

$\mathbf{T}$ now represents the non-hydrostatic stress tensor. Density $\rho$ and gravity $g$ refer to the deformed state, and $\mathbf{e}$ is a unit vector with radial component $\mathbf{e}_r$ and tangential component $\mathbf{e}_t$. The potential $U$ is the sum of two parts: $U_1$, the potential of the external force $\mathbf{F}$, and $U_2$, the non-hydrostatic potential of self-attraction after deformation. Inside the body $U = U_1 + U_2$ is subject to Poisson's equation,

$$\nabla^2 U = -4\pi G\nabla \cdot (\rho\mathbf{d}). \quad (2.1.2)$$

Partial solutions to these equations can be found if $U$ is harmonic and of frequency $\sigma$, i.e. if

$$U = \sum_n U'_n(r) S_n\, \mathrm{e}^{j\sigma t}, \quad (2.1.3)$$

where $S_n$ is a surface harmonic of degree $n$. A. E. H. Love showed that in this case the solution of (2.1.1) and (2.1.2) can be written in the form

$$\mathbf{d} = \sum_n [V_n(r) S_n \mathbf{e}_r + W_n(r)\nabla S_n \mathbf{e}_t]\, \mathrm{e}^{j\sigma t}, \quad (2.1.4)$$

where $V_n(r)$ and $W_n(r)$ are unknown functions defining the radial and tangential deformations. For realistic Earth models the solution is further facilitated by transforming (2.1.1) with (2.1.2) and (2.1.4) into six first-order differential equations of a form, first given by Alterman, Jarosch & Pekeris (1959),

$$\mathrm{d}y_\alpha/\mathrm{d}r = a_{\alpha\beta}y_\beta, \qquad \alpha, \beta = 1, \ldots 6 \tag{2.1.5}$$

The six parameters, $y_\alpha$, represent the radial factors in the following quantities:

$\alpha = 1$; radial displacement, $y_1 = V_n(r)$
$\alpha = 2$; radial stress
$\alpha = 3$; tangential displacement, $y_3 = W_n(r)$
$\alpha = 4$; tangential stress
$\alpha = 5$; potential perturbation, $y_5 = U_n'(r)$
$\alpha = 6$; perturbation in potential gradient, $Y_6 = \partial U_n'/\partial r - 4\pi G\rho V_n$.

The $a_{\alpha\beta}$ are functions of the Lamé constants, $\lambda(r)$ and $\mu(r)$, and of $\rho(r)$, $g(r)$, the harmonic degree $n$, and the frequency $\sigma$ of the deformation. The equations (2.1.5) are solved with boundary conditions relevant to the particular problem discussed. They include (i) regularity at the origin, (ii) that stresses vanish across free surfaces, (iii) continuity of deformation and stress across internal surfaces of discontinuity, and (iv) that internal and external gravitational potentials and their respective gradients must be equal at free surfaces and across surfaces of discontinuity (see, for example, Chinnery 1975).

A number of geophysical problems can be resolved with the above formulation. These include free oscillations, tidal deformations, response to loading of the Earth by surface or internal loads, and rotational deformations. Free oscillations represent solutions of (2.1.5) if there is no external force acting on the body. The so-modified equations give non-trivial solutions for only certain values of $\sigma$; the eigenvibration or free-oscillation frequency. Measurements at the surface of the Earth give the frequencies and relative amplitudes of these oscillations when they are excited by large earthquakes. The problem is discussed by Alterman *et al.* (1959), Gilbert (1972), and others. In the problem of the tidal deformations, the equations (2.1.5) are solved for a planet subject to a known external potential that does not load the surface. Stresses across the free surface must vanish. The observed quantities are the deformation of this surface $r = R$ and the change in the gravitational potential at $r = R$. This problem has been discussed for realistic Earth models by Takeuchi (1950), Pekeris & Accad (1972) and others. Peltier (1974) has extended the theory to a Maxwell rheology. Rotational deformation problems are very similar to the tidal problems, the tidal potential being replaced by the potential of the centrifugal force. When the Earth is loaded at its free surface, the potential $U_1(r)$ represents the gravitational potential of the load. The boundary conditions differ from those of the tide problem in that the stress is now continuous across the loaded surface $r = R$. Longman (1963) and Farrell (1972) discuss the problem. A related question is for internal loading. The Earth's gravity field indicates that important lateral density anomalies occur within the mantle. These will stress the Earth relative to the hydrostatic equilibrium state. Kaula (1963) has discussed the problem of estimating the stress and density in the Earth corresponding to a known external gravitational potential.

### 2.1.2 *Love numbers*

In the tidal, loading, and rotation problems with which we are concerned here, the response to an applied potential of the form (2.1.3) is assumed to be linear: the deformations are also harmonic with the same degree $n$ as $U_n$. Then with

$$U_1 = \sum_n U_{1,n} = \sum_n U'_{1,n}(r) S_n \mathrm{e}^{\mathrm{j}\sigma t},$$

$$\begin{pmatrix} V_n(r) \\ W_n(r) \\ U'_{2,n}(r) \end{pmatrix} = U'_{1,n}(r) \begin{pmatrix} h_n(r)/g \\ l_n(r)/g \\ k_n(r) \end{pmatrix}.$$

In terms of the $y_\alpha$ of (2.1.5)

$$\left. \begin{aligned} y_1(r) &= h_n(r) U'_{1,n}(r)/g(r), \\ y_3(r) &= l_n(r) U'_{1,n}(r)/g(r), \\ y_5(r) &= [1 + k_n(r)] U'_{1,n}(r). \end{aligned} \right\} \qquad (2.1.6)$$

On the surface $r=R$,

$$\left.\begin{aligned} \mathbf{d}_r(R) &= \mathbf{e}_r y_1(R) S_n e^{j\sigma t} = [h_n(R)/g] U_{1,n}(R)\mathbf{e}_r, \\ \mathbf{d}_t(R) &= \mathbf{e}_t y_3(R)(\nabla S_n) e^{j\sigma t} = [l_n(R)/g] \nabla U_{1,n}(R)\mathbf{e}_t, \\ \Delta U(R) &= y_5(R) S_n e^{j\sigma t} = [1+k_n(R)] U_{1,n}(R). \end{aligned}\right\} \quad (2.1.7)$$

The constants $h_n(R)$, $k_n(R)$ and $l_n(R)$, or simply $h_n$, $k_n$, $l_n$ are referred to as Love numbers of degree $n$. The $h$ and $k$ were first introduced by Love in 1909 and the $l$ was introduced by T. Shida in 1912. This notation is usually reserved for Love numbers that define the deformation of a radially symmetric and elastic Earth in the absence of loading. If the potential does load the Earth, we denote the appropriate Love numbers by $h'$, $k'$, $l'$. These are often referred to as load Love numbers or load deformation coefficients. Similar parameters, $h''_n$, $k''_n$, $l''_n$, can be introduced to define the elastic deformations due to tangential stresses applied at the Earth's surface. These Love and load numbers are not all independent, Molodensky (1977) having demonstrated that for any radially symmetrical model and for any degree $n$, there exist three relations between them. They are

$$\left.\begin{aligned} k'_n &= k_n - h_n \\ h''_n &= 3n[(n+1)/(2n+1)](l-l') \\ 1+k''_n &= 3n[(n+1)/(2n+1)]l. \end{aligned}\right\} \quad (2.1.8)$$

Amongst the early studies of the Earth's tidal deformations, the work of Kelvin is most important. In his well-known estimate that the Earth's mean rigidity is greater than that of steel, Kelvin treated the Earth as homogeneous and incompressible. Inherent in Kelvin's treatment are two parameters that describe the deformation at the surface of the body in response to a harmonic potential of degree 2, and that relate to the density and rigidity. These correspond to the Love numbers $h_2$ and $k_2$ for the Kelvin Earth model, the homogeneous incompressible sphere. The three Love numbers for this model are

$$h_2(R) = \frac{5/2}{1+\tilde{\mu}}, \qquad k_2(R) = \frac{3/2}{1+\tilde{\mu}}, \qquad l_2(R) = \frac{3/4}{1+\tilde{\mu}}, \quad (2.1.9a)$$

where

$$\tilde{\mu} = (19/2)\mu/\rho g R. \tag{2.1.9b}$$

Also

$$h_2'(R) = -\frac{5/3}{1+\tilde{\mu}}, \qquad k_2'(R) = -\frac{1}{1+\tilde{\mu}}. \tag{2.1.9c}$$

The unrealistic nature of these assumptions was well understood but was necessary to solve an already complex problem. More detailed Earth models were studied by G. Herglotz, S. Hough, Love and others early in this century, but the solution for realistic models in which $\rho$, $\mu$ and $K$ vary with radius has only become possible with the introduction of electronic computers. This was first attempted by Takeuchi (1950) and later by Takeuchi, Saito & Kobayashi (1962), Longman (1963), Farrell (1972) and Pekeris & Accad (1972).

Seismological observations provide the principal information from which the elastic structure of the Earth's interior can be deduced. These data include free oscillations, surface waves and body wave information. Inversion of the ensemble determines the velocities $V_p$, $V_s$ of compressional and shear waves and the density, as a function of depth. The velocities relate to $K(r)$, $\mu(r)$ and $\rho(r)$ by

$$V_p^2(r) = [K(r) + \tfrac{4}{3}\mu(r)]/\rho(r), \qquad V_s^2(r) = \mu(r)/\rho(r).$$

Early Earth models by Bullen, Gutenberg and others were based on body wave data only. These data do not give a unique determination of $K(r)$, $\mu(r)$ and $\rho(r)$, and either an Adams–Williamson relation between the velocities and density or an equation of state is required (see, for example, Bullen 1963). In more recent models, free-oscillation observations have also been used which impose further restrictions on the distribution of $\rho$, $\mu$ and $K$ although the inversion of the data remains non-unique (Cleary & Anderssen 1979). Two recent Earth models are illustrated in figure 2.1. The model 1066A by Gilbert & Dziewonski (1975) is based on a large number of free-oscillation periods and to a lesser degree on differential travel times of body waves. $V_p$, $V_s$, and $\rho$ vary quite smoothly with radius in this model due to the inability to resolve fine detail with these global data. The second model C2 by Anderson & Hart (1976) uses, in addition to these data, high-resolution body

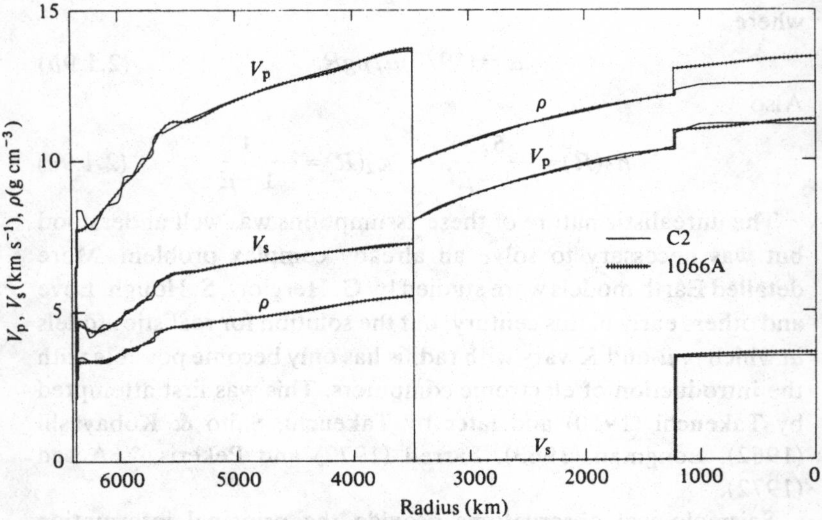


Figure 2.1. Models C2 from Anderson & Hart (1976) and 1066A from Gilbert & Dziewonski (1975) of the Earth's radial seismic velocity and density structure.

wave information that is more path dependent than the normal modes. These models do not differ in their general form from the earlier Bullen and Gutenberg models. Combined with laboratory data and with petrological and geochemical arguments, these velocity and density profiles provide constraints on the composition of the various parts of the solid Earth. Recent discussions are provided by Ringwood (1975) and Anderson (1977).

Table 2.1 summarizes some results by Farrell (1972) for the Love number calculations. Three Earth models have been considered: (i) a Gutenberg–Bullen A model, (ii) the same basic model, but with the upper 1000 km replaced by a typical oceanic structure with a pronounced low-velocity layer similar to that in the model C2 illustrated in figure 2.1, and (iii) the same, but with a continental shield structure for the upper 1000 km. The low-degree Love numbers differ at most by 0.5% for the oceanic and continental upper mantle models: they do not appear to be very sensitive to mantle structure and Farrell's results are unlikely to be much modified if the more recent Earth models are used. Load deformation coefficients computed by Farrell for the Gutenberg–Bullen

Table 2.1. *Theoretical Love numbers of degree n computed by Farrell (1972) for three different Earth models: Gutenberg–Bullen (G–B) Earth model (first line for each n), an Earth model with a typical oceanic upper-mantle structure (second line for each n), and an Earth model with a typical shield upper mantle (third line for each n)*

| | $n$ | $h_n$ | $l_n$ | $k_n$ |
|---|---|---|---|---|
| G–B Earth model | 2 | 0.6114 | 0.0832 | 0.3040 |
| Oceanic mantle | | 0.6149 | 0.0840 | 0.3055 |
| Shield mantle | | 0.6169 | 0.0842 | 0.3062 |
| | 3 | 0.2891 | 0.0145 | 0.0942 |
| | | 0.2913 | 0.0145 | 0.0943 |
| | | 0.2923 | 0.0147 | 0.0946 |
| | 4 | 0.1749 | 0.0103 | 0.0429 |
| | | 0.1761 | 0.0103 | 0.0424 |
| | | 0.1771 | 0.0104 | 0.0427 |

model are given in table 2.2. They agree with the relations (2.1.8). Also indicated are the tangential stress coefficients $h''_n$ and $k''_n$ deduced from the relations (2.1.8).

## 2.2 Anelastic deformation

Numerous geophysical observations indicate that the Earth's response to forces is not purely elastic: seismic waves decay with time as does the Chandler wobble, the tidal response lags the stress-producing potential, and the Earth's topography is generally close to isostatic balance. Thus the actual response of the Earth to a stress field will be partly elastic, partly inelastic, and rheological laws other than the simple linear stress–strain, or Hookean, behaviour need to be considered. The rate of decay of seismic waves with distance and time is the principal source of information for the mantle's anelastic behaviour and can yield information about the thermal and crystalline structures of the planet. A number of mechanisms by which seismic energy may be dissipated have been proposed, but further experimentation under high pressure and

Table 2.2. *Load deformation coefficients of degree n computed by Farrell (1972) for a Gutenberg–Bullen Earth model, and the tangential stress deformation coefficients based on the relations (2.1.8)*

| $n$ | $-h'_n$ | $nl'_n$ | $-nk'_n$ | $h''_2$ | $-k''_2$ |
|---|---|---|---|---|---|
| 1 | 0.290 | 0.113 | 0 | — | — |
| 2 | 1.001 | 0.059 | 0.615 | 0.192 | −0.948 |
| 3 | 1.052 | 0.223 | 0.585 | −0.306 | −0.926 |
| 4 | 1.053 | 0.247 | 0.527 | −0.345 | −0.931 |

temperature relevant to mantle conditions is necessary to decide upon their relative importance. Various mechanisms have been reviewed by Gordon & Nelson (1966), Anderson (1967), Jackson & Anderson (1970), Weertman (1970) and others. Gordon & Nelson suggest that at seismic frequencies, the mechanisms in order of increasing importance are (i) viscous-grain boundary sliding, (ii) stress-induced atomic ordering, and (iii) dislocation damping. Stress-induced fluid flow in a partially molten material, thermoelastic damping, and high-temperature internal-friction background may also contribute to the overall damping of seismic waves. The first two mechanisms are characteristically frequency dependent and amplitude independent, while dislocation damping is frequency dependent for low-amplitude stresses and becomes frequency independent for large-amplitude stresses. For most of the above-mentioned mechanisms the dissipation behaviour is essentially linear, and the energy loss is conveniently discussed in terms of the specific dissipation function $Q^{-1}$ defined, for large $Q$, as

$$Q^{-1} = (1/2\pi)\Delta E/E, \tag{2.2.1a}$$

where $E$ is the maximum value of the elastic energy stored per unit cycle per unit volume and $\Delta E$ is the amount of energy dissipated per unit cycle per unit volume. In any non-linear system a detailed consideration of the energy loss is required to compute $Q^{-1}$ and this function then loses much of its usefulness.

The definition (2.2.1*a*) is not always very useful since there is no convenient way for determining the elastic energy. An alternative definition is

$$Q^{-1} = (1/2\pi)/\Delta E/2\langle E\rangle, \qquad (2.2.1b)$$

where $\langle E\rangle$ is the average stored energy. Another useful definition is

$$Q^{-1} = (1/2\pi)\Delta E_k/E_k, \qquad (2.2.1c)$$

where $E_k$ is the peak kinetic energy in the cycle and $\Delta E_k$ is the change in $E_k$ over one cycle. For large $Q$ these definitions are equivalent (O'Connell & Budiansky 1978).

The most general linear relation between stress, strain and their first derivatives is (Fung 1965; Malvern 1969)

$$T + \tau_T \dot{T} = E_R(\varepsilon + \tau_\varepsilon \dot{\varepsilon}), \qquad (2.2.2)$$

where $E_R$ is a deformation modulus, $\tau_T$ is the stress relaxation time under constant strain, and $\tau_\varepsilon$ is the strain relaxation time under constant stress. A solid described by such a relation is called a 'standard linear solid'. If the stress is harmonic, $T = T_0 e^{j\sigma t}$, the deformation of a linear solid is also harmonic, $\varepsilon = \varepsilon_0 e^{j\sigma t}$. Substituting into (2.2.2) gives

$$T_0/\varepsilon_0 = E_R(1 + j\tau_\varepsilon\sigma)/(1 + j\tau_\varepsilon\sigma) = \mathcal{M}^* = B\, e^{j\theta},$$

where $\mathcal{M}^* = [\mathcal{M}_0 + \mathcal{M}_1(\sigma)] + j\mathcal{M}_2(\sigma)$ is a complex modulus. Also

$$B^2 = E_R(1 + \tau_\varepsilon^2\sigma^2)/(1 + \tau_T^2\sigma^2).$$

For small dissipation

$$\tan\theta = Q^{-1} = \frac{\mathrm{Im}(\mathcal{M}^*)}{\mathrm{Re}(\mathcal{M}^*)} = \frac{\sigma(\tau_\varepsilon - \tau_T)/(1 + \sigma^2\tau_\varepsilon^2)}{1 - \sigma^2\tau_\varepsilon(\tau_\varepsilon - \tau_T)/(1 + \sigma^2\tau_\varepsilon^2)}$$
$$= (\tau_\varepsilon - \tau_T)\sigma/(1 + \tau_T\tau_\varepsilon\sigma^2). \qquad (2.2.3)$$

Strain lags stress by an angle $\theta$. For small dissipation

$$\mathcal{M}_1(\sigma) \ll \mathcal{M}_0,\ \mathcal{M}_2(\sigma) \ll \mathcal{M}_0,$$

and the magnitude of the modulus at frequency $\sigma$ compared with the elastic modulus $\mathcal{M}_0$ is

$$\mathcal{M}(\sigma)/\mathcal{M}_0 \simeq 1 + \mathcal{M}_1(\sigma)/\mathcal{M}_0. \qquad (2.2.4)$$

With $(\tau_\varepsilon\tau_T)^{1/2} = \sigma_0^{-1}$ and

$$\alpha = \tfrac{1}{4}(\tau_\varepsilon - \tau_T)/\tau_\varepsilon\tau_T$$

(2.2.3) becomes

$$Q^{-1} = 4\alpha\sigma/(\sigma^2 + \sigma_0^2), \qquad (2.2.5)$$

which is the usual definition corresponding to the mechanical linearly damped oscillator model in which damping is proportional to velocity (see, for example, Feynman *et al.* 1963); $\alpha^{-1}$ is the relaxation time and $\sigma_0$ the resonance frequency of the model. If the frequency of the applied stress equals the resonance frequency,

$$Q^{-1} = Q_0^{-1} = 2\alpha/\sigma_0. \qquad (2.2.6)$$

It is readily shown that, for small damping,

$$Q_0^{-1} = \Delta\sigma/\sigma_0, \qquad (2.2.7)$$

where $\Delta\sigma$ is the width of the peak in $B^2$ at the half-power points. With (2.2.5)

$$Q_0^{-1}(\sigma) = 2Q^{-1}\sigma\sigma_0/(\sigma^2 + \sigma_0^2) \qquad (2.2.8)$$

and dissipation is strongly frequency dependent, with the maximum occurring at resonance.

Many of the frequency-dependent dissipation processes can be described by relations of this type (Liu, Anderson & Kanamori 1976; O'Connell & Budiansky 1977). For a continuous distribution of relaxation times, the internal friction for the standard linear solid becomes (Liu *et al.* 1976)

$$\tan\theta = \frac{\displaystyle\int_0^\infty \mathrm{d}\tau_T \int_0^\infty \mathrm{d}\tau_\varepsilon D(\tau_T, \tau_\varepsilon)\frac{\sigma(\tau_\varepsilon - \tau_T)}{1 + \sigma^2\tau_\varepsilon^2}}{1 - \displaystyle\int_0^\infty \mathrm{d}\tau_T \int_0^\infty \mathrm{d}\tau_\varepsilon D(\tau_T, \tau_\varepsilon)\frac{\sigma^2\tau_\varepsilon(\tau_\varepsilon - \tau_T)}{1 + \sigma^2\tau_\varepsilon^2}}, \qquad (2.2.9)$$

where $D(\tau_T, \tau_\varepsilon)$ is the distribution function of the mechanism.

Grain boundary and defect phenomena, as well as the inter-atomic forces, are critically temperature dependent, a rise in temperature causing a strong decrease in the relaxation time. Pressure also plays an important role – at increased pressure the relaxation time increases – and the combined temperature–pressure effect on the specific dissipation for a single thermally activated process will lead to a variation of the dissipation with depth. Furthermore, depending on the actual mechanism, damping is also sensitive to grain size, to the orientation and densities of dis-locations and liquid-filled cavities, and to the thermal and elastic contrasts between grains. A consequence is that several mechanisms may contribute to the dissipation at any one time; the dominance of one mechanism over another may vary with depth as the material characteristics change with temperature, pressure, composition and mineralogy. Thus, rather than a single absorption band, the Earth may possess many bands or a single, very broad, band representing the coalescence of many individual bands. This is indeed suggested by the seismic data (Anderson *et al.* 1977). Seismic observations, the absorption of body waves and the decay of free oscillations, indicate that over a wide range of periods from a few seconds to an hour, $Q$ can be interpreted as being frequency independent but variable with depth (Anderson 1967; S. W. Smith 1972). This interpretation is not, however, unique and the data can be equally satisfied with a strongly frequency-dependent $Q$. This frequency independence is quite different from the strong frequency dependence that would be expected if a single linear mechanism dominated. Anderson *et al.* suggest that one or several dissipation mechanisms occur whose resonance frequencies vary with depth, and that the absorption band is sufficiently broad to encompass the seismic frequencies (see also Kanamori & Anderson 1977). If, in (2.2.9), the relaxation times vary between the limits $\tau_1$ and $\tau_2$ according to $\tau^{-1}$, then $\tan\theta$ is very nearly constant between these limits (Liu *et al.* 1976). Thus the suggested frequency independence of $Q^{-1}$ does not require an intepretation in terms of a non-linear model but can be considered as a consequence of the super-positioning of a large number of linear models. For this constant-$Q$ model and to first order in $Q^{-1}$ the modulus dependence on frequency (2.2.4) reduces to (Kolsky 1956; Kanamori & Anderson 1977)

$$\mathcal{M}(\sigma_2)/\mathcal{M}(\sigma_1) = \mu(\sigma_2)/\mu(\sigma_1) = 1 + (2/\pi Q)\ln(\sigma_2/\sigma_1). \tag{2.2.9a}$$

Experiments by Brennan & Stacey (1977) on granites and basalts bear out the validity of this equation at low temperatures and pressures. Equation (2.2.9$a$) determines, for example, the correction to $\mu$ that should be applied if Love numbers at frequency $\sigma_2$ are computed from body wave or surface wave seismic data of frequency $\sigma_1$; $\mu$ decreases with increasing period. Thus, for the

Kelvin Earth model (2.1.9), the Love number $k_2$ at frequency $\sigma_2$ becomes

$$k_2(\sigma_2) = \frac{3/2}{1+(19/2)[\mu(\sigma_1)/\rho g R][1+(2/\pi Q)\ln(\sigma_2/\sigma_1)]}, \tag{2.2.10a}$$

the magnitude of the Love number increasing with decreasing frequency.

Figure 2.2 illustrates recent estimates of the $Q$-structure of the Earth at seismic frequencies. These results are for shear waves, dissipation by compression being small throughout the mantle. Sailor & Dziewonski (1978) propose generally lower shear $Q$-values than Anderson & Hart (1978*a*). In a later paper, Anderson & Hart (1978*b*) revised their model significantly and find results similar to the Sailor & Dziewonski model in that the lower-mantle shear $Q$ of their earlier model is now much reduced (figure 2.2). The models are characterized by a low-$Q$ zone underneath the lithosphere and, in the case of the Anderson & Hart models, by a second low-$Q$ zone at the base of the mantle. For the spheroidal mode of degree 2, the models give a $Q$ of about 600 compared with observed values ranging from 350 to 600. This oscillation samples preferentially the lower mantle and the $Q$-value is only partly dependent upon the upper mantle. This harmonic is most comparable with dissipation of tidal and rotational energy as these are also defined by second-degree harmonics. Whether the broad absorption band encompasses these lower frequencies is unknown and of considerable interest in understanding the dissipation mechanisms of the Earth. Anderson & Hart suggest that at the lower frequencies, the low-$Q$ zone in the upper mantle may move higher into the lithosphere, but this is unlikely to change the average mantle $Q$ significantly.

Table 2.3 summarizes percentage changes in $k_2$ for various tidal and rotational frequencies and for different values of $Q$. For $Q \simeq 600$, $\Delta k_2 \simeq 1\%$ for diurnal frequencies and 1.4% for annual frequencies. For $Q \simeq 300$, $\Delta k_2 \simeq 2\%$ and 3% for these two frequencies. Clearly, dispersion effects may be important if we can extrapolate the seismic absorption band to these frequencies. The load deformation coefficients will also be frequency dependent.

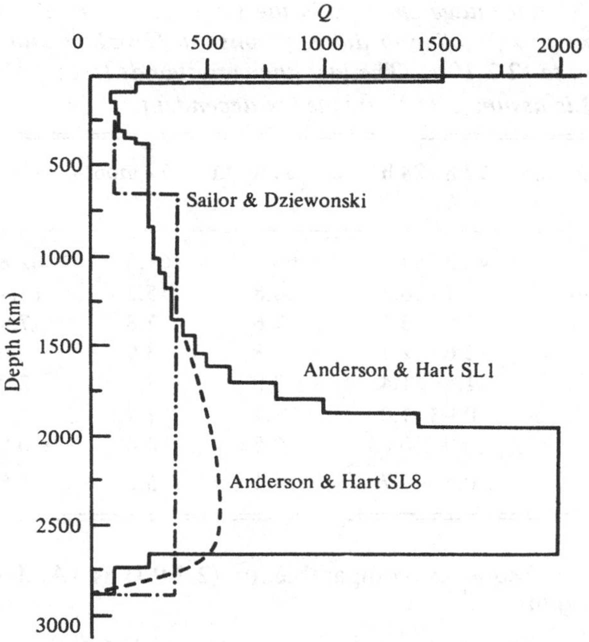


Figure 2.2. The $Q$-structure of the Earth for shear waves at seismic frequencies: model SL1 of Anderson & Hart (1978*a*), SL8 of Anderson & Hart (1978*b*), and the model by Sailor & Dziewonski (1978).

The seismic frequencies cover nearly four decades, from 1 Hz to $3\times10^{-4}$ Hz. Tidal frequencies between $2\times10^{-5}$ Hz and $10^{-5}$ Hz are still relatively near to the lower part of the seismic spectrum and here the dissipation mechanisms may be similar. More problematical is dissipation at the Chandler wobble frequency of $3\times10^{-8}$ Hz, some four decades lower than the seismic free-oscillation frequencies. Anderson & Minster (1979) suggest that high-temperature background may be a dominant dissipation mechanism over a wide frequency range in regions of the Earth where the temperatures exceed about one-half of the melting point temperature. For this mechanism, $Q$ is proportional to $(\text{frequency})^{\alpha}$, where $\alpha$ is of the order of 0.2 to 0.4 and, if valid, the $Q$ at the wobble frequency may be significantly less than the seismic $Q$s. For such a frequency-dependent model the dependence of

Table 2.3. *Percentage changes in the Love number $k_2$ due to dispersion as a function of different constant Q-values and period according to* (2.2.10*a*). *The last line corresponds to* (2.2.10*b*) *in which Q is assumed to be frequency dependent.*

| *Q* / Period | 12 h | 24 h | 6 month | 12 month | 430 d |
|---|---|---|---|---|---|
| 10 | 42.3 | 44.5 | 60.6 | 61.6 | 62.8 |
| 50 | 9.7 | 10.3 | 14.8 | 15.2 | 15.4 |
| 100 | 5.0 | 5.2 | 7.6 | 7.8 | 7.9 |
| 200 | 2.6 | 2.7 | 3.8 | 3.9 | 4.0 |
| 400 | 1.3 | 1.4 | 2.0 | 2.0 | 2.0 |
| 600 | 0.93 | 1.0 | 1.4 | 1.4 | 1.4 |
| 1000 | 0.60 | 0.63 | 0.80 | 0.80 | 0.80 |
| 600 | 0.5 | 0.7 | 4.1 | 5.2 | 5.5 |

rigidity on frequency comparable to (2.2.9*a*) is (Anderson & Minster 1979)

$$\mu(\sigma_2)/\mu(\sigma_1) = 1 + \cot\left(\tfrac{1}{2}\alpha\pi\right)(\sigma_2/\sigma_1)^{\alpha} Q_1^{-1}, \qquad (2.2.9b)$$

where $\sigma_1$ and $Q_1^{-1}$ are the values appropriate to the seismic frequencies. The Love number $k_2$ follows as

$$k_2(\sigma_2) \simeq \frac{3/2}{1 + (19/2)[\mu(\sigma_1)/\rho g R][1 + \cot(\tfrac{1}{2}\alpha\pi)(\sigma_2/\sigma_1)^{\alpha} Q_1^{-1}]} \qquad (2.2.10b)$$

The last line in table 2.3 summarizes the percentage change in $k_2(\sigma_2)$ for different values of frequency $\sigma_2$, using for $Q_1$ the spheroidal free-oscillation value of 600. At very low frequencies, information on anelasticity comes from post-glacial rebound studies; the isostatic re-adjustment of the crust and mantle after the melting of the last Pleistocene ice sheets of the northern regions of Europe and North America provides the main observational evidence for mantle viscosity. Recent discussions are by Cathles (1975), Peltier (1976), Peltier & Andrews (1976) and Kaula (1979). At these low frequencies, anelastic deformation probably follows a steady-state response: if the temperatures equal or exceed one-half the melting point temperature, the deformation of the material subject to a constant stress consists of an initial instantaneous response followed by a nearly constant deformation rate (Stacey 1977). Such a response can be approximated by a Maxwell rheology, which is a special case of (2.2.2), or

$$\dot{\varepsilon} = 1/\mu \dot{T} + 1/\eta T, \tag{2.2.11}$$

in which the rigidity $\mu$ governs the initial elastic response and the dynamic viscosity $\eta$ governs the long-term behaviour. $\tau = \mu/\eta$ is the characteristic relaxation time. Initially the response is elastic and asymptotically approaches that of a Newtonian viscous fluid. Visco-elastic problems described by linear rheologies such as (2.2.2) or (2.2.11) are solved by a formalism that is very similar to that of the equivalent elastic problems, by using the correspondence principle (Fung 1965; Peltier 1974). Applying it to a spherical and homogeneous Maxwell body that is subject to a periodic potential, one obtains complex Love numbers

$$\left.\begin{aligned} \mathbf{k}_2^{\mathrm{M}}(\sigma) &\simeq k_2\{1 + \mathrm{j}[\tilde{\mu}/(1+\tilde{\mu})](1/\sigma\tau)\}, \\ \mathbf{h}_2^{\mathrm{M}}(\sigma) &\simeq h_2\{1 + \mathrm{j}[\tilde{\mu}/(1+\tilde{\mu})](1/\sigma\tau)\}, \end{aligned}\right\} \tag{2.2.12}$$

where $k_2$ and $h_2$ are the Love numbers for the equivalent elastic model; $\tilde{\mu}$ is defined by (2.1.9). The lag of the response $\theta$ relates to the Maxwell-body $Q$ according to

$$Q_{\mathrm{M}}^{-1} \simeq \tan(-\theta) = [\tilde{\mu}/(1+\tilde{\mu})](1/\sigma\tau)$$

and $Q_{\mathrm{M}}$ varies linearly with frequency while the dissipation remains small. Similar relations can be obtained for other linear solids. For a Kelvin–Voigt solid, for example,

$$Q_{\mathrm{K-V}}^{-1} = [\tilde{\mu}/(1+\tilde{\mu})]\sigma\tau$$

and $Q_{\mathrm{K-V}}$ varies inversely with frequency.

## 2.3 The core

The Earth's core consists of two main regions: an outer core, 3485 km $\gtrsim r \gtrsim$ 1215 km, and an inner core, $r \lesssim$ 1215 km. From the absence of transmitted shear waves and from the nature of the reflection coefficients of ScS and PcP waves, the outer core is generally interpreted as being completely molten. The inner core is believed to be solid. Seismic body wave data are ambiguous on this point but the inversion of free-oscillation data yields Earth models that seem to require a solid inner core (Gilbert & Dziewonski 1975). A review of the core structure and properties is given by Jacobs (1975). Anderson (1977) reviews recent evidence for the core composition. Apart from the electromagnetic properties, two aspects of the core are of some importance in studies of the Earth's rotation: the degree to which the core is stratified and the core viscosity.

### 2.3.1 *Density stratification*

A useful measure of stratification in the core is provided by the Brunt–Vaisala frequency $N(r)$, defined as

$$N^2(r) = -g(r)\{[1/\rho(r)]\partial\rho(r)/\partial r + \rho(r)g(r)/K(r)\}, \quad (2.3.1)$$

or equivalently, by a dimensionless parameter $\beta(r)$ introduced by Pekeris & Accad (1972):

$$\begin{aligned}\beta(r) &= [K(r)/g(r)\rho^2(r)]\partial\rho(r)/\partial r + 1 \\ &= [-K(r)/g^2(r)\rho(r)]N^2(r). \quad (2.3.2)\end{aligned}$$

Physically, $N^2(r)$ represents the frequency with which a small element of fluid will oscillate adiabatically when slightly displaced radially from its rest position. If the core is chemically homogeneous with vanishing $\mu$, and the density stratification is a consequence of adiabatic compression, the Adams–Williamson relation,

$$(1/\rho)\partial\rho/\partial r = -\rho g/K,$$

holds, and $N^2(r) = 0$. The core is then said to be neutrally stratified. For a stably stratified core $N^2 > 0$ while for an unstable stratification $N^2 < 0$. Figure 2.3 illustrates $N^2(r)$ for the two Earth models considered in section 2.1: model 1066A of Gilbert & Dziewonski (1975) and model C2 of Anderson & Hart (1976). $N^2(r)$ depends critically on the density gradient, but, as knowledge of the density is not independent of $\mu$ and $K$ and as resolution in the core is poor, $N^2(r)$ remains poorly known. Only in the outer part of the core is its value likely to be significant. Both models suggest that $N^2(r)$ is negative in the outer part of the core and that its sign may change. The earlier Gutenberg model shows a similar behaviour. This suggestion of unstable stratification is what may be expected from geodynamo considerations since, in the alternative case, radial mantle convection and dynamo action would be inhibited. Pekeris

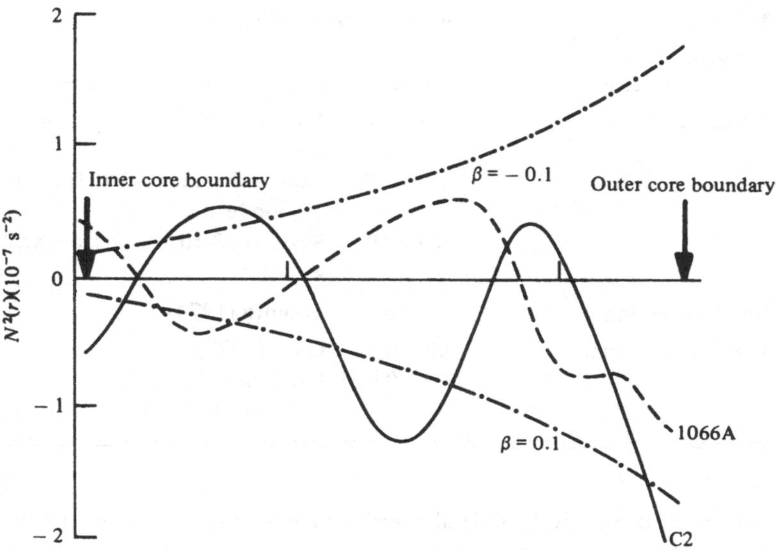


Figure 2.3. The Brunt–Vaisala frequency $N^2(r)$ for core models C2 and 1066A; $\beta$ is the dimensionless parameter defined by (2.3.2).

& Accad suggest that $\beta(r) = -0.2$, apparently because such a value yields a period for the $S_2^0$ spheroidal mode that still agrees with observations. Such a value, however, is quite different from that suggested by the 1066A and C2 models of the core.

Pekeris & Accad (1972) have investigated the core dynamics for three cases of stratification. They find that in all cases the Love numbers vary slightly with frequency according to

$$h_2(\sigma) = h_2(\sigma = 0) + \alpha\sigma^2,$$

and similarly for $k_2$ and $l_2$. Such a frequency dependence was suggested earlier by Jeffreys & Vicente (1966). For $N^2 = 0$, the Love numbers $h_2$, $k_2$ decrease by about 0.5% for periods of 12 h to infinity. The unstable and neutrally stable cores do not exhibit free oscillations with periods longer than about 54 min, corresponding to the period of the fundamental spheroidal mode of degree 2 of the Earth. For the stable stratified model, additional core oscillations occur with periods ranging from 54 min up to infinity, and the Love numbers will exhibit resonance at these periods. Pekeris & Accad, however, have neglected the Coriolis force in their equations of

Table 2.4. *Summary of recent estimates of the kinematic core viscosity* ($cm^2\ s^{-1}$)

| | | |
|---|---|---|
| Seismic estimates | $<5\times10^7$ | Jeffreys (1970) from body waves |
| | $10^8$–$10^{10}$ | Sato & Espinosa (1967) from ${}_0T_2$ mode |
| | $<7\times10^9$ | Suzuki & Sato (1970) from ScS waves |
| Nutation estimate | $\leq10^5$ | Toomre (1974) |
| Theoretical estimates | 0.04–0.20 | Gans (1972) |
| | 0.003–0.005 | Gubbins (1976) |
| | $>1$ | Bukowinski & Knopoff (1976) |

motion and Smylie (1974) showed that with the inclusion of this force the periods of the core modes are compressed to below 1.5 d (see also Crossley 1975; Johnson & Smylie 1977).

### 2.3.2 *Core viscosity*

Table 2.4 summarizes some recent estimates of core viscosity. Rochester (1970) summarized earlier estimates. Basically two observations give constraints on the core viscosity: the damping of seismic waves, and the precession and nutations of the Earth in space. These give only upper limits. Other estimates, based on theories of the behaviour of liquid iron at core pressure and temperatures, give much lower values. Seismic observations indicate an upper limit of about $10^8$–$10^9$ cm$^2$ s$^{-1}$ for the kinematic viscosity. These include observations of the propagation of body waves through the core and of the reflection coefficients of ScS waves. The nutation observations indicate an upper limit of the order $10^5$ cm$^2$ s$^{-1}$, from the phase lead of the forced 18.6-yr nutation (Toomre 1974). The theoretical estimates, in table 2.4, are based upon different hypotheses. Gans's (1972) estimate of $\nu$ is obtained by assuming that the boundary between the inner and outer cores represents a melting transition and by using the Andrade hypothesis, in which the behaviour of a liquid at melting point is assumed to be much like a solid, and the viscosity relates to the mass per molecule, the interatomic spacing, and the characteristic lattice vibration frequency of the solid. Higgins & Kennedy (1971) have argued that the entire outer core may lie close to the melting curve, suggesting that the Andrade analysis may be relevant to the outer core (see Jacobs 1975). Gubbins (1976) estimates $\nu$ using Chapman's corresponding-state theorem in which viscosity is expressed as a function of density, temperature and parameters describing the inter-molecular forces. Bukowinski & Knopoff (1976) estimate $\nu$ from the electronic structure of iron at high pressures and temperatures.

## 2.4 Gravity field

### 2.4.1 *The inertia tensor*

A planet's external gravitational potential is conveniently expressed in a spherical harmonic expansion as

$$U(r, \phi, \lambda) = \frac{GM}{r}\left[1 + \sum_{n=2}^{\infty}\sum_{m=0}^{n}\left(\frac{R_e}{r}\right)^n \times (C_{nm}\cos m\lambda + S_{nm}\sin m\lambda)P_{nm}(\sin\phi)\right], \quad (2.4.1)$$

where $r$ is the geocentric distance, $\phi$ the latitude, and $\lambda$ the longitude of the point at which $U$ is determined. $G$ is the gravitational constant, $M$ the mass and $R_e$ the equatorial radius of the Earth. $C_{nm}$, $S_{nm}$ are the Stokes coefficients of degree $n$ and order $m$. They relate to the elements of the $n$th-order inertia tensor according to

$$\left.\begin{matrix} C_{nm} \\ S_{nm} \end{matrix}\right\} = \frac{1}{MR_e^n}\frac{2(n-m)!}{(n+m)!}\int_M (r')^n P_{nm}(\sin\phi')\left\{\begin{matrix} \cos m\lambda' \\ \sin m\lambda' \end{matrix}\right\} \mathrm{d}M, \quad (2.4.2)$$

where $(r', \phi', \lambda')$ are the coordinates of the mass element d$M$. The $P_{nm}(\sin\phi)$ are the Legendre polynomials. In accordance with the definition (2.4.2), the $P_{nm}(\sin\phi)$ are unnormalized so that the surface integral

$$\frac{1}{4\pi}\int_S\left[P_{nm}(\sin\phi)\left\{\begin{matrix} \cos m\lambda \\ \sin m\lambda \end{matrix}\right\}\right]^2 \mathrm{d}S = \frac{(n+m)!}{(2n+1)(n-m)!(2-\delta_{0m})},$$

where

$$\delta_{0m} = \begin{cases} 1 & \text{if} \quad m = 0 \\ 0 & \text{if} \quad m \neq 0 \end{cases}. \quad (2.4.3)$$

Fully normalized Legendre polynomials $\bar{P}_{nm}(\sin\phi)$ are defined such that

$$\frac{1}{4\pi}\int_S\left[\bar{P}_{nm}(\sin\phi)\begin{Bmatrix}\cos m\lambda\\ \sin m\lambda\end{Bmatrix}\right]^2 dS=1.$$

The corresponding Stokes coefficients $\bar{C}_{nm}$, $\bar{S}_{nm}$ relate to $C_{nm}$, $S_{nm}$ by

$$\begin{Bmatrix}\bar{C}_{nm}\\ \bar{S}_{nm}\end{Bmatrix}=N_{nm}^{-1}\begin{Bmatrix}C_{nm}\\ S_{nm}\end{Bmatrix},$$

with

$$N_{nm}=[(2n+1)(2-\delta_{0m})(n-m)!/(n+m)!]^{1/2}. \quad (2.4.4)$$

Of the Stokes coefficients, the most important is $C_{20}$, which is of the order $10^{-3}$. All others are of the order $(C_{20})^2$. Table 2.5 summarizes values for the important second-degree terms. More complete sets of values to degree 16 and beyond are given by Gaposchkin & Lambeck (1971), Gaposchkin (1974) and Smith *et al.* (1976).

The second-degree Stokes coefficients relate to the inertia tensor $I_{ij}(i,j=1,2,3)$ of order 2 according to

$$\left.\begin{aligned}C_{20}&=-[I_{33}-\tfrac{1}{2}(I_{11}+I_{22})]/MR_e^2,&&\\ C_{21}&=-I_{13}/MR_e^2, & S_{21}&=-I_{23}/MR_e^2,\\ C_{22}&=-(I_{11}-I_{22})/4MR_e^2, & S_{22}&=-I_{12}/2MR_e^2.\end{aligned}\right\} \quad (2.4.5)$$

Recent estimates from gravity and satellite orbit calculations give

$$C_{20}=-(1082.63\pm0.01)10^{-6}.$$

This includes a small term resulting from a permanent tidal correction (Romanowicz & Lambeck 1977).

$$\Delta C_{20}=+9.30\,k_f\times10^{-8},$$

where $k_f$ is the fluid Love number.† With $k_f\simeq0.93$ (equation 2.4.12), $\Delta C_{20}=+8.69\times10^{-8}$, and the corrected value is

$$C'_{20}=-1082.54\times10^{-6}.$$

For an Earth in hydrostatic equilibrium,

$$(C_{20})_{HE}=-(1072.19\pm0.45)10^{-6},$$

according to Jeffreys (1963), or

$$(C_{20})_{HE}=-1073.65\times10^{-6},$$

† The fluid Love number $k_f$ should be used, not the elastic value $k_2=0.30$ as did Romanowicz & Lambeck.

Table 2.5. *Constants defining the form and gravity of the Earth*

| | | |
|---|---|---|
| Gravitational constant | $G$ | $(6.672 \pm 0.004)\ 10^{-8}\ \text{cm}^3\ \text{g}^{-1}\ \text{s}^{-2}$ |
| Mass of Earth | $M$ | $(5.974 \pm 0.004)\ 10^{27}\ \text{g}$ |
| | $GM$ | $3.982 \times 10^{20}\ \text{cm}^3\ \text{s}^{-2}$ |
| Mean radius of Earth | $R$ | $(6.371012(1 \pm 5 \times 10^{-6})\ 10^{8}\ \text{cm}$ |
| Equatorial radius | $R_e$ | $(6.378140 \pm 5 \times 10^{-6})\ 10^{8}\ \text{cm}$ |
| Mean equatorial gravity | $g_e$ | $978.03\ \text{cm s}^{-2}$ |
| | $C_{20}$ | $(1082.63 \pm 0.01)\ 10^{-6}$ |
| Stokes coefficients (unnormalized) | $C_{22}$ | $(1.54 \pm 0.04)\ 10^{-6}$ |
| | $S_{22}$ | $(-8.84 \pm 0.04)\ 10^{-7}$ |
| Polar moment of inertia | $\frac{I_{33}}{MR_e^2}$ | 0.3306 |
| Mean equatorial moment of inertia | $\frac{I_{11}}{MR_e^2}$ | 0.3295 |
| Mean rotational velocity of Earth | $\Omega$ | $7.292115 \times 10^{-5}\text{s}^{-1}$ |
| Flattening of equipotential surface | $F_o$ | 1/298.26 |

according to Caputo (1965). The difference $C'_{20} - (C_{20})_{HE}$, after the tidal correction, is $-10.35 \times 10^{-6}$ for Jeffreys' value or $-8.98 \times 10^{-6}$ for Caputo's value. This is also of the order $(C_{20})^2$. Some early studies (see, for example, Munk & MacDonald 1960; MacDonald 1966) attached considerable significance to this difference and attributed it to a delay in the Earth's response to the change in the rotational velocity. Goldreich & Toomre (1969), however, showed that the departure of $C_{20}$ from hydrostatic equilibrium is no more significant than the other terms $C_{nm}$, $S_{nm}$ in the potential. This is discussed further in chapter 11.

The precession of the Earth's orbit on the ecliptic determines the constant

$$\mathcal{H} = [I_{33} - \tfrac{1}{2}(I_{11} + I_{22})]/I_{33} = 0.003275 \pm 3.3 \times 10^{-5} \quad (2.4.6)$$

(Romanowicz & Lambeck 1977). With (2.4.5) this gives

$$-C_{20}/\mathcal{H} = I_{33}/MR_e^2 = 0.3306. \quad (2.4.7a)$$

For a rotationally symmetric body

$$I_{11}/MR_e^2 = I_{22}/MR_e^2 = (-C_{20}/\mathscr{H})(1-\mathscr{H}) = 0.3295. \tag{2.4.7b}$$

The trace of the inertia tensor $I_{ij}$ is conserved under a large variety of deformations (Rochester & Smylie 1974). Thus for any change $\Delta I_{ii}$ in $I_{ii}$

$$\sum_i \Delta I_{ii} = 0.$$

For a rotationally symmetric body $\Delta I_{11} = \Delta I_{22}$ and $\Delta I_{33} = -2\Delta I_{11}$. The corresponding change in $C_{20}$ is therefore

$$-\Delta C_{20} = \tfrac{3}{2}\Delta I_{33}/MR_e^2 \tag{2.4.8}$$

### 2.4.2 *Fluid Love numbers*

For a rotating Earth in hydrostatic equilibrium, the free surface is given by

$$r = R[1 - \tfrac{2}{3}F_h P_{20} + O(F_h^2)],$$

where the hydrostatic flattening $F_h$ is given by

$$F_h = \frac{5}{2}\frac{\Omega^2 R_e}{g_e} \Big/ \left[1 + \left(\frac{5}{2} - \frac{15 I_{33}}{4MR_e^2}\right)^2\right] + O(F_h^2).$$

$\Omega$ is the velocity of rotation and $g_e$ is the mean gravity on the equator. With (2.4.7)

$$F_h = 0.967(\Omega^2 R_e/g_e).$$

The radial deformation is

$$d_r = -\tfrac{2}{3}F_h R P_{20}. \tag{2.4.9}$$

The harmonic part of the potential of the centrifugal force is (equation 3.4.1*b*)

$$\Delta U_2 = -\tfrac{1}{3}\Omega^2 R^2 P_{20}. \tag{2.4.10}$$

The surface deformation can also be expressed with the aid of a Love number, $h_f$ (of degree 2), for a fluid Earth as

$$d_r = h_f \Delta U_2/g. \tag{2.4.11}$$

Equating (2.4.9) and (2.4.11) gives

$$h_f = 1.934.$$

The potential of the free surface will be

$$(1+k_f)\Delta U_2,$$

where $k_f$ (also of degree 2) is the fluid potential Love number. The free surface is raised with respect to the centre of the Earth by

$$(1+k_f)\Delta U_2/g = h_f \Delta U_2/g.$$

Hence

$$k_f = 0.934. \tag{2.4.12}$$

The observed flattening, $F_o$, of the equipotential surface relates to $C_{20}$ by

$$F_o = \tfrac{3}{2}(-C_{20} + \tfrac{1}{3}\Omega^2 R_e/g_e) + O(F_o^2), \tag{2.4.13}$$

and using this instead of $F_h$ in the above analysis gives

$$k_o = 3(C-A)G/R^5\Omega^2 = 0.942. \tag{2.4.14}$$

Munk & MacDonald interpret $k_o$ as a secular Love number, the difference between $k_f$ and $k_o$ reflecting the difference between the hydrostatic flattening and the observed flattening. This difference is of the same order as the neglected terms in (2.4.9) and (2.4.13) and is not very significant. Here $k_o$ is merely a convenient constant.

An example of the use of the fluid Love number is the computation of the variation of the polar moment of inertia with time due to a secular change in $\Omega$, assuming that the planet remains in hydrostatic equilibrium. The potential of the deformation by the centrifugal force is

$$\Delta U_2 = -\tfrac{1}{3}k_o\Omega^2(t)R_e^2 P_{20}(\sin\phi)$$

which may also be written in the form (2.4.1), or

$$\Delta U_2 \simeq (GM/R)C_{20}P_{20}.$$

Equating these two expressions, and using (2.4.5) and the definition

$$I = \tfrac{1}{3}\sum_i I_{ii} = \text{constant},$$

(see Rochester & Smylie (1974) for further discussion) gives

$$I_{33}(t) = I[1 - \tfrac{2}{9}k_o R_e^5 \Omega^2(t)/GI]. \tag{2.4.15}$$

THREE

# ROTATIONAL DYNAMICS

## 3.1 Rigid body rotation

The fundamental equations governing the rotation of a body are Euler's dynamical equations. These describe the rotational response of a body to an applied torque **L** by

$$\dot{\mathbf{H}} = \mathbf{L}, \tag{3.1.1}$$

where **H** is the angular momentum of the body. This equation describes the motion in an inertial reference frame **X**. For many problems it is more convenient to express forces, velocities and torques with respect to axes that are fixed in the rotating body. These body-fixed axes are denoted by **x** and their angular velocity by **ω**; the components $\omega_i$ $(i = 1, 2, 3)$ of **ω** specify the angular velocities with which the axes $x_i$ turn about themselves. Euler's equations, referred to these axes, are (see, for example, Goldstein 1970)

$$\mathrm{d}\mathbf{H}/\mathrm{d}t + \boldsymbol{\omega} \wedge \mathbf{H} = \mathbf{L}, \tag{3.1.2}$$

where the components of **H** and **L** are now along the moving axes. For a body of mass $M$

$$\begin{aligned} \mathbf{H} &= \int_M \mathbf{x} \wedge (\boldsymbol{\omega} \wedge \mathbf{x})\, \mathrm{d}M \\ &= \int_M (\mathbf{x} \cdot \mathbf{x}\mathbf{1} - \mathbf{x} \cdot \mathbf{x}^{\mathrm{T}})\, \boldsymbol{\omega} \mathrm{d}M \qquad (3.1.3) \\ &= \mathbf{I} \cdot \boldsymbol{\omega}, \end{aligned}$$

where $\mathbf{1}(\equiv \delta_{ij}$, with $i, j = 1, 2, 3$; $\delta_{ij} = 1$ if $i = j$, otherwise $\delta_{ij} = 0)$ is a unit matrix, $\mathbf{x}^{\mathrm{T}}$ denotes the transpose of **x** and

$$\mathbf{I} = \int_M (\mathbf{x} \cdot \mathbf{x}\mathbf{1} - \mathbf{x} \cdot \mathbf{x}^{\mathrm{T}})\, \mathrm{d}M \tag{3.1.4}$$

is the second-degree symmetric inertia tensor. With (3.1.3) Euler's equation becomes

$$(\mathrm{d}/\mathrm{d}t)(\mathbf{I} \cdot \boldsymbol{\omega}) + \boldsymbol{\omega} \wedge (\mathbf{I} \cdot \boldsymbol{\omega}) = \mathbf{L}. \tag{3.1.5}$$

For axes $x_i$ fixed in a rigid body, the inertia tensor does not vary with time and the axes can be chosen such that the products of inertia vanish, or

$$I_{ij} = -\int_M x_i x_j \, \mathrm{d}M = 0 \quad \text{for} \quad i \neq j. \tag{3.1.6}$$

A further simplification results if the body is rotationally symmetric about $x_3$, for then $I_{11} = I_{22} \equiv A$, $I_{33} \equiv C$ and

$$\left.\begin{aligned} \mathrm{d}\omega_1/\mathrm{d}t + [(C-A)/A]\omega_2\omega_3 &= L_1/A, \\ \mathrm{d}\omega_2/\mathrm{d}t - [(C-A)/A]\omega_1\omega_3 &= L_2/A, \\ \mathrm{d}\omega_3/\mathrm{d}t &= L_3/C. \end{aligned}\right\} \tag{3.1.7}$$

For zero torques, the solution of these equations is

$$\left.\begin{aligned} \omega_1 &= a_0 \cos \sigma_r t + b_0 \sin \sigma_r t \\ \omega_2 &= a_0 \sin \sigma_r t - b_0 \cos \sigma_r t \\ \omega_3 &= \text{constant} = \Omega, \end{aligned}\right\} \tag{3.1.8}$$

where $a_0$, $b_0$ and $\Omega$ are constants of integration and the frequency of the motion $\sigma_r$ is defined by

$$\sigma_r = [(C-A)/A]\Omega. \tag{3.1.9}$$

Thus, for a small displacement of the principal axis $x_3$ from the rotation axis $\boldsymbol{\omega}$, the latter rotates in a circular path about $x_3$, according to (3.1.8) with a frequency $\sigma_r$ that is a function of the mass distribution in the body, and with an amplitude and phase that are determined by the initial displacement. This motion is referred to as the free Eulerian precession or free nutation† of the rigid body. For values of $A$, $C$ and $\Omega$ that are appropriate to the Earth (section 2.4), $\sigma_r \simeq 1/306 \text{ rev d}^{-1}$.

The equations (3.1.7) and their solution (3.1.8) provide a useful zeroth-order approximation to any discussion of the Earth's rotation. Their solution is stable if the motion of $\boldsymbol{\omega}$ is about a principal axis that is either one of maximum or minimum inertia. If $\boldsymbol{\omega}$ rotates about an intermediate principal axis, the motion becomes unstable; the rotation axis begins to wander away to a point diametrically opposite before returning to the original position. The Earth rotates about an axis that lies close to the mean position of the greatest

† Euler used the term 'precession' to describe this motion but it is now more usual to describe it as a nutation (see, for example, Routh 1905; Lambert, Schlesinger & Brown 1931).

principal axis and in most studies the motion can be considered as stable.

Equations (3.1.5) are valid for an instant $t$ when the inertial system $\mathbf{X}$ coincides with the moving axes $\mathbf{x}$. For a subsequent instant $t'$ the equations remain valid provided that a new inertial system $\mathbf{X}'$ is defined that coincides with the rotating axis at this time. Thus a complete description of the motion also requires the geometric relation between the positions of the axes $\mathbf{X}, \mathbf{X}'\ldots$ at times $t, t'\ldots$ This relation can be defined with the aid of the three Eulerian angles $\alpha_i$ (Woolard 1953). The inertial system $\mathbf{X}$ is defined by the plane of the mean ecliptic (the $X_1X_2$-plane) and the mean equinox ($X_1$) for the epoch $T_0$. $X_2$ lies $\frac{1}{2}\pi$ to the East of $X_1$, and $X_3$ is normal to the ecliptic; $x_3$ is directed to the mean position of the rotation axis, $x_1$ is directed towards the mean meridian of Greenwich, and $x_2$ is $\frac{1}{2}\pi$ East of $x_1$. Then, with Woolard's definition,

$\alpha_1$ = inclination of the $x_1x_2$-plane on the mean ecliptic.

$\alpha_2$ = angle in the $X_1X_2$-plane between $X_1$ and the descending node of the $x_1x_2$-plane on the ecliptic. $\alpha_2$ is measured positive eastward from $X_1$.

$\alpha_3$ = angle in the $x_1x_2$-plane between the descending node and the $x_1$-axis. $\alpha_3 = \omega_3 T \simeq \Omega T$.

$$\mathbf{X} = \mathcal{R}_3(-\alpha_2)\mathcal{R}_1(\alpha_1)\mathcal{R}_3(-\alpha_3)\mathbf{x},$$

where $\mathcal{R}_i(\alpha_j)$ denotes an anti-clockwise rotation through an angle $\alpha_j$ about an axis $x_i$. The time derivatives of $\alpha_j$ represent the motion of the $\mathbf{x}$-axes with respect to the inertial frame. Resolving these velocities into components along the $x_i$-axes and equating them to $\omega_i$ gives (Woolard 1953; Webster 1959)

$$\begin{pmatrix} \dot{\alpha}_1 \\ \dot{\alpha}_2 \sin \alpha_1 \\ \dot{\alpha}_3 + \dot{\alpha}_2 \cos \alpha_1 \end{pmatrix} = R(-\alpha_3) \begin{pmatrix} \omega_1 \\ \omega_2 \\ \omega_3 \end{pmatrix} \qquad (3.1.10)$$

with

$$\mathcal{R}(-\alpha_3) = \begin{pmatrix} \cos \alpha_3 & -\sin \alpha_3 & 0 \\ \sin \alpha_3 & \cos \alpha_3 & 0 \\ 0 & 0 & 1 \end{pmatrix}.$$

These are Euler's kinematic equations of motion. With (3.1.5) they describe completely the motion of $\boldsymbol{\omega}$ both in space and with respect to the body. Studies of polar motion and changes in the speed of rotation usually involve (3.1.5); studies of precession and nutation usually involve (3.1.10). The two sets of equations indicate that, for every oscillation of $\boldsymbol{\omega}$ with respect to $\mathbf{x}$, there will be an associated nutation in space. Since $\dot{\alpha}_3 \simeq \Omega$ the frequency of the nutation in space of the Eulerian motion is

$$\sigma_r + \Omega \tag{3.1.11}$$

and the free wobble is associated with a nearly diurnal oscillation in space. A complete discussion of the forced precession and nutations of a rigid body is given by Woolard (1953) and Kinoshita (1977). (See also Murray 1978.)

The kinematic representation of the motion of $\boldsymbol{\omega}$, originally due to Poinsot in 1851, is illustrated in figure 3.1. If no torques act on the body, $\mathbf{H}$ is fixed in space and $\boldsymbol{\omega}$ traces out a cone about the principal axis $\mathbf{x}_3$. This is the free Eulerian nutation or wobble, given by (3.1.8), with frequency $\sigma_r$; $\boldsymbol{\omega}$ also traces out a smaller cone about $\mathbf{H}$ and this is the associated nutation described by (3.1.10). The axes $\boldsymbol{\omega}$, $\mathbf{H}$ and $\mathbf{x}_3$ are coplanar, with $\mathbf{H}$ lying between the other two. The frequency (3.1.11) of this nutation is approximately diurnal. Its amplitude relative to the wobble amplitude is of the order $(C-A)/A$ or 1/300.

## 3.2 Non-rigid body rotation: general formulation

Apart from the ratios $(C-A)/C$ or $(C-A)/A$, no geophysical parameters enter into the discussion of rigid body motion and, in the absence of external torques, the motion is completely predictable once the initial conditions have been established. Such motion would be of no further geophysical interest. The Earth's actual motion differs from rigid body motion in two respects: the inertia tensor $\mathbf{I}$ is time dependent, and motion occurs relative to the axes $\mathbf{x}$. Hence the total angular momentum $\mathbf{H}$ must be written as

$$\mathbf{H}(t) = \mathbf{I}(t)\boldsymbol{\omega}(t) + \mathbf{h}(t), \tag{3.2.1}$$

where

$$\mathbf{h} = \int_M (\mathbf{x} \wedge \mathbf{u})\, \mathrm{d}M, \tag{3.2.2}$$

is the angular momentum vector due to motion, with velocity $\mathbf{u}$,

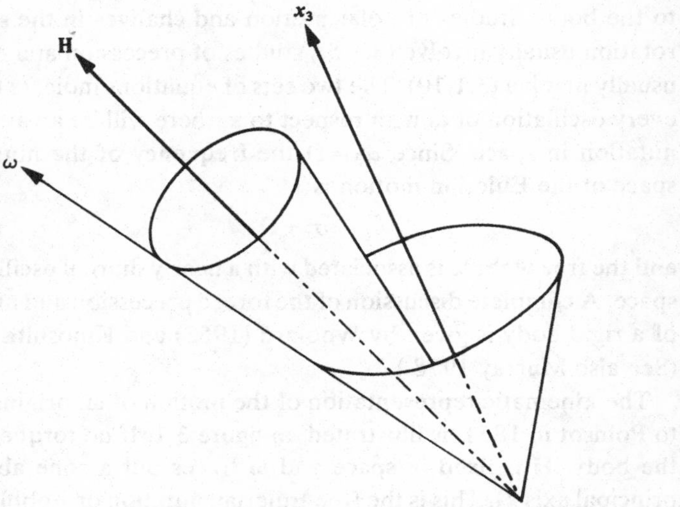


Figure 3.1. The Poinsot representation of rigid body rotation. **H** is the angular momentum axis, **ω** the instantaneous rotation axis, and $\mathbf{x}_3$ the principal axis. In the absence of external torques, **H** is fixed in space; **ω** describes a periodic nearly diurnal motion in space about **H** and a much larger motion, the free Eulerian nutation, about $\mathbf{x}_3$.

relative to **x**. Substituting (3.2.1) into (3.1.5) gives

$$(\mathrm{d}/\mathrm{d}t)[\mathbf{I}(t)\boldsymbol{\omega}+\mathbf{h}(t)]+\boldsymbol{\omega}\wedge\mathbf{I}(t)\boldsymbol{\omega}+\mathbf{h}(t)=\mathbf{L}. \qquad (3.2.3)$$

These are the Liouville equations (Routh 1905; Munk & MacDonald 1960).

In most discussions on non-rigid rotation, the excursions from uniform rotation are small and it is convenient to write

$$\omega_1=\Omega m_1, \qquad \omega_2=\Omega m_2, \qquad \omega_3=\Omega(1+m_3), \qquad (3.2.4)$$

where $\Omega$ is the mean angular velocity of the Earth. The $m_i$ are small dimensionless quantities, of the order of $10^{-6}$ or less. The $m_1$, $m_2$, $1+m_3$ represent the direction cosines of **ω** relative to $x_3$; $m_1$ and $m_2$ are the components of polar motion or wobble and $\Omega\mathrm{d}m_3/\mathrm{d}t$ is nearly the acceleration in diurnal rotation. Changes in the inertia tensor and the relative angular momenta $h_i$ are also small in most problems and it is convenient to write (3.2.3) in a perturbation form. We write

$$I_{11}=A+\Delta I_{11}(t) \qquad I_{22}=A+\Delta I_{22}(t), \qquad I_{33}=C+\Delta I_{33}(t),$$
$$I_{ij}=\Delta I_{ij}(t), \qquad i\neq j, \tag{3.2.5}$$

where $\Delta I_{ij}(t)/C$ are small quantities. Substituting (3.2.4) and (3.2.5) into (3.2.3), and neglecting the squares and products of small quantities, the equations of motion reduce to

$$\dot{m}_1/\sigma_r+m_2=\psi_2, \qquad \dot{m}_2/\sigma_r-m_1=-\psi_1, \tag{3.2.6a}$$
$$\dot{m}_3=\dot{\psi}_3, \tag{3.2.6b}$$

with

$$\begin{aligned}\psi_1&=[\Omega^2\Delta I_{13}+\Omega\Delta\dot{I}_{23}+\Omega h_1+\dot{h}_2-L_2]/\Omega^2(C-A)\\ \psi_2&=[\Omega^2\Delta I_{23}-\Omega\Delta\dot{I}_{13}+\Omega h_2-\dot{h}_1+L_1]/\Omega^2(C-A)\\ \psi_3&=[-\Omega^2\Delta I_{33}-\Omega h_3+\Omega\textstyle\int_0^t L_3\,\mathrm{d}t]/\Omega^2 C,\end{aligned} \tag{3.2.7a}$$

and

$$\sigma_r=[(C-A)/A]\Omega. \tag{3.2.7b}$$

The omission of the squares and products of small quantities is adequate in most studies. For example, the period of the free nutation is given with a precision of about 1 d by the above development and this compares with an observational precision of about 2 d (chapter 5). For other investigations, such as the ellipticity of the wobble, some of the smaller terms must be retained.

Complex notation gives a more compact form of the equations of motion. Writing

$$\left.\begin{aligned}\mathbf{m}&=m_1+\mathrm{j}m_2,\\ \boldsymbol{\psi}&=\psi_1+\mathrm{j}\psi_2,\\ \boldsymbol{\Delta}\mathbf{I}&=\Delta I_{13}+\mathrm{j}\Delta I_{23},\\ \mathbf{h}&=h_1+\mathrm{j}h_2,\\ \mathbf{L}&=L_1+\mathrm{j}L_2,\end{aligned}\right\} \tag{3.2.8}$$

then

$$\left.\begin{aligned}\mathrm{j}(\dot{\mathbf{m}}/\sigma_r)+\mathbf{m}&=\boldsymbol{\psi},\\ \dot{m}_3&=\dot{\psi}_3,\end{aligned}\right\} \tag{3.2.9}$$

and

$$\boldsymbol{\psi}=[1/\Omega^2(C-A)][\Omega^2\boldsymbol{\Delta}\mathbf{I}-\mathrm{j}\boldsymbol{\Delta}\dot{\mathbf{I}}+\Omega\mathbf{h}-\mathrm{j}\dot{\mathbf{h}}+\mathrm{j}\mathbf{L}]. \tag{3.2.10}$$

The solution for $\boldsymbol{\psi}=0$ is

$$\mathbf{m}=\mathbf{m}_0 e^{j\sigma_r t}. \quad (3.2.11)$$

Otherwise

$$\mathbf{m}=e^{j\sigma_0 t}\left[\mathbf{m}_0-j\sigma_0\int_{-\infty}^{t}\boldsymbol{\psi}(\tau)e^{-j\sigma_r\tau}\,d\tau\right]. \quad (3.2.12)$$

The $\psi_i$ are referred to as excitation functions and include all factors that perturb the rotational motion from that defined by the rigid body motion discussed above (equations 3.1.4–9). They include, for example, changes in $\Delta I_{ij}$ and $h_i$ arising from the atmospheric and oceanic circulation, the Earth's elastic deformation due to centrifugal and tidal forces, large-scale mass redistribution associated with mantle convection, and electromagnetic coupling of core motions to the mantle. Equations (3.2.6) clearly separate the astronomical and geophysical problems: astronomers observe the motion of the rotation axis with respect to the Earth's surface; geophysical observations and theory provide the information for evaluating the excitation functions $\psi_i$. The problem is to explain the $m_i$ in terms of the available geophysical data, or, failing that, to deduce certain geophysical properties from the observed rotation variations.

Two aspects of the choice of the terrestrial system need further elaboration. (i) The $x_i$ are assumed to be fixed to the Earth, but this is incompatible with the notion of a non-rigid Earth. High-frequency deformations of the solid Earth occur due to tidal forces, for example, while at the other end of the spectrum, slow secular displacements occur due to tectonic upheaval and continental drift. The former can often be monitored and corrections may be applied to the observations, while the latter, being secular and slow, do not disrupt greatly the higher-frequency irregularities in rotation. A pragmatic choice of a fixed reference frame is one that is defined by the coordinates of observatories on tectonically stable continents, whose motions are partly predictable by current global plate tectonics models. Unfortunately, many of the astronomical observatories have not always been placed with this condition in mind: the International Latitude Service, which has monitored polar motion since 1900, has four of its five stations in tectonically complex areas and these stations may have been subject to important local or regional displacements (section 5.3.2).

(ii) In theoretical discussions of the excitation functions, other choices of axes are sometimes convenient. Routh (1905) discusses some possibilities. One choice is such that $h_i = 0$ at all times. These are the *mean* axes introduced by Tisserand in his *Mécanique Céleste*. Such axes are fixed in the body only when all relative motion ceases. Another choice is to select $x_i$ such that $I_{ij} = \delta_{ij} I_{ij}$ at all times. These axes were introduced by G. H. Darwin in his study of geological influences on the rotation axis. Choices such as these often simplify the discussion of the geophysical influences on rotation, but they do not permit a ready comparison of the excitation functions with the astronomical observations that relate the motion of the rotation axis with respect to the station positions.

In many studies it is possible to separate the wobble equations (3.2.6*a*) from the l.o.d. equation (3.2.6*b*) since the excitation functions $\psi_1$ and $\psi_2$ do not contain the same elements of the inertia tensor and angular momenta and torque vectors as does $\psi_3$. Some cross-coupling may occur in certain situations. One example is the pole tide. The periodic wobble (3.2.11) sets up an oscillation in the oceans which, due to the continent distribution, is not symmetrical about the instantaneous rotation axis. This pole tide contributes to $\Delta I_{33}$ and introduces a perturbation in $m_3$, of the same frequency as the wobble (chapter 6). But this is small and negligible. A case where the cross-coupling may be much more important is discussed in section 3.5.

## 3.3 Rotation of spheroid with fluid core

A special case of non-rigid body rotation is that of a rigid shell containing a fluid-filled cavity. In the initial state the shell and fluid rotate together about an axis $\boldsymbol{\omega}_0$ that is also a principal axis $\mathbf{x}_3$. The cavity is spherical and the fluid is homogeneous, incompressible and inviscid. At time $t_0$ an impulse torque $\mathbf{L} = \mathbf{L}\delta(t - t_0)$, $L_3 = 0$ acts on the shell, and the instantaneous rotation axis now wobbles about $\mathbf{x}_3$. For this idealized case the core and shell are completely decoupled: the shell motion is not transmitted to the core and the latter continues to rotate as a rigid body about its original axis. The frequency of the wobble is now

$$\sigma = [(C_m - A_m)/A_m]\Omega,$$

where $C_m, A_m, A_m$ are the principal moments of the shell or mantle. If $C_c, A_c, A_c$ denote the principal moments of the core,

$$\sigma \simeq [(C-A)/(A-A_c)]\Omega = (A/A_m)\sigma_r, \tag{3.3.1}$$

and the effect of the liquid cavity is to shorten the wobble period. The frequency of the motion of $\boldsymbol{\omega}$ in space is

$$(A/A_m)\sigma_r + \Omega$$

and is almost unchanged from the value (3.1.11) for the rigid body. The axes $\mathbf{x}_3$, $\boldsymbol{\omega}$ and $\mathbf{H}$ remain coplanar and the ratio of the amplitudes of wobble and nearly diurnal nutation is only marginally modified: Poinsot's representation (figure 3.1) remains essentially unmodified.

For a spheroidal cavity, the interior flow field departs from a rigid body rotation. Kelvin showed that the relative motion consists of a rigid rotation and a deformational motion due to the flow of the fluid past the spheroidal boundary. This secondary flow exerts asymmetrically distributed pressures over the boundary that couple the core to the shell. This form of coupling is referred to as pressure coupling or inertial coupling and has been studied by W. Hopkins in 1839, by Kelvin, Hough and F. Sludskii in 1896 and by Poincaré in 1910. The equations of motion for the shell and the core have been derived by Lamb (1932) using the Helmholtz equations to describe the flow in the fluid of uniform vorticity. The shell motion is defined by the direction cosines $m_1$, $m_2$, and that of the core by $n_1$, $n_2$. Both motions are with respect to axes fixed in the shell. The equations of motion now reduce to

$$\left.\begin{aligned} \dot{\mathbf{m}} - \mathrm{j}\Omega[(C-A)/A_m]\mathbf{m} &= -\mathrm{j}\Omega[(C-A)/A_m]\boldsymbol{\psi}, \\ \dot{\mathbf{n}} - \mathrm{j}\Omega(1+f_c)(\mathbf{m}-\mathbf{n}) &= 0, \end{aligned}\right\} \tag{3.3.2}$$

where $f_c$ is the flattening of the core. (See Lamb (1932) p. 726, but note that in (3.3.2) $\mathbf{n}$ is the motion with respect to mantle-fixed axes). For the Earth $f_c \simeq 1/400$. The excitation function $\boldsymbol{\psi}$ acts on the mantle only and the core responds passively. In the case $\boldsymbol{\psi} = 0$, the free-oscillation frequency of the system is

$$\sigma = (A/A_m)(1+f_c)\sigma_r. \tag{3.3.3}$$

The coupling is quite weak and the core mainly abstains from Eulerian nutation. The solution also points to the existence of a second free-oscillation mode or wobble whose frequency is

$$\sigma = -\Omega[1+(A/A_{\mathrm{m}})f_{\mathrm{c}}] \tag{3.3.4}$$

and which represents a retrograde motion of the pole about $x_3$. Its period differs from a sidereal day by $Af_{\mathrm{c}}/A_{\mathrm{m}}$, or by about 4 min. This is the diurnal free wobble and has again been the subject of attention in the recent literature (Rochester, Jensen & Smylie 1974; Toomre 1974; M. L. Smith 1977). The associated free nutation of $\boldsymbol{\omega}$ in space about $\mathbf{H}$ has the same frequency

$$\sigma+\Omega \simeq -(Af_{\mathrm{c}}/A_{\mathrm{m}})\Omega \tag{3.3.5}$$

and is also retrograde. $\mathbf{H}$, $\boldsymbol{\omega}$ and $\mathbf{x}_3$ remain coplanar, as for the Eulerian wobble, but $\mathbf{H}$ and $\boldsymbol{\omega}$ lie on opposite sides of $\mathbf{x}_3$ (figure 3.2). The ratio of this wobble and associated nutation is $Af_{\mathrm{c}}/A_{\mathrm{m}}$ and now the nutation amplitude is much larger than the wobble amplitude, by a factor of about 400 for the Earth. Thus, if the Earth has an observable nearly diurnal wobble – sometimes claimed to have an amplitude of 0.″01 (see Yatskiv (1972) and Rochester *et al.* (1974) for a summary of results) – the associated wobble in space must have an amplitude about 400 times larger and a period of about 400 d. The astronomically observed quantity would be the angle $\boldsymbol{\omega}-\mathbf{H}$ but it is not seen in the data (Rochester *et al.* 1974; Capitaine 1975).

## 3.4 Rotational deformation

The Earth's principal departure from rigid body rotation is its deformation due to the variable centrifugal force, and it is convenient to describe other perturbations in the motion with respect to this deformed state. The potential of the centrifugal force at point $P$, distant $l$ from the instantaneous rotation axis, is

$$U_{\mathrm{c}} = \tfrac{1}{2}\omega^2 l^2.$$

The direction cosines of $\boldsymbol{\omega}$ are $m_i = \omega_i/\omega$ so that

$$l^2 = \sum_i x_i^2 - \left(\sum_i \frac{\omega_i x_i}{\omega}\right)^2$$

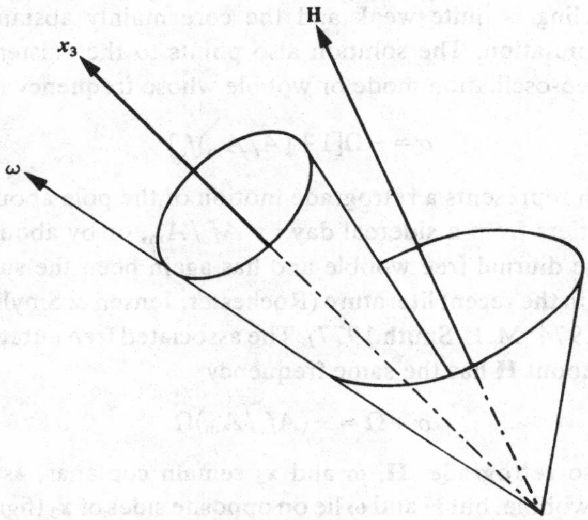


Figure 3.2. The Poinsot representation of the nearly diurnal free wobble for a body containing a spheroidal, liquid-filled cavity. The motion of **ω** about **x**$_3$ describes a nearly diurnal free wobble, and the much larger motion of **ω** about **H** describes the associated nutation in space.

and

$$U_c = \tfrac{1}{2}\omega^2 r^2 - \tfrac{1}{2}\left(\sum_i \omega_i x_i\right)^2,$$

with

$$r^2 = \sum_i x_i^2,$$

$$\omega^2 = \sum_i \omega_i^2 \simeq \Omega^2(1+2m_3) \simeq \Omega^2.$$

This potential can be further arranged as

$$U_c = \tfrac{1}{3}\omega^2 r^2 + \Delta U_c, \qquad (3.4.1a)$$

where

$$\Delta U_c = \tfrac{1}{6}\sum_i \omega_i^2 (x_{i+1}^2 + x_{i+2}^2 - 2x_i^2) - \sum_i \omega_i \omega_{i+1} x_i x_{i+1}.$$

The first term in (3.4.1) results in a small, purely radial, deformation (about 0.004 cm at the Earth's surface for typical variations in $m_3$ of $10^{-8}$ (Lambeck 1973)). The second term is harmonic in degree 2 and can be written in terms of spherical coordinates ($r$, $\phi$, $\lambda$) for the point $P$ as

$$\Delta U_c = (R^2/6)(\omega_1^2+\omega_2^2-2\omega_3^2)(r/R)^2 P_{20}(\sin\phi) - (R^2/3)(r/R)^2 \times(\omega_1\omega_3\cos\lambda+\omega_2\omega_3\sin\lambda)P_{21}(\sin\phi)+(R^2/12)(r/R)^2 \times[(\omega_2^2-\omega_1^2)\cos 2\lambda - 2\omega_1\omega_2\sin 2\lambda]\times P_{22}(\sin\phi). \tag{3.4.1b}$$

This potential deforms the Earth and, for an elastic body, the further change can be described at the Earth's surface, $r=R$, by

$$\Delta U_c'(R) = k_2 \Delta U_c(R).$$

Outside the Earth,

$$\Delta U_c'(r) = (R/r)^3 k_2 \Delta U_c(R).$$

This can be written in the form (2.4.1) as

$$\Delta U_c'(R) = \frac{GM}{r}\left(\frac{R}{r}\right)^3 \sum_m (C_{2m}^* \cos m\lambda + S_{2m}^* \sin m\lambda) P_{2m}(\sin\phi), \tag{3.4.1c}$$

with

$$C_{20}^* = (R^3/6GM)(\omega_1^2+\omega_2^2-2\omega_3^2)k_2,$$
$$C_{21}^* = -(R^3/3)(\omega_1\omega_3/GM)k_2, \qquad S_{21}^* = -(R^3/3)(\omega_2\omega_3/GM)k_2;$$
$$C_{22}^* = (R^3/12)[(\omega_2^2-\omega_1^2)/GM]k_2,$$
$$S_{22}^* = -(R^3/6)(\omega_1\omega_2/GM)k_2.$$

Equating $C_{2m}^*$ and $S_{2m}^*$ with the appropriate elements in the second-degree inertia tensor (2.4.5) gives the following changes $\Delta I_{ij}(t)$ in the Earth's inertia tensor due to the rotational deformation:

$$\left.\begin{aligned} \Delta I_{13} &= \frac{k_2 R^5 \omega_1\omega_3}{3G} = \frac{k_2 R^5 \Omega^2 m_1(1+m_3)}{3G} \\ &\simeq \frac{k_2 R^5 \Omega^2 m_1}{3G}, \\ \Delta I_{23} &= \frac{k_2 R^5 \omega_2\omega_3}{3G} = \frac{k_2 R^5 \Omega^2 m_2(1+m_3)}{3G} \\ &\simeq \frac{k_2 R^5 \Omega m_2}{3G}. \end{aligned}\right\} \tag{3.4.2}$$

The excitation functions $\psi_1$ and $\psi_2$ become

$$\left.\begin{aligned}\psi_1&=\frac{\Omega^2}{C-A}k_2\frac{R^5}{3G}\left(m_1+\frac{\dot m_2}{\Omega}\right)=\frac{k_2}{k_o}\left(m_1+\frac{\dot m_2}{\Omega}\right),\\ \psi_2&=\frac{\Omega^2}{C-A}k_2\frac{R^5}{3G}\left(m_2-\frac{\dot m_1}{\Omega}\right)=\frac{k_2}{k_o}\left(m_2-\frac{\dot m_1}{\Omega}\right),\end{aligned}\right\}\qquad(3.4.3)$$

with

$$k_o=3(C-A)G/\Omega^2R^5=3GMC_{20}/\Omega^2R^3=0.942.\qquad(3.4.4)$$

The $m_i$ are of the order $10^{-6}$ with a period of about 430 d so that $|\dot{\mathbf{m}}|/\Omega\simeq 5\times10^{-9}\ll m$. Hence $\psi_i\simeq k_2m_i/k_o$. Writing

$$\sigma_0=\sigma_r(1-k_2/k_o),\qquad(3.4.5)$$

and substituting the excitations $\psi_i$ into (3.2.6) gives (in the absence of all other excitations)

$$\dot m_1/\sigma_0+m_2=0,\qquad \dot m_2/\sigma_0-m_1=0.\qquad(3.4.6)$$

The solution is

$$\left.\begin{aligned}m_1&=m_0\cos(\sigma_0t+\theta),\\ m_2&=m_0\sin(\sigma_0t+\theta),\end{aligned}\right\}\qquad(3.4.7)$$

and the motion is circular, as for the rigid body motion, but the frequency is reduced from $\sigma_r$ to $\sigma_0$. Observations of the body tide indicate that $k_2\simeq0.29$–$0.30$ (chapter 6) so that $\sigma_0\simeq 2\pi/445$ rev d$^{-1}$, close to the observed frequency of about $\sigma_o\simeq 2\pi/435$ rev d$^{-1}$ (chapter 8). Elasticity lengthens the period of the Eulerian precession from about 305 d to 435 d.

Figure 3.3($a$) illustrates the motion $\mathbf{m}$ of the rotation axis and of the excitation axis $\boldsymbol{\psi}$ relative to the body-fixed axes $\mathbf{x}$. Rigid body rotation is illustrated by case ($a$) in which the $\mathbf{x}$-axes are also principal axes. Initially $\mathbf{m}$ is aligned with $\mathbf{x}_3$ and $\boldsymbol{\psi}=0$, $\mathbf{m}=0$. A disturbance shifts the excitation axis $\boldsymbol{\psi}_r$ for this rigid body and now $\mathbf{m}$ turns about this new position with frequency $\sigma_r$ and with an amplitude $m_0$ that is a function of the initial displacement. For the elastic case, the rotation axis wobbles freely about a mean position. The excitation function is given by (3.4.3) or

$$\boldsymbol{\psi}_D=(k_2/k_o)\mathbf{m}$$

and is a consequence of the bulge adjusting itself to the continuously

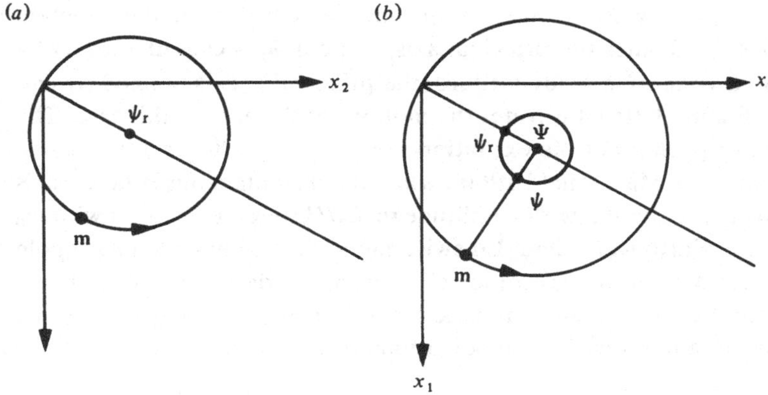


Figure 3.3. Motion of the rotation axis with respect to Earth-fixed axes. (*a*) The rotation for the rigid Earth. In the initial state $\mathbf{m}=0$, $\boldsymbol{\psi}=0$. A perturbation shifts the excitation pole to $\boldsymbol{\psi}_r$ and $\mathbf{m}$ now moves about $\boldsymbol{\psi}_r$ with frequency $\sigma_r$ and amplitude $\mathbf{m}_0$. (*b*) The motions for an elastic Earth. For the same initial displacement, the mean excitation pole $\boldsymbol{\Psi}$ now moves by an amount $k_o/(k_o-k_2)\boldsymbol{\psi}$. Both $\mathbf{m}$ and the instantaneous excitation pole $\boldsymbol{\psi}=\boldsymbol{\psi}_r+\boldsymbol{\psi}_D$ move about this position.

changing position of $\mathbf{m}$. The equations of motion now are

$$\mathrm{j}(\dot{\mathbf{m}}/\sigma_r)+\mathbf{m}=\boldsymbol{\psi}_D$$

or

$$\mathrm{j}(\dot{\mathbf{m}}/\sigma_0)+\mathbf{m}=0, \tag{3.4.8}$$

where $\sigma_0=\sigma_r(1-k_2/k_o)$ is the frequency of the free oscillation of the elastic Earth. From (3.4.2) it follows that $\boldsymbol{\psi}_D$ represents the orientation of the principal axis of the Earth with respect to $\mathbf{x}$.

If the Earth is subject to a forced excitation, the total excitation $\boldsymbol{\psi}$ is the sum of two parts: the forcing function $\boldsymbol{\psi}_r$, evaluated as if the Earth were rigid, and $\boldsymbol{\psi}_D$. The equations of motion are

$$\mathrm{j}(\dot{\mathbf{m}}/\sigma_r)+\mathbf{m}=\boldsymbol{\psi}_r+\boldsymbol{\psi}_D$$

or

$$\mathrm{j}(\dot{\mathbf{m}}/\sigma_0)+\mathbf{m}=[k_o/(k_o-k_2)]\boldsymbol{\psi}_r. \tag{3.4.9}$$

The elastic yielding of the Earth therefore modifies the amplitude of the excitation function by a factor $k_o/(k_o-k_2)$; the rotational deformation increases the amplitude of the wobble by about 1.4 from what it would be if the Earth were rigid. If the excitation $\boldsymbol{\psi}_r$ is due entirely to changes in the inertia tensor then the function $\boldsymbol{\psi}_r+\boldsymbol{\psi}_D$ defines the principal axis, while if $\boldsymbol{\psi}_r$ is caused entirely by relative motion or by torques, the principal axis is defined by $\boldsymbol{\psi}_D$.

Figure 3.3(*b*) illustrates the motions of the axes in this case. The mean position of the excitation axis is given by $\boldsymbol{\Psi}=k_o/(k_o-k_2)\boldsymbol{\psi}_r$ (Munk & MacDonald 1960), while the instantaneous rotation axis $\mathbf{m}$ moves about it with amplitude $\mathbf{m}_0 k_o/(k_o-k_2)$ compared with $\mathbf{m}_0$ if the Earth were rigid. Likewise the instantaneous excitation pole $\boldsymbol{\Psi}$ rotates about this position with an amplitude given by $\boldsymbol{\psi}_D$. The $\mathbf{m}$, $\boldsymbol{\psi}$ and $\boldsymbol{\Psi}$ are coplanar, and the separation $\mathbf{m}-\boldsymbol{\psi}$ between $\mathbf{m}$ and the instantaneous excitation pole remains the same.

## 3.5 Damped linear motion

The Earth's response to a disturbing potential is not purely elastic but lags it. This anelastic response can be expressed in a very general way by the complex Love number $\mathbf{k}_2$ (chapter 2). Writing $\mathbf{k}_2=k_2+\mathrm{j}\kappa$, the excitation functions (3.4.3) are

$$\boldsymbol{\psi}\simeq[(k_2+\mathrm{j}\kappa)/k_o]\mathbf{m},$$

and the equations of motion are

$$\mathrm{j}\dot{\mathbf{m}}/\boldsymbol{\sigma}_0+\mathbf{m}=0$$

with

$$\boldsymbol{\sigma}_0=\sigma_r(1-k_2/k_o)-\mathrm{j}\sigma_r\kappa/k_o. \tag{3.5.1}$$

That is, to allow for anelasticity it suffices to introduce the complex frequency $\boldsymbol{\sigma}_0=\sigma_0+\mathrm{j}\alpha$, This is valid in so far as the complex form of the Love numbers is valid. The solution of these equations is, for $\boldsymbol{\psi}=0$.

$$\mathbf{m}=\mathbf{m}_0\mathrm{e}^{\mathrm{j}\boldsymbol{\sigma}_0 t}=\mathbf{m}_0\mathrm{e}^{-\alpha t}\mathrm{e}^{\mathrm{j}\sigma_0 t}, \tag{3.5.2}$$

and the amplitude of the free wobble is damped by the factor $\mathrm{e}^{-\alpha t}$. The relaxation time is

$$\tau=1/\alpha,$$

and the specific dissipation is (equation 2.2.6)

$$1/Q=2\alpha/\sigma_0. \tag{3.5.3}$$

The spectral density of the damped motion is proportional to

$$\{[(\sigma-\sigma_0)/\sigma_0]^2+(1/2Q_0)^2\}^{-1}. \tag{3.5.4}$$

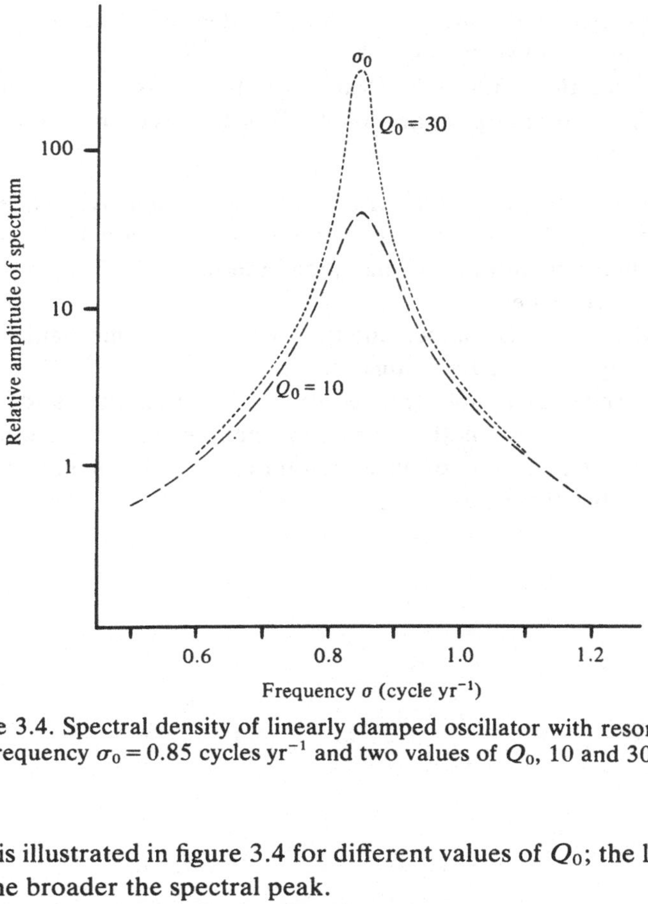


Figure 3.4. Spectral density of linearly damped oscillator with resonance frequency $\sigma_0 = 0.85$ cycles $yr^{-1}$ and two values of $Q_0$, 10 and 30.

This is illustrated in figure 3.4 for different values of $Q_0$; the lower $Q_0$ the broader the spectral peak.

The linearly damped wobble also illustrates an interesting situation in which the wobble and l.o.d. equations (3.2.6) cannot be treated separately. Due to tidal friction (chapter 10), the Earth undergoes a slow secular acceleration with consequence that $\Omega$ was greater in the past than it is now and that $\sigma_0$ was also greater. This raises the possibility that at some time earlier, $\sigma_0$ equalled the frequency of the seasonal forcing function. At such a resonance, the spectral density of the damped motion is proportional to $(2Q_0)^2$, some 500 times larger than at present if $Q_0 \simeq 60$. If the present rate of deceleration has remained constant, this resonance would have occurred some $2 \times 10^8$ yr ago. Cannon (1974; but see Lambeck

1975*b*) has speculated on possible thermal and geological consequences of such an event.

The solution (3.5.2) illustrates the three essential problems related to the interpretation of the Eulerian free precession (see chapter 8):

(i) The elongation of the period, although qualitatively explained through the use of the Love number $k_2$ (equation 3.4.5), needs a more quantitative explanation in terms of the Earth's physical properties.
(ii) The motion is damped and the mechanism, or mechanisms, of dissipation need investigation.
(iii) Despite damping, the astronomical observations do not indicate a systematic decrease in amplitude of the free wobble. This is indicative of an excitation process that maintains the wobble against damping.

FOUR

# EXCITATION FUNCTIONS

The geophysical discussion of the Earth's rotation centres around the evaluation of the excitation functions $\psi_i$ defined by (3.2.7), while the astronomical discussion centres around the determination of the motion of the instantaneous rotation axis, $\omega_i$ or $m_i$. A primary goal in studying these problems is to understand the geophysical processes that excite the irregular rotation and conversely, to invert the astronomical observations for geophysical parameters describing some global properties of the Earth. Two approaches are possible. One is to make a detailed evaluation of the geophysical mechanisms perturbing the rotation, and this is done in the following chapters. The other is to compare the properties of the observed time series and power spectra with those that would be expected if the excitation functions had some simple mathematical form. A number of such forms and related formulae are discussed in this chapter and from them considerable insight into the irregular rotation can already be gained.

## 4.1 Alternative formulations

The excitation functions (3.2.7) consist of contributions from (i) a redistribution of mass, (ii) relative motion of matter, and (iii) torques. Thus

$$\psi_i = \psi_i(\text{matter}) + \psi_i(\text{motion}) + \psi_i(\text{torques}),$$

with

$$\left.\begin{aligned} \boldsymbol{\psi}(\text{matter}) &= [1/(C-A)]\boldsymbol{\Delta}\mathbf{I}, \\ \boldsymbol{\psi}(\text{motion}) &= (1/\Omega(C-A))[-\mathrm{j}\,\boldsymbol{\Delta}\dot{\mathbf{I}} + \mathbf{h} - (\mathrm{j}/\Omega)\dot{\mathbf{h}}], \\ \boldsymbol{\psi}(\text{torque}) &= \mathrm{j}\mathbf{L}/\Omega^2(C-A), \end{aligned}\right\} \quad (4.1.1)$$

and

$$\left.\begin{aligned}\psi_3(\text{matter}) &= -\Delta I_{33}/C,\\ \psi_3(\text{motion}) &= -h_3/\Omega C,\\ \psi_3(\text{torque}) &= \Omega\int_0^t L_3\,\mathrm{d}t.\end{aligned}\right\} \tag{4.1.2}$$

Each of these terms is summarized in table 4.1 for both Cartesian and spherical coordinates. In these expressions $\rho(x_i, t)$ is the density of matter taking part in the motion and $\Delta\rho(x_i, t)$ is the change in the density distribution. Velocities are denoted by $u_i$, accelerations by $\dot{u}_i$. In spherical coordinates, the velocities $u_r$, $u_\phi$, $u_\lambda$ are positive radially outwards, polewards and eastwards. The torques may be of two types: torques due to a body force **f**, and the torque due to surface forces $p_{nk}$, in the direction **k**, on a unit surface whose normal is **n**. An example of the former is the lunar attraction on the Earth's bulge; an example of the latter is the action of surface winds. The volume integrals are over a volume $V$ enclosed by a surface $S$. The actual choice of this surface depends on the problem studied. $V$ could include the fluid parts of the Earth as well as the solid Earth. In this case effects of winds and ocean currents are evaluated in terms of relative motion or relative angular momenta. Alternatively, $V$ could include only the solid Earth and now wind and ocean effects are evaluated in terms of surface stresses integrated over the surface $S$ of the solid Earth. The choice of method is, in most cases, decided by the available observations, that is, whether it is simpler or more precise to measure relative motion or surface stresses.

The elements of the inertia tensor $\Delta I_{ij}$ follow from (3.1.4), or they may be expressed in spherical coordinates as

$$\Delta I_{ij} = \int_V \Delta\rho\, r^2 \Pi_{ij}\,\mathrm{d}V, \tag{4.1.3a}$$

where

$$\left.\begin{aligned}\Pi_{11} &= 1-\cos^2\phi\cos^2\lambda,\\ \Pi_{22} &= 1-\cos^2\phi\sin^2\lambda,\\ \Pi_{33} &= \cos^2\phi,\\ \Pi_{12} &= -\tfrac{1}{2}\cos^2\phi\sin 2\lambda = \Pi_{21},\\ \Pi_{13} &= -\sin\phi\cos\phi\cos\lambda = \Pi_{31},\\ \Pi_{23} &= -\sin\phi\cos\phi\sin\lambda = \Pi_{32}.\end{aligned}\right\} \tag{4.1.3b}$$

Table 4.1. *Excitation functions $\psi_i$ due to a redistribution of matter, motion (velocity and acceleration), and* [illegible] *Cartesian and spherical coordinates*

| | | | |
|---|---|---|---|
| | | Cartesian coordinates $(x_i)$ | |
| Matter | $\frac{-\Delta\rho}{C-A} x_1 x_3 \, \mathrm{d}V$ | $\frac{-\Delta\rho}{C-A} x_2 x_3 \, \mathrm{d}V$ | $\frac{-\Delta\rho}{C}(x_1^2+x_2^2) \, \mathrm{d}V$ |
| Velocity | $\frac{-2\rho}{\Omega(C-A)} x_3 u_2 \, \mathrm{d}V$ | $\frac{2\rho}{\Omega(C-A)} x_3 u_1 \, \mathrm{d}V$ | $\frac{\rho}{\Omega C}(-x_1 u_2 + x_2 u_1) \, \mathrm{d}V$ |
| Acceleration | $\frac{\rho}{\Omega^2(C-A)}(x_3 \dot{u}_1 - x_1 \dot{u}_3) \, \mathrm{d}V$ | $\frac{\rho}{\Omega^2(C-A)}(x_3 \dot{u}_2 - x_2 \dot{u}_3) \, \mathrm{d}V$ | 0 |
| Torque | $\frac{1}{\Omega^2(C-A)}[\rho(x_2 f_3 - x_3 f_2) \, \mathrm{d}V$ $+ (x_2 p_{n3} - x_3 p_{n2}) \, \mathrm{d}S]$ | $\frac{-1}{\Omega^2(C-A)}[\rho(x_1 f_2 - x_2 f_1) \, \mathrm{d}V$ $+ (x_1 p_{n3} - x_3 p_{n1}) \, \mathrm{d}S]$ | $\frac{1}{\Omega^2 C}[\rho(x_1 f_2 - x_2 f_1) \, \mathrm{d}V$ $+ (x_1 p_{n2} - x_2 p_{n1}) \, \mathrm{d}S]$ |
| | | Spherical coordinates $(r, \phi, \lambda)$ | |
| Matter | $\frac{-\Delta\rho \, r^2}{C-A} \cos\phi \sin\phi \cos\lambda \, \mathrm{d}V$ | $\frac{-\Delta\rho \, r^2}{C-A} \cos\phi \sin\phi \sin\lambda \, \mathrm{d}V$ | $\frac{-\Delta\rho \, r^2}{C} \cos^2\phi \, \mathrm{d}V$ |
| Velocity | $\frac{-2\rho r}{\Omega(C-A)} \sin\phi \, (u_\lambda \cos\lambda$ $- u_\phi \sin\phi \sin\lambda$ $+ u_r \cos\phi \sin\lambda) \, \mathrm{d}V$ | $\frac{+2\rho R \sin\phi}{\Omega(C-A)} (-u_\lambda \sin\lambda$ $- u_\phi \sin\phi \cos\lambda$ $+ u_r \cos\phi \cos\lambda) \, \mathrm{d}V$ | $\frac{-\rho}{\Omega C} \cos\phi \, u_\lambda \, \mathrm{d}V$ |
| Acceleration | $\frac{\rho r}{\Omega^2(C-A)} (\dot{u}_\lambda \sin\phi \sin\lambda$ $- \dot{u}_\phi \cos\lambda) \, \mathrm{d}V$ | $\frac{\rho r}{\Omega^2(C-A)} (\dot{u}_\lambda \sin\phi \cos\lambda$ $- \dot{u}_\phi \sin\lambda) \, \mathrm{d}V$ | 0 |
| Torques due to body force | $\frac{\rho r}{\Omega^2(C-A)} [f_\phi \sin\lambda -$ $f_\lambda \sin\phi \cos\lambda] \, \mathrm{d}V$ | $\frac{-\rho r}{\Omega^2(C-A)} [f_\phi \cos\lambda +$ $f_\lambda \sin\phi \sin\lambda] \, \mathrm{d}V$ | $\frac{\rho r}{\Omega^2 C} f_\lambda \cos\phi \, \mathrm{d}V$ |

## 4.2 Surface loading

In numerous instances, the excitation function is a consequence of a redistribution of mass over the Earth's surface: for example, an exchange of mass between the polar ice caps and the oceans, a redistribution of water between oceans, atmosphere and ground-water, or a redistribution of mass within the atmosphere. In all cases it is convenient to express the change in mass distribution as a surface load $q(\phi, \lambda : t)$ and to expand it into surface spherical harmonics. That is

$$q(\phi, \lambda : t) = \sum_{n=0}^{\infty} \sum_{m=0}^{n} (q'_{nm}(t) \cos m\lambda + q''_{nm}(t) \sin m\lambda) P_{nm}(\sin \phi).$$

The excitation functions are, with $\rho \, \mathrm{d}V = q \, \mathrm{d}S$ (see table 4.1),

$$\boldsymbol{\psi}(\text{matter}) = -\frac{R^2}{C-A} \int_S q \cos \phi \sin \phi \, \mathrm{e}^{\mathrm{j}\lambda} \, \mathrm{d}S,$$

$$\psi_3(\text{matter}) = -\frac{R^2}{C} \int_S q \cos^2 \phi \, \mathrm{d}S.$$

With

$$\cos \phi \sin \phi \, \mathrm{e}^{\mathrm{j}\lambda} = \tfrac{1}{3} P_{21}(\sin \phi) \, \mathrm{e}^{\mathrm{j}\lambda},$$
$$\cos^2 \phi = \tfrac{2}{3}(P_{00} - P_{20}),$$

and with the orthogonality properties of the surface harmonics, these excitation functions reduce to

$$\boldsymbol{\psi} = -\tfrac{4}{5}\pi[R^4/(C-A)](q'_{21} + \mathrm{j}q''_{21}),$$
$$\psi_3 = -\tfrac{8}{3}\pi(R^4/C)(q'_{00} - \tfrac{1}{5}q'_{20}).$$

Only zeroth- and second-degree harmonics in the surface load enter into the excitations. If the mass of the load is conserved, $q'_{00} = 0$. Under a surface load, the elastic Earth deforms and further modifies the excitation function by an amount $\boldsymbol{\psi}_\mathrm{L}$. This is most readily accounted for with the aid of the load deformation coefficients $k'_n$, since only low-degree harmonics enter into the discussion. Then

$$\boldsymbol{\psi} = \boldsymbol{\psi}_\mathrm{r} + \boldsymbol{\psi}_\mathrm{L}$$
$$= -\tfrac{4}{5}\pi[R^4/(C-A)](1+k'_2)(q'_{21} + \mathrm{j}q''_{21}), \qquad (4.2.1)$$

$$\psi_3 = \psi_{\mathrm{r},3} - \psi_{\mathrm{L},3}$$
$$= -\tfrac{8}{3}\pi(R^4/C)[(1+k'_0)q'_{00} - \tfrac{1}{5}(1+k'_2)q'_{20}]. \qquad (4.2.2)$$

The rotational response of an excitation function $\boldsymbol{\psi}_r$ that loads the Earth therefore differs from the same function acting on a rigid Earth in that there is (i) an additional rotational deformation leading to $\boldsymbol{\psi}_D$ and (ii) a load deformation leading to $\boldsymbol{\psi}_L$. The effective excitation is, in consequence,

$$\begin{aligned}\boldsymbol{\psi} &= \boldsymbol{\psi}_r + \boldsymbol{\psi}_D + \boldsymbol{\psi}_L \\ &= (1+k_2')\boldsymbol{\psi}_r + (k_2/k_o)\mathbf{m}.\end{aligned} \tag{4.2.3}$$

The wobble equations (3.2.6) now are

$$\dot{\mathbf{m}}/\sigma_r + \mathbf{m} = \boldsymbol{\psi} \tag{4.2.4}$$

or, with (4.2.3) and (3.4.5),

$$\dot{\mathbf{m}}/\sigma_0 + \mathbf{m} = [k_o/(k_o - k_2)](1+k_2')\boldsymbol{\psi}_r. \tag{4.2.5}$$

Numerically $k_o = 0.942$, $k_2 \simeq 0.30$, $k_2' \simeq -0.30$ and the right-hand side of (4.2.5) is approximately equal to $\boldsymbol{\psi}_r$. To evaluate the effective wobble excitation function we first determine $\boldsymbol{\psi}_r$ as if the Earth were rigid and multiply it by

$$\text{(i)} \quad X_{\text{wobble}} = k_o/(k_o - k_2) \tag{4.2.6a}$$

if $\boldsymbol{\psi}_r$ does not load the Earth, or

$$\text{(ii)} \quad X_{\text{wobble}} = [k_o/(k_o - k_2)](1+k_2') \tag{4.2.6b}$$

if $\boldsymbol{\psi}_r$ does load the Earth.

For $\psi_3$ we apply the operation (4.2.2) if the excitation function loads the Earth's surface. In many problems mass will be conserved and

$$\psi_3 = (1+k_2')\psi_{r,3} \equiv X_{\text{l.o.d.}}\psi_{r,3}. \tag{4.2.7}$$

If no loading occurs $X_{\text{l.o.d.}} = 1$. Munk & MacDonald refer to $X_{\text{wobble}}$ and $X_{\text{l.o.d.}}$ as the 'transfer function'.

In other problems the load or stress is limited to either the oceans or the continents. Ocean tides constitute an example of the former. In situations of this type we introduce the ocean function $\mathscr{C}(\phi, \lambda)$, defined such that

$$\left.\begin{aligned}\mathscr{C}(\phi,\lambda) &= 1 \quad \text{where there are oceans,} \\ \mathscr{C}(\phi,\lambda) &= 0 \quad \text{where there is land.}\end{aligned}\right\} \tag{4.2.8}$$

Likewise a continent function $\mathscr{C}'(\phi, \lambda)$ can be introduced, such that

$$\mathscr{C}'(\phi,\lambda) = 1 - \mathscr{C}(\phi,\lambda). \tag{4.2.9}$$

In terms of spherical harmonics the ocean and continent functions are

$$\mathscr{C}(\phi, \lambda) = \sum_{nm} (a_{nm} \cos m\lambda + b_{nm} \sin m\lambda) P_{nm}(\sin \phi), \tag{4.2.10}$$

$$\mathscr{C}'(\phi, \lambda) = \sum_{nm} (a'_{nm} \cos m\lambda + b'_{nm} \sin m\lambda) P_{nm}(\sin \phi), \tag{4.2.11}$$

with

$$a'_{00} = 1 - a_{00}$$

and for all other $n$, $m$

$$a'_{nm} = -a_{nm}, \qquad b'_{nm} = -b_{nm}.$$

The ocean function coefficients $a_{nm}$, $b_{nm}$ have been evaluated by Balmino, Lambeck & Kaula (1973) up to degree and order 36. Table 4.2 gives the normalized coefficients $\bar{a}_{nm}$, $\bar{b}_{nm}$ for some low-degree coefficients of the spherical harmonic expansion of the topography and of the ocean function. The unnormalized coefficients are given by (equation 2.4.4)

$$\left.\begin{matrix} a_{nm} \\ b_{nm} \end{matrix}\right\} = N_{nm} \left\{\begin{matrix} \bar{a}_{nm} \\ \bar{b}_{nm} \end{matrix}\right. .$$

## 4.3 Schematic excitation functions

Many of the geophysical phenomena perturbing the Earth's rotation can be approximated by simple excitations such as step, delta, ramp or periodic functions. The general properties of the solutions for such excitations are discussed in this section, while the discussion of specific geophysical excitations is deferred to later chapters. Consider first a step function in $\boldsymbol{\psi}$:

$$\boldsymbol{\psi}(t) = \Delta\boldsymbol{\psi}\, H(t),$$

where

$$\begin{aligned} H(t) &= 0 \quad \text{for } t < 0 \\ &= 1 \quad \text{for } t \geq 0. \end{aligned}$$

This represents an instantaneous shift in the position of the excitation pole at time $t = 0$. If, at $t < 0$, $m = 0$, the solution (3.2.12) of the

Table 4.2. *Coefficients $\bar{a}_{nm}$, $\bar{b}_{nm}$ (normalized) for the spherical harmonic expansion of the Earth's topography and of the ocean function.* (*From Balmino, Lambeck & Kaula, 1973.*)

| | | *Topography* | | *Ocean function* | |
|---|---|---|---|---|---|
| Degree | Order | $\bar{a}_{nm}$ (m) | $\bar{b}_{nm}$ (m) | $\bar{a}_{nm}$ | $\bar{b}_{nm}$ |
| 0 | 0 | −2300 | | 0.697 | |
| 1 | 0 | 639 | | −0.126 | |
| 1 | 1 | 591 | 409 | −0.108 | −0.056 |
| 2 | 0 | 515 | | −0.060 | |
| 2 | 1 | 336 | 306 | −0.040 | −0.051 |
| 2 | 2 | −415 | −80 | 0.040 | 0.002 |
| 3 | 0 | −174 | | 0.045 | |
| 3 | 1 | −135 | 115 | 0.044 | −0.032 |
| 3 | 2 | −446 | 458 | 0.070 | −0.089 |
| 3 | 3 | 116 | 545 | −0.016 | −0.089 |
| 4 | 0 | 373 | | −0.024 | |
| 4 | 1 | −222 | −249 | 0.036 | 0.030 |
| 4 | 2 | −382 | 56 | 0.090 | −0.021 |
| 4 | 3 | 360 | −118 | −0.053 | 0.005 |
| 4 | 4 | −57 | 449 | 0.014 | −0.101 |

equations of motion becomes

$$m(t) = -j\sigma_0 e^{j\sigma_0 t} \int_0^t \Delta\psi\, e^{-j\sigma_0 \tau}\, d\tau$$

$$= \Delta\psi(1 - e^{j\sigma_0 \tau}). \qquad (4.3.1)$$

At time $t = 0$, the excitation pole has jumped from $\psi = 0$ to a new position, $\psi = \Delta\psi$, about which the rotation axis now turns. If, at a later instant, the excitation pole jumps from $\Delta\psi$ to $\Delta\psi + \Delta\psi'$, the rotation pole moves about this new position without there being a break in the pole path (figure 4.1). An excitation function that consists of a sequence of step functions can be represented as the sum of a number of rectangular functions of variable length $a^{-1}$.

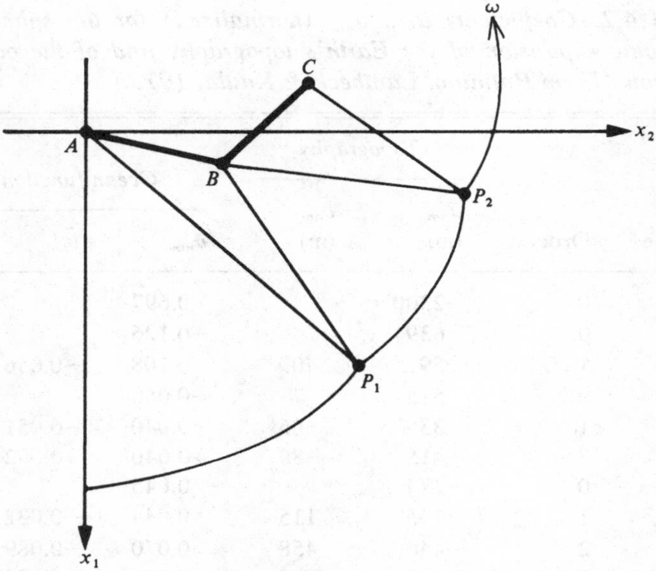


Figure 4.1. Motion of rotation axis $\boldsymbol{\omega}$ due to step function changes in $\boldsymbol{\psi}$. At time $T_1$, when $\boldsymbol{\omega}$ is at $P_1$, the excitation pole is shifted from $A$ to $B$; $\boldsymbol{\omega}$ now turns about $B$ with radius $BP_1$ until time $T_2$, when $\boldsymbol{\omega}$ is at $P_2$, and a second shift in the excitation pole occurs from $B$ to $C$.

The amplitude spectrum of one such rectangular function is

$$\mathcal{S}_{\psi}(f)=\left[\frac{\sin \pi f/a}{\pi f/a}\right]^2 \tag{4.3.2}$$

and falls off rapidly as the frequency increases, until it becomes zero at $f = a$. Beyond this point the spectrum consists of small-amplitude ripples. The longer the time constant $a^{-1}$, the more pronounced will be the spectrum at low frequencies. If $\boldsymbol{\psi}$ consists of a series of rectangular pulses, with variable time constants, the spectrum will be smoothed and the ripples vanish. The characteristic spectrum will be one of decreasing power with increasing frequency. This is in agreement with the result (4.3.1) that the mean excitation function shifts with each jump and the total motion after a sequence of jumps is a two-dimensional random walk.

The step function change in $\psi_3$ likewise leads to a spectrum of $m_3$ in which the signal decreases with increasing frequency. The spectrum in $\dot{m}_3$ varies with frequency according to

$$\mathscr{S}_{\dot{m}_3}(f) = (2\pi f)^2 \mathscr{S}_{\psi}(f) = (2\pi f)^2 \left[\frac{\sin \pi f/a}{\pi f/a}\right]^2 \tag{4.3.3}$$

and vanishes at zero frequency. Astronomers do not observe $m_3$ directly but measure the integrated amount by which the Earth is slow after an interval of a few days (chapter 5). That is, they measure

$$\text{Л} = -\int_t m_3 \, \mathrm{d}t. \tag{4.3.4}$$

A step function change in $\psi_3$ therefore causes a change in slope of Л (figure 4.2), and the resulting spectrum of Л is

$$\mathscr{S}_{\text{Л}}(f) = [1/(2\pi f)^2] \mathscr{S}_{\psi}(f). \tag{4.3.5}$$

A sudden impulse in the excitation $\boldsymbol{\psi}$ is described by the delta function

$$\delta(t) = 0, \qquad t \gtrless 0$$

$$\int_{-\infty}^{\infty} \delta(t) \, \mathrm{d}t = 1,$$

as

$$\boldsymbol{\psi}(t) = \boldsymbol{\Delta\psi} \, \delta(t).$$

In this case the partial solution of the wobble equations follows from (3.2.12) as

$$\begin{aligned} \mathbf{m}(t) &= -\mathrm{j}\sigma_0 \, \mathrm{e}^{\mathrm{j}\sigma_0 t} \, \boldsymbol{\Delta\psi} \int_{-\infty}^{\infty} \delta(t) \, \mathrm{e}^{-\mathrm{j}\sigma_0 \tau} \, \mathrm{d}\tau \\ &= -\sigma_0 \, \boldsymbol{\Delta\psi} \, \mathrm{e}^{\mathrm{j}(\sigma_0 t + \pi/2)}. \end{aligned} \tag{4.3.6}$$

Now the instantaneous rotation pole undergoes a jump of magnitude $\sigma_0 \, \boldsymbol{\Delta\psi}$, while the excitation pole remains at its original position. The jump of $\mathbf{m}$ is shifted $\frac{1}{2}\pi$ in phase from the direction of the impulse (figure 4.3). The power spectrum of a delta function is independent of frequency and a sequence of irregularly spaced delta functions also produces a white noise spectrum. The delta functions do not cause a shift in the position of the mean excitation pole. The corresponding spectrum of $\dot{m}_3$ is proportional to $(2\pi f)^2$ and vanishes at very low frequencies. For delta function impulses in $\psi_3$, the changes in the observed Л are represented by step functions,

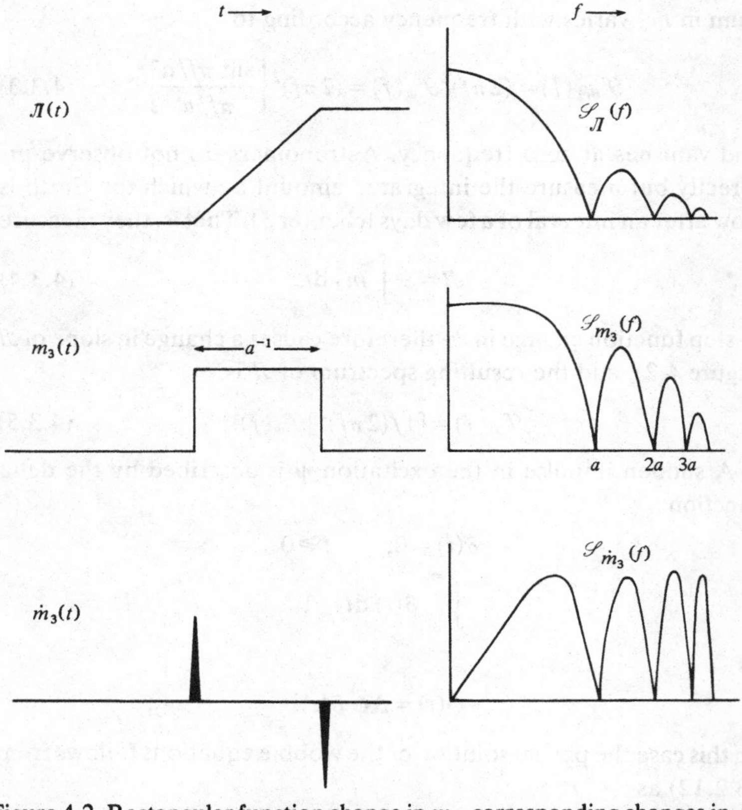


Figure 4.2. Rectangular function change in $m_3$, corresponding changes in Л and $\dot{m}_3$, and their power spectra.

and the spectrum of Л is enhanced at low frequencies according to $(2\pi f)^{-2}$. Figure 4.4 illustrates the case when Л$(t)$ is a rectangular function.

Geophysical problems that involve step function type changes in the inertia tensor have wobble excitation functions that consist of both a step and a delta function. That is,

$$\Delta\mathbf{I} = \Delta\mathbf{I}\, H(t),$$

$$\Delta\dot{\mathbf{I}} = \Delta\mathbf{I}\, \delta(t),$$

$$\boldsymbol{\psi} = \frac{1}{C-A}\left(\Delta\mathbf{I} - \frac{\mathrm{j}}{\Omega}\Delta\dot{\mathbf{I}}\right) = \frac{\Delta\mathbf{I}}{C-A}\left[H(t) - \frac{\mathrm{j}}{\Omega}\delta(t)\right].$$

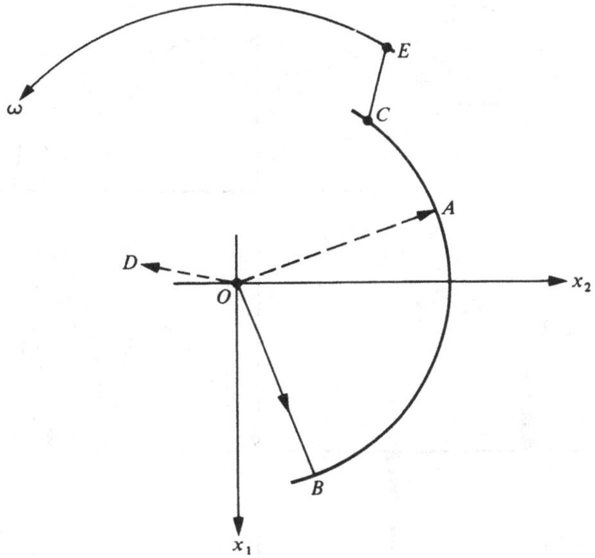


Figure 4.3. Motion of the rotation axis $\boldsymbol{\omega}$ due to delta function changes in $\boldsymbol{\psi}$. Initially $\boldsymbol{\omega}=0$, $\boldsymbol{\psi}=0$. At time $T_1$ a delta function excitation occurs and momentarily displaces $\boldsymbol{\psi}$ from $O$ to $A$. $\boldsymbol{\omega}$ jumps to $B$, $\frac{1}{2}\pi$ behind $A$ in phase, and now rotates about $O$. At time $T_2$, $\boldsymbol{\omega}$ is at $C$ and a second impulse occurs in the direction $OD$; $\boldsymbol{\omega}$ now moves to $E$ and continues to rotate about $O$ but with a modified radius.

With (4.3.1) and (4.3.6)

$$\mathbf{m}=\frac{\mathbf{\Delta I}}{C-A}\left[(1-\mathrm{e}^{\mathrm{j}\sigma_0 t})+\mathrm{j}\frac{\sigma_0}{\Omega}\,\mathrm{e}^{\mathrm{j}(\sigma_0 t+\pi/2)}\right]. \qquad (4.3.7)$$

But $\sigma_0/\Omega \simeq 1/430$, the last term is negligible compared with the first term, and it is the step function change that dominates $\boldsymbol{\psi}$. An example of an excitation of this type is the mass redistribution associated with earthquakes (chapter 8).

A secular change in the excitation function can be represented by a ramp function

$$R(t)=tH(t),$$

as

$$\boldsymbol{\psi}(t)=\boldsymbol{\Delta\psi}\; tH(t).$$

The solution of the equations of motion now is

$$\mathbf{m}(t)=[t-(\mathrm{j}/\sigma_0)]\,\boldsymbol{\Delta\psi}+(\mathrm{j}/\sigma_0)\,\boldsymbol{\Delta\psi}\,\mathrm{e}^{\mathrm{j}\sigma_0 t}. \qquad (4.3.8)$$

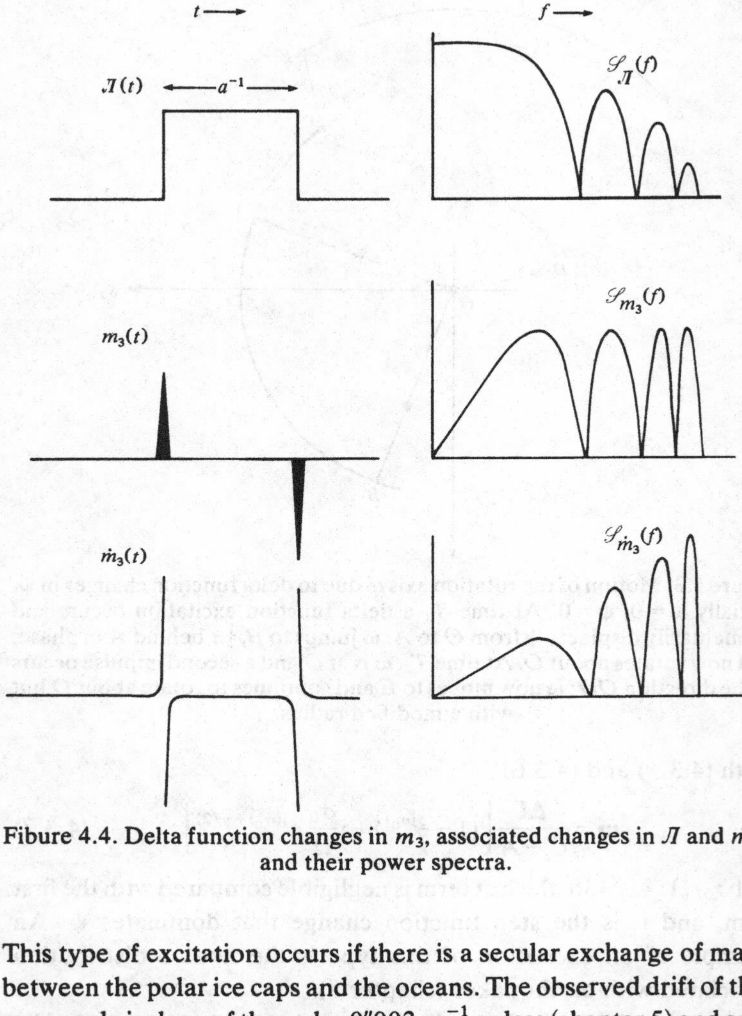


**Fibure 4.4. Delta function changes in $m_3$, associated changes in $\mathcal{I}$ and $\dot{m}_3$, and their power spectra.**

This type of excitation occurs if there is a secular exchange of mass between the polar ice caps and the oceans. The observed drift of the mean pole is slow, of the order 0.″003 yr$^{-1}$ or less (chapter 5) and any excitation of the wobble, the second term in (4.3.8), is small indeed. The ramp function gives a spectrum whose power at low frequencies is much enhanced compared with the previous spectra: for a step function the power spectrum is proportional to $(\frac{1}{2}\pi f)^2$ and the ramp function, being the integral of the former, has a spectrum that is proportional to $(\frac{1}{2}\pi f)^4$. A ramp function change in $\dot{m}_3$ implies a step function change in $\dot{m}_3$ or a quadratic variation in $\mathcal{I}$ (figure 4.5).

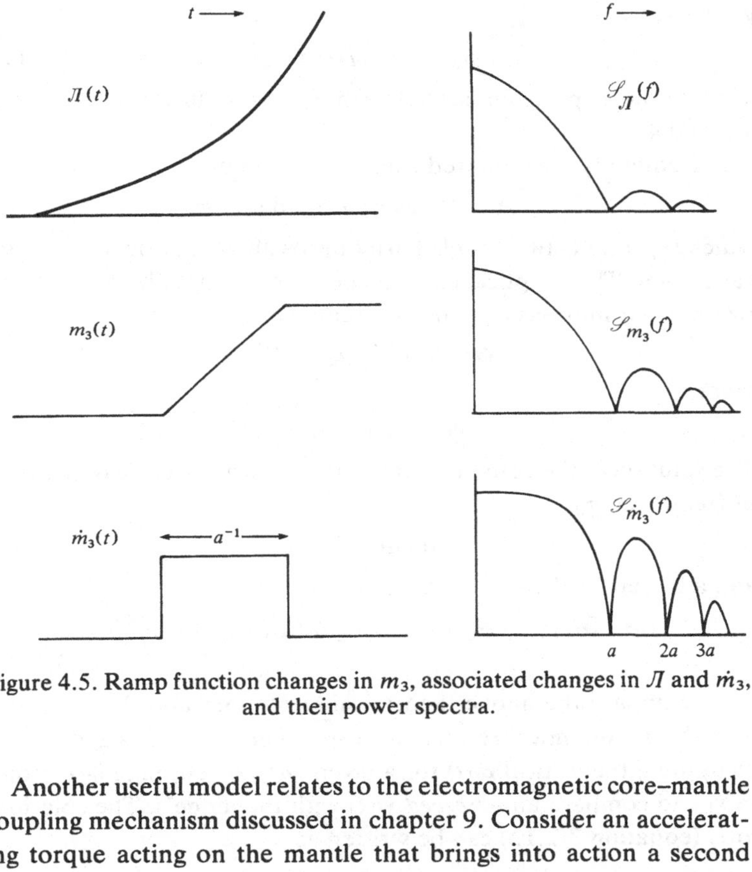


Figure 4.5. Ramp function changes in $m_3$, associated changes in Л and $\dot{m}_3$, and their power spectra.

Another useful model relates to the electromagnetic core–mantle coupling mechanism discussed in chapter 9. Consider an accelerating torque acting on the mantle that brings into action a second torque, opposite to the first, and growing in magnitude with time until a new equilibrium state is reached. That is,

$$\boldsymbol{\psi} = \Delta\boldsymbol{\psi}\, H(t)\, \mathrm{e}^{-at} \tag{4.3.9}$$

and

$$\mathbf{m}(t) = [\mathrm{j}\sigma_0/(a + \mathrm{j}\sigma_0)]\, \Delta\boldsymbol{\psi}\, (\mathrm{e}^{-at} - \mathrm{e}^{\mathrm{j}\sigma_0 t}). \tag{4.3.10}$$

At time $t$, the excitation pole jumps from $\boldsymbol{\psi} = 0$ to a new position $\mathrm{j}\sigma_0\, \Delta\boldsymbol{\psi}/(a + \mathrm{j}\sigma_0)$, then it drifts back to the original position with a time constant $a^{-1}$. The amplitude spectrum of $\boldsymbol{\psi}$ is proportional to

$$1/[a^2 + (2\pi f)^2],$$

and approaches $1/a^2$ at low frequencies. For $m_3$, a torque of this kind results in

$$\dot{m}_3 = \Delta\psi_3\, H(t)\, e^{-at}, \tag{4.3.11}$$

and its power spectrum lies between the cases illustrated in figures 4.2 and 4.5.

A frequently encountered excitation function is

$$\boldsymbol{\psi} = \boldsymbol{\psi}^c \cos \sigma t + \boldsymbol{\psi}^s \sin \sigma t,$$

which represents two simple harmonic oscillators at right angles to each other. This elliptical path can be considered as the resultant of two circular motions of opposite frequency or

$$\boldsymbol{\psi} = \boldsymbol{\psi}^+ e^{j\sigma t} + \boldsymbol{\psi}^- e^{-j\sigma t}, \tag{4.3.12a}$$

where

$$\boldsymbol{\psi}^{\pm} = \tfrac{1}{2}(\boldsymbol{\psi}^c \mp j\boldsymbol{\psi}^s).$$

The solution of the equations of motion consists of a free oscillation of frequency $\sigma_0$,

$$\mathbf{m} = \mathbf{m}_0\, e^{j\sigma_0 t},$$

and a forced oscillation of frequency $\sigma$,

$$\mathbf{m} = [\sigma_0/(\sigma - \sigma_0)]\boldsymbol{\psi}^+ e^{j\sigma t} + [\sigma_0/(\sigma_0 - \sigma)]\boldsymbol{\psi}^- e^{-j\sigma t}. \tag{4.3.12b}$$

To compare the above theoretical excitations and their spectra with the astronomical results, one can either proceed as above and compute a theoretical $\mathbf{m}(t)$ for a given $\boldsymbol{\psi}(t)$, or estimate $\boldsymbol{\psi}(t)$ from $\mathbf{m}(t)$ and compare an *observed* $\boldsymbol{\psi}(t)$ with the models. The solution $\mathbf{m}(t)$ (equation 3.2.12) can be written as

$$\mathbf{m}(t) = \int_{-\infty}^{t} e^{j\sigma_0(t-\tau)}\boldsymbol{\psi}(\tau)\, d\tau$$

$$= \int_{0}^{\infty} e^{j\sigma_0\tau}\boldsymbol{\psi}(t-\tau)\, d\tau,$$

or

$$\mathbf{m}(t) = \int_{-\infty}^{\infty} \mathbf{g}(\tau)\boldsymbol{\psi}(t-\tau)\, d\tau,$$

where

$$\mathbf{g}(\tau) = \begin{cases} 0 & t<0 \\ e^{j\sigma_0 t} & t>0 \end{cases}$$

is the wobble impulse response function. Thus $\mathbf{m}(t)$ is the result of a

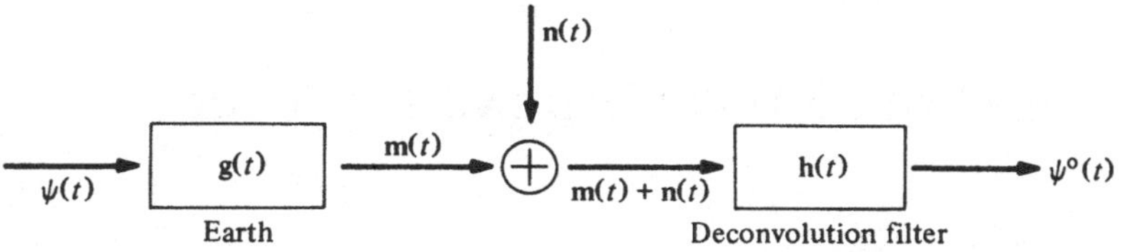


Figure 4.6. Schematic representation of the deconvolution of polar motion $\mathbf{m}(t)$ for estimating the excitation function $\boldsymbol{\psi}(t)$.

convolution of the excitation $\boldsymbol{\psi}(t)$ with this response function; the problem of deducing the excitation from the astronomically observed $\mathbf{m}(t)$ is one of deconvolution. The essential aspects of this deconvolution process are illustrated in figure 4.6. The excitation function acts on the Earth, whose wobble response is given by $\mathbf{g}(t)$. The resulting wobble is $\mathbf{m}(t)$ but, in the measuring process, noise $\mathbf{n}(t)$ is added. One now seeks a deconvolution filter whose response $\mathbf{h}(t)$ is such that its output $\boldsymbol{\psi}^0(t)$, caused by the input $\mathbf{m}(t)+\mathbf{n}(t)$, provides a good estimate of the true excitation function. The quality of the filter approximation is given by the quantity

$$\boldsymbol{\varepsilon}(t)=\boldsymbol{\psi}(t)-\boldsymbol{\psi}^0(t).$$

Smylie, Clarke & Mansinha (1970) use the optimum Wiener filter which minimizes the mean square of $\boldsymbol{\varepsilon}(t)$, and they discuss it in relation to the problem where the excitation function can be characterized by a two-dimensional random walk. A simpler deconvolution process that does not attempt to separate noise from signal in $\mathbf{m}(t)$ is to write, for a discrete observation series,

$$\mathbf{m}(t)=\mathrm{e}^{\mathrm{j}\sigma_0 t}\left[\mathbf{m}_0-\mathrm{j}\sigma_0\sum_0^t \boldsymbol{\psi}(\tau)\,\mathrm{e}^{-\mathrm{j}\sigma_0\tau}\,\Delta t\right]$$

and for the instant $t-\Delta t$

$$\mathbf{m}(t-\Delta t)=\mathrm{e}^{\mathrm{j}\sigma_0(t-\Delta t)}\left[\mathbf{m}_0-\mathrm{j}\sigma_0\sum_0^{t-\Delta t} \boldsymbol{\psi}(\tau)\,\mathrm{e}^{-\mathrm{j}\sigma_0\tau}\,\Delta t\right].$$

Thus

$$\mathbf{m}(t)=\mathbf{m}(t-\Delta t)\,\mathrm{e}^{\mathrm{j}\sigma_0\Delta t}-\mathrm{j}\sigma_0\boldsymbol{\psi}(t)\,\Delta t \qquad (4.3.13a)$$

and

$$\boldsymbol{\psi}(t)=-(1/\mathrm{j}\sigma_0\,\Delta t)[\mathbf{m}(t)-\mathrm{e}^{\mathrm{j}\sigma_0\Delta t}\mathbf{m}(t-1)]. \qquad (4.3.13b)$$

# FIVE

# THE ASTRONOMICAL EVIDENCE

## 5.1 Observations of the Earth's rotation

### 5.1.1 *The astronomical observations*

Precise measurements of the time elapsed between two consecutive transits of a star across a meridian determine the l.o.d. with respect to some uniform time scale. Measurement of l.o.d. therefore involves two processes: the astronomical observation of the star transits, and the establishment of a reference time. Changes in the l.o.d. are small, of the order of $10^{-8}$, and observations of numerous stars from a number of observatories, over several nights, are required in order that the signal rises above the noise of the measuring process. Thus, what is observed by astronomers is the integrated amount by which the Earth is in advance or behind after a number of days, compared with the uniform time scale. Time kept by the Earth is referred to as universal time (UT)). Strictly speaking, the time interval between successive star transits defines sidereal time, whereas UT is a mean solar time. The relation between these two systems is quite complex and is discussed in detail in most textbooks on spherical astronomy (see, for example, Smart 1962; Woolard & Clemence 1966). For geophysical purposes, the observed quantity can be considered to be the universal time. Star transits are observed with respect to an Earth-fixed meridian, defined by the station coordinates and the body-fixed **x**-axes. But the rotation axis moves with respect to this system, the polar motion, and small corrections to the transit times are required in order to determine the rotation about the instantaneous axis. UT corrected for this effect is denoted by UT1. The uniform reference time $\mathcal{U}$ can be defined and measured in several ways, but for the moment it is useful to consider it as the time that would be kept if the Earth rotated uniformly with angular velocity $\Omega$ equal to its mean velocity.

The observed quantity then is

$$\text{Л}_{\Delta T} = -(\text{UT1} - \text{И}),$$

the amount by which the Earth is slow after an interval $\Delta T$. Conventional usage is that $\text{Л}$ is positive if the Earth is slow relative to $\text{И}$. The change in the l.o.d., $\Delta$(l.o.d.), and the instantaneous rotational velocity $\omega_3$, or the perturbation in velocity $m_3$, are related by

$$m_3 = (\omega_3 - \Omega)/\Omega = -\Delta(\text{l.o.d.})/\text{l.o.d.} = -\mathrm{d}\text{Л}/\mathrm{d}t = (\mathrm{d}/\mathrm{d}t)(\text{UT1} - \text{И}). \tag{5.1.1}$$

Astronomers observe the station latitudes by measuring the zenith distances of stars crossing the meridian. For a star of declination $\delta_\mathrm{p}$, transiting at zenith distance $z_\mathrm{p}$ to the poleward side of the vertical, the astronomical latitude of the station is (figure 5.1)

$$\phi^\mathrm{a} = \delta_\mathrm{p} - z_\mathrm{p},$$

while for a star crossing to the equatorward side of the vertical

$$\phi^\mathrm{a} = \delta_\mathrm{e} - z_\mathrm{e},$$

and

$$\phi^\mathrm{a} = \tfrac{1}{2}(\delta_\mathrm{p} + \delta_\mathrm{e}) + \tfrac{1}{2}(z_\mathrm{e} - z_\mathrm{p}). \tag{5.1.2}$$

Thus, for two stars of known declination, the observed difference in zenith distance at meridian transit provides an estimate of the astronomical latitude. This is the basis of the Horrebow–Talcott method of positional astronomy. If, with time, the rotation axis changes its position relative to the Earth's surface, $\phi^\mathrm{a}$ also varies. Polar motion, the motion of the rotation axis with respect to the solid Earth, is therefore also referred to as the variation of (astronomical) latitude. In terms of $\mathbf{m}$, the change in astronomical latitude of a station $P_n$ at longitude $\lambda_n$ is

$$\Delta\phi_n^\mathrm{a}(t) = m_1(t)\cos\lambda_n + m_2(t)\sin\lambda_n. \tag{5.1.3}$$

Observations from two or more stations, well separated in longitude, determine the complete motion of the Earth's pole of rotation.

Latitude observations consist of measuring the positions of stars; l.o.d. measurements consist of recording the times at which stars occupy certain positions. Observations of both position and time provide all three components $m_1$, $m_2$ and $m_3$ of the instantaneous

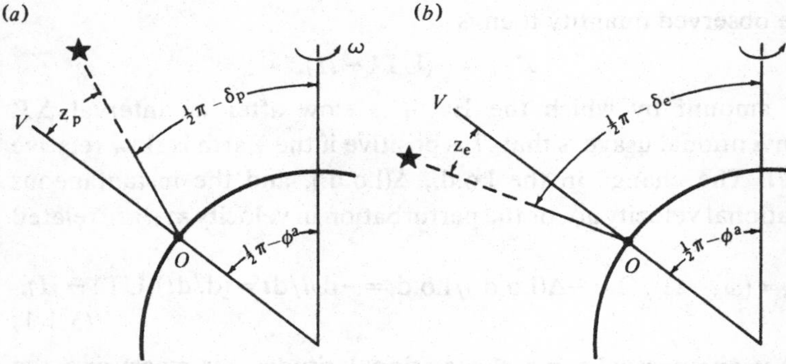


Figure 5.1. Geometry of latitude observations. The observer at $O$ observes ($a$) the zenith angle $z_p$ of a star transiting to the poleward side of the vertical $OV$ and ($b$) the zenith angle $z_e$ of a second star to the equatorward side of $OV$. These two observations determine the astronomical latitude $\phi^a$ of $O$ if the declinations $\delta_p$ and $\delta_e$ of the two stars are known.

rotation axis. Measuring these quantities is a delicate operation beset with numerous problems: instrumental limitations, atmospheric refraction, the definition of the meridian or vertical, and the non-uniformity and imprecision of the reference time scale all contaminate the results.

The early astronomical observations useful for studying the Earth's rotation consisted of occultations and transits of the Moon, Sun and planets, and were carried out with telescopes not specifically designed for measuring time and latitude. Only with the discovery of polar motion, at the close of the nineteenth century, were special instruments designed for these purposes. The principal instruments are (i) zenith telescopes, with which near-vertical transits of stars are observed, and which include visual telescopes and photo-zenith tubes, and (ii) prismatic astrolabes. The visual zenith telescopes, introduced in 1899 for latitude observations, use the Horrebow–Talcott method (equation 5.1.2). Two known stars are selected which cross the meridian within a few minutes of each other on opposite sides of, and close to, the vertical. The small difference in their zenith distances is measured with the aid of a micrometer, rather than by vertical circle readings. Precise levels are mounted on the instrument to relate the telescope axis to the vertical. The precision of the latitude measurement depends on instrumental and environmental factors, on the precision and calibration of the micrometer, on the precision with which the local vertical can be defined, and on the accuracy with which the star positions are known. Most of these error sources also perturb the other observation techniques discussed below. Perhaps the most important source of error is associated with the micrometer calibration. This is not known *a priori* with sufficient accuracy and has to be determined from the observations themselves. Any uncertainty in its value has immediate consequences on the pole path and, in particular, may introduce large errors in the annual component of the pole path: G. Cecchini (Melchior 1957) found that the introduction of improved values for the micrometer calibration modified the annual term in the pole path by as much as $0''.05$, which is comparable with the amplitude of the annual term itself! Melchior (1972) dwells further on this matter. Yumi (1970) estimates that the precision of a single night's latitude observation is of the order $0''.2$–$0''.3$.

Photo-zenith tubes (PZT), developed in 1911, represent some important improvements over the visual telescopes. The horizontal levels of the latter have been replaced by a mercury pool from which the star images are reflected on to a photographic plate in the telescope's focal plane. The image of a star in the zenith coincides with the lens's inner nodal point and neither a tilt nor a translation of the lens will alter the position of the zenith on the plate. Neither will images near the zenith be sensibly displaced by such motions. The photographic plate can be rotated about a vertical axis, and multiple exposures are made, before and after rotating the plate through 180°. The relative positions of the images determine the zenith distance of the star at the moment it transits the meridian (Markowitz 1960). If the instants of exposure are recorded with precision, the times of meridian transit and UT $-$ $\mathcal{H}$ are also determined. Tests from 1911 to 1914 indicated that these instruments, originally developed by the US Coast and Geodetic Survey for latitude observations, were superior to the visual telescopes already employed. In 1923 the PZT was used for time observations as well, and by 1934 the instrument was routinely used at the US Naval Observatory (USNO). At present some 10 PZT's contribute to the latitude and time determination by the Bureau International de l'Heure. Markowitz (1960) concludes that the standard deviation for one night's observations is $0^s.006$ in time and $0''.07$ in latitude.

An instrument widely used for latitude and time determination since its introduction in 1958 is the impersonal prismatic astrolabe of Danjon, developed at the Paris Observatory (Danjon 1958, 1960*a*). The time is recorded at which a star reaches a fixed altitude, 60° in the case of Danjon's instrument. To minimize personal errors in timing this instant, an impersonal micrometer is placed in front of the focal plane. By observing transits across the 60°-altitude circle, many more stars can be observed, particularly brighter stars, which are generally known with higher accuracy than the faint stars observed in the restricted field of the zenith telescopes. Débarbat & Guinot (1970) discuss the method of observing stars at fixed altitudes. Guinot (1958) discusses in detail the observation process, reductions and early results obtained with the astrolabe at the Paris Observatory, and concludes that, from one night's observations, the standard deviation of time determination is $0\overset{s}{.}0045$ and of latitude is $0\overset{''}{.}05$.

### 5.1.2 *Time*

Until about 40 yr ago, the Earth's rotation had been assumed to be constant, apart from possible secular or very long-period accelerations, and clocks served to subdivide the Earth's time into convenient units. Pendulum clocks have been the principal time-keeping device at observatories for several centuries. Joost Bürgi, an assistant of Tycho Brahe, was apparently the first to design a pendulum clock although the invention was virtually lost and rediscovered a century later by Christiaan Huygens in 1656. Jean Ricard soon introduced regular time observations at the Paris Observatory. Pendulum clocks were brought to a high state of precision by Rieffer in about 1890 and by Shortt in 1921, and by the mid-1930s these clocks were sufficiently precise to detect seasonal changes in the l.o.d. No longer could UT be considered a uniform time. Quartz crystal clocks, with greater stability, superseded pendulum clocks by about 1940. These clocks are essentially frequency standards in which an alternating electric current is applied across a crystal and the frequency of this current is adjusted until it resonates with the crystal's natural frequency. The frequency at which this resonance occurs depends upon the ambient pressure and temperature, and upon the elastic properties and the shape and size of the crystal. Stringent environmental control of the clock is essential to maintain high stability. Ageing of the crystals results in a drift of the resonance frequency that, in the past, could be determined only by comparing it with the Earth's rotation. Thus only changes in rotation with time scales comparable with the interval over which the crystals behave 'normally' can be observed with these clocks. Typically this is not much beyond a year. The dependence of the frequency and the drift rate on the crystal properties implies that no two clocks have identical natural frequencies, and no one clock can be exactly reproduced. These characteristics make the crystal clocks of only limited value in establishing a uniform time scale over periods longer than a few years. By 1950 the crystal clocks confirmed that the l.o.d. varied annually and led to the detection of a semi-annual oscillation and to suggestions of monthly and fortnightly variations in the l.o.d.

Crystal clocks do have very high stabilities over short time intervals, and they are now used in combination with stabler atomic frequency standards to provide a measure of uniform time. Atomic frequency standards use the natural vibrations of atoms or molecules: atoms in an initially excited state drop to a lower-energy state and radiate energy at a frequency that is characteristic of the atom and that defines a spectral line in the spectrum of the radiated energy. Caesium atoms provide the principal standard in defining atomic frequencies. The atoms are excited by electromagnetic energy, but only at resonance, when the frequency of the applied energy equals the frequency of the emitted energy, will the transition in energy state and the emission of the characteristic spectral line occur. A quartz crystal oscillator controls the applied electromagnetic field, and by regulating its frequency until emission occurs, the oscillator frequency is calibrated with respect to the standard atomic frequency. It is this combination of atomic frequency standard and quartz crystal oscillator that forms the atomic clock and defines the atomic time scale introduced in 1955. The importance of these clocks is that they are extremely stable and quite insensitive to temperature and other environmental effects. Frequency stability of caesium standards is of the order of 2 parts in $10^{13}$ $yr^{-1}$, and the uncertainty of the frequency is a few parts in $10^{12}$ (Terrien 1976). These stabilities far exceed the stability of the Earth as a time keeper, and the irregularities in rotation, both at high and low frequencies, can now be observed with accuracy and reliability.

The uncertainty in the l.o.d. determination is now entirely dominated by the astronomical part of the observation.

*The atomic time scale.* Since 1955, a number of atomic time scales have been used, the principal ones being A1, A3 and AT. The last (or TAI, Temps Atomique International) is a scale established by the Bureau International de l'Heure (BIH) in 1969, and represents an average of standards kept at several laboratories in Western Europe and North America. Each of these stations maintains two or more caesium standards – 16 standards are kept at the US Naval Observatory in Washington alone. They are related and compared by means of the Loran C navigation system and by transporting portable standards between observatories. Guinot, Feissel & Granveaud (1971) discuss the formation of AT in detail. From 1958 to 1969 the time scale kept by the BIH was A3, of which AT is a prolongation without discontinuity. The definition of A3 is given by Guinot & Feissel (1969). A1 is a scale kept by the US Naval Observatory and is also based on numerous frequency standards kept in North America and Western Europe. A1 differs from A3 by a nearly constant amount of −34.4 ms. Fluctuations between AT and A1 have not exceeded a small fraction of a millisecond since 1969.

*Ephemeris time.* To study the irregularities in the Earth's speed of rotation back beyond 1955 when atomic time first became available, a time scale determined from the astronomical data itself must be introduced. This is ephemeris time (ET), a uniform time scale that satisfies the Newtonian equations of motion of the Sun, Moon and planets and that is deduced from a comparison of astronomical observations with gravitational theories of these motions. Observed positions, with times measured by pendulums, refer to universal time, due to the inability of these clocks to maintain a time scale that is independent of the Earth's rotation for much longer than a year. Computed positions are based on the assumption that the Earth keeps a uniform time. Discrepancies between the two sets of positions determine the relation between the time scales, within the limitations of the observations and of the orbital theory. In principle, ET is defined in terms of the orbital motion of the Earth about the Sun, but in practice it is determined mainly from observations of the motion of the Moon since the latter's mean motion, as seen from the Earth, is the most rapid of all celestial bodies. The history of the discrepancies between the observed and computed positions goes back some three centuries. Edmund Halley in 1695 was apparently the first to suggest that the Moon may be accelerated in longitude, using as evidence the difficulty he experienced in reconciling certain ancient eclipse observations with the then available lunar theory. The subsequent contributions to an improved lunar theory by Lagrange, Laplace, Hansen, Euler, Adams, Delaunay and others are well known (Berry 1961; Munk & MacDonald 1960) and led to the conclusion, unaccepted at that time by many astronomers, that the Moon was being accelerated by an unknown force. Immanuel Kant, in 1754, had already proposed that tidal friction could explain such an acceleration, but this suggestion received little attention, mainly because it implied that the Sun should also be accelerated in longitude; this had not been observed. William Ferrel, in 1853, and Delaunay attempted to calculate the effects of tidal friction and stressed that all other planetary motions, observed with respect to the Earth's rotational time scale, should appear to be accelerated as well. Only in 1905 did Cowell discover the requisite solar acceleration.

The planetary observations usually consist of the UT of transits of the Sun, Moon or planets across the observer's meridian, the times of occultations of stars by the Moon, or transits of Mercury and Venus across the Sun's disc. Discrepancies between the observed and computed positions are most marked in the longitudes. If UT departs from ET due to an acceleration $\dot{\omega}_3$ of the Earth, then, with (5.1.1),

$$\mathcal{T}(t) = -(\mathrm{UT}-\mathrm{ET}) = -\int_t\int_t \frac{\dot{\omega}_3}{\Omega}\,\mathrm{d}t\,\mathrm{d}t \qquad (5.1.4a)$$

and in time $t$ the Sun will have moved through an angle

$$\Delta\lambda'_{\odot} = -n_{\odot}\int_t\int_t \frac{\dot{\omega}_3}{\Omega}\,\mathrm{d}t\,\mathrm{d}t, \qquad (5.1.4b)$$

where $n_{\odot}$ is the Sun's mean motion in longitude. We consider $\dot{\omega}_3$ to consist of a constant but small acceleration $\dot{\Omega}$ and a fluctuating part $\Delta\dot{\omega}_3$. That is,

$$\dot{\omega}_3 = \dot{\Omega} + \Delta\dot{\omega}_3.$$

Then

$$\Delta\lambda'_\odot = a'_\odot + b'_\odot t + \tfrac{1}{2}c'_\odot t^2 + \beta(t), \qquad (5.1.5a)$$

where

$$\beta(t) = \frac{n_\odot}{\Omega}\int_t\int_t \frac{\Delta\dot{\omega}_3}{\Omega}\,dt\,dt$$

represents the irregularly fluctuating part in the longitude discrepancy. If the Earth undergoes an acceleration in its *orbital* motion that has not been included in the theory, or, equivalently, the Sun is accelerated by $\dot{n}_\odot$ in its apparent motion around the Earth, the computed longitude is further in error by

$$\Delta\lambda''_\odot = a''_\odot + b''_\odot t + \tfrac{1}{2}c''_\odot t^2.$$

The total longitude discrepancy is

$$\Delta\lambda_\odot = \Delta\lambda'_\odot + \Delta\lambda''_\odot = a_\odot + b_\odot t + \tfrac{1}{2}c_\odot t^2 + \beta(t), \qquad (5.1.5b)$$

with

$$\begin{pmatrix} a_\odot \\ b_\odot \\ c_\odot \end{pmatrix} = \begin{pmatrix} a'_\odot \\ b'_\odot \\ c'_\odot \end{pmatrix} + \begin{pmatrix} a''_\odot \\ b''_\odot \\ c''_\odot \end{pmatrix}.$$

The constant $c_\odot$ is a function of the secular acceleration of the Sun, $\dot{n}_\odot$, and of the Earth, $\dot{\Omega}$. For the planets, the longitude discrepancy can be expressed by a relation similar to (5.1.5*b*) but only the inner planets have sufficiently rapid mean motions to provide useful observations. Spencer Jones (1939) found for Mercury and Venus, within observational errors, longitude discrepancies proportional to $\Delta\lambda_\odot$. Thus, for Mercury,

$$\Delta\lambda_{☿} = (n_{☿}/n_\odot)[a_\odot + b_\odot t + \tfrac{1}{2}c_\odot t^2 + \beta(t)]. \qquad (5.1.6)$$

The more recent analyses of the Venus observations by Duncombe (1958) and of the Mercury observations by Morrison & Ward (1975) confirm this. Also, R. R. Newton's (1976) study of the ancient planetary conjunctions and occultations (chapter 10) indicates that, within observational errors, (5.1.6) is valid. This implies that either (i) the Sun and planets do not undergo a significant secular acceleration and their longitude discrepancies are entirely caused by $\dot{\omega}_3$, or (ii) the Sun and planets undergo accelerations, not included in the theory, that are proportional to their mean motions. A lack of physically plausible mechanisms for the latter interpretation leads to its rejection in favour of the first. Hence,

$$\dot{\omega}_3(t) = -(\Omega/n_\odot)[c_\odot + \ddot{\beta}(t)]. \tag{5.1.7}$$

Figure 5.2 illustrates Spencer Jones's results for the solar declination and right ascension observations, the Mercury transits and the Venus longitudes. The quantities plotted are $(n_☾/n_\odot)[\frac{1}{2}c_\odot t^2 + \beta(t)]$ for the Sun and $(n_☾/n_{\rm p})[\frac{1}{2}c_\odot t^2 + \beta(t)]$ for the planets. The Moon, unlike Mercury or Venus, does not appear to follow the above rule (5.1.6), as is illustrated in figure 5.2 in which the quantity $[\frac{1}{2}c_☾ t^2 + \beta(t)n_☾/n_\odot]$ is plotted from the data given by Spencer Jones. Writing for the lunar longitude discrepancy,

$$\Delta\lambda_☾ = a_☾ + b_☾ t + \tfrac{1}{2}c_☾ t^2 + (n_☾/n_\odot)\beta(t), \tag{5.1.8}$$

the difference

$$\Delta\lambda_☾ - (n_☾/n_\odot)\,\Delta\lambda_\odot \tag{5.1.9}$$

is suggestive of a quadratic function (figure 5.2). This is compatible with an acceleration of the Moon in its orbit and a longitude discrepancy of

$$\Delta\lambda_☾'' = a_☾'' + b_☾'' t + \tfrac{1}{2}c_☾'' t^2. \tag{5.1.10}$$

Equation (5.1.10) plus the longitude discrepancy $(n_☾/n_\odot)\,\Delta\lambda_\odot$ caused by the Earth's rotation yield the total observed discrepancy given by (5.1.8) in which

$$\begin{pmatrix} a_☾ \\ b_☾ \\ c_☾ \end{pmatrix} = \begin{pmatrix} a_☾'' \\ b_☾'' \\ c_☾'' \end{pmatrix} + \frac{n_☾}{n_\odot}\begin{pmatrix} a_\odot \\ b_\odot \\ c_\odot \end{pmatrix}. \tag{5.1.11}$$

Tidal friction (chapter 10), by transferring angular momentum from the Earth's spin to the Moon's motion, is the mechanism accelerating the Moon in its orbit as well as causing a major part of the Earth's secular acceleration $\dot{\Omega}$. Such a transfer between the Earth and Sun makes an insignificant contribution to the Earth's motion about the Sun (chapter 10). This is also indicated by the observation that $\dot{n}_\odot \simeq 0$ and $\dot{n}_☿ \simeq 0$.

Observations of the Sun and inner planets determine ET by (5.1.4) and (5.1.5*a*). Lunar and solar observations provide the Moon's orbital acceleration $\dot{n}_☾$ not accounted for in the lunar

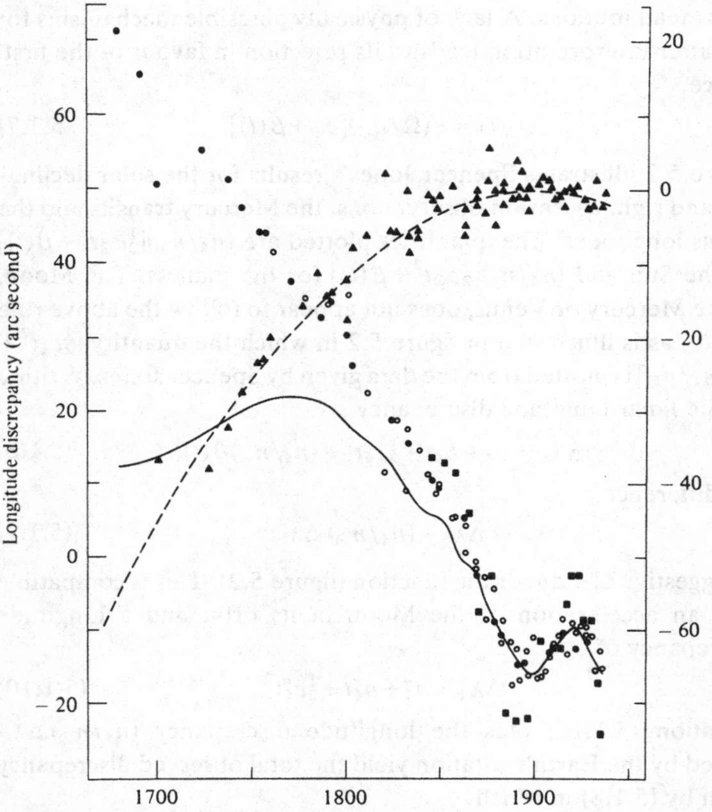


Figure 5.2. Longitude discrepancies of the Sun (○), Mercury (●) and Venus (■), expressed in the form $(n_☾/n_\odot)[\frac{1}{2}c_\odot t^2+\beta(t)]$ for the Sun and $(n_☾/n_p)\cdot[\frac{1}{2}c_\odot t^2+\beta(t)]$ for the planets. Also shown is the longitude discrepancy of the Moon (solid line) in the form $[\frac{1}{2}c_☾ t^2+\beta(t)n_☾/n_\odot]$. All data are from Spencer Jones (1939). The differences $(\Delta\lambda_☾ - \Delta\lambda_\odot n_☾/n_\odot)$ (indicated by ▲) suggest a quadratic function (broken line) and determine the acceleration of the Moon in its orbit.

theory, since, with (5.1.9),

$$\dot{n}_☾ = c_☾ - (n_☾/n_\odot)c_\odot = (\mathrm{d}^2/\mathrm{d}t^2)[\Delta\lambda_☾ - (n_☾/n_\odot)\,\Delta\lambda_\odot]. \tag{5.1.12}$$

But solar and planetary observations are generally less precise and less frequent than lunar observations. Also, a long time interval of data is necessary in order to separate the secular part $\dot{\Omega}$ from the long-period fluctuations $\beta(t)$: at least several centuries of data are required. This is seen from Morrison's (1978) analysis of the telescope observations; even 300 yr of data appears insufficient to separate the two. The first difficulty is overcome by estimating $\dot{\omega}_3(t)$ from the lunar data, once $\dot{n}_☾$ has been determined. That is, from the difference (5.1.9) $\dot{n}_☾$ is estimated (equation 5.1.12) and

$$\Pi(t) = -(\mathrm{UT}-\mathrm{ET}) = -(\Omega/n_☾)[\Delta\lambda_☾ - (a_☾ + b_☾ t + \tfrac{1}{2}\dot{n}_☾ t^2)].$$

The second difficulty can be resolved by introducing ancient eclipse observations into the solution for the accelerations (chapter 10).

## 5.2 Length-of-day

### 5.2.1 *The decade fluctuations*

Telescope observations of the l.o.d. go back to the seventeenth century and the observed changes are characterized by time constants of the order of 10–20 yr. These changes are loosely referred to as the *decade fluctuations.* The principal sources of data are the studies of Spencer Jones (1932, 1939), and Brouwer (1952) and they include solar, lunar and planetary observations. The most important planetary observations are the Mercury transits across the Sun's face, available from 1667 to the present. The time of contact between Mercury and the Sun's disc is recorded with respect to UT while the computed time of contact is based on ET. Mercury's orbit is inclined at some 7° to the ecliptic and transits can occur only when the Earth is near the planet's line of nodes. This occurs twice each year, in May at the descending node and in November at the ascending node. Mercury passes through the line of nodes during these months only once every 6–7 yr and the occurrence of the transits is rather infrequent. Contact is established four times during the transit: two external contacts – when Mercury lies outside the Sun's disc – and two internal contacts. Only the latter two are sufficiently precise. Wittman (1974) gives a recent discussion of observational problems, a subject with an extensive literature of its own. Clemence (1943) has analysed meridian observations of Mercury, in addition to the transits, but the former do not add any further significant information. Morrison & Ward (1975) have rediscussed the Mercury transits from 1677 to 1973. Transit observations of Venus across the Sun are less frequent than Mercury transits and are also, according to Spencer Jones, subject to large systematic errors and he does not use them in his solution. Right ascension observations of Venus, taken after 1835, are included, but their average scatter with respect to the Mercury transits is large (figure 5.2). Solar declination observations from 1760 onwards and right ascension observations since 1835 have also been used by Spencer Jones.

The lunar observations used by Spencer Jones consist of occultations rather than meridian observations as the latter can be subject to serious systematic errors. Brouwer (1952) has re-analysed these data, both occultations and meridian passages, to establish $\dot{\omega}_3(T)$ from 1621 to 1950 using Spencer Jones's results for the Sun's longitude discrepancy. The older data from 1663 to 1862 have been revised more recently by Martin (1969), and Morrison (1973) has compiled Brouwer's and Martin's data, together with modern observations, into as homogeneous a series as possible, referenced to the same star catalogues and astronomical and orbital constants. For the years 1621–1820 only isolated values for $\dot{\omega}_3$ are available but after 1820 the observations are sufficiently frequent to give annual values. Morrison's results after 1862 remain essentially those of Brouwer.† Stoyko's (1969) results for 1900–1955 also differ little from Brouwer's values for this period, any differences being a consequence of minor revisions of the ET scale and of different smoothing of the data. Oesterwinter & Cohen's (1972) solution for the Earth's irregular rotation is one result in their general adjustment for the orbital elements of the Moon and planets. Their results for the years 1912–1954 are independent of those by Brouwer in the sense that they use an independent orbital theory and that only a subset of the lunar observations, the transit observations of the US Naval Observatory, has been used. But they are less complete.

Of greater interest than the $\varPi(t)$ is the proportional change in l.o.d., or $m_3(t)$ defined by (5.1.1), and $\dot{m}_3(t)$. These have been determined by fitting a spline function through $\varPi(t)$ and differentiating the smoothed values, once to obtain $m_3(t)$ and twice to obtain $\dot{m}_3(t)$. The results based on Morrison's compilation are

† Morrison (1979) has re-analysed the lunar observations in the period 1861–1978. The results do not differ greatly from those in Figure 5.3.

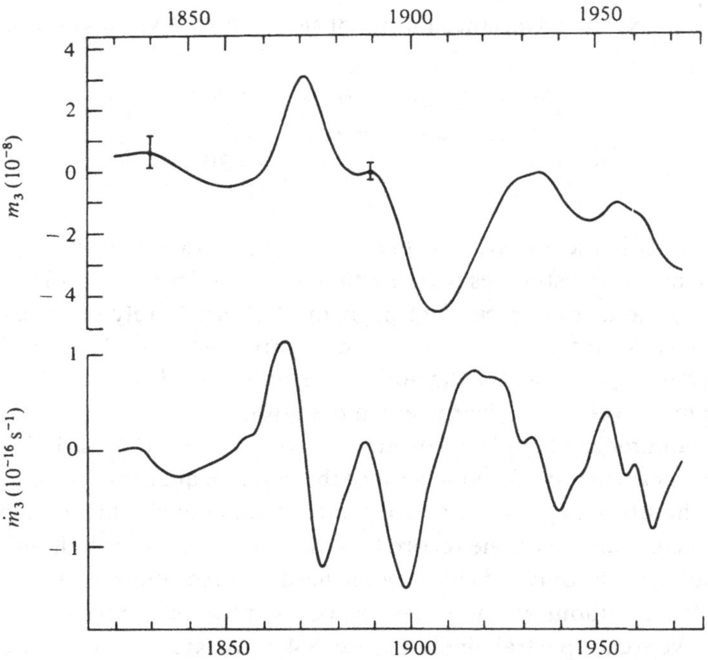


Figure 5.3. Values of $m_3$ and $\dot{m}_3$ from 1820 to 1975 based on the compilation of *Л*$(t)$ by Morrison (1973). A spline function has been fitted to *Л* using Brouwer's uncertainty estimates.

illustrated in figure 5.3. Brouwer estimates that the mean error $\varepsilon_{Л}$ of the annual values is of the order 1 s for the years 1820–1850, $0^{s}.4$ for 1850–1923 and $0^{s}.2$ for 1923–1950. It is unlikely that successive values of *Л* are uncorrelated and Brouwer's study indicates that differences between the occultation and meridian transits are frequently of a systematic nature with the same sign persisting for several years. If the correlated observations are assumed to follow a Markoff process, the autocorrelation function is reduced to $e^{-1}$ of its value after $t'$ yr, and the covariance matrix of successive 5-yr means of $m_3$ is given, with $t' = 5$ yr, by

$$\frac{v_{Л}^2}{2.5\times 10^{16}}\begin{pmatrix} 0.70 & -0.10 & -0.15 & -0.07 & -0.01 & \cdots \\ & 0.70 & -0.10 & -0.15 & -0.07 & \cdots \\ & & 0.70 & -0.10 & -0.15 & \cdots \\ & & & 0.70 & -0.10 & \cdots \\ & & & & 0.70 & \cdots \end{pmatrix}.$$

For $t' = 5$ yr the covariance matrix of successive 10-yr means of $m_3$ is

$$\frac{v_{\Pi}^2}{10^{17}}\begin{pmatrix} 0.88 & -0.30 & -0.13 & 0.00 & \cdots \\ & 0.88 & -0.30 & -0.13 & \cdots \\ & & 0.88 & -0.30 & \cdots \\ & & & 0.88 & \cdots \end{pmatrix},$$

where $v_{\Pi}^2$ is the variance of the annual mean values. Comparing these accuracy estimates with the time series of figure 5.3 indicates that (i) the astronomical data prior to 1850 are barely significant when 10-yr averages are computed, (ii) from 1850 to 1923 these 10-yr averages lie above the noise level, and (iii) only after 1923 do 5-yr mean values lie clearly above the noise in the data.

In a number of studies, spectral analyses of the l.o.d. variations have been attempted, but in view of the variable quality of the data, and the obviously long time constants of some of the fluctuations when compared with the record length, such analyses are of limited value: they are only indicative of the nature of the continuum of the spectrum without any implications of the physical significance of any eventual spectral lines. Figure 5.4 illustrates the smoothed power spectrum of the l.o.d. observations. If the variance of $\Pi$ is $v_{\Pi}^2$, and successive values of $\Pi$ are uncorrelated, the power spectrum of the observational errors is a white noise spectrum

$$\mathscr{E}_{\Pi}(f) = (v_{\Pi}^2/f_{\mathrm{N}})\ \mathrm{s}^3, \tag{5.2.1}$$

where $f_{\mathrm{N}} = (2\ \Delta t)^{-1}$ is the Nyquist frequency, and $\Delta t$ the sampling interval. For correlated observations following a Markoff process with $t' = 2\pi/a$,

$$\mathscr{E}_{\Pi}(f) = \frac{4\pi a}{a^2 + (2\pi f)^2} v_{\Pi}^2\ \mathrm{s}^3. \tag{5.2.2}$$

The error spectrum of $m_3$, the time derivative of $\Pi$, is

$$\mathscr{E}_{m_3}(f) = (2\pi f)^2 \mathscr{E}_{\Pi}(f)\ \mathrm{s}, \tag{5.2.3}$$

and that of $\dot{m}_3$ is

$$\mathscr{E}_{\dot{m}_3}(f) = (2\pi f)^4 \mathscr{E}_{\Pi}(f)\ \mathrm{s}^{-1}. \tag{5.2.4}$$

These latter spectra lead to an emphasis of the high-frequency contents of the spectrum. Figure 5.4 illustrates the error spectra of $m_3$ and $\dot{m}_3$ for the different values of $v_{\Pi}^2$ discussed earlier. The

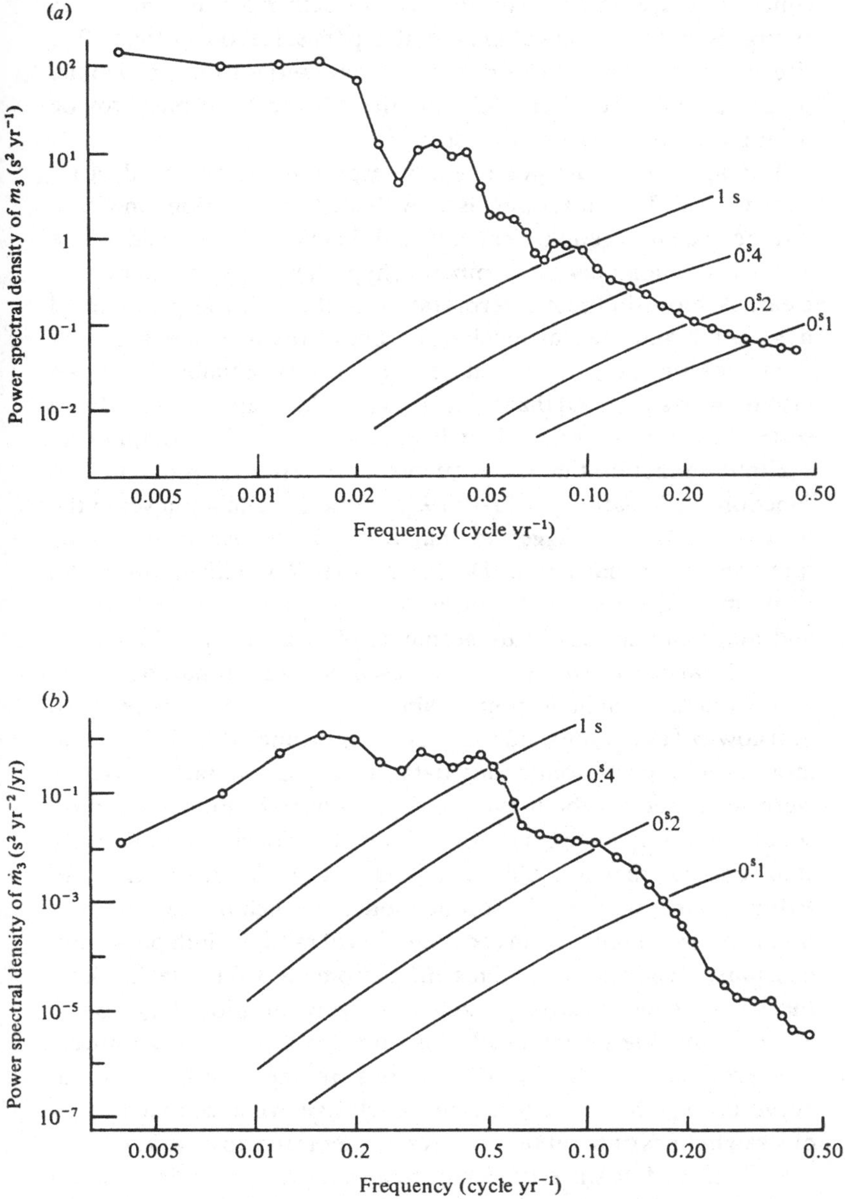


**Figure 5.4. Smoothed amplitude spectrum of (*a*) $m_3$ and (*b*) $\dot{m}_3$, and error spectra for different values of $v_{\Pi}$.**

conclusions are similar to those drawn directly from the time series comparison. Only changes in l.o.d. that persist for longer than 10 yr rise above the noise level for the entire data series from 1820 to the present, and only after 1923 do the ET observations provide information for periods down to 5 yr.

Two approaches are possible in interpreting the observed variations in l.o.d. The most direct is to evaluate the excitation functions, but the available geophysical data and theory are often inadequate and a distinction between competing hypotheses may not always be clear. A more indirect interpretation is through the geometrical description and the statistical properties of the time series. Such a discussion on the geometrical description was initiated by Brown (1926), who suggested that the $\Pi$-curve could be approximated by a series of segments of straight lines (figure 4.2). This implies an excitation function that is a sequence of step functions due to step function modifications of $\Delta I_{33}$ or $h_3$, or due to delta impulses in the torque $L_3$. Brown suggested that the changes occurred in time intervals of less than a year. De Sitter, in 1927, modified this model by permitting the changes to occur over an interval of several years, and approximated $\Pi(T)$ by segments of straight lines joined by parabolic arcs. Brouwer (1952) represented $\Pi(T)$ by parabolic arcs with common tangent points about every 10 yr (figure 4.5). Markowitz (1970) pursued this further and concluded that, for the more precise post-atomic time data, the successive tangent points were separated by about every 4–5 yr. Brouwer's model implies a sequence of step changes in $\dot{m}_3$ and this could be a consequence of step function changes in $L_3$ or of changes in $\mathrm{d}h_3/\mathrm{d}t$ or $\mathrm{d}\Delta I_{33}/\mathrm{d}t$. Brouwer also considered a second model in which the variations in $m_3$ are a consequence of an accumulation of random, independently distributed, changes in $\dot{m}_3$. This differs from his first model in which the $\dot{m}_3$ are constant over periods of varying duration. Figures 4.2 and 4.5 illustrate schematically the models of Brown and Brouwer. The problem is to fit the observations by one or other of these curve-fitting schemes. There is no conclusive evidence that a series of straight lines or parabolas better represents the observed $\Pi(T)$. If anything, the data since 1955 suggest an irregular periodic variation with a period of perhaps 6–8 yr superimposed on very long-period fluctuations (figure 5.5). Also, Brouwer's study of the pre-atomic

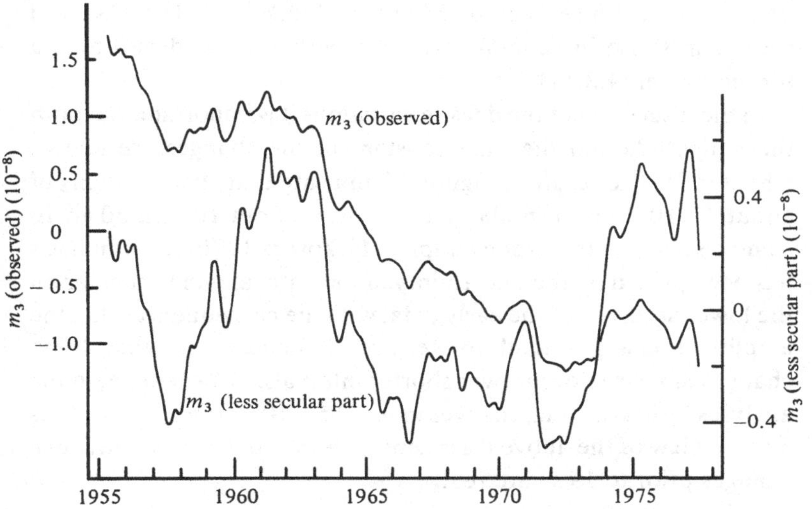


Figure 5.5. Time series of $m_3$ from 1955 to 1977 after removal of the 6-month and 12-month oscillations and after some smoothing to remove high-frequency (>6 cycle yr$^{-1}$) fluctuations.

time data and Markowitz's study of the recent data indicate that the more precise the data the shorter becomes the period over which the curves can be fitted with satisfactory precision, and the more frequent become the postulated changes in velocity or in acceleration.

Munk & MacDonald attempted to distinguish between these models by investigating the amplitude spectra. That is, they attempted to match the observed spectrum of $m_3$ or $\dot{m}_3$ by the spectra appropriate to the curve-fitting scheme and to measurement noise. Error spectra are given by (5.2.1–4). The power spectra appropriate to Brown's model are sketched on the right of figure 4.2, and those appropriate to Brouwer's model are sketched on the right of figure 4.5. The latter leads to a much more pronounced signal at low frequencies than does Brown's model. Brouwer's second model leads to a white noise spectrum for $\dot{m}_3$ and differs from Brown's model only near zero frequency. The observed spectrum for $\dot{m}_3$ is devoid of low-frequency information (figure 5.4($b$)), and indicates that Brown's model may be the most appropriate description of the l.o.d. decade fluctuations. This conclusion had already been reached by Munk & MacDonald. The observed spectrum is also in general agreement with that predicted by the torque model (4.3.11).

In the discussion of the decade variations it is important that both the magnitude and the time constant of the changes are known (chapter 9). The results in figure 5.3 indicate that changes in $m_3$ of about $4\times10^{-8}$ occur in about 10 yr, or of about $10^{-7}$ in 30 yr. In some discussions (see, for example, Markowitz 1970) less emphasis has been placed on the uncertainty of the data, and in particular on the lower accuracy of the early data, with the consequences that the amplitude changes tend to be over-estimated and that these changes appear to occur over shorter intervals. For example, in the analyses by Markowitz, changes in $m_3$ of $5\times10^{-8}$ occur in as little as 5 yr. In view of the above discussion, it is improbable that any such changes prior to 1923 are real.

### 5.2.2 *Higher-frequency variations*

By 1955 the quality of the astronomical observations had improved sufficiently so that, when the atomic time scale was introduced, an entirely new part of the l.o.d. spectrum rose unambiguously above the noise level. The main sources of results are the observations from individual observatories, in particular from the US Naval Observatory's PZT observations (Markowitz 1970), or from the Bureau International de l'Heure (BIH). An advantage of the former source is that the results are more homogeneous but they may also be subject to greater systematic errors than the BIH data, which are based on observations from many stations with different instruments, observing methods and star catalogues. The principal instruments contributing to the BIH results are the PZTs, astrolabes and photo-electric transit instruments used mainly in the USSR. In 1975 some 50 stations participated in the BIH program but only about 20 of them made a significant contribution. Since the PZTs and astrolabes also provide information on station latitude, the BIH determines time and latitude simultaneously. No homogeneous BIH data set exists at present, and from the various lists of results published at different times, the following compilation is considered to be the best available:

(i) For the years 1955.5–1962.0, revised values of UT2−AT are given in BIH (1965). UT2 is UT1 corrected for the seasonal changes in the Earth's rotation and was initially introduced in an attempt to establish a uniform time scale from UT1. It is of no interest now, except that we have to remove these seasonal terms to re-establish UT1. The corrections to be applied are

$$-\text{Л} = \text{UT1} - \text{AT} = (\text{UT2} - \text{AT}) - (\text{UT2} - \text{UT1})$$

with

$$\text{UT2} - \text{UT1} = (22 \sin 2\pi t - 17 \cos 2\pi t - 7 \sin 4\pi t + 6 \cos 4\pi t) \text{ ms},$$

where $t$ is the fraction of the year. For the interval 1955.5–1958.0 the atomic time was based on the unweighted mean of all frequency standards communicated to the BIH. After 1958.0 the A3 time scale was adopted and the passage from one to the other is assumed to have been smooth. The data points are given at 10-d intervals and have been subject to considerable smoothing. Their accuracy $v_{\text{Л}}$ is of the order of 3 ms.

(ii) For the years 1962.0–1967.0, the revised values published in the BIH annual report of 1970 are adopted. These values are given at 5-d intervals but they have also been subject to a very considerable smoothing, to the extent that their power spectrum will contain little significant power at frequencies above about 6 cycle $\text{yr}^{-1}$. The system of station longitudes, defining the terrestrial reference system, as well as the star catalogues, were revised at the beginning of 1962 and this may have introduced some discontinuity in the data at this time. For this interval $v_{\text{Л}} \simeq 2$ ms.

(iii) For the later years the unsmoothed observations published in the BIH annual reports are adopted. These values are given at 5-d intervals and their accuracy increased from about 1.5 ms from 1967 to 1972 to 1 ms for the more recent data (Guinot 1970*a*). The atomic time scale was changed from A3 to AT in 1969 with the transition being smooth. The l.o.d. data are subject to errors associated with the tidal deflections of the vertical (Guinot 1970*b*), errors which should be clearly distinguished from the real tidal fluctuations in rotation. The BIH annual reports tabulate the sum of the two effects but for geophysical work only the deflection corrections should be applied. The high-frequency tidal effects on polar motion are insignificant.

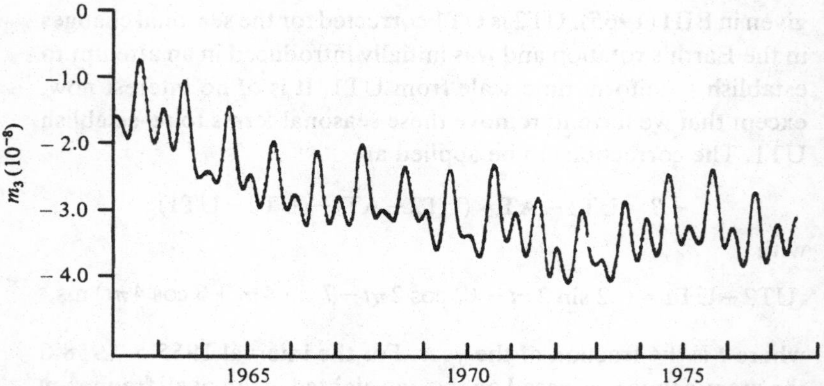


Figure 5.6. Time series of $m_3$ from 1962 to 1977 based on the smoothed values of $\Lambda$ published by the BIH.

The data since 1962 are homogeneous in that special attention has been given to maintaining a constant definition of the BIH reference system, even though some stations have been added or subtracted at different times and observation programs have been modified. The process is described by Guinot (1970*a*).

In the above-mentioned BIH reports, smoothed values are also given from 1962 to the present and these may be more useful for studies of seasonal and longer-period fluctuations for the entire data set. These smoothed values are given at 5-d intervals. Figure 5.6 illustrates the $m_3$, computed from $\Lambda$ in the same way as before (section 5.2.1). The seasonal terms are clearly evident in $m_3$. For $v_\Lambda = 3$ ms, the error spectrum of $\Lambda$ in

$$\mathscr{E}_\Lambda(f) = (0.003)^2/(2\times 10\times 86400)^{-1}\ \mathrm{s}^3$$
$$\simeq 15.0\ \mathrm{s}^3$$

for values of $\Lambda$ at 10-d intervals. The amplitude of the annual term in $\Lambda$ is about 20 ms and its spectral density is $(0.020)^2/(2\times 10\times 86400)^{-1}$. Hence the signal-to-noise ratio at the annual frequency is about 40 db for the data from 1955 to 1962. For the data since 1972, $\mathscr{E}_\Lambda(f) \simeq 1\ \mathrm{s}^3$ and the signal-to-noise ratio is about 60 db. For the semi-annual term, the signal-to-noise ratio is likewise high. Because of the large number of stations contributing to the l.o.d. determination since 1955, many error sources that have a seasonal character for individual stations – such as refraction, star catalogues and instruments – become less important. One possible exception

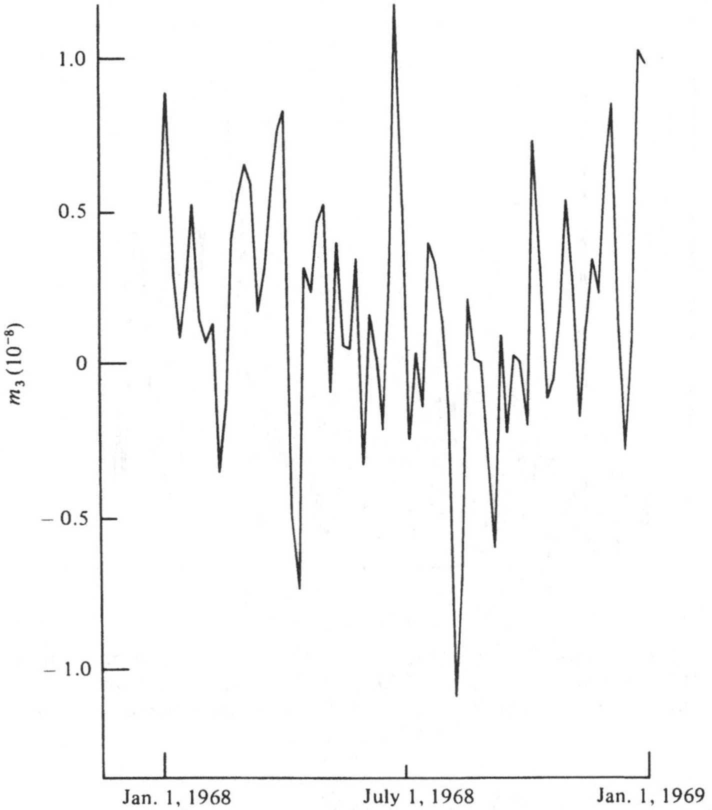


Figure 5.7. Time series of $m_3$ for 1968 based on the unsmoothed values of Л published by the BIH.

occurs in early 1962, due to the change in the stellar reference frame from the FK3 to the FK4 system. Okazaki (1975) has investigated this potential error source and concludes that its effect on the annual term is small, but that a modification of the amplitude of the semi-annual term can be expected, and that this term has been systematically underestimated prior to 1962.0. This complication was not recognized by Lambeck & Cazenave (1973) in their comparison of the seasonal wind excitation functions with the astronomical data for the years 1957–1963. Figure 5.5 illustrates the residual in $m_3$ after the removal of the annual and semi-annual oscillations.

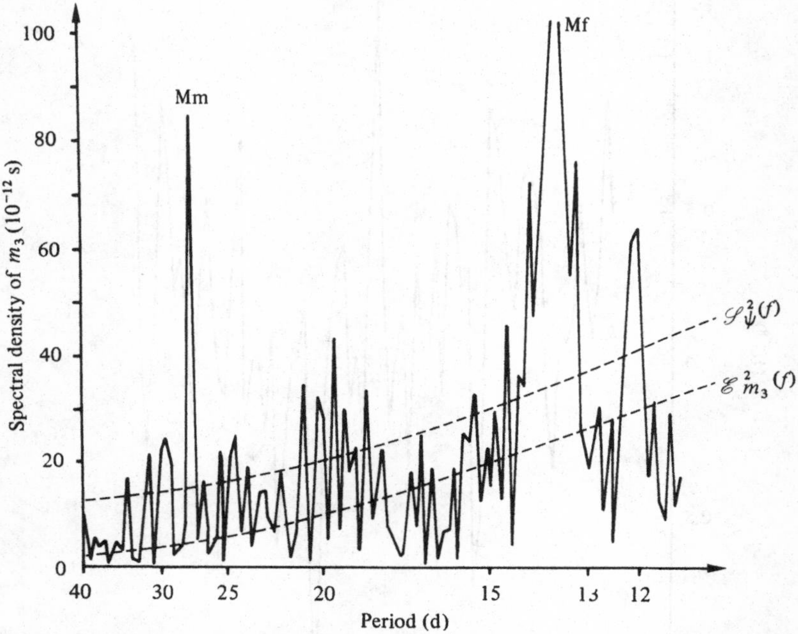


Figure 5.8. Unsmoothed amplitude spectrum of $m_3$. Tidal terms occur near 14 d (Mf) and near 27 d (Mm). The estimated error spectrum $\mathscr{E}_{m_3}(f)$ is based on $\nu_{\Pi} = 1$ ms. $\mathscr{S}_{\psi}(f)$ is the estimated continuum of the meteorological excitation spectrum.

The unsmoothed character of the post-1967 data is clearly evident from the time series in figure 5.7. The power spectrum of $m_3$ for these data is illustrated in figure 5.8. Tidal terms near 14 d and 27 d (chapter 6) rise well above the noise but the apparent noise level is greater than expected from the observational errors alone. With $\sigma_{\Pi} \simeq 1$ ms, $\mathscr{E}_m(f) = (2\pi f)^2 \mathscr{E}_{\Pi} = 0.9(2\pi f)^2$ s. This represents about one-half of the observed continuum at high frequencies and the difference is a consequence of meteorological noise, irregular and high-frequency changes in the zonal wind circulation (Lambeck & Cazenave 1974). Figure 5.9 illustrates a particularly abrupt change in $\Pi$ of some 10 ms within a 5-d interval (Guinot 1970*a*). The reality of such jumps is difficult to assess and their history is reminiscent of Hamlet's ghost: *'Tis here! 'Tis there! 'Tis gone!*

Guinot (personal communication, 1973) concluded that this particular discontinuity is probably real as its presence is suggested

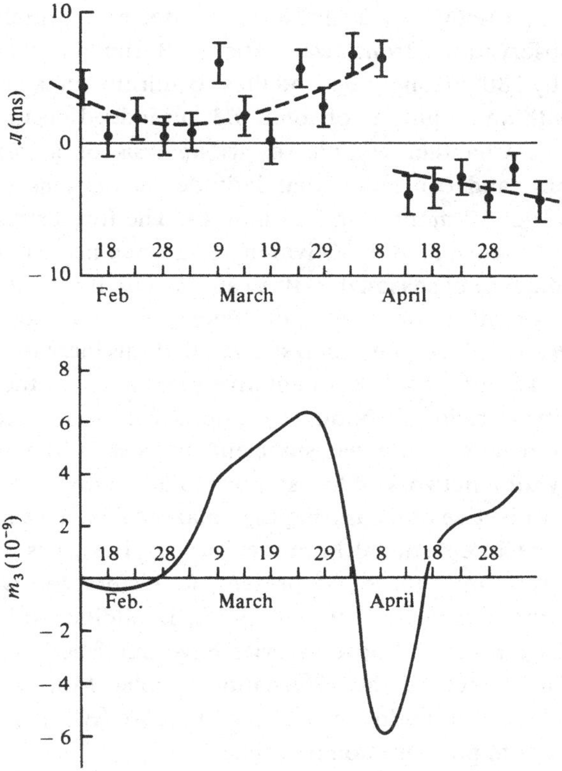


Figure 5.9. Variation in $\Lambda$ and $m_3$, after removal of seasonal terms, in early 1968.

in a number of the individual station observations, but latterly he has been more cautious. Possible explanations for these abrupt changes are discussed in chapter 7.

## 5.3 Polar motion

Euler, in 1765, concluded from his analysis of the rotational motions of rigid bodies that the Earth's pole of rotation could undergo a 305-d oscillation with respect to the crust. Lambert *et al.* (1931) have reviewed the unsuccessful search by astronomers for evidence for such a variation in astronomical latitude. Only in 1884–1885 did the observations yield evidence for it: observations taken at Berlin indicated minute changes in the astronomical latitude. This discovery led, in 1891, to a series of simultaneous latitude observations from two stations, Berlin and Honolulu, separated by 180° in longitude, and these confirmed that changes in latitude, with an amplitude of some 0″.3, do indeed occur. Meanwhile, S. C. Chandler completed his analysis of astronomical observations, and concluded that latitude fluctuations occurred with periods of 12 months and 14 months. The former result had already been anticipated by Kelvin in 1862 when he discussed the possible influence of seasonal shifts in air mass on the rotation axis, but the 14-month period was quite unexpected and contrary to Euler's theory. S. Newcomb soon showed that this increase from 10 months to 14 months could be qualitatively explained by the Earth's departure from rigidity (chapter 8). The Berlin observations and Chandler's results stimulated sufficient interest in the scientific community for a network of five stations to be set up in 1899, with the special objective of measuring the polar motion. This network, known as the International Latitude Service (ILS), has operated continuously from 1900 to the present. Its history and work has been recently reviewed by Stoyko (1973). In addition to the ILS, other organizations and observatories have published pole coordinates. These include the International Polar Motion Service (IPMS) and the Bureau International de l'Heure (BIH). The latter determines both polar motion and time.

### 5.3.1 *Data*

*International Latitude Service.* The ILS has observed the polar motion since 1900 using visual zenith telescopes and the Horrebow–Talcott method. The program consists of five stations at the same latitude, 39°08′ North, so that during the course of a year all stations observe the same stars. Observations each night consist of six to eight pairs of stars, and throughout the year 12 such groups are observed. To minimize the errors associated with the micrometer readings, these groups are selected so that the sum of the micrometer measures for a group is very nearly zero. But as precession slowly moves the stars out of the zeniths of the latitude stations, this condition is no longer satisfied after some years, and a new group of stars must be selected. Such changes in the star catalogues were made in 1906.0, 1912.7, 1935.0, 1955.0 and

1967.0. Uncertainties in the star positions may then introduce discontinuities in the pole path. To allow for catalogue errors, for errors in the fundamental constants defining nutation and aberration and for the neglect of certain parallax terms, the equation (5.1.3) expressing the relation between a change in latitude $\Delta\phi(t)$ and the pole coordinates is modified to

$$\Delta\phi_i(t) = m_1 \cos \lambda_i + m_2 \sin \lambda_i + z(t), \qquad (5.3.1)$$

where the correction term $z(t)$ absorbs all errors that are common to the stations. This term is referred to as the Kimura term. If, for example, there is an annual exchange of ocean and atmospheric mass between the northern and southern hemispheres then the effect on latitude is absorbed by the Kimura term. To absorb star catalogue errors, a correction term $z(t)$ is introduced for each group of stars. Recent attempts to interpret the $z(t)$ in terms of geophysical phenomena include the studies by Ishii & Naito (1974), Naito & Ishii (1974) and Naito (1974), but such interpretations remain uncertain due to the multitude of astronomical, observational and geophysical factors that contribute to them.

Discontinuities in the data occur with changes in the directors of the ILS who find it necessary to revise the reduction procedures. Thus Munk & MacDonald refer to the German (1900–1922), Japanese (1923–1934) and Italian (1935–1966) eras of the ILS, to which can now be added a second Japanese era from 1967.0 to the present under the direction of S. Yumi. For each of these periods, different star programs have been used and the pole positions have been referred to different origins due to changes in the adopted mean latitudes of the participating stations. Finally, although the ILS has provided a continuous service throughout 75 yr of international cooperation, the participation of the individual stations has been intermittent. Only two, Mizusawa (Japan) and Ukiah (California) have observed continuously. Carloforte (Sardinia) did not observe from 1943 to 1946; Gaithersburg (Maryland) did not observe from 1914 to 1931, while the original ILS station Tchardjui (USSR) was moved some 3° in longitude to Kitab with a gap in the observational record from about 1920 to 1930. A sixth station, Cincinnati (Ohio), operated early in this century for only a few years.

A number of ILS data sets have been compiled by various authors from the original data. One set, compiled by Walker and Young in 1957 and updated by Jeffreys (1968*a*) has had, until quite recently, widespread use. Other compilations include those by E. Proverbio, by Gaposchkin (1972) and by Stoyko (1973). The most complete revisions are those by Vicente & Yumi (1969, 1970) for the years 1900–1969. Despite these efforts, the data are still in an unsatisfactory state. The uniform revision of all ILS results by Melchior, announced by Munk & MacDonald in 1960, is still awaited, although some important preliminary steps have been taken (Melchior 1972). S. Yumi and R. Vicente anticipate that the final revisions will be completed by early 1980.

*Bureau International de l'Heure.* Since 1955, the BIH has routinely computed the position of the instantaneous rotation axis in an adjustment that also determines the rate of rotation. For a station at $\phi_n$, $\lambda_n$, the apparent changes in astronomical latitude and longitude due to $m_1$, $m_2$ and $\tau$ are

$$\begin{pmatrix}\Delta\phi_i\\ \Delta\lambda_i\end{pmatrix}=\begin{pmatrix}\cos\lambda_1 & \sin\lambda_1 & 0\\ \sin\lambda_1 & -\cos\lambda_1 & -2\pi/\text{l.o.d.}\end{pmatrix}\begin{pmatrix}m_1\\ m_2\\ \tau\end{pmatrix}.$$

PZTs, astrolabes, and visual zenith telescopes, including the five that constitute the ILS system, are used. In 1975 a total of 38 stations contributed to the pole position solutions with weights ranging from 1 to 100. Of these, about 28 dominate the final solution. As for the l.o.d. data, no homogeneous data set exists since the inception of the service. The data sources are the same as those for the l.o.d. discussed in section 5.2.2. Since 1967.0, the pole positions are unsmoothed and referred to a uniform system, the BIH system of 1968 (Guinot 1970*b*), and have been corrected for tidal deflections of the vertical and for the forced diurnal nutations. The resulting values are given in the annual reports of the BIH. Prior to 1967.0 only smoothed values are available. Accuracy estimates of the unsmoothed 5-d pole positions at successive 5-d intervals are of the order $0''.015$ (Guinot 1970*b*). Yumi (1970) estimates that the standard deviation of a monthly mean value of the ILS pole position is of the order of $0''.02$–$0''.03$. A comparison of

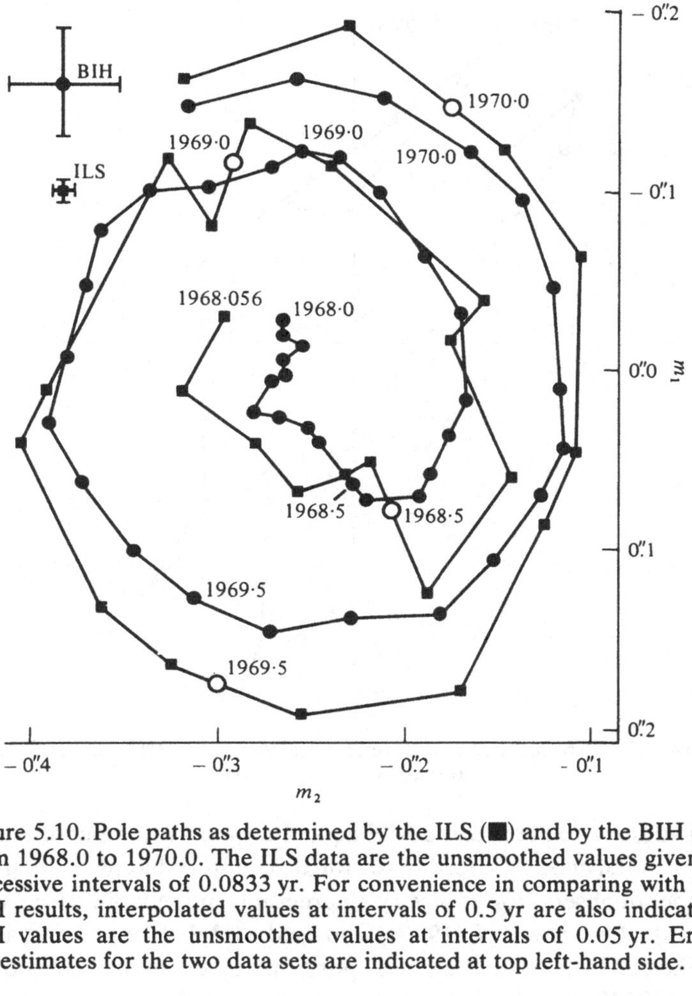


Figure 5.10. Pole paths as determined by the ILS (■) and by the BIH (●) from 1968.0 to 1970.0. The ILS data are the unsmoothed values given at successive intervals of 0.0833 yr. For convenience in comparing with the BIH results, interpolated values at intervals of 0.5 yr are also indicated. BIH values are the unsmoothed values at intervals of 0.05 yr. Error estimates for the two data sets are indicated at top left-hand side.

the BIH and ILS results reveals significant differences. Typical results are illustrated in figure 5.10. Part of the discrepancy is a consequence of the uncertainty in the ILS seasonal terms (section 5.3.2). Non-seasonal differences of 0″.1 or more also occur and may persist for several months. Furthermore, there appear to be systematic differences: annual mean differences between the ILS and BIH pole positions from 1962 to 1975 show fluctuations of the order 0″.02 in both $m_1$ and $m_2$ (figure 5.11).

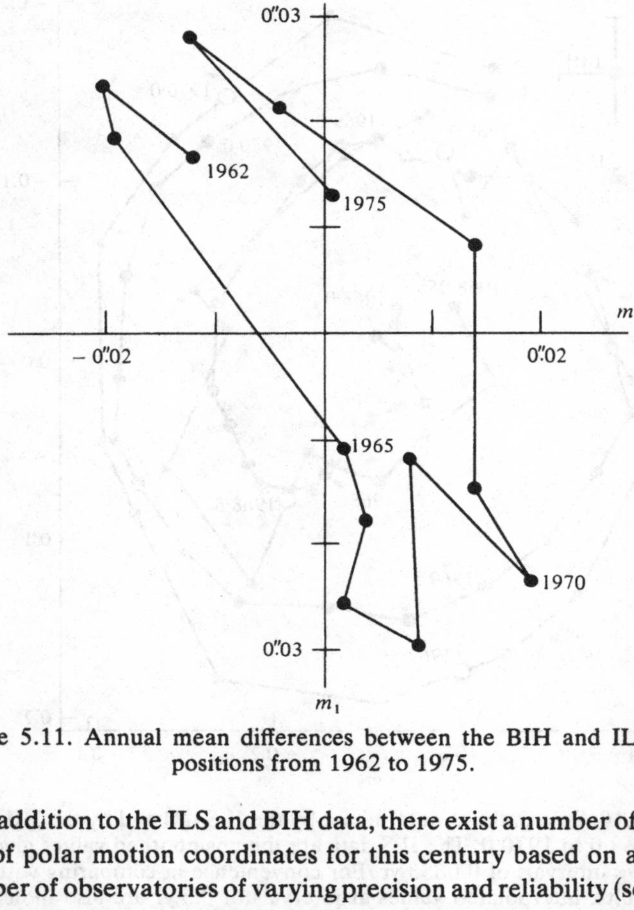


Figure 5.11. Annual mean differences between the BIH and ILS pole positions from 1962 to 1975.

In addition to the ILS and BIH data, there exist a number of other sets of polar motion coordinates for this century based on a large number of observatories of varying precision and reliability (see, for example, Federov *et al.* (1972), who used a total of 72 stations, and Stoyko (1973)). Since 1962, pole positions have also been provided by the IPMS, based on observations from 26 stations. These observatories include the five ILS stations and also contribute to the BIH results (see, for example, Yumi 1970).

### 5.3.2 *The wobble spectrum*

*Low-frequency variations.* Numerous analyses of the ILS data have indicated that the mean rotation pole experiences a slow secular drift of about 0″.003 $yr^{-1}$ in a direction 65°–75° West longitude (table 5.1). The uncertainty of this drift is about 20% in rate

Table 5.1. *Estimates of the secular polar motion from the ILS data*

| Author | Rate (arc second $yr^{-1}$) | Direction | |
|---|---|---|---|
| Markowitz (1970) | 0.0035 | 65° West | |
| Yumi & Wako (1970) | 0.0022 | 77° West | |
| Stoyko (1973) | 0.0040 | 73° West | (five ILS stations) |
| Stoyko (1973) | 0.0032 | 70° West | (three ILS stations) |

and 10° in direction. Whether this secular motion is apparent (a consequence of the measuring process, local crustal deformations and deflections of the vertical) or real (due to continental drift or a secular motion of the axis of rotation) is difficult to ascertain, as many of the studies use essentially the same data. The studies by Markowitz (1970), Yumi & Wako (1970) and Stoyko (1973) all use the ILS data from 1900 to about 1965. Any differences in the secular shifts are due to (i) small differences in the data sets, (ii) differences in the estimation process, and (iii), in the case of Yumi & Wako, to a removal of what are believed to be local latitude shifts of the Mizusawa and Ukiah stations. Stoyko (1973) has estimated the secular drift from only the three ILS stations that have provided a (nearly) continuous service since the inception of the ILS. This test is of some importance as it has sometimes been suggested that the secular pole shift may be a consequence of variations in the stations participating in the ILS. These results do not differ significantly from those based on all five stations, although her results do indicate rates that are very different for the last 10–20 yr of her analysis. A further independent estimate of the drift has been made by McCarthy (1972), who analysed the Washington PZT observations. His estimate of the change in latitude is comparable with that predicted by the above results.

The power spectrum of the polar motion at low frequencies does not shed much light on the nature of the secular excitation, for, with only 70 yr of data, the number of spectral estimates is insufficient. If the secular motion is a consequence of a two-dimensional walk, its spectrum will vary with frequency according to $f^{-2}$ (section 4.3) and according to $f^{-4}$ if it is due to a systematic motion of the pole in the same direction. If the excitation function takes the form of delta function impulses, the spectrum will vary with frequency according to $f^{-0}$. Mandelbrot & McCamy (1970) have attempted to distinguish between these models using a technique they call cumulative range analysis. Their conclusions are, however, very data dependent. Using the ILS data set compiled by Walker & Young, they concluded that the secular motion follows a process between white noise and Brownian motion, while in a second analysis (see their note added in proof) of an improved ILS data set, they concluded that below about 0.15 cycle $yr^{-1}$ the spectrum is characteristic of a Brownian motion process, in which the pole wanders because of consecutive and independent jumps in the excitation function. A new application of this method to more recent compilations of the ILS data may be of some interest.

Several attempts have been made to deduce the rates of motion of the continents from the polar motion data by assuming that the station motion is representative of the motion of the tectonic unit to which it is attached (Arur & Mueller 1971; Proverbio & Quesada 1973). In particular, attempts have been made to test the so-called plate tectonics hypothesis, the modern version of the older concept of continental drift. An essential aspect of the plate tectonics model is that it represents a time-averaged situation: paleomagnetic observations provide relative motions that are averages over $10^6$ yr and longer, and seismic observations along the plate boundaries indicate that at least 100 yr of averaging are required in order to deduce motions that will be representative of the motions of the plates as a whole. A second aspect of the model is that the paleomagnetic data provide only relative displacements; absolute motions, relative to the deep interior of the Earth, can only be surmised from additional observations, for example the *hot spots* model of Minster *et al.* (1974). Globally, the plate tectonics hypothesis is based on a small number of plates: six in Le Pichon's (1968) model, 10 in Minster *et al.*'s model. On this macroscopic scale the boundaries are relatively simple zones of compression, extension or lateral motion. Viewed microscopically the boundaries are much more complicated, with the deformation extending over large areas and often with the formation of small boundary plates. In the

Mediterranean, for example, a number of small plates exist that apparently move orthogonally to, and at much higher rates than, the main Eurasian and African plates between which they are sandwiched.

In view of the above few comments, the use of the astronomical data to test the plate tectonics hypothesis is quite limited. In the first place the astronomical data can give average motions over only some 70 yr and, secondly, four of the five ILS stations lie near the plate boundaries and may be subject to deformations that will not be representative of the plates as a whole. Mizusawa lies on the Eurasian plate, close to the subduction zone forming this plate's margin with the Pacific plate. Kitab and Carloforte also lie on the Eurasian plate, close to the southern compressive margins bordering on the Indian–Arabian plates and the African plate. Ukiah lies barely on the North American plate, very close to its demarcation line with the Pacific plate. All four stations are located in regions of high seismic activity. Only Gaithersburg is located in a relatively stable region away from plate boundaries. Two plates enter into the discussion, the Eurasian plate and the North American plate, two of the slower-moving plates in Minster *et al.*'s model of absolute plate motions.

Rather than solving for the plate motions, we attempt a direct calculation of the contribution that these motions may make to the pole path. We compute only the geometric effect and do not consider the changes in the inertia tensor that may be associated with the tectonic motion and mantle convection. The latter calculation makes little sense while neither the mechanism driving the plates nor the mantle flow at depth is known.

The plate motion is represented by a rigid body rotation on the sphere about a pole $P'(\phi', \lambda')$, through an angle $\mathrm{d}A$ measured positive if the rotation is clockwise. The station $P_i(\phi_i, \lambda_i)$ is assumed to move with this plate. The rotation $\mathrm{d}A$ displaces $P_i$ in latitude by

$$\mathrm{d}\phi_i = \frac{\cos A \sin l \cos \phi_\mathrm{p} \sin (\lambda_i - \lambda_\mathrm{p})}{\sin \phi_\mathrm{p} \cos \phi_i \cos (\lambda_i - \lambda_\mathrm{p}) - \sin \phi_i \cos \phi_\mathrm{p}} \mathrm{d}A,$$

with

$$\cos A = (\sin \phi_\mathrm{p} \cos l - \sin \phi_i)/(\cos \phi_\mathrm{p} \sin l),$$

$$\cos l = \sin \phi_\mathrm{p} \sin \phi_i + \cos \phi_\mathrm{p} \cos \phi_i \cos (\lambda_i - \lambda_\mathrm{p}),$$

where $A$ is the angle subtended at $P$ by $P_i$ and the Earth's rotation axis, and $l$ is the distance $PP_i$. Minster *et al.* give $\phi_p = 38.1°$, $\lambda_p = 110.5°$ West, $dA = -0°.12 \times 10^{-6}$ yr$^{-1}$ for the Eurasian plate, and $\phi_p = -48.1°$, $\lambda_p = 82.1°$ West, $dA = -0°.24 \times 10^{-6}$ yr$^{-1}$ for the North American plate. With a weight for each station that is proportional to the period of observation, the apparent shift in the pole position is

$$\dot{m}_1 = -1''.7 \times 10^{-4}\ \text{yr}^{-1},$$

$$\dot{m}_2 = -0''.4 \times 10^{-4}\ \text{yr}^{-1}.$$

This is a magnitude smaller than the observed drift. The observed polar motion cannot be explained as an apparent motion resulting from the station displacements, *if* the model proposed by Minster *et al.* is representative of the plate motions over the last 70 yr and *if* the stations move with the plates. Dickman (1977) obtains comparable results.

Superimposed on the ILS secular pole shift is a quasi-long-period oscillation which Markowitz (1970) interprets as an approximately linear motion in a direction 60° East–120° West longitudes. The two results (figure 5.12) by Markowitz and by Stoyko (1968) are based on the same basic data but to which different corrections and smoothing techniques have been applied. Only from about 1927 onwards are the two curves similar. The significance of these fluctuations since 1927 is dubious. Some rapid changes coincide with changes in the star catalogues (for example, 1923 and 1935). Another change near 1946 occurred at the time when the Carloforte station was re-introduced into the ILS after a 3-yr absence. The important change in direction in 1950 coincides with a time when Cecchini made some modifications in the reduction process. In view of these coincidences, Melchior's (1957, and elsewhere) warning that these oscillations may well be a consequence of changes in star catalogue and instrumental errors, needs heeding. A comparison of figures 5.11 and 5.12 further confirms that the long-period wobble may be an artifact of the observation and reduction processes since the differences between the annual mean pole positions of the BIH and the ILS from 1962 to 1968 reveal a drift that is in the same general direction and magnitude as the secular ILS pole determined by Markowitz for the same period.

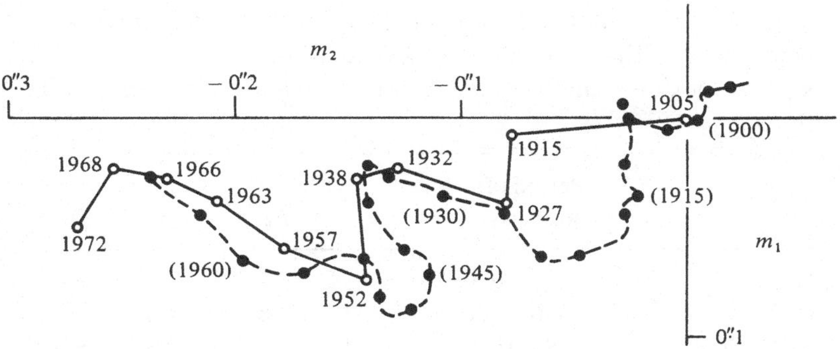


Figure 5.12. Motion of the mean pole according to Stoyko (●, dates in parentheses). Pole positions represent running means based on 6-yr intervals. Positions every 3 yr are indicated. The motion of the mean pole according to Markowitz is indicated by ○; these positions also represent 6-yr means.

*Higher-frequency variations.* The principal characteristics of the polar motion are the 12-month and 14-month terms. The former is a forced oscillation due to seasonal shifts in the mass distribution in the atmosphere, oceans and groundwater, while the latter, the Chandler wobble, is a free oscillation of the Earth. A semi-annual oscillation in polar motion, also associated with meteorological, oceanographic and hydrological excitations, is very small since its frequency is much further from the resonant Chandler frequency than the annual oscillation (equation 4.3.12). Observations of the annual wobble of the rotation axis are beset with a number of difficulties. Optical refraction remains an important error source in individual station latitude results, even for near-zenith observations. Anomalous conditions can introduce errors of as much as 0″.1–0″.2 and may persist for several weeks, while seasonal variations in the atmospheric conditions may introduce apparent annual oscillations in latitude. Possibly more important are the annual errors introduced by erroneous micrometer calibrations for the visual zenith telescope observations of the ILS. A comparison of the five ILS station latitudes with those predicted by the BIH pole path shows, in fact, quite large discrepancies that follow a seasonal pattern: for 1975 the root-mean-square value of the discrepancy in amplitude is nearly as large as the amplitude of the annual term

Table 5.2. *Estimates of the annual components $m_1$ and $m_2$ in the polar motion. The $m_i$ are expressed in the form* $m_i = a_i \cos \odot + b_i \sin \odot$, *where* $\odot$ *is the longitude of the mean Sun, in units of* $10^{-2}$ s

| | Period of analysis | $a_1$ | $b_1$ | $c_1$ | $d_1$ |
|---|---|---|---|---|---|
| ILS | 1956–1970 | −54 | −95 | −70 | −38 |
| BIH | 1956–1970 | −32 | −87 | −74 | −27 |
| Gaposchkin (1972) ILS | 1891–1970 | −43 | −80 | 58 | −34 |
| Jeffreys (1968) ILS | 1899–1961 | −66 | −62 | 60 | −45 |
| | 1899–1905 | −26 | −50 | 59 | −10 |
| | 1906–1912 | −50 | −61 | 60 | −33 |
| | 1913–1919 | −61 | −62 | 77 | −66 |
| | 1920–1926 | −54 | −89 | 86 | −49 |
| | 1927–1933 | −78 | −80 | 94 | −70 |
| | 1934–1940 | −74 | −56 | 60 | −50 |
| | 1941–1947 | −40 | −58 | 52 | −18 |
| | 1948–1954 | −107 | −67 | 57 | −62 |
| | 1954–1961 | −102 | −39 | −7 | −45 |

itself. Wells (1972) has analysed the variation in latitude for a number of stations and concludes that, because of these discrepancies, the annual polar motion term estimated from the ILS data does not accurately reflect the true annual wobble of the Earth.

The situation for the BIH results may not be much better for, despite the greater number of stations contributing to the pole determination, the annual term is given primarily by those stations that define the 1968 BIH system. Table 5.2 summarizes some recent estimates of the annual wobble based on ILS and BIH data. The first two results are for 15 yr of BIH and ILS data. The agreement between the two is perhaps not as poor as implied by the above comparisons over shorter periods: possibly, systematic errors in the ILS data vary sufficiently with time for the total effect over several decades to be much reduced, particularly if several observation programs are covered in the interval. The two estimates based on longer but different versions of the ILS data are in

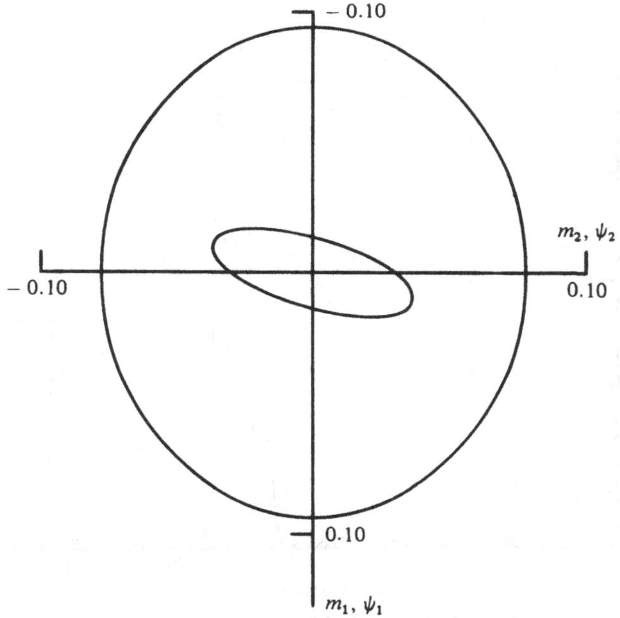


Figure 5.13. Average annual wobble of the Earth's axis of rotation as estimated from the BIH data from 1956 to 1970. The smaller ellipse represents the inferred annual excitation function required to maintain this wobble.

reasonable agreement. Jeffreys (1968*a*) has estimated the average annual wobble for 7-yr periods from the ILS data from 1899 to 1961 and finds a variation in amplitude by a factor of 2 (table 5.2). Whether or not all of this fluctuation can be attributed to the observation and reduction processes, or whether it is due to changes in the annual excitation function, is not clear and is further discussed in chapter 7. Figure 5.13 illustrates the average annual pole path as determined from the BIH data (line 2 of table 5.2). The motion is slightly elliptic with the major axis lying some 10° east of Greenwich. The inferred annual excitation function (equation 4.3.13) is a nearly rectilinear oscillation in the meridian planes 90° East and 90° West.

Figure 5.14 illustrates the power spectrum of 70 yr of ILS polar motion data. The spectrum in figure 5.14(*a*) is unsmoothed and shows a characteristic double-peak structure, and it has been

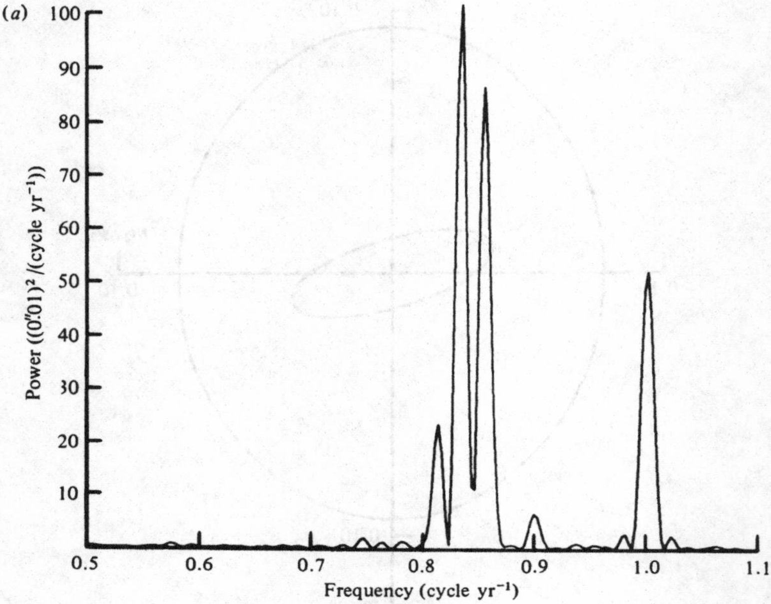
(a)
Power ((0″.01)² /(cycle yr⁻¹))
100
90
80
70
60
50
40
30
20
10
0.5
0.6
0.7
0.8
0.9
1.0
1.1
Frequency (cycle yr⁻¹)

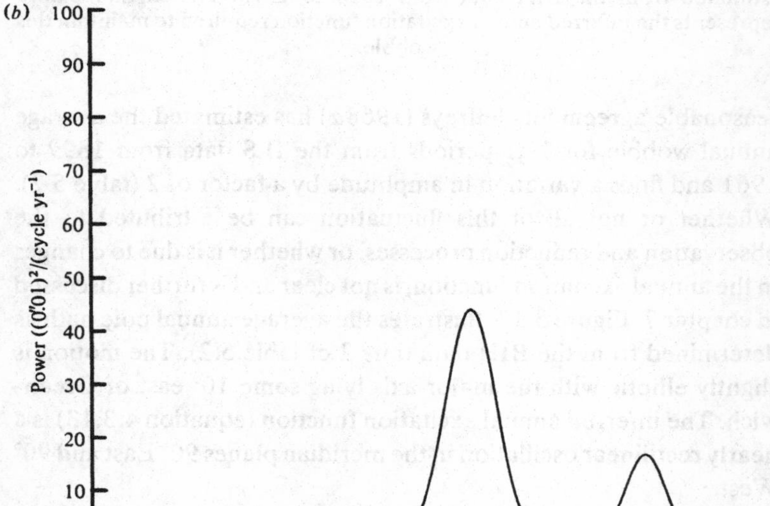
(b)
Power ((0″.01)²/(cycle yr⁻¹))
100
90
80
70
60
50
40
30
20
10
0.5
0.6
0.7
0.8
0.9
1.0
1.1
Frequency (cycle yr⁻¹)

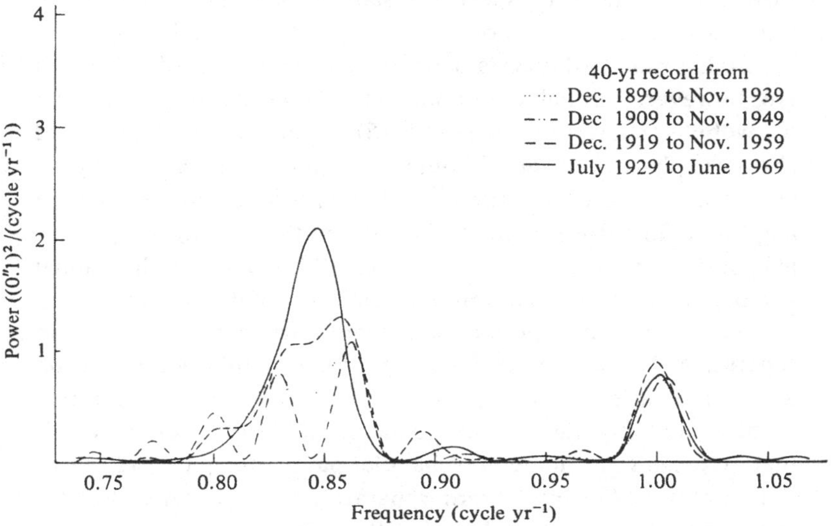


Figure 5.14. The power spectrum of the Chandler wobble. (*a*) The unsmoothed spectrum from 70 yr of ILS data; (*b*) smoothed spectrum using a Parzen window of length 35 yr; (*c*) spectra for four subsets of data each of 40-yr length. All spectra are from Pedersen & Rochester (1972).

suggested that instead of one free oscillation, the Chandler wobble actually consists of two components separated in frequency by about 0.02 cycle $\mathrm{yr}^{-1}$ (Colombo & Shapiro 1968; Gaposchkin 1972). The finite record length, the discrete sampling and the noise in the data, however, make an unsmoothed spectrum a quite unreliable estimate of the spectral structure of the motion, and some smoothing is essential. Thus reliability is enhanced at the cost of detail in the spectra. The spectrum in figure 5.14(*b*) has been smoothed using a Parzen window of length about one-half of the record length. The two peaks have now merged. The unsmoothed spectra in figure 5.14(*c*) are for four subsets of data, each of 40-yr length. They show great dissimilarity at the Chandler frequency $\sigma_0$. In contrast, the annual term is comparatively stable. These spectra illustrate a number of problems associated with the Chandler wobble. First, the spectrum at 14 months is quite broad, more so than at 12 months. This suggests a damping mechanism, and that the motion may be described by a damped oscillator. Secondly, the variability of the four spectra based on subsets of the data is suggestive of a non-stationary process (Pedersen & Rochester 1972). Unsmoothed spectra based on a short interval of noisy data are, however, unreliable. Guinot's (1972) results better illustrate the wobble behaviour (see figure 5.15). From 1925 to 1929 a rapid change in phase occurred without there being a significant change in amplitude, while during the preceding and following periods the amplitude fluctuated significantly, without there being any noticeable shift in phase (see also Brillinger 1973). Thirdly, the annual component of the wobble is amplified because of its proximity to the Chandler resonance period and some care must be taken in separating the two components (see, for example, Korsun *et al.* 1974). The variable nature of the annual term in the five ILS station latitudes and the inhomogeneous nature of the time series suggest that, in order to estimate the Chandler parameters, it is preferable to first remove the annual term separately for each station and for each series of observations based on the same reduction process and star positions, before combining the results for all five stations. Such a procedure has been adopted by Guinot (1972) with the results summarized in table 5.3. Wells & Chinnery (1973) have also advocated this approach. Fourthly, the statistical properties of the noise in the observations should be known. The estimation process is essentially one of matching the observed spectrum of the wobble by a theoretical spectrum for a damped oscillator plus a noise spectrum, and any such analysis for the optimum wobble parameters must take into account the above four factors at some point.

Consider a discrete series of $\mathbf{m}$ sampled at intervals $\Delta t$. The linearly damped oscillator model gives (equation 4.3.13*a*)

$$\mathbf{m}(t) - \mathrm{e}^{\mathrm{j}(\sigma_0 + \mathrm{j}\alpha)\,\Delta t}\mathbf{m}(t-\Delta t) = -\mathrm{j}\sigma_0 \boldsymbol{\psi}(t)\,\Delta t.$$

The observed $\mathbf{m}(t)$ are subject to measurement noise, or

$$\mathbf{m}(t) = \mathbf{m}'(t) + \boldsymbol{\varepsilon}_m(t)$$

where the $\mathbf{m}'(t)$ satisfy the above model. Then

$$\mathbf{m}(t) - \mathrm{e}^{\mathrm{j}(\sigma_0 + \mathrm{j}\alpha)\Delta t}\mathbf{m}(t-\Delta t) = -\mathrm{j}\sigma_0 \boldsymbol{\psi}(t)\Delta t + \boldsymbol{\varepsilon}_m(t) - \boldsymbol{\varepsilon}(t-\Delta t)\mathrm{e}^{-\alpha\Delta t}\mathrm{e}^{\mathrm{j}\sigma_0\Delta t}$$

This equation completely specifies the observed motion due to the excitation $\boldsymbol{\psi}(t)$ and observation noise $\boldsymbol{\varepsilon}_m(t)$, always assuming that

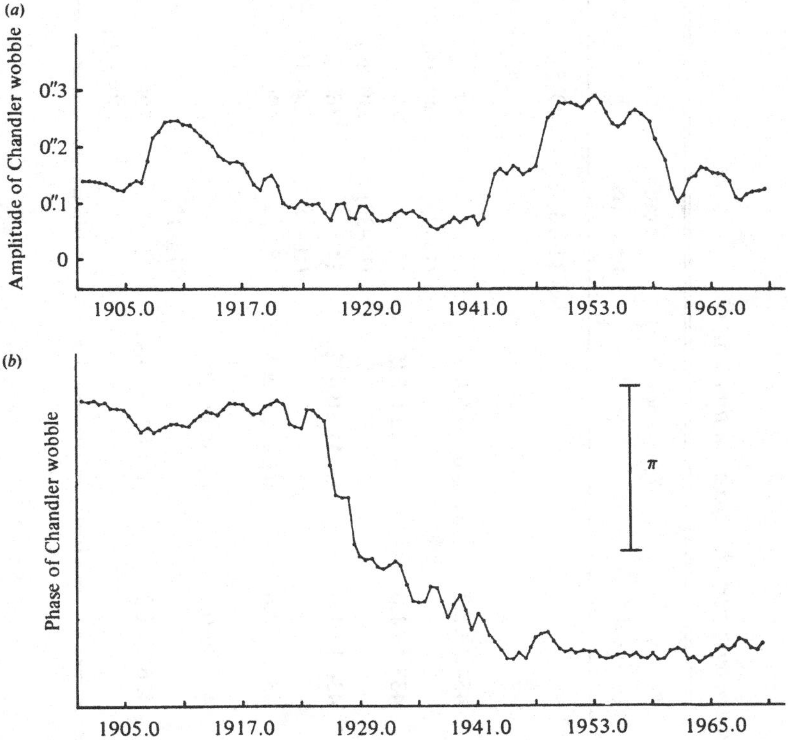


Figure 5.15. Variation in ($a$) the amplitude of the Chandler wobble with time from the ILS and additional latitude stations, and ($b$) the phase of the wobble. From 1925 to 1928 a change in phase of almost $\pi$ occurred. (From Guinot 1972.)

the linearly damped model is valid. The simplest assumption that can be made is that $\boldsymbol{\psi}(t)$ is an independent stationary process, so that the amplitude spectrum of $\boldsymbol{\psi}$ is $\mathcal{S}_\psi(\sigma) = v_\psi^2 = \text{constant}$, and that the error spectrum $\mathscr{E}_m(\sigma) = v_m^2 = \text{constant}$. Then the amplitude spectrum of $\mathbf{m}$ follows as

$$\mathcal{S}_m(\sigma) = \tfrac{1}{2}\pi\left[v_m^2 + \frac{(\sigma\,\Delta t)^2 v_\psi^2}{1 - 2\mathrm{e}^{-\alpha\,\Delta t}\cos\left[(\sigma - \sigma_0)\,\Delta t\right] + \mathrm{e}^{-2\alpha\,\Delta t}}\right].$$

The unknown parameters, to be determined by matching this spectrum to the observed spectrum, are $v_m^2$, $v_\psi^2$, the damping constant $\alpha$ and the free-oscillation frequency $\sigma_0$. The usual

Table 5.3. *Estimates of the Chandler wobble parameters: period, relaxation time and Q*

| Author | Data | Period (sidereal day) | Frequency (cycle $yr^{-1}$) | Relaxation time (yr) | $Q$ |
|---|---|---|---|---|---|
| Mandelbrot & McCamy (1970) | ILS 1900–1954 | | | 11–13 | 30–5 |
| Jeffreys (1968) | ILS 1899–1961 | $434.3 \pm 2.2$ | $0.843 \pm 0.004$ | 23 (14–73) | 60 (40–190) |
| Brillinger (1973) | ILS 1900–1970 | $433.4 \pm 1.6$ | $0.845 \pm 0.025$ | 17 (11–30) | 45 (30–80) |
| Currie (1974) | ILS 1900–1973 | $434.1 \pm 1.0$ | $0.844 \pm 0.002$ | 14 (10–18) | 36 (24–46) |
| Wilson & Haubrich (1976) | ILS 1901–1970 | $435.2 \pm 2.6$ | $0.841 \pm 0.005$ | 38 (20–150) | 100 (50–400) |
| Guinot (1972) | ILS + 8 others 1900–1970 | $436.9 \pm 0.7$ | $0.838 \pm 0.001$ | 16 (15–17) | 40 |
| Ooe (1978) | ILS 1900–1975 | $436.2 \pm 2.2$ | $0.8400 \pm 0.004$ | 38 | 100 (50–300) |

procedure is to solve for these parameters using the maximum likelihood approach (Jeffreys 1940; Walker & Young 1955; Brillinger 1973).

Mandelbrot & McCamy (1970) and Brillinger concluded that the observed spectrum of the polar motion, near the Chandler frequency, has the characteristics of the linearly damped oscillator (table 5.3). The spectrum of the excitation function deduced from the astronomical data (figure 8.9), using (4.3.13*b*) after removal of the seasonal terms, is reasonably constant over a wide frequency range about the Chandler frequency, indicating that the assumed spectrum $\mathscr{S}_\psi(\sigma)=v_\psi^2$ provides a reasonable approximation to the real spectrum. Jeffreys (1968*a*) and Wilson & Haubrich (1976*a*) have used a procedure similar to that outlined above, with the latter inferring the maximum likelihood estimators from only a narrow band of the spectrum about the Chandler frequency. From (3.5.4)

$$\mathscr{E}_\psi(\sigma)=\{[(\sigma_0-\sigma)/\sigma_0)]^2+1/4Q^2\}\mathscr{E}_m(\sigma)$$

and is a minimum at $\sigma=\sigma_0$. These two studies lead to similar results for wobble and $Q$ (table 5.3). Currie (1974) has estimated the parameters using the maximum-entropy spectral analysis technique (see, for example, Smylie *et al.* 1973), a procedure that had already been applied to the ILS data by Claerbout (1969) who obtained $Q\simeq 50$. Ooe (1978) used a mixed autoregressive–moving average model.

The total power of the Chandler wobble generated by $\mathscr{S}_\psi(\sigma)$ follows from (3.5.4) as

$$\mathscr{S}_m=\int_0^\infty \mathscr{S}_m(\sigma)\,d\sigma=\int_0^\infty \frac{\mathscr{S}_\psi(\sigma)\,d\sigma}{[(\sigma_0-\sigma)/\sigma_0]^2+1/4Q^2}.$$

With $\mathscr{S}_\psi(\sigma)$ nearly uniform about $\sigma_0$ (figure 8.9), we can write $\mathscr{S}_\psi(\sigma)=\mathscr{S}_\psi(\sigma_0)$ and, for $Q$ relatively large,

$$\mathscr{S}_m=\sigma_0 Q\mathscr{S}_\psi(\sigma_0)\ \text{rad}^2.$$

From figure 8.9, $\mathscr{S}_\psi(\sigma_0)\simeq 14\times10^{-16}\ \text{rad}^2/(\text{cycle yr}^{-1})$. This value is a function of $Q$ and $\sigma_0$ for which the values $Q\simeq 100$, $\sigma_0\simeq 0.8415\times 2\pi\ \text{rad yr}^{-1}$ of Wilson & Haubrich (1976*a*) have been adopted. Then $\mathscr{S}_m\simeq 74\times10^{-14}\ \text{rad}^{-2}$. Ooe (1978) obtained $\mathscr{S}_\psi(\sigma_0)\simeq 21\times10^{-16}\ \text{rad}^2/(\text{cycle yr}^{-1})$ and Munk & MacDonald, using ILS data from 1899 to 1954 and $Q\simeq 60$, obtained $\mathscr{S}_m=$

$24\times10^{-14}\,\mathrm{rad}^{-2}$. $\mathscr{S}_m$ also includes a contribution from observational noise. If $v_m^2$ is the variance of the monthly values $\mathscr{E}_m(\sigma)=v_m^2/2\pi f_N$, where $f_N=6$ cycle yr$^{-1}$ is the Nyquist frequency. Integrating from $f=0$ to $f=6$ cycle yr$^{-1}$ yields $\mathscr{E}_m=6v_m^2/$(cycle yr$^{-1}$). With $v_m\simeq0''.02$ for the ILS data, $\mathscr{E}_m\simeq5\times10^{-14}\,\mathrm{rad}^2$.

### 5.4 New observation techniques

The present astrometric methods of observing the Earth's irregular rotation have probably reached their limits in precision and resolution and any major improvements will have to come from quite different techniques. To provide useful results, any such technique must (i) give improved precision, (ii) give improved resolution, (iii) give long-term stability to permit secular and long-period oscillations to be studied with less ambiguity than at present, and (iv) permit a long-term observation program. Three main techniques have been developed in recent years and involve (i) satellite tracking, (ii) lunar laser ranging, and (iii) long-baseline radio interferometry. Lambeck (1979*b*) gives a review of these methods.

The first results for pole positions from satellite orbits were obtained by Anderle & Beuglass (1970) from the Doppler tracking of the US Navy navigation satellites. Since 1969, Doppler pole solutions have been routinely computed and results have also been obtained *a posteriori* for the years 1964–1969. Hence a continuous record of some 15 yr is now available (Anderle 1973, 1976). A great advantage of the Doppler method is that the observations are relatively precise: the observations are not dependent on weather conditions and the US Navy navigation system consists of a large number of stations (19 in 1973) that permanently track the satellites and provide extensive orbital coverage. Pole positions are determined at intervals of 2 d. Results from 1964 to 1967 are of limited quality, but those since 1968 are comparable in precision with the astronomical estimates. Particularly from 1970 to the present, the agreement with the BIH data is excellent (Anderle 1973, 1976). Since 1972 the Doppler solutions have been introduced into the BIH results (BIH Annual Report for 1972). Laser ranging to satellites has, so far, only provided occasional results (see, for example, Smith *et al.* 1972).

Satellite methods suffer a drawback similar to the astronomical observations of the Earth's rotation, namely, that it is difficult to maintain a homogeneous reference system. The parameters defining the Earth's gravity field and the tracking station coordinates are subject to frequent revision and, each time they are changed, discontinuities in the pole path result. Other unmodelled or inadequately modelled forces acting on the satellite will also introduce errors in the pole positions (Lambeck 1971; Anderle 1973). Satellite methods do not yet provide useful estimates of the Earth's irregular rate of rotation. Only the longitude of the ascending node $\Phi$ is perturbed by errors in the sidereal angle $\theta$ but this orbital element is difficult to compute with precision: the zonal harmonics in the gravity field cause secular and periodic perturbations, and air drag and radiation pressure also make major contributions to $\Phi(t)$. Separation of these effects from errors in $\theta$ is not possible in the way it is for polar motion, where a distinctly periodic signal is introduced into the orbital elements. Possibly, satellite observations may provide useful estimates of high-frequency and irregular changes in l.o.d.

A number of attempts have been made to range to the Moon but only the McDonald observatory in Texas has so far yielded useful results† (Bender *et al.* 1973). Accuracies of a few tens of centimetres are routinely achieved. Scientific results include improved lunar libration parameters, orbital constants, and initial results for the Earth's rotation. Williams (1977) reviewed the present achievements. A disadvantage of the lunar laser method of determining the Earth's rotation is that around new Moon it is difficult to range successfully and there is usually a period of a few days when no observations are made. This disadvantage can be overcome by combining lunar and artificial satellite observations. The former gives a direct determination of the sidereal angle $\theta$ but with interruptions during which $\theta$ could be interpolated, using observations of the node of satellites. At the same time, the satellite observations provide a more direct determination of latitude changes than does the lunar method. Preliminary lunar laser results for the Earth's rate of rotation, obtained by Harris & Williams

† The Australian station at Orroral Valley near Canberra has been in operation since late 1978.

(1977) and by King, Counselman & Shapiro (1978), indicate accuracies comparable with those obtained by conventional astronomical observations.

The technique of long-baseline interferometry using stellar radio sources has been reviewed by Shapiro & Knight (1970) and by Counselman (1976). Essentially, the orientation of the vector joining two antennae is determined relative to the stellar reference frame, and the Earth's rotation can be deduced from fluctuations in the orientation of this vector. Shapiro *et al.* (1974) present initial results for $m_i$ that are comparable in accuracy with those obtained by the BIH. More recent results have been given by Robertson *et al.* (1979) for both polar motion and UT1.

As with the other high-precision observations, not only changes in rotation introduce discrepancies between the observed and theoretical signals. Periodic tidal deformations, crustal and tectonic motions and precession and nutation all contribute to these discrepancies and it is not clear whether they can all be separated with any one technique. Furthermore, these new techniques are not designed just for rotation studies: satellite tracking provides information on the Earth's gravity field, lunar laser ranging is of primary interest in studies of the Moon's motion and long-baseline radio interferometry is of importance in the study of the stellar sources. In all cases, the irregularities in rotation are just one other perturbation. Hence the final solution for future rotation studies may well be a combination of all of these techniques.

# SIX

# TIDES

Apart from precession and nutation, the consequences of gravitational lunar and solar attraction on rotation would be of little interest if the Earth were truly rigid. But due to the departure from absolute rigidity, these forces deform the Earth and modify the rotation. Observations of these rotational perturbations provide a global measure of the Earth's elastic and non-elastic response to periodic forces, in the frequency range from 2 cycle $d^{-1}$ to 2 cycle $yr^{-1}$. Tidal deformations occur in the solid Earth, in the oceans and in the atmosphere, and all three interact in different ways with the Earth's rotation. The interactions are of three types. First, the periodic tidal deformations induce temporal changes in the planet's inertia tensor and lead to perturbations in both the polar motion and the rate of rotation. The solid tides will be most important, although, as ocean and atmospheric tides have the same frequencies as the solid tide, all three contribute to the variations in the inertia tensor. For most tidal frequencies,

$$\Delta I_{ij}(\text{solid tide})/\Delta I_{ij}(\text{ocean tide}) \simeq 10\text{–}15.$$

Atmospheric tides are much smaller and, except for the solar semi-diurnal tide, can be ignored. For this tide

$$\Delta I_{ij}(\text{solid tide})/\Delta I_{ij}(\text{atmospheric tide}) \simeq 100.$$

Secondly, the tides deflect the observer's meridian and periodically displace the position of the observer, resulting in perturbations of the observation record, either by modifying the apparent amplitudes of tidal perturbations in the rotation spectrum or by introducing additional periodicities. Thirdly, the anelastic response results in a transfer of angular momentum from the Earth to the Moon and leads to secular changes in the motion of the Moon and in the Earth's spin. In this problem, ocean tides are most important, while the atmospheric tide plays a small but interesting role.

Tides are also raised by periodic changes in the centrifugal force and the *pole tide*, the 14-month period tide in the oceans, has been much discussed. There will also be annual and semi-annual ocean tides caused by the forced component in the polar motion and by the variations in the speed of rotation. These are less frequently discussed, although they are more important than the solar annual ocean tide Sa.

In this chapter we consider mainly the first problem and lay some of the basic groundwork for the discussion of the pole tide and secular accelerations. The complete discussion of these problems is deferred to chapters 8 and 10. One of the first references to tidal variations in the Earth's rotation was by Jeffreys in 1928, who estimated the theoretical amplitudes of the semi-annual tide fluctuations in the l.o.d. and concluded that they were insignificant compared with the then available l.o.d. data. The subsequent discussion until about 1960 has been reviewed by Munk & MacDonald (1960), who conclude that the only tidal term of importance is the solar semi-annual effect on l.o.d., with the lunar monthly Mm and fortnightly Mf tides rising only marginally above the noise level. With the establishment of the precise atomic time scale, these tidal frequencies are now clearly evident in l.o.d. observations (figure 5.8) and the Love numbers $k_2$ can be estimated with some confidence for the Mm and Mf frequencies (Guinot 1974).

The tidal deformations result in the observer's meridian not being fixed, and stars transit the meridian earlier or later than they would if the meridian was not deflected. A principal consequence is an aliasing between the diurnal tidal deflections and the nearly diurnal pattern followed in the observing program at the astronomical observatories. Guinot (1970*b*) has shown that this will produce perturbations in the l.o.d. observations at frequencies near 14 d. As these do not signify a change in rotation, they should be removed from the record by correcting each station's observations for the tidal deflections of the meridian.

New methods for determining the Earth's rotation will not be free from these indirect tidal deformations either: satellite orbits are perturbed by the tide potential; Earth-based observatories, whether they be satellite tracking stations, lasers for ranging to the

Moon or long-baseline interferometers, will be periodically displaced with respect to the Earth's centre of mass by amounts greater than the observational precision.

## 6.1 Earth tides

### 6.1.1 *Love numbers*

A potential $U(\mathbf{r})$ acts on the Earth, deforms it, and modifies the external gravitational potential. A most convenient way of defining this deformation is through the Love number approach (equations 2.1.7) in which the Earth's response is evaluated in the frequency domain; the forcing function is expanded in spherical harmonics and each harmonic is multiplied by the corresponding Love number. This method is most efficient when the forcing function is described by low-degree harmonics. This is the case for the tidal and centrifugal forces, where nearly all the energy of the forcing function is concentrated in the second-degree harmonics. Closely related is the deformation problem in which the Earth is subject to a spatially and temporally varying load. These problems often involve a forcing function whose energy is contained in high-degree harmonics. Examples are the loading of coastal sea floors by ocean tides or the variable ice mass stored in the polar caps. These loads are concentrated in limited areas, but any change in their redistribution can have global repercussions; thus a partial melting of the ice caps results in a global redistribution of water over the oceans and in a global loading of the sea floor. The Love number approach to computing the response to the load at any point requires that the load be known all over the planet so that it can be decomposed into its harmonic components. The load is usually known with precision only around the point in question, and as it is this nearby load that contributes most to the deformation, the approach is clearly not ideal. A Green's function procedure is more efficient (Longman 1962; Farrell 1972; Peltier 1974). Nearly all surface loading problems that occur in the discussion of the rotation of the Earth concern low-degree harmonics, since the inertia tensor entering into the excitation functions is of second degree (section 3.4), and the Love number method is adequate and efficient.

The usefulness of the Love numbers lies in their convenient description of the deformations of a planet due to an applied harmonic potential; they form convenient parameters to relate the results of complex theories of the Earth's response to refined measurements taken on the surface. Their intepretation is not without its own difficulties, however.

Consider the response of the Earth to a perturbing potential. This response can be considered as the sum of several terms and can be written as

$$\mathcal{R} = \mathcal{R}_0 + \sum_i \Delta\mathcal{R}_i,$$

where $\mathcal{R}_0$ is the elastic response of a radially symmetric, oceanless, but otherwise realistic, Earth model, and $\Delta\mathcal{R}_i$ are perturbations from this state. $\mathcal{R}_0$ is now best calculated using Earth models derived from seismic observations, since the latter give directly the distribution of density and elastic moduli with depth. These calculations show that the Love numbers are quite insensitive to the choice of Earth model (table 2.1) and are of only limited interest for planets whose seismic structure is already known. The first perturbation to be considered, $\Delta\mathcal{R}_1$, is an eventual departure of this radial Earth model from elasticity. For real Earth models, the response, and hence the Love numbers, can be expected to depend on both the amplitude and the frequency of the perturbing potential since different rheologies may dominate at different magnitudes or durations of the applied forces. Over the frequency range usually encountered in rotational studies, $\Delta\mathcal{R}_1$ is likely to be small since dissipation in the solid Earth appears to be small (section 2.2). An observation of $\Delta\mathcal{R}_1$ is nevertheless of considerable interest for discussions of the dependence of $Q^{-1}$ on frequency. A second contribution, $\Delta\mathcal{R}_2$, due to the asymmetrical distribution of the ocean tides and tidal loading of the Earth, is generally the most important perturbation, and its phase may differ considerably from that of $\Delta\mathcal{R}_1$. A smaller contribution, $\Delta\mathcal{R}_3$, may result from anomalous mechanical properties of local and regional geologic structure, and will in most cases be difficult to distinguish from $\Delta\mathcal{R}_2$. Dynamic interactions between the mantle and fluid core may lead to a further term, $\Delta\mathcal{R}_4$. $\Delta\mathcal{R}_2$ and $\Delta\mathcal{R}_3$ are responses that vary locally, while $\Delta\mathcal{R}_1$ and $\Delta\mathcal{R}_4$ are global responses.

Observational evidence for the low-degree Love numbers comes from measurements of the Earth's response to external potentials that do not load the Earth. No load Love numbers have been observed experimentally for at least two reasons: unlike the tidal and centrifugal force functions, the potential of the surface load is seldom known with precision, and whereas the former are strictly periodic, with known astronomical frequencies, the surface loads are not always so. The subject of Earth tides has been recently reviewed by Slichter (1972), Jobert (1973*b*) and Melchior (1978). Melchior, Kuo & Ducarme (1976), Baker & Lennon (1976), Ostrovsky (1976) and Pertsev (1977) present recent results for the gravimetric and tilt tides for Europe and Asia. The amplitudes of the gravimetric and tilt tides show a considerable dispersion that is mainly a consequence of the interference of the ocean tide with the body tide. Even observations in the middle of continents appear to be perturbed by the ocean tide (Pertsev 1969; Kuo *et al.* 1970). Observations of the solid Earth phase lags are generally unsatisfactory due to instrumental problems and due to the ocean-loading contributions to the tide potential. Early gravity measurements by Harrison *et al.* (1963) gave lags ranging from $-20°$ to $+24°$ for the $M_2$ tide with an average of 3.5°. Farrell (1970) has considered the ocean-loading contributions to the observations of Harrison *et al.* and finds that the average residual lag is essentially zero. Gravimetric tide observations in the interior of the USSR indicate a lag of about 0.5°, part of which can be attributed to ocean loading even though the stations lie far from the coast. The gravimetric tide lags for western Europe are also attributed to tidal loading and the residual lags are small (Melchior *et al.* 1976; Pertsev 1977). Observations by Lambert (1970), Kuo *et al.* (1970) and Baker & Lennon (1976) confirm that the solid Earth lag is less than a fraction of a degree. Measurements of the tilt tide are even less satisfactory for estimating the solid Earth tide lag due to instrumental problems and environmental perturbations.

Ocean tides are in general not sufficiently well known to eliminate all of the ocean-loading contribution. Hendershott & Munk (1970), in a review of ocean tides, conclude that improvements in the solid tide results can only come about from major improvements in our knowledge of the ocean tide. Recent results for marine loading, such as those by Warburton, Beaumont & Goodkind (1975), Baker & Lennon (1976) and Moens (1976), are encouraging and indicative of progress in this area. Kuo & Jachens (1977) report initial results on the inversion of the ocean-loading contributions for estimating the ocean tides.

### 6.1.2 *Tide-generating potential*

The potential $U$ of the gravitational attraction at $\mathbf{r}$ due to a mass $m^*$ at $\mathbf{r}^*$ is (see, for example, Brouwer & Clemence 1961)

$$U(\mathbf{r}) = Gm^*[1/|\mathbf{r}-\mathbf{r}^*| - \mathbf{r}\cdot\mathbf{r}^*/r^{*3}]. \tag{6.1.1}$$

The geocentric angle $S$ between the position $\mathbf{r}$ and $\mathbf{r}^*$ is given by

$$\cos S = \mathbf{r}\cdot\mathbf{r}^*/rr^* \tag{6.1.2}$$

$$= \sin\phi \sin\phi^* + \cos\phi \cos\phi^* \cos(\lambda - \lambda^*). \tag{6.1.3}$$

The $(r^*, \phi^*, \lambda^*)$ and $(r, \phi, \lambda)$ are the spherical coordinates of the mass $m^*$ at $\mathbf{r}^*$ and of the position $\mathbf{r}$, respectively. If $\lambda$ and $\lambda^*$ are longitudes with respect to an Earth-fixed meridian this potential is defined in a reference frame that rotates with the Earth. With (6.1.2) the potential (6.1.1) can be expressed by Legendre polynomials $P_{l0}(\cos S)$ as

$$U(r) = \frac{GM^*}{r^*} \sum_{l=2}^{\infty} \left(\frac{r}{r^*}\right)^l P_{l0}(\cos S). \tag{6.1.4}$$

With (6.1.3) and the addition theorem of spherical harmonics (see, for example, Jeffreys & Jeffreys 1962, p. 646), the polynomials can be expanded as

$$P_{l0}(\cos S) = \sum_{m=0}^{l} (2-\delta_{0m}) \frac{(l-m)!}{(l+m)!} \times P_{lm}(\sin\phi) P_{lm}(\sin\phi^*) \cos m(\lambda - \lambda^*)$$

and

$$U(r) = \frac{GM^*}{r^*} \sum_{l} \left(\frac{r}{r^*}\right)^l \sum_{m} (2-\delta_{0m}) \frac{(l-m)!}{(l+m)!} \times P_{lm}(\sin\phi) P_{lm}(\sin\phi^*) \cos m(\lambda - \lambda^*). \tag{6.1.5}$$

The motion of the attracting mass, Moon or Sun, is more conveniently expressed by orbital elements than by the $r^*$, $\phi^*$, $\lambda^*$ and the time derivatives $\dot{r}^*$, $\dot{\phi}^*$, $\dot{\lambda}^*$. A common choice, useful for the subsequent discussion on the tidal evolution of the lunar orbit, consists of the instantaneous Keplerian elements of the Moon (or Sun). These elements $\kappa_i$ are defined as follows (see, for example, Brouwer & Clemence 1961; Kaula 1966):

$\kappa_1 = a^* =$ semi-major axis of the orbit,
$\kappa_2 = e^* =$ eccentricity,
$\kappa_3 = I^* =$ inclination of the orbital plane on the equator,
$\kappa_4 = M^* =$ mean anomaly and $\dot{M}^* = n^*$, where $n^*$ is the mean motion of the body,
$\kappa_5 = \omega^* =$ argument of perigee,
$\kappa_6 = \Phi^* =$ longitude of the ascending node.

The appropriate transformation from spherical coordinates to these elements is discussed by Kaula (1966). The result is

$$\left(\frac{1}{r^*}\right)^{l+1} P_{lm}(\sin\phi^*)[\cos m\lambda^* + \mathrm{j}\sin m\lambda^*] = \left(\frac{1}{a^*}\right)^{l+1} \sum_{p=0}^{l} F_{lmp}(I^*)$$
$$\times \sum_{q=-\infty}^{\infty} G_{lpq}(e^*) \begin{bmatrix} \cos v^*_{lmpq} + \mathrm{j}\sin v^*_{lmpq} \\ \sin v^*_{lmpq} - \mathrm{j}\cos v^*_{lmpq} \end{bmatrix}_{l-m\ \text{odd}}^{l-m\ \text{even}}, \qquad (6.1.6)$$

with $v^*_{lmpq} = (l-2p)\omega^* + (l-2p+q)M^* + m(\Phi^* - \theta)$. The sidereal angle is denoted by $\theta$. The $F_{lmp}(I^*)$ and $G_{lpq}(e^*)$ are polynomials in $\sin I^*$ and $e^*$ and are summarized in table 6.1 for the most important terms.† More complete tabulations are given by Kaula. For small $I$ the $F_{lmp}$ are proportional to $I^{(l-2p-m)}$ and only those indices that result in small values for the combination $(l-2p-m)$ of indices need be considered. The $G_{lpq}(e^*)$ are proportional to $e^{|q|}$ and, as the eccentricities of both the lunar and Earth orbits are small, the summation over the index $q$ needs to be carried out only over a small number of terms; that is, $q = 0, \pm 1, \pm 2$. Also, because the factor $(R/a^*)^l \simeq (1/60)^l$ is small for the Moon, and very much smaller for the Sun, only terms with $l = 2$ are important. With the

† $G_{lpq}(e)$ is itself an infinite summation of terms proportional to $e^{k'}$ with $k' = |q|$ to $\infty$. The inclination function can also be expressed in terms of the Clebsch–Gordan or Wigner 3-$j$ coefficients (see also B. Jeffreys 1965; Izsak 1964).

transformation (6.1.6) the potential $U(r)$ becomes

$$U(r)=\frac{Gm^*}{a^*}\sum_l\left(\frac{r}{a^*}\right)^l\sum_m(2-\delta_{0m})\frac{(l-m)!}{(l+m)!}P_{lm}(\sin\phi)$$

$$\times\sum_p F_{lmp}(I^*)\sum_q G_{lpq}(e^*)\begin{bmatrix}\cos\\ \sin\end{bmatrix}_{l-m\text{ odd}}^{l-m\text{ even}}(v^*_{lmpq}-m\lambda). \qquad (6.1.7)$$

Table 6.1. *Inclination functions $F_{lmp}(i)$ for $l=2$ and eccentricity functions $G_{lpq}(e)$ for $l=2$ and $-2<q<2$ to order $e^2$ (from Kaula 1966)*

| $l$ | $m$ | $p$ | $F_{lmp}(i)$ |
|---|---|---|---|
| 2 | 0 | 0 | $-\frac{3}{8}\sin^2 i$ |
| 2 | 0 | 1 | $\frac{3}{4}\sin^2 i-\frac{1}{2}$ |
| 2 | 0 | 2 | $-\frac{3}{8}\sin^2 i$ |
| 2 | 1 | 0 | $\frac{3}{4}\sin i\,(1+\cos i)$ |
| 2 | 1 | 1 | $-\frac{3}{2}\sin i\cos i$ |
| 2 | 1 | 2 | $-\frac{3}{4}\sin i\,(1-\cos i)$ |
| 2 | 2 | 0 | $\frac{3}{4}(1+\cos i)^2$ |
| 2 | 2 | 1 | $\frac{3}{2}\sin^2 i$ |
| 2 | 2 | 2 | $\frac{3}{4}(1-\cos i)^2$ |

| $l$ | $p$ | $q$ | $l$ | $p$ | $q$ | $G_{lpq}(e)$ |
|---|---|---|---|---|---|---|
| 2 | 0 | $-2$ | 2 | 2 | 2 | 0 |
| 2 | 0 | $-1$ | 2 | 2 | 1 | $-\frac{1}{2}e+\ldots$ |
| 2 | 0 | 0 | 2 | 2 | 0 | $1-\frac{5}{2}e^2+\ldots$ |
| 2 | 0 | 1 | 2 | 2 | $-1$ | $\frac{7}{2}e+\ldots$ |
| 2 | 0 | 2 | 2 | 2 | $-2$ | $\frac{17}{2}e^2+\ldots$ |
| 2 | 1 | $-2$ | 2 | 1 | 2 | $\frac{9}{4}e^2+\ldots$ |
| 2 | 1 | $-1$ | 2 | 1 | 1 | $\frac{3}{2}e+\ldots$ |
| | | | 2 | 1 | 0 | $(1-e^2)^{-3/2}$ |

In a first approximation, the only time-dependent variable is the sidereal angle $\theta$, and the potential exhibits three main periodicities: long periods when $m=0$, nearly diurnal periods when $m=1$ and nearly semidiurnal periods when $m=2$. In a second approximation, the position of the Moon or Sun also varies with time and additional periods occur in the potential for different values of $p$ and $q$. These group about the three fundamental periods. Further periods result due to small variations in $\omega^*$ and $\Phi^*$ with time, and the total spectrum of the potential is very rich indeed, particularly when the total tide-raising potential, that due to the Sun and Moon, is considered.

Table 6.2. *Definition of the fundamental frequencies in the Earth–Moon–Sun motion*

| | |
|---|---|
| $f_1^{-1}$ | period of the Earth's rotation relative to the Moon, equal to 1 lunar day |
| $f_2^{-1}$ | mean period of the Moon's orbital motion, equal to 1 lunar month |
| $f_3^{-1}$ | mean period of the Sun's orbital motion, equal to 1 yr |
| $f_4^{-1}$ | the mean period of the lunar perigee, equal to 8.85 yr |
| $f_5^{-1}$ | the mean period of regression of the lunar nodes, equal to 18.61 yr |
| $f_6^{-1}$ | the period of perihelion, equal to 20 940 yr |
| $2\pi f_1 t$ | local mean lunar time, equal to $\tau+\lambda$, where $\lambda$ is longitude |
| $2\pi f_2 t$ | mean longitude of the Moon, equal to $s$ |
| $2\pi f_3 t$ | mean longitude of the Sun, equal to $h$ |
| $2\pi f_4 t$ | longitude of the lunar perigee, equal to $p'$ |
| $2\pi f_5 t$ | $\Omega=-N'$, where $\Omega$ is the longitude of the ascending node of the Moon |
| $2\pi f_6 t$ | longitude of perihelion, equal to $p''$ |

Analytical developments of the equations of motion of the Earth–Moon–Sun system rarely use the equatorial coordinates defined above. Instead they employ ecliptic elements as these facilitate the integration of the equations of motion. Widely used are the Brown variables, which define the motion of the three bodies in terms of six fundamental frequencies $f'_i$, defined in table 6.2. The potential can be expressed as a sum of terms $U_\beta$ where

$$U_\beta(r)=(r/R)^l A_\beta P_{lm}(\sin\phi)\cos(2\pi f_\beta t+m\lambda), \qquad (6.1.8)$$

with

$$A_\beta \simeq \frac{Gm^*}{a^*}\left(\frac{R}{a^*}\right)^l (2-\delta_{0m})\frac{(l-m)!}{(l+m)!}F_{lmp}(I^*_e)G_{lpq}(e^*),$$

and where $I_e^*$ is the inclination of the orbit of the tide-raising body on the ecliptic, $t$ is universal time, and the $f_\beta$ are linear relations of $f_i$. Table 6.3 summarizes $f_\beta$, amplitude factors $A_\beta$, the corresponding indices $lmpq$ for the equivalent expansion in terms of the Keplerian equatorial elements, and the relations between $v^*_{lmpq}$ and $2\pi f_\beta t$ in the form

$$2\pi f_\beta t = -v^*_{lmpq} + \pi(\tfrac{1}{2}r_\beta + m), \tag{6.1.9}$$

where $r_\beta$ is an integer and depends on the harmonic considered (see Lambeck 1977). Traditionally, each frequency $f_\beta$ is associated with a symbol, following Darwin's nomenclature, as indicated in table 6.3. The complete development of (6.1.8) is given by Doodson (1921) and more recently by Cartwright & Edden (1973). In the astronomical literature, time $t$ is usually referred with respect to mean noon, December 31, 1899. In tidal developments the origin is usually adopted as $0^{\text{h}}.00$ January 1, 1900 (Doodson & Warburg 1941). The relation (6.1.9) takes this into consideration.

### 6.1.3 *The solid tide potential*

The tide-generating potential (equation 6.1.7 or 6.1.8) can be expressed as

$$U_l(r) = r^l S_l(\phi, \lambda), \tag{6.1.10}$$

where $S_l(\phi, \lambda)$ is a surface harmonic. At $r = R$ this potential is

$$U_l(R) = \left(\frac{R}{r}\right)^l U_l(r).$$

The additional potential $\Delta U_l(R)$ due to the deformation of the Earth is defined as $k_l U_l(R)$. Outside the Earth, by Dirichlet's theorem,

$$\begin{aligned}\Delta U_l(r) &= k_l(R/r)^{l+1} U_l(R)\\ &= k_l(R/r)^{2l+1} U_l(r).\end{aligned} \tag{6.1.11}$$

With (6.1.7)

$$\begin{aligned}\Delta U_l(r) = \frac{Gm^*}{a^*} \sum_{m=0}^{l} \left(\frac{R}{r}\right)^{l+1} \left(\frac{R}{a^*}\right)^{l} k_l(2-\delta_{0m}) \frac{(l-m)!}{(l+m)!}\\ \times P_{lm}(\sin\phi) \sum_{p=0}^{l} F_{lmp}(I^*) \sum_{q=-\infty}^{+\infty} G_{lpq}(e^*)\\ \times \begin{bmatrix}\cos\\ \sin\end{bmatrix}_{l-m \text{ odd}}^{l-m \text{ even}} (v^*_{lmpq} - m\lambda).\end{aligned} \tag{6.1.12}$$

Table 6.3. *Arguments $2\pi f_\beta \ell$ and amplitudes $A_\beta$ of the tide-raising potential for the major constituents. The lmpq denote the indices that give the approximate equivalence between the two definitions (6.1.7) and (6.1.8) of the potential; $r_\beta$ is defined by (6.1.9)*

| Tide symbol | *lmpq* | Origin L, Lunar S, solar | Period | $2\pi f_\beta T$ | $r_\beta$ | $A_\beta$ ($10^3$) |
|---|---|---|---|---|---|---|
| Semi-diurnal tides: $P_{22}(\sin\phi) = 3\cos^2\phi$ | | | | | | |
| $M_2$ | 2200 | L | 12.42h | $30°\ell + 2h - 2s$ | | 7.958 |
| $S_2$ | 2200 | S | 12.00h | $30°\ell$ | | 3.708 |
| $N_2$ | 2201 | L | 12.66h | $30°\ell + 2h - 3s + p'$ | | 1.525 |
| $K_2$ | 2210 | L+S | 11.97h | $30°\ell + 2h$ | | 1.010 |
| $L_2$ | 220−1 | L | | $30°\ell + 2h - s - p' + \pi$ | 2 | 0.226 |
| $T_2$ | 2201 | S | | $30°\ell - h + p''$ | | 0.221 |
| $2N_2$ | 2202 | L | | $30°\ell + 2h - 4s + 2p'$ | | 0.200 |
| Diurnal tides: $P_{21}(\sin\phi) = 3\sin\phi\cos\phi$ | | | | | | |
| $K_1$ | 2110 | L+S | 23.93h | $15°\ell + h + \frac{1}{2}\pi$ | 1 | 9.310 |
| $O_1$ | 2100 | L | 25.82h | $15°\ell + h - 2s - \frac{1}{2}\pi$ | −1 | 6.611 |
| $P_1$ | 2100 | S | 24.07h | $15°\ell - h - \frac{1}{2}\pi$ | −1 | 3.087 |
| $Q_1$ | 2101 | L | | $15°\ell + h - 3s - p' - \frac{1}{2}\pi$ | −1 | 1.262 |
| Long-period tides: $P_{20}(\sin\phi) = \frac{3}{2}\sin^2\phi - \frac{1}{2}$ | | | | | | |
| 18.6-yr | — | L | 18.6yr | $N'$ | | −1.735 |
| Sa | 2011, 201−1 | S | 1.0yr | $h$ | | −0.316 |
| Ssa | 2000, 2020 | S | 0.5yr | $2h$ | | −1.920 |
| MSm | | L+S | 31.85d | $s - h + p'$ | | −0.421 |
| Mm | 2011, 201−1 | L | 27.55d | $s - p'$ | | −2.183 |
| MSf | | L+S | 14.77d | $2s - 2h$ | | −0.368 |
| Mf | 2000, 2020 | L | 13.66d | $2s$ | | −4.105 |
| | | L | 13.63d | $2s + N'$ | | −1.710 |

This expression for the tide potential assumes an instantaneous or elastic response. The consequence of anelasticity is to delay the response of the Earth by an amount $\Delta t$. The maximum deformation is reached at a time $\Delta t$ after the Moon or Sun has passed through the observer's meridian. During this interval the Earth has rotated through an angle $\dot{\theta}\,\Delta t \equiv \Omega\,\Delta t$ while the Moon has moved through $n_{☾}\,\Delta t$, $n_{☾}$ being the mean lunar motion. Viewed from space (figure 6.1), the tidal bulge is ahead of the Moon by approximately $(\Omega - n_{☾})\,\Delta t \simeq \Omega\,\Delta t$ since $n_{☾} \ll \Omega$. This delay is most conveniently introduced into the above potential by considering a fictitious position $\tilde{r}^*$, immediately above the bulge of the tide-raising body, and then by assuming that the response is instantaneous. That is, we are concerned with the potential at $(r, \phi, \lambda)$ raised by the Moon at a fictitious position $(\tilde{r}^*, \tilde{\phi}^*, \tilde{\lambda}^*)$. The latter position can also be transformed into fictitious Keplerian elements $\tilde{\kappa}_i$ defined by $\kappa_i + \dot{\kappa}_i\,\Delta t$, but in the short time interval – about 10–20 min – it takes for the Earth to respond, the only elements that will have significantly changed are the sidereal angle, by $\Omega\,\Delta t$, and the mean motion $M^*$, by $n^*\,\Delta t$. To a lesser extent $\Phi^*$ and $\omega^*$ will have changed by small amounts due to the secular rates of these elements; $a^*$, $e^*$ and $I^*$ will not have undergone any perceptible change. Thus to introduce the lag into equation (6.1.7) we need only substitute $\tilde{v}^*_{lmpq}$ for $v^*_{lmpq}$ where

$$\tilde{v}_{lmpq} = (l-2p)\omega^* + (l-2p+q)M^* + m(\Phi^* - \theta) + \varepsilon^*_{lmpq}, \tag{6.1.13a}$$

with

$$\begin{aligned}\varepsilon^*_{lmpq} &= [(l-2p)\dot{\omega} + (l-2p+q)\dot{M}^* + m(\dot{\Phi}^* - \Omega)]\,\Delta t \\ &\simeq [(l-2p+q)n^* - m\Omega]\,\Delta t,\end{aligned} \tag{6.1.13b}$$

$\Delta t$ is negative for a delayed response and $\varepsilon$ will be positive if $m\Omega > (l-2p+q)n^*$. In terms of the ecliptic variables the potential of the tidal deformation becomes

$$\Delta U_\beta = (R/r)^{l+1} P_{lm}(\sin\phi) k_l A_\beta \cos(2\pi f_\beta t + \varepsilon_\beta + m\lambda), \tag{6.1.14a}$$

with

$$\varepsilon_\beta = (2\pi f_\beta - m\Omega)\,\Delta t. \tag{6.1.14b}$$

It should be stressed that the phase lag $\varepsilon_\beta$ is not directly comparable with the phase lag observed at the Earth's surface by gravity meters

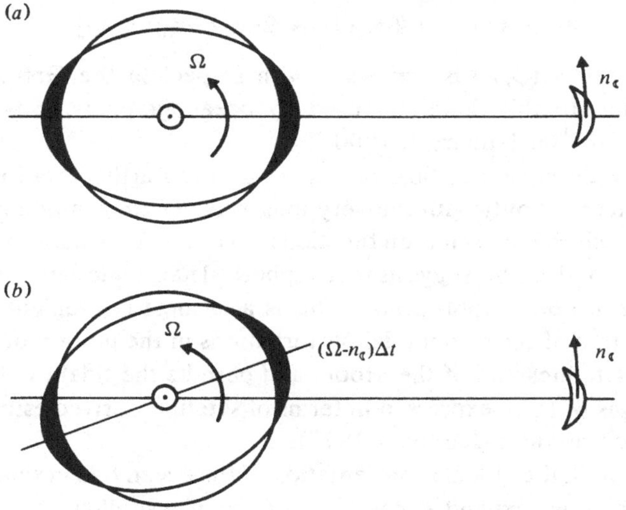


Figure 6.1. The tidal bulge (*a*) for an elastic Earth and (*b*) for the case where there is a delay $\Delta t$ in the Earth's response.

or by tilt meters. In the former case, if $\Gamma_\beta$ is the observed lag,

$$\varepsilon_\beta \simeq [(1 - \tfrac{3}{2}k_2 + h_2)/(-\tfrac{3}{2}k_2 + h_2)]\Gamma_\beta,$$

while if $\Gamma_\beta$ is the observed lag of the tilt tide,

$$\varepsilon_\beta \simeq [(1 + k_2 - h_2)/(k_2 - h_2)]\Gamma_\beta.$$

If the seismic $Q$ of the second-degree spheroidal mode is applicable to the tidal frequencies, then the anticipated phase lag $\varepsilon_\beta$, corrected for ocean effects, is about $0.^\circ 1$, or $\Gamma_\beta \sim 0.^\circ 01$. For $Q \sim 100$ at tidal frequencies $\varepsilon_\beta \sim 0.^\circ 6$ and $\Gamma_\beta \sim 0.^\circ 1$.

## 6.2 Ocean and atmospheric tides

### 6.2.1 *The ocean tide potential*

The ocean tide raised by the potential (6.1.7) or (6.1.8) contains a large number of frequencies $f_\beta$. For each component the tide is given at any position on the Earth by an amplitude $\xi^0_\beta(\phi, \lambda)$ and a phase $\chi_\beta(\phi, \lambda)$ that both vary over the Earth's surface. Thus (Hendershott & Munk 1970)

$$\xi_\beta(\phi, \lambda : T) = \xi^0_\beta(\phi, \lambda)\cos[2\pi f_\beta \ell - \chi_\beta(\phi, \lambda)].$$

The phase $\chi_\beta(\phi, \lambda)$ is expressed with respect to the Greenwich meridian. In this definition used by oceanographers, $\ell$ is with respect to $0^h_.00$ January 1, 1900.

In the discussion of tidal influences on the Earth's rotation, we are concerned only with the very long-wavelength components in the tide, since the excitation functions $\psi_i$ contain only harmonics of degree 0 and 2. This suggests that a spherical harmonic expansion of the ocean tide is appropriate. This is also most convenient for a description of the ocean tidal perturbations in the motion of close Earth satellites and of the Moon, and permits the tidal excitation functions $\psi_i$ to be expressed in terms of satellite-derived estimates of the ocean tides (Lambeck 1977).

To obtain the global representation of the ocean tide in spherical harmonics, we expand $\xi^0_\beta \cos \chi_\beta$ and $\xi^0_\beta \sin \chi_\beta$ as follows:

$$\left.\begin{aligned} \xi^0_\beta \cos \chi_\beta &= \sum_{s=1}^{\infty} \sum_{t=0}^{s} \\ &\quad \times (a'_{\beta,st}\cos t\lambda + b'_{\beta,st}\sin t\lambda) P_{st}(\sin\phi), \\ \xi^0_\beta \sin \chi_\beta &= \sum_{s=1}^{\infty} \sum_{t=0}^{s} \\ &\quad \times (a''_{\beta,st}\cos t\lambda + b''_{\beta,st}\sin t\lambda) P_{st}(\sin\phi). \end{aligned}\right\} \tag{6.2.1a}$$

On the continents $\xi^0_\beta = 0$. Then

$$\xi_\beta = \sum_s \sum_t \sum_{+}^{-} D^{\pm}_{\beta,st} \cos[2\pi f_\beta \ell \pm t\lambda - \varepsilon^{\pm}_{\beta,st}] P_{st}(\sin\phi), \tag{6.2.1b}$$

with

$$\left.\begin{aligned} D^{\pm}_{\beta,st} \cos \varepsilon^{\pm}_{\beta,st} &= \tfrac{1}{2}(a'_{\beta,st} \mp b''_{\beta,st}), \\ D^{\pm}_{\beta,st} \cos \varepsilon^{\pm}_{\beta,st} &= \tfrac{1}{2}(a''_{\beta,st} \pm b'_{\beta,st}). \end{aligned}\right\} \tag{6.2.1c}$$

The summation of the form $\sum_{+}^{-} D^{\pm}\cos(\alpha \pm \beta - \varepsilon^{\pm})$ implies $D^{+}\cos(\alpha + \beta - \varepsilon^{+}) + D^{-}\cos(\alpha - \beta - \varepsilon^{-})$. The potential of this layer outside the Earth is

$$\Delta U_\beta(r) = 4\pi G R \rho_w \sum_s \sum_t \sum_{+}^{-} \frac{1+k'_s}{2s+1}\left(\frac{R}{r}\right)^{s+1} \times D^{\pm}_{\beta,st} \cos(2\pi f_\beta t \pm t\lambda - \varepsilon^{\pm}_{\beta,st}) P_{st}(\sin\phi), \tag{6.2.2}$$

where the factor $1+k'_s$ allows for the Earth's elastic yielding under the variable ocean load; $\rho_w$ is the density of the oceans.

*Equilibrium tide.* The parameters $D^{\pm}_{st}$ and $\varepsilon^{\pm}_{st}$ are known for only a few of the principal tidal frequencies, and for the remainder an order-of-magnitude estimate is possible only by using an equilibrium tide theory. For a tide-raising potential of degree $l$, order $m$ and frequency $f_\beta$ (equation 6.1.8)

$$U_\beta = A_\beta \cos(2\pi f_\beta t + m\lambda) P_{lm}(\sin\phi),$$

the equilibrium tide is

$$\xi'_\beta = [(1+k_2-h_2)/g] U_\beta \mathscr{C}(\phi,\lambda), \tag{6.2.3a}$$

where $\mathscr{C}(\phi,\lambda)$ is the ocean function. Expanding the latter into spherical harmonics, as in section 4.2, gives, writing $\sigma_\beta$ for $2\pi f_\beta t$,

$$\xi'_\beta = \frac{1}{2}\frac{1+k_2-h_2}{g} A_\beta \sum_{i=0}^{\infty}\sum_{j=0}^{i}\sum_{+}^{-}[(a_{ij}\cos\sigma_\beta + b_{ij}\sin\sigma_\beta) \times \cos(m\pm j)\lambda + (a_{ij}\sin\sigma_\beta \mp b_{ij}\cos\sigma_\beta)\sin(m\pm j)\lambda] \times P_{lm}(\sin\phi) P_{ij}(\sin\phi). \tag{6.2.3b}$$

Terms like $\sum_{+}^{-}[\cos(m\pm j)$ or $\sin(m\pm j)\lambda] P_{lm}(x) P_{ij}(x)$ can be expanded into linear series of polynomials in the form

$$\sum_u \sum_{+}^{-} Q^{\pm}_{lmiju} P_{u,|m\pm j|}(x)\, [\cos(m\pm j)\lambda \text{ or } \sin(m\pm j)\lambda] \tag{6.2.4}$$

with (Balmino 1978)

$$Q^{\pm}_{lmiju} = \frac{2u+1}{2}\frac{(u-|m\pm j|)!}{(u+|m\pm j|)!}\int_{-1}^{+1} P_{lm}(x) P_{ij}(x) P_{u,|m\pm j|}(x)\, dx.$$

The $Q^{\pm}_{lmiju}$ relate to the Clebsch–Gordan coefficients and the

Wigner 3-$j$ coefficients encountered in quantum mechanics. They are non-zero for even $(l+u+i)$ and for max $(|m\pm j|, |l-i|) \leq u \leq l+i$. The equilibrium tide becomes, after further re-arrangement,

$$\xi'_\beta = \frac{1}{2}\frac{1+k_2-h_2}{g} A_\beta \sum_i \sum_j \sum_u \sum_+^- \{a_{ij} \cos[\sigma_\beta + (m\pm j)\lambda]$$

$$\pm b_{ij} \sin[\sigma_\beta + (m\pm j)\lambda]\} Q^{\pm}_{lmiju} P_{u,|m\pm j|}(\sin\phi). \qquad (6.2.5)$$

In order that mass is conserved, the terms with $u=0$ and $|m+j|=0$ should vanish, but in (6.2.5) this condition gives

$$\Delta\xi'_\beta \equiv \xi'_\beta(u=0, |m+j|=0) \qquad (6.2.6)$$

$$= \frac{1}{2}\frac{1+k_2-h_2}{g} A_\beta (a_{lm} \cos\sigma_\beta - b_{lm} \sin\sigma_\beta) \frac{(l+m)!}{2l+1}$$

and we have to apply a correction term: the equilibrium tide becomes

$$\xi_\beta = \xi'_\beta - \Delta\xi'_\beta = \sum_{u=1}^{\infty} \xi_{\beta,u}. \qquad (6.2.7)$$

The surface load due to this tide is found by multiplying each harmonic of degree $u$ by $\rho_w g(1+k'_u)$.

The equilibrium tide (6.2.7) does not allow for the effect of the ocean loading on the tide itself, and the complete expression for the tide is, with $\bar{\rho}$ equal to the mean density of the Earth,

$$\xi_\beta = \frac{1+k_2-h_2}{g} U_\beta \mathscr{C}(\phi,\lambda) + \sum_u (1+k'_u-h'_u) \frac{3}{2u+1} \frac{\rho_w}{\bar{\rho}} \xi_{\beta,u}. \qquad (6.2.8)$$

The solution of this equation for $\xi_{\beta,u}$ is discussed by Dahlen (1976) in connection with the pole tide but, since the diurnal and semi-diurnal tides depart significantly from the equilibrium condition, and as we are only interested in the magnitudes of a few of the low-degree terms in these tides, the simplified formalism (equation 6.2.7 with equations 6.2.5 and 6.2.6) is sufficient. To compare the equilibrium tide with the development (6.2.1), we compare those terms in (6.2.5) that give $u=s$ and $|m\pm j|=t$ with the corresponding terms in (6.2.1). Furthermore, $l=2$, and we consider only those terms for which $s=2, 4$ and $t=m$. Then

$$\left.\begin{aligned}
D^{+}_{\beta,st}\cos\varepsilon^{+}_{\beta,st} &= \frac{1}{2}\frac{(1+k_2-h_2)}{g}A_\beta\sum_i a_{i0}Q^{+}_{2mi0s},\\
D^{+}_{\beta,st}\sin\varepsilon^{+}_{\beta,st} &= 0,\\
D^{-}_{\beta,st}\cos\varepsilon^{-}_{\beta,st} &= \frac{1}{2}\frac{(1+k_2-h_2)}{g}A_\beta\sum_i a_{i0}Q^{-}_{2mi(2m)s},\\
D^{-}_{\beta,st}\sin\varepsilon^{-}_{\beta,st} &= \frac{1}{2}\frac{(1+k_2-h_2)}{g}A_\beta\sum_i b_{i0}Q^{-}_{2mi(2m)s}.
\end{aligned}\right\}\qquad(6.2.9)$$

The discussion on equilibrium tides goes back to Laplace. Wunsch (1967) reviewed the history of the arguments for and against the theory. Laplace argued that the long-period tides would very nearly follow an equilibrium theory; for small amplitudes and slow changes the ocean adjusts itself to the changing potential and only a small amount of friction is necessary to damp out any non-equilibrium deviations. Kelvin adopted this viewpoint for the Mf tide in an attempt to estimate the elastic factor $(1+k_2-h_2)$ appearing in the equilibrium tide, for example in (6.2.5). Subsequent analyses of the ocean tides in many ports apparently verified Kelvin's argument: G. H. Darwin found $0.67\pm0.07$ and W. Schweydar found $0.63\pm0.04$ for this factor, in quite good agreement with recent determinations from the solid tide observations and from elastic Earth model calculations. Both Darwin and S. S. Hough attempted theoretical analyses of the long-period tides on a water-covered globe and concluded that the amplitudes would be far from equilibrium due to the establishment of strong zonal currents, but, as pointed out by Rayleigh, such currents would be effectively blocked by the north–south continental barriers. More recently Proudman (1960), investigating frictional effects on long-period tides, concluded that the 18.6-yr tide follows an equilibrium theory as probably do the semi-annual and annual tides, although the Mm and Mf tides probably do not. Rossiter (1967) estimated the average 18.6-yr tide for European stations and found that it is in phase with the equilibrium tide although its amplitude is somewhat reduced. Currie (1976), applying the maximum-entropy method to 33 long-interval records, found that this tide follows closely the equilibrium values. The annual (Sa) and semi-annual

(Ssa) tides are dominated by seasonal variations of meteorological and oceanographical origin and cannot be used to deduce the validity of the equilibrium theory at these frequencies. For Mm and Mf, Wunsch (1967) has analysed in detail the tide records for a number of Pacific stations and concludes that they are substantially non-equilibrium, that the average reduction of the amplitudes is about 50% compared with their equilibrium values, and that the observed tide lags the equilibrium tide by some 20° on the average.

### 6.2.2 *Ocean tide models*

Knowledge of the world's ocean tides is rather sparse. Long records of tides exist along many parts of the shoreline and are extremely valuable for predicting the tides locally, but these tides are most often influenced by the coastline geometry and by the shallow coastal sea floor where frictional forces will significantly modify the tides from their open-sea behaviour. Thus the coastal tides that sometimes reach more than 5 m are very local effects and are not characteristic of the global ocean tide. The best observations of the open-ocean tides come from the island stations, particularly from those islands that rise up steeply from the deep-sea floor, such as volcanic islands. All such island records show that the ocean tide does not in general exceed 1 m. The limited availability of unperturbed tidal stations means that the global tidal patterns cannot be established with reliability from measurements alone, and one has to resort to theory for estimating the global tides. The development of bottom pressure gauges for measuring the tides in the open sea has led to important improvements in the knowledge of regional ocean tides, as demonstrated by Munk, Snodgrass & Wimbush (1970) and Filloux (1971) for tides off the Californian coast, by Luther & Wunsch (1975) for the Central Pacific, and by D. E. Cartwright and colleagues for the North Atlantic. But the information is still too sparse to be useful for global models. Complementary reviews of ocean tides are given by Hendershott & Munk (1970), by Hendershott (1973) and by Cartwright (1977).

This lack of adequate observational data means that the prediction of open-ocean tides is largely based on theory, on the solution of the Laplace tidal equations (see, for example, Lamb 1932; Platzman 1971). Complete solutions are complex and must include the following factors:

(i) the land–sea distribution and the ocean depth;
(ii) energy dissipation, usually assumed to occur only in shallow seas and along coastlines;
(iii) the Earth, being elastic, is itself subjected to a solid tide which will work on the sea floor and modify the ocean tide;
(iv) the ocean tide, representing a variable load, will deform the Earth and modify the ocean tide;
(v) the computed tidal patterns must be in agreement with the observed tides for those stations where the tides are observed free from local disturbing effects.

Miles (1974) has investigated the various approximations made in deriving the classical Laplace tidal equations. Further complications, as yet not evaluated, may arise from the generation and dissipation of baroclinic tides.

Numerical solutions for the global $M_2$ tide have been published by Bogdanov & Magarik (1967), Tiron, Sergeev & Michurin (1967), Pekeris & Accad (1969), Zahel (1970, 1977), Hendershott (1972), and Gordeyev, Kagan & Polyakov (1977). Pekeris & Accad (1969) and Zahel (1970) have attempted to solve the Laplace tide equations using only a knowledge of the ocean–continent distribution, bathymetry and the tide-generating force. Both consider the tide on a rigid Earth. Pekeris & Accad (1969) introduce a frictional force that is proportional to the tidal velocity rather than to the square of this velocity if dissipation is by bottom friction. This is comparable with the Guldberg–Mohn assumption about *virtual* internal friction introduced into some oceanographic problems (see, for example, Neumann 1968). This friction model, and the definition of the boundaries of the oceans as the 100-m isobath, results in dissipation that is fairly uniformly distributed along coastlines. The authors determine a global friction coefficient by adopting a value that results in the best agreement between computed and observed tides at island stations. Zahel (1970, 1977) introduces a bottom friction force that is proportional to the square of the velocity and also allows for dissipation by turbulent friction, using a constant eddy viscosity coefficient for the world's oceans. The former allows for dissipation in shallow seas, the latter for dissipation in the open oceans.† The ocean geometry is defined by the 50-m isobath. Zahel's 1977 model differs from this earlier model in that (i) it considers the solid tide, (ii) the computational grid size has been reduced from 4° to 1°, and (iii) the eddy viscosity coefficient has been reduced. In his models, flow is not permitted across the coastlines since specific dissipation mechanisms have been introduced. Thus dissipation by bottom friction in shallow (less than 50 m) seas is excluded. Bogdanov & Magarik (1967) and Hendershott (1972) solve the Laplace tidal equations by imposing specific tide values along the coastlines. For several extensive coastlines no tide observations are available and interpolated values have been used. No frictional forces are introduced in these models and currents are allowed to flow across the boundaries. Thus dissipation is assumed to occur in the shallow seas, although the actual mechanism is not specified. Implicit in these studies is that the coastal observations are representative of the nearby deep-water tides. Tiron, Sergeev & Michurin (1967) have published models for $M_2$, $S_2$, $K_1$ and $O_1$, but these models differ considerably from the others. This is possibly a consequence of their treatment of the dissipation in that they appear to impose the condition that observed tides equal computed tides where the former are reliably known and that flow normal to the coastlines vanishes in regions where the tide is unknown.

Hendershott (1972) and Zahel (1977) consider the interaction between the solid tide and ocean tide, although models such as that of Bogdanov & Magarik apparently give realistic results due to their introduction of observed tides as constraints. Allowance for the Earth's tidal deformation effectively reduces the tidal potential $\Delta U_2$ to $(1+k_2-h_2)\Delta U_2$ and presumably the tidal amplitudes will be reduced by the factor $(1+k_2-h_2)$. But if dissipation is explicitly introduced in the models through mechanisms that depend on the tidal velocities, the rate of dissipation, and hence the tidal amplitudes, may be further modified. By an iterative procedure, Hendershott evaluates the interaction of the ocean floor's elastic yielding under the variable tide load. His resulting model is

† The introduction of a non-linear frictional force implies that the different tidal frequencies cannot be modelled separately.

inconclusive, but does indicate that this interaction may be at least as important as the solid tide effect. Presumably, by imposing the boundary conditions, this effect is already partly included in his and in Bogdanov & Magarik's models. The recent solution by Gordeyev *et al.* (1977) for $M_2$ is the most complete as it includes the self-loading effect that eluded Hendershott.† Dissipation is modelled by both a virtual friction coefficient and turbulent friction plus a boundary condition of no slip along coastlines (see also Gordeyev, Kagan & Rivkind 1973; Kagan 1977). The spatial structure of their models, with and without self-loading, does not differ greatly, but the amplitudes and phase do differ in important respects.

Hendershott (1973) compares various solutions. Agreement is generally good in the North Atlantic but elsewhere considerable discrepancies exist: in the Indian and Pacific oceans differences in amplitude of a factor of 2 occur between the solution of Pekeris & Accad and that of Hendershott. A comparison with the recent tidal chart by Luther & Wunsch (1975) indicates that the Pekeris & Accad model overestimates the tides in the Central Pacific. In the Indian Ocean it is the Hendershott model that overestimates the tides. We cannot consider the available models to be adequate as far as the detail is concerned. Available models for the other frequencies in the tidal potential must be considered even less adequate or non-existent. Bogdanov & Magarik (1967) published a solution for $S_2$ and in a later paper (Bogdanov & Magarik 1969) they give models for $K_1$ and $O_1$. Zahel (1973) gives a model for $K_1$. Empirical charts for the phase of the global $K_1$ and $O_1$ tides are given by Dietrich (1944).

For the discussion on the fluctuations in the rate of rotation we require the coefficients $D^{\pm}_{\beta,20}$, $\varepsilon^{\pm}_{\beta,20}$; for the polar motion we require $D^{+}_{\beta,21}$, $\varepsilon^{+}_{\beta,21}$; for the discussion on the tidal dissipation we require the coefficients $D^{+}_{\beta,lm}$, $\varepsilon^{+}_{\beta,lm}$, with the same degree and order as the tide-raising potential. All of the above-mentioned models have been harmonically analysed for the coefficients $D^{\pm}_{\beta,st}$ and $\varepsilon^{\pm}_{\beta,st}$, and the relevant terms for the discussion on the rotation are summarized in table 6.4 (for further details see Lambeck, Cazenave & Balmino

† Zahel (1978) has also solved this problem in a preliminary model for $M_2$, as has Parke (1978). See also Accad & Pekeris (1978).

Table 6.4. *Estimates of the* $D^+_{22}$ sin $\varepsilon^+_{22}$ *or* $D^+_{21}$ cos $\varepsilon^+_{21}$ *for different ocean tide models*

| | Tide | $D^+_{22}$ or $D^+_{21}$ (cm) | $\varepsilon^+_{22}$ or $\varepsilon^+_{21}$ (degree) | $D^+_{22}$ sin $\varepsilon^+_{22}$ or $D^+_{21}$ cos $\varepsilon^+_{21}$ (cm) | $D^+_{22}$ sin $\varepsilon^+_{22}$ or $D^+_{21}$ cos $\varepsilon^+_{21}$ (corrected) (cm) |
|---|---|---|---|---|---|
| $M_2$ | Bogdanov & Magarik (1967) | 4.33 | 126 | 3.51 | |
| $M_2$ | Pekeris & Accad (1969) | 4.57 | 110 | 4.37 | 3.00[a] |
| $M_2$ | Zahel (1970) | 4.90 | 105 | 4.73 | 3.30[a] |
| $M_2$ | Hendershott (1972) | 3.61 | 105 | 3.48 | |
| $M_2$ | Zahel (1977) | 4.66 | 110 | 4.38 | |
| $M_2$ | Equilibrium tide | 4.00 | 0 | | |
| $S_2$ | Bogdanov & Magarik (1967) | 1.87 | 140 | 1.20 | 1.42[b] |
| $S_2$ | Equilibrium tide | 1.86 | 0 | | |
| $S_2$ | Atmospheric tide | 0.34 | 292 | −0.32[c] | |
| $K_1$ | Dietrich (1944) | 2.34 | 222 | −1.74 | |
| $K_1$ | Bogdanov & Magarik (1969) | 1.65 | 250 | −0.56 | |
| $K_1$ | Zahel (1973) | 6.64 | 221 | −5.01 | 3.51[a] |
| $K_1$ | Equilibrium tide | 2.34 | 180 | | |
| $O_1$ | Dietrich (1944) | 1.66 | 38 | 1.31 | |
| $O_1$ | Bogdanov & Magarik (1969) | 0.67 | 61 | 0.32 | |
| $O_1$ | Equilibrium tide | 1.66 | 0 | | |

[a] Reduced by factor $(1+k_2-h_2)$ to allow for yielding of Earth
[b] Corrected for atmospheric loading
[c] Without ocean response to atmospheric load

1974; Lambeck 1977). The most important effect of the tides on the rotation is associated with the tidal dissipation problem discussed in chapter 10, and for the other effects an order-of-magnitude calculation is adequate. These can be obtained from an equilibrium theory (section 6.2.1). Such a theory also enables the tidal coefficients for lesser-known ocean tides to be estimated.

For the $M_2$ tide the coefficients for the several ocean models discussed above can be compared. The Pekeris & Accad (1969) and Zahel (1970) coefficients tend to be larger than those for the other $M_2$ solutions due, in part at least, to these models having been computed for a rigid Earth. The 'corrected' values in table 6.4 correspond to the tide coefficients on the rigid Earth reduced by the factor $(1+k_2-h_2)$, or by about 70%. The uncorrected amplitudes $D^+_{M_2,22}$, range from about 3.6 to 4.9 cm and the phase varies by about 20°. The agreement appears satisfactory despite the differences in the manner in which the Laplace equations are resolved and the way in which dissipation is treated. The only other tide for which a comparison is possible is the $K_1$ tide, for which we have the numerical models by Bogdanov & Magarik and by Zahel and the empirical model of Dietrich. The comparison of these three $K_1$ models reveals the unsatisfactory status of the global diurnal tide models. Zahel's corrected amplitude $D^+_{K_1,21}$ is about three times greater than the other two values and may be a consequence of his using a value for the effective eddy viscosity that is too large (W. Zahel, personal communication). All diurnal tides summarized in table 6.4 indicate that $|\cos \varepsilon^+_{21}| < |\sin \varepsilon^+_{22}|$, suggesting that the diurnal tides are closer to their equilibrium values than the semi-diurnal tides.

Table 6.5 summarizes the 'best' estimates of the $D^+_{\beta,lm}$ and $\varepsilon^+_{\beta,lm}$ for the tides that enter into the tidal dissipation discussion. They are from Lambeck (1977) and are based on (i) numerical models of the ocean tides, (ii) ages of the semi-diurnal tides† and (iii) equilibrium theory.

### 6.2.3 *Atmospheric tides*

In many discussions on the secular deceleration of the Earth's rotation, the atmospheric tide has played a role that is well beyond its significance in the total tidal effect; neglected ocean tides such as $O_1$, $K_1$, and the virtually unknown $N_2$, are much more important (Lambeck 1975*a*). This apparent importance is undoubtedly related to the fact that the atmospheric tide leads the Sun and hence

† The age of the tide denotes the interval between the time of maximum equilibrium tide and the time of local spring tide.

Table 6.5. *Summary of 'best estimates' of ocean tide parameters deduced from tide models, age of tide observations and equilibrium theory (from Lambeck 1977)*

| | $D^+_{22}$ or $D^+_{21}$ (cm) | $\varepsilon^+_{22}$ or $\varepsilon^+_{21}$ (degree) | $D^+_{22}\sin\varepsilon^+_{22}$ or $D^+_{21}\cos\varepsilon^+_{21}$ |
|---|---|---|---|
| Ocean model results | | | |
| $M_2$ | 3.64 | 112 | $3.37\pm0.30$ |
| $S_2$[c] | 1.66 | 134 | $1.19\pm0.25$ |
| $N_2$ | 0.69 | 100 | $0.68\pm0.20$ |
| $K_2$(lunar) | 0.32 | 136 | $0.22\pm0.07$ |
| $K_2$(solar) | 0.14 | 136 | $0.10\pm0.03$ |
| $L_2$ | 0.10 | −56 | $-0.08\pm0.02$ |
| $2N_2$ | 0.09 | 88 | $0.09\pm0.03$ |
| $T_2$ | 0.10 | 133 | $0.07\pm0.02$ |
| $K_1$(lunar) | 1.37 | 236 | $-0.77\pm0.15$ |
| $K_1$(solar) | 0.63 | 236 | $-0.35\pm0.07$ |
| $O_1$ | 1.17 | 50 | $0.75\pm0.15$ |
| $P_1$ | 0.54 | 55 | $0.31\pm0.10$ |
| $Q_1$ | 0.22 | 47 | $0.15\pm0.05$ |
| Atmospheric tides | | | |
| $S_2$[a] | | | $-0.10\pm0.05$ |
| $S_2$[b] | | | $-0.32\pm0.10$ |

[a] To be used with the ocean model results
[b] To be used with the satellite results
[c] Equivalent phase angle for combined ocean and atmosphere

gives a positive acceleration to the Earth, in contrast to the ocean's decelerating effect. This has led to speculation on the possible role it may have played in the past (Holmberg 1952), but an atmosphere similar to that of Venus, and thermally driven, is required if it is to cancel the effects of the present ocean tides.

Chapman & Lindzen (1970) review the subject of atmospheric tides. Of the numerous frequencies in the atmospheric tide, the only one of some importance is the solar tide at the $S_2$ frequency. Unlike the ocean tide, its global distribution is reasonably well known from ground level pressure records and, because it is less influenced by the ocean–continent distribution, its spatial configuration is more uniform. In its harmonic expansion, proportionally more power is contained in the low-degree than high-degree harmonics and the ratio of the 2, 2-terms in the atmospheric and oceanic expansions will be larger than may be expected from simply comparing the tidal amplitudes at any one island station.

The gravitational potential of the atmosphere can be computed in the same way as the ocean tide. Atmospheric pressure $p(\phi, \lambda)$ at each point on the Earth's surface is harmonically analysed into its constituent frequency components $p_\beta(\phi, \lambda)$. Each such component $p_\beta$ is expressed by a series analogous to the ocean tide expansion (6.2.1), and the surface load is $p_\beta/g$. In computing the potential outside the Earth due to the layer, the effective depth of the atmosphere is assumed to be small compared with the Earth's radius. The elastic yielding of the Earth under this variable load is allowed for by introducing the load deformation coefficients $k'_s$. More troublesome is the question of how to allow for the ocean response to this load. Does the ocean respond as an inverted barometer or not (chapter 7)? A study by Cartwright (1968) of the radiational, or meteorological, tides around the British Isles found an average ratio (radiational tide)/(gravitational tide) of 0.18 for $S_2$ and an average phase difference between these tides of 130°. Zetler (1971), in a similar study for stations along the US coast, finds comparable results. This radiational tide appears to be largely excited by the atmospheric tidal load on the ocean surface, but the response is very different from that of an inverted barometer. In some cases the radiational tide has an amplitude that is larger than the predicted amplitude by a factor of 10 (see also Cartwright & Edden 1977). This may be partially due to a direct effect of solar radiation on the sea surface or due to the offshore–onshore wind cycle. The first effect would not load the crust; the second effect will modify only the coastal tide, not the mid-ocean tide.

The $S_2$ ocean tide solution of Bogdanov & Magarik (1967) may include the ocean response to the atmospheric tide since it has been constrained by tide observations along the coastlines, and presumably no atmospheric correction has been made. A 'corrected' $S_2$ ocean tide coefficient then would be

$$D_{22}^{+} \sin \varepsilon_{22}^{+} = (D_{22}^{+} \sin \varepsilon_{22}^{+})_{\text{observed ocean}} - a_{00}(D_{22}^{+} \sin \varepsilon_{22}^{+})_{\text{atmosphere}}.$$

A more satisfactory solution, also relevant to the discussion of meteorological influences on the rotation of the Earth, would be along the following lines. If the ocean response is static, the total potential due to the atmospheric layer and the deformed ocean is

$$U = \sum_n U_n = \frac{3}{\bar{\rho}} \sum_n \left(\frac{1+k_n'}{2n+1}\right)(p_n + g\rho_w \xi \mathscr{C}), \qquad (6.2.10)$$

where $\xi$ is the pressure-induced ocean tide. The equilibrium surface will be deformed by

$$N = \sum_n \left[ (1 + k_n' - h_n') \frac{U_n}{g} - \frac{p_n \mathscr{C}}{\rho_w g} \right]$$

which is defined over the entire globe. Over the oceans $N = \xi$, and with (6.2.10),

$$N = \sum_n \left[ \frac{3}{\bar{\rho}(2n+1)g} (1 + k_n')(1 + k_n' - h_n')(p_n + g\rho_w N \mathscr{C}) - \frac{p_n \mathscr{C}}{\rho_w g} \right]. \qquad (6.2.11)$$

This equation, when solved for $N$, gives the ocean tide $\xi = N\mathscr{C}$. Mass must be conserved in the solution. That is,

$$\int_S N\mathscr{C} \, dS = 0. \qquad (6.2.12)$$

The actual solution of (6.2.11) is complicated by the fact that, for a given frequency, $p$ will consist of numerous harmonics, each of which will contribute to the entire spectrum of $N$.

## 6.3 Satellite observations of tidal parameters†

### 6.3.1 *Methodology*

The tidal potential (6.1.12) will act on a satellite orbiting the Earth and cause perturbations in its motion. These perturbations are of interest in that they permit some of the tidal parameters to be estimated from precise analyses of orbits and because the tidal effects on the lunar motion represent a special case (chapter 10). To

† A recent review of results is given by Felsentreger *et al.* (1979).

study these perturbations in the orbits of close Earth satellites, the coordinates $(r, \phi, \lambda)$ in potential (6.1.12) are also transformed into the Keplerian elements $\kappa_i$ of the satellite, using the transformation (6.1.6). The potential now becomes

$$\Delta U(r) = \sum_{l=2}^{\infty} \sum_{m=0}^{l} \sum_{p=0}^{l} \sum_{q=-\infty}^{\infty} \sum_{j=0}^{l} \sum_{g=-\infty}^{\infty} \Delta U_{lmpqjg}$$

with

$$\Delta U_{lmpqjg} = \frac{Gm^*}{a^*} k_l \left(\frac{R}{a}\right)^l \left(\frac{R}{a}\right)^{l+1} (2-\delta_{0m}) \frac{(l-m)!}{(l+m)!} F_{lmp}(I^*) F_{lmj}(I) \times G_{lpq}(e^*) G_{ljg}(e) \cos(v^*_{lmpq} - v_{lmjg} + \varepsilon_{lmpq}), \tag{6.3.1}$$

where the lag $\varepsilon_{lmpq}$ is defined by (6.1.13*b*).

This form of $\Delta U$ is equivalent in all respects to that developed by Kaula (1964, 1969) and used by Goldreich (1966), Peale (1973), Lambeck *et al.* (1974), Lambeck (1977) and others in studies of tidal effects on the Moon itself or on the motion of close Earth satellites.

We require the acceleration of the satellite due to the force $\nabla \Delta U(r)$, and this is most conveniently found by using the Lagrangian form for the equations of motion (see, for example, Brouwer & Clemence 1961; Kaula 1966). These equations are

$$\left.\begin{aligned}
\frac{\mathrm{d}a}{\mathrm{d}t} &= \frac{2}{na}\frac{\partial \Delta U}{\partial M},\\
\frac{\mathrm{d}e}{\mathrm{d}t} &= \frac{(1-e^2)^{1/2}}{na^2 e}\frac{\partial \Delta U}{\partial M} - \frac{(1-e^2)^{1/2}}{na^2 e}\frac{\partial \Delta U}{\partial \omega},\\
\frac{\mathrm{d}\omega}{\mathrm{d}t} &= -\frac{\cos I}{na^2(1-e^2)^{1/2}\sin I}\frac{\partial \Delta U}{\partial I} + \frac{(1-e)^{1/2}}{na^2 e}\frac{\partial \Delta U}{\partial e},\\
\frac{\mathrm{d}I}{\mathrm{d}t} &= \frac{\cos I}{na^2(1-e^2)^{1/2}\sin I}\frac{\partial \Delta U}{\partial \omega}\\
&\quad - \frac{1}{na^2(1-e^2)^{1/2}\sin I}\frac{\partial \Delta U}{\partial \Phi},\\
\frac{\mathrm{d}\Phi}{\mathrm{d}t} &= \frac{1}{na^2(1-e)^{1/2}\sin I}\frac{\partial \Delta U}{\partial I},\\
\frac{\mathrm{d}M}{\mathrm{d}t} &= n - \frac{1-e^2}{na^2 e}\frac{\partial \Delta U}{\partial e} - \frac{2}{na}\frac{\partial \Delta U}{\partial a}.
\end{aligned}\right\} \tag{6.3.2}$$

They express the time rate of change of the orbital elements due to the tidal potential. For the inclination of the satellite orbit, for example,

$$\left.\frac{\mathrm{d}I}{\mathrm{d}t}\right|_{lmpqjg} = \frac{GM^*[(l-2p)\cos I - m]}{na^2(1-e^2)^{1/2}\sin I} k_l \frac{1}{a^*}\left(\frac{R}{a^*}\right)^l\left(\frac{R}{a}\right)^{l+1}$$
$$\times(2-\delta_{0m})\frac{(l-m)!}{(l+m)!}$$
$$\times F_{lmj}(I)F_{lmp}(I^*)G_{ljg}(e)$$
$$\times G_{lpq}(e^*)\sin(v^*_{lmpq} - v_{lmjg} + \varepsilon_{lmpq}).$$

The integration of this, and similar expressions for the other elements, is carried out either numerically or analytically by assuming that the only time-varying elements are (i) the mean motions of the satellite and of the tide-raising body, and (ii) the linear rates in $\dot{\Phi}$, $\dot{\omega}$, $\dot{\Phi}^*$, $\dot{\omega}^*$ due principally to interactions of the motions with the Earth's flattening and the solar attraction (see, for example, Kaula 1966; Gaposchkin 1973). The results for the variation in semi-major axis, eccentricity and inclination are

$$\left.\begin{aligned}
\Delta a_{lmpq} &= \frac{2}{na}\frac{(l-2j+g)}{(\dot{v}^*_{lmpq} - \dot{v}_{lmjg})} A_{lmpqjg}\cos\gamma_{lmpqjg}\\
\Delta e_{lmpq} &= \frac{(1-e^2)^{1/2}}{na^2 e}\frac{[(l-2j+g)-(l-2j)]}{(\dot{v}^*_{lmpq} - \dot{v}_{lmjg})} A_{lmpqjg}\cos\gamma_{lmpqjg}\\
\Delta I_{lmpq} &= \frac{1}{na^2(1-e^2)^{1/2}\sin I}\frac{[\cos I(l-2j)-m]}{(\dot{v}^*_{lmpq} - \dot{v}_{lmjg})}\\
&\quad\times A_{lmpqjg}\cos\gamma_{lmpqjg}
\end{aligned}\right\}\qquad(6.3.3a)$$

with

$$A_{lmpqjg} = k_l\left(\frac{R}{a^*}\right)^l\left(\frac{R}{a}\right)^{l+1}\frac{Gm^*}{a^*}(2-\delta_{0m})\frac{(l-m)!}{(l+m)!}$$
$$\times F_{lmp}(I^*)F_{lmj}(I)G_{lpq}(e^*)G_{ljg}(e), \qquad (6.3.3b)$$

and

$$\gamma_{lmpqjg} = (v^*_{lmpq} - v_{lmjg}).$$

These and the corresponding perturbations in the other elements have periods longer than 1 d only for those terms for which the

Table 6.6. *Theoretical amplitudes of tidal perturbations in the inclination and right ascension in the orbits of artificial Earth satellites GEOS 1 and STARLETTE*

| | STARLETTE | | | GEOS 1 | | |
|---|---|---|---|---|---|---|
| Tide | Period (d) | $\Delta i$ (arc second) | $\Delta\Omega$ (arc second) | Period (d) | $\Delta i$ (arc second) | $\Delta\Omega$ (arc second) |
| $M_2$ | 11 | 0.19 | 0.20 | 12 | 0.17 | 0.12 |
| $S_2$ | 36 | 0.21 | 0.43 | 56 | 0.41 | 0.47 |
| $K_2$ | 46 | 0.11 | 0.16 | 80 | 0.20 | 0.25 |
| $O_1$ | 11 | 0.09 | 0.05 | 13 | 0.06 | 0.07 |
| $P_1$ | 60 | 0.18 | 0.22 | 85 | 0.15 | 0.30 |
| $K_1$ | 90 | 0.95 | 1.47 | 160 | 0.95 | 2.64 |

combination of indices $(l-2j+g)$ vanishes. For all other terms not satisfying this condition, the perturbations will be of small amplitudes since their frequencies entering into the denominator of equations such as (6.3.3) are large. The frequencies of the long-period tidal terms are governed by both the lunar and the satellite motions around the Earth, and the spectrum will differ quite considerably from that of the tidal variations observed at the Earth's surface in, for example, gravity. Also, as the amplitude of the orbital perturbations depends upon frequency, tidal terms observed at the Earth's surface to be of small amplitude may become important in the satellite spectrum if the lunar and satellite motions become commensurable, when $\dot{v}^*_{lmpq} \simeq \dot{v}_{lmjg}$. Thus by a careful selection of elements for an orbit, different fundamental tidal frequencies can be made to have more or less important effects on the satellite. Table 6.6 illustrates two such theoretical spectra for perturbations in the inclination of the satellite STARLETTE (7501001) and GEOS 1 (6508901). The $S_2$ tide perturbs the satellite motion considerably more than does the $M_2$ tide, while at the Earth's surface it is only about one-half of $M_2$. In the case of GEOS 1, the $K_1$ tide perturbation in inclination is about three times greater than the perturbations due to $M_2$, while on the Earth's surface it represents only about 60% of $M_2$ in amplitude.

The amplitudes of the orbit perturbations are proportional to the Love numbers $k_2$, and the phases lag the direct attraction of the Sun or Moon on the satellite by an amount $\varepsilon_{lmpq}$. Observations of the perturbations permit these parameters to be deduced, and initial attempts were made by Kozai (1968) and R. R. Newton (1968). A review is given by Lambeck *et al.* (1974). If only the solid tide potential were of importance, any differences in the Love number and phase lags from theoretical values would be small indeed, but ocean tides also perturb the satellite motions, and, if not allowed for in the orbital theory, introduce significant variations in the tidal effective Love numbers with frequency. The ocean tide perturbations are found by applying the transformation (6.1.6) to the ocean tide potential (6.2.2) and substituting the resulting expression into the Lagrange equations (6.3.2). For the satellite inclination, for example,

$$\left.\frac{\mathrm{d}I}{\mathrm{d}t}\right|_{\beta,stuv} = \frac{4\pi GR^2}{a}\rho_{\mathrm{w}}\frac{1+k'_s}{2s+1}\left(\frac{R}{a}\right)^s \frac{1}{na^2(1-e^2)^{1/2}\sin I} \times D^{\pm}_{\beta,st}F_{stu}(I)G_{suv}(e)\,[(s-2u)\cos I - t] \times \begin{bmatrix} \mp\sin \\ \cos \end{bmatrix}^{s-t \text{ even}}_{s-t \text{ odd}} \gamma^{\pm}_{\beta,stuv}, \tag{6.3.4}$$

with

$$\gamma^{\pm}_{\beta,stuv} = v_{stuv} \pm 2\pi f_\beta \ell + \varepsilon^{\pm}_{\beta,st}.$$

Integrating these expressions with respect to time, assuming that all time-dependent variables are contained in $\gamma^{\pm}_{\beta,stuv}$, gives for $a$, $e$ and $I$

$$\left.\begin{aligned}
\Delta a_{\beta,stuv} &= \frac{2}{na}A'_{\beta,stuv}\frac{(s-2u+v)}{\dot{\gamma}^{\pm}_{\beta,stuv}}\begin{bmatrix} \pm\cos \\ \sin \end{bmatrix}^{s-t \text{ even}}_{s-t \text{ odd}} \gamma^{\pm}_{\beta,stuv}, \\
\Delta e_{\beta,stuv} &= \frac{(1-e^2)^{1/2}}{na^2 e}A'_{\beta,stuv}\frac{[(s-2u+v)-(s-2u)]}{\dot{\gamma}^{\pm}_{\beta,stuv}} \\
&\quad \times \begin{bmatrix} \pm\cos \\ \sin \end{bmatrix}^{s-t \text{ even}}_{s-t \text{ odd}} \gamma^{\pm}_{\beta,stuv}, \\
\Delta I_{\beta,stuv} &= \frac{1}{na^2(1-e^2)^{1/2}\sin I}\frac{[\cos I(s-2u)-t]}{\dot{\gamma}^{\pm}_{\beta,stuv}}A'_{\beta,stuv} \\
&\quad \times \begin{bmatrix} \pm\cos \\ \sin \end{bmatrix}^{s-t \text{ even}}_{s-t \text{ odd}} \gamma^{\pm}_{\beta,stuv},
\end{aligned}\right\} \tag{6.3.5}$$

with

$$A'_{\beta,stuv} = \frac{4\pi G R^2 \rho_w}{a} \frac{(1+k'_s)}{2s+1} \left(\frac{R}{a}\right)^s D^{\pm}_{\beta,st} F_{stu}(I) G_{suv}(e).$$

Long-period perturbations, longer than 1 d, occur only when $\gamma^{\pm}_{\beta,stuv}$ does not contain the sidereal angle $\theta$. As $2\pi f_\beta t \equiv m\Omega + 2\pi f'_\beta t$ (table 6.3), diurnal and semi-diurnal tides do not give rise to long-period perturbations for those terms in (6.3.5) containing $\gamma^{-}_{\beta,stuv}$. For $m = 2$, the coefficients $D^{+}_{\beta,st}$ with $s$, $t = 2,2$; 4,2; 6,2; . . . give rise to long-period perturbations, as do the coefficients $D^{+}_{\beta,st}$ with $s$, $t = 3,2$; 5,2; 7,2; . . .But for these $v = \pm 1$ and the amplitudes are smaller by a factor $e$ than those due to the former coefficients. Thus, unless the satellite orbit is very eccentric, these perturbations are quite small. For $m = 1$ only coefficients $D^{+}_{\beta,st}$ with $s$, $t = 2,1$; 4,1; 6,1; . . .give rise to long-period perturbations for which $v = 0$. Also the amplitudes of the perturbations are proportional to $(R/a)^{s+1}$, and orbital perturbations due to coefficients with $s > 4$ tend to be small. Furthermore, the perturbations (6.3.5) due to $D^{+}_{\beta,21}$ or $D^{+}_{\beta,22}$ have the same dependence on the orbital elements as do the perturbations due to the solid tide (6.3.3) with the same frequency $f_\beta$. The two cannot be separated.

The $D^{\pm}_{\beta,42}$ occur only in the ocean tide potential. Because of its different inclination function $F_{stu}(I)$, it can be separated from the leading term in the ocean tide expansion, even though the two have the same frequency – if at least two orbital elements of two different orbits of close Earth satellites are available for analysis.

### 6.3.2 *Results*

Satellite results for some of the tide coefficients $D^{\pm}_{\beta,st}$, $\varepsilon^{\pm}_{\beta,st}$, given in table 6.7, were found by assuming that the solid tide is described by $k_2 = 0.30$ and $\varepsilon_2 = 0°$. With the present accuracy of observations and of orbital theory, only a few terms in the ocean tide expansion can be estimated. For a semi-diurnal tide, $l$, $m = 2,2$, these are $D^{+}_{22}$, $D^{+}_{42}$ and $\varepsilon^{+}_{22}$, $\varepsilon^{+}_{42}$; for the diurnal tide they are $D^{+}_{21}$, $D^{+}_{41}$ and $\varepsilon^{+}_{21}$, $\varepsilon^{+}_{41}$. But these are just the coefficients that play the most important role in tidal effects upon the Earth's rotation and on the lunar motion. The results are based on solutions by Cazenave, Daillet & Lambeck

Table 6.7. *Estimates of ocean tide parameters deduced from orbit analysis*

| Tide symbol | $D_{22}^{+}$ (cm) | $\varepsilon_{22}^{+}$ (degree) | |
|---|---|---|---|
| $M_2$ | 3.21 | 98 | Daillet (1977) |
| $M_2$ | 3.14 | 105 | Cazenave & Daillet (1977) |
| $S_2$ | 1.74 | 126 | Daillet (1977) |
| $S_2$ | 1.45 | 121 | (Oceanic and atmospheric tide) |

(1977), Daillet (1977) and Cazenave & Daillet (1977).† The $M_2$ tide solution by Daillet is based on observations of three TRANSIT satellites and of GEOS 3. The solution by Cazenave and Daillet is based on perturbations in the orbit of the satellite STARLETTE. Agreement between the two solutions is satisfactory and gives $D_{22}^{+} \sin \varepsilon_{22}^{+} \simeq 3.1$ cm. This value is somewhat lower than the $M_2$ model result of $3.37 \pm 0.30$ (table 6.5). This may be a consequence of the fact that the tide models do not take into account the loading of the sea floor by the tide itself. The $S_2$ solution by Daillet is based on the TRANSIT satellites and GEOS 1. A correction for the atmospheric tide has been applied. The ratio $(D_{22}^{+} \sin \varepsilon_{22}^{+})M_2/(D_{22}^{+} \sin \varepsilon_{22}^{+})S_2$ for the satellite solutions is 2.2 in agreement with the equilibrium ratio of 2.3. The difference is not significant. Separation of the atmospheric tide from the ocean $S_2$ tide will not be possible from orbit perturbation analysis alone, and when applying the results to the lunar motion, it will be more useful to estimate combined fluid tide parameters for this frequency.

## 6.4 Tidal perturbations in rotation

### 6.4.1 *Rotational changes of tidal origin*

*Solid tides.* The solid Earth tide potential (6.1.14) can be written in the form used for the Earth's self-attraction (2.4.1) as

$$\Delta U_{\beta,lm} = \frac{GM}{r} \sum_{m} \left(\frac{R}{r}\right)^{l} P_{lm} (\sin \phi)\, (C_{\beta,lm}^{*} \cos m\lambda + S_{\beta,lm}^{*} \sin m\lambda),$$

† Further solutions for the $M_2$ tides are given by Goad & Douglas (1978) and Felsentreger, Marsh & Williamson (1978).

with

$$\left.\begin{matrix} C^*_{\beta,lm} \\ S^*_{\beta,lm} \end{matrix}\right\} = \frac{R}{GM} k_l A_{\beta,lm} \left\{ \begin{matrix} \cos(2\pi f_\beta t + \varepsilon_{\beta,lm}) \\ -\sin(2\pi f_\beta t + \varepsilon_{\beta,lm}) \end{matrix}\right. \tag{6.4.1}$$

and where the Stokes coefficients $C^*_{\beta,lm}$ and $S^*_{\beta,lm}$ depend on the tidal frequency and on time. These coefficients relate to the inertia tensor of order $l$ and, for the relevant elements entering into the excitation functions (4.1.1) and (4.1.2) (equations 2.4.5 and 2.4.8),

$$\left.\begin{aligned} \Delta I_{\beta,13} &= -MR^2 C^*_{\beta,21}, \\ \Delta I_{\beta,23} &= -MR^2 S^*_{\beta,21}, \\ \Delta I_{\beta,33} &= -\tfrac{2}{3} MR^2 C^*_{\beta,20}. \end{aligned}\right\} \tag{6.4.2}$$

That is,

$$\psi = X_{\text{wobble}}(C^*_{\beta,21} + \mathrm{j}S^*_{\beta,21})/C_{20}, \tag{6.4.3a}$$

where $C_{20}$ is the second-degree zonal harmonic in the geopotential. Also

$$\psi_3 = X_{\text{l.o.d.}}(2MR^2/I_{33})C^*_{\beta,20} \simeq 2C^*_{\beta,20}. \tag{6.4.3b}$$

Substituting (6.4.1) and the results of table 6.3. into these excitation functions gives the theoretical amplitudes summarized in table 6.8. For the tidal frequencies that are high compared with the frequency $\sigma_0$ of the free nutation, the motion of the rotation axis has an amplitude that is about $\sigma_0/2\pi f_\beta$ times the amplitudes of the excitation function (equation 4.3.12), and will generally be unimportant.

The semi-diurnal body tides cause no perceptible perturbation in rotation since the corresponding inertia tensor elements, $\Delta I_{22}$ and $\Delta I_{23}$, enter into the excitation function only as a second-order effect (equations 3.2.7). Diurnal body tides cause small, nearly diurnal, perturbations in the direction of the axis of rotation but do not affect the l.o.d. The amplitudes of these wobbles are quite unimportant. Zonal tides affect l.o.d. only, and these perturbations are the most important of all tidal perturbations in rotation. Lunar tides cause principal perturbations near 13.5 d and 27 d, solar tides cause perturbations near 6 months and 12 months. Longer perturbations occur with periods of 8.8 yr and 18.6 yr. The theoretical amplitudes of the tidal changes in $m_3$ given in table 6.8 are based on a nominal Love number of $k_2 = 0.30$. They are unlikely to be in error by more than a few per cent *if* dissipation in the mantle is small (section 2.2).

Table 6.8. *Tidal excitation functions and perturbations in wobble and l.o.d.*

| Tide symbol | $\lvert\bar{\psi}\rvert$ | $\lvert\bar{m}\rvert$ | $m_3 \cos(2\pi f_\beta t)$ | $\lvert \Lambda_3 \rvert$ (ms) |
|---|---|---|---|---|
| Solid tide | $(10^{-6})$ | | $(10^{-8})$ | |
| $K_1$ | 4.1 | 9.6 | | |
| $O_1$ | 2.9 | 6.8 | | |
| $P_1$ | 1.4 | 3.2 | | |
| $Q_1$ | 5.6 | 1.3 | | |
| 18.6-yr | | | −0.167 | 156 |
| Sa | | | −0.030 | 1.53 |
| Ssa | | | −0.185 | 4.64 |
| MSm | | | −0.041 | 0.18 |
| Mm | | | −0.210 | 0.80 |
| Msf | | | −0.036 | 0.07 |
| Mf | | | −0.395 | 0.74 |
| 13.63-d | | | −0.165 | 0.30 |
| Ocean tide | $(10^{-9})$ | | | |
| 18.6-yr | 2.0 | 2.0 | −0.016 | 14.7 |
| Ssa | 2.2 | 0.9 | −0.018 | 0.44 |
| Mm | 2.5 | 0.2 | −0.020 | 0.08 |
| Mf | 4.7 | 0.2 | −0.037 | 0.07 |

Comparison of these amplitudes with the error spectrum of $m_3$ (figures 5.4 and 5.8) indicates that the tidal-induced changes at frequencies of the Ssa, Mm and Mf lie well above the noise level of the astronomical observations, but at the lower-frequency lunar tides, the signal falls below the noise level of the presently available data.

*Fluid tides.* The contribution of the ocean and atmospheric tides to the excitation functions can be computed in a manner similar to (6.4.1–3) by substituting the appropriate harmonics of the fluid tide expansions into the excitation functions (4.2.1) for a surface load. Now the semi-diurnal tides will induce a small semi-diurnal wobble in the orientation of the rotation axis since the ocean tides contain harmonics of degree 2 and order 1. But the effect is negligibly small.

More important are zonal ocean tide effects on the l.o.d. No realistic ocean models are available and an equilibrium tide is assumed. From the general equilibrium tide expansion (6.2.5) we require those terms that give $u=2$ and $|m+j|=0$. Thus with $l,m=2,0$ the appropriate part of the expansion is

$$\xi'_{\beta,(2,0)}=\frac{1}{2}\frac{(1+k_2-h_2)}{g}A_\beta\sum_i(Q^+_{20i02}+Q^-_{20i02})$$
$$\times a_{i0}\cos 2\pi f_\beta t\cdot P_{20}(\sin\phi),$$

in which

$$Q^+_{20i02}=Q^-_{20i02},$$

and

$$\sum_i a_{i0}Q^+_{20i02}=a_{00}+\tfrac{2}{7}a_{20}+\tfrac{2}{7}a_{40}.$$

Thus

$$\xi'_{\beta(2,0)}=\frac{1}{2}\frac{(1+k_2-h_2)}{g}A_\beta(a_{00}+\tfrac{2}{7}a_{20}+\tfrac{2}{7}a_{40})P_{20}(\sin\phi)\cos 2\pi f_\beta t.$$

The surface load is $\rho_w(1+k'_2)\xi'_{\beta(2,0)}$. As the tide does not contain a harmonic of degree and order zero because of the correction (6.2.7), the excitation function is

$$\psi_3=\frac{4\pi}{15}\frac{R^4}{C}\frac{(1+k_2-h_2)}{g}X_{\text{l.o.d.}}$$
$$\times A_\beta(a_{00}+\tfrac{2}{7}a_{20}+\tfrac{2}{7}a_{40})\rho_w, \qquad (6.4.4)$$

and represents about 10% of the solid tide perturbation (table 6.8). This important contribution of the ocean tide to the total tidal excitation will prevent meaningful estimates being made of the solid Earth Love numbers from the l.o.d. observations until appropriate tide models are available. In particular, any lag between the lunar attraction and the solid Earth's response will be impossible to estimate without further *a priori* information on the long-period zonal ocean tides. A more useful approach may be to assume that the solid tide is known and to interpret any difference between the theoretical response and the observed response as a measure of the departure of the ocean tide from an equilibrium theory at these frequencies.

The long-period ocean tides also contain terms of degree 2 and order 1, and diurnal perturbations in the direction of the rotation axis can be expected. The appropriate coefficients in the equilibrium tide expression are

$$\xi_{\beta(2,1)}=\frac{(1+k_2-h_2)}{g}A_\beta\{(\tfrac{1}{7}a_{21}+\tfrac{10}{21}a_{41})\cos 2\pi f_\beta t\cos\lambda$$
$$+(\tfrac{1}{7}b_{21}+\tfrac{10}{21}b_{41})\cos 2\pi f_\beta t\sin\lambda)\}P_{21}(\sin\phi),$$

yielding the excitation function

$$\psi_{\beta,1}=-\frac{4}{5}\pi\frac{R^4}{C-A}X_{\text{wobble}}\,\rho_w\frac{(1+k_2-h_2)}{g}$$
$$\times A_\beta(\tfrac{1}{7}a_{21}+\tfrac{10}{21}a_{41})\cos 2\pi f_\beta t$$

$$\psi_{\beta,2}=-\frac{4}{5}\pi\frac{R^4}{C-A}X_{\text{wobble}}\,\rho_w\frac{(1+k_2-h_2)}{g}$$
$$\times A_\beta(\tfrac{1}{7}b_{21}+\tfrac{10}{21}b_{41})\cos 2\pi f_\beta t.$$

The motion of the rotation axis follows from (4.3.12). The amplitude of the motion barely exceeds $2\times10^{-9}$ rad for the 18.6-yr tide and is less for the other zonal tides.

### 6.4.2 *Tides of rotational origin*

Apart from the lunar and solar attraction, tides are also raised in the oceans by variations in the centrifugal force. The principal tide raised in this manner is the pole tide with a 14-month period and its interactions with the Earth's rotation are discussed in chapter 8. The pole tide also contains an annual term which, as Munk & Haubrich (1958) point out, is more important than the solar annual tide Sa. A further tide is raised by changes in the speed of rotation and this aspect has been briefly discussed by Lambeck & Cazenave (1976) in connection with the decade changes in l.o.d. These l.o.d. tides are considerably smaller than the already small pole tide.

That part of the centrifugal potential responsible for the pole tide is given by (3.4.1*b*). The equilibrium tide (6.2.3*a*) becomes

$$\xi'=-(R^2\Omega^2/3g)(1+k_2-h_2)$$
$$\times(m_1\cos\lambda+m_2\sin\lambda)P_{21}(\sin\phi)\mathscr{C}(\phi,\lambda).\quad(6.4.6)$$

We ignore the loading term for the moment. The tide is then expanded according to (6.2.3–5)

$$\xi'=-\frac{1}{6}\frac{R^2\Omega^2}{g}(1+k_2-h_2)\sum_i\sum_j\sum_u\sum_{\pm}[(m_1a_{ij}\mp m_2b_{ij})\cos\lambda(j\pm i)$$
$$+(m_2a_{ij}\pm m_1b_{ij})\sin\lambda(j\pm 1)]Q^{\pm}_{21iju}P_{u,|j\pm1|}(\sin\phi).$$

Of the coefficients $a_{ij}$ and $b_{ij}$ occurring in $\mathscr{C}(\phi, \lambda)$, the most important is $a_{00}$, all others are smaller by a factor of 7 or more. Thus the equilibrium tide will be expressed to a good approximation by those terms in (6.4.6) for which $i, j = 0, 0$. This occurs only for $|j \pm 1| = 1$ and $u = 2$ since max $(|2 \pm j|, |2 - i|) < u < (2 + i)$. Then

$$\xi' = -\tfrac{1}{6}(R^2\Omega^2/g)$$
$$\times (1 + k_2 - h_2)(2a_{00}m_1 \cos \lambda$$
$$+ 2a_{00}m_2 \sin \lambda)P_{21} (\sin \phi)$$
$$+ O(\xi a_{ij}/a_{00};\ i, j = 0).$$

As the equilibrium tide can be expressed to a high degree of accuracy by a single harmonic despite the complicated ocean–continent distribution, the effect on the tide of the deformation of the Earth under the ocean can be readily approximated by writing the total equilibrium tide as (cf. equation 6.2.8)

$$\xi' = -\frac{R^2\Omega^2}{6g}(1 + k_2 - h_2) \sum_i \sum_j \sum_{\pm} \sum_u \{[(m_1 a_{ij} \mp m_2 b_{ij}) \cos \lambda\, (j \pm 1)$$
$$+ (m_2 a_{ij} \pm m_1 b_{ij}) \sin \lambda\, (j \pm 1)] Q^{\pm}_{21 iju} P_{u,|j \pm 1|} (\sin \phi)$$
$$+ \tfrac{6}{5} a_{00}(1 + k_2' - h_2')(m_1 \cos \lambda + m_2 \sin \lambda)(\rho_w/\bar{\rho}) P_{21} (\sin \phi)\} \tag{6.4.7}$$

This is the pole tide. Its frequency is determined by the polar motion **m**. It will be 14 months for the Chandler pole tide and 12 months for the annual pole tide.

To evaluate the excitation functions we require only the terms of degree 2. Specifically, for the wobble, we require only the terms for which $u = 2$ and $|j \pm 1| = 1$. The relevant components in the ocean tide are

$$\xi_{21} = -(R^2\Omega^2/6g)(1 + k_2 - h_2)[(A_1 m_1 + A_2 m_2)$$
$$\times \cos \lambda + (B_1 m_1 + B_2 m_2) \sin \lambda] P_{21} (\sin \phi), \tag{6.4.8a}$$

with

$$A_1 = 2(a_{00} + \tfrac{1}{7}a_{20} - \tfrac{4}{21}a_{40}) + \tfrac{12}{7}a_{22} + \tfrac{40}{7}a_{42} + \tfrac{6}{5}(1 + k_2' - h_2')(\rho_w/\bar{\rho})a_{00}$$
$$A_2 = \tfrac{12}{7}b_{22} + \tfrac{40}{7}b_{42} = B_1 \tag{6.4.8b}$$
$$B_2 = 2(a_{00} + \tfrac{1}{7}a_{20} - \tfrac{4}{21}a_{40}) - \tfrac{12}{7}a_{22} - \tfrac{40}{7}a_{42} + \tfrac{6}{5}(1 + k_2' - h_2')(\rho_w/\bar{\rho})a_{00}.$$

As a first approximation,

$$|\xi|_{\text{pole tide}} = -\tfrac{1}{3}(R^2\Omega^2/g)(1+k_2-h_2)a_{00}P_{21}(\sin\phi)|\mathbf{m}|$$

and

$$|\xi'_{\max}|_{\text{pole tide}} = 0.5 \text{ cm} \tag{6.4.9}$$

for latitude 45° and $|\mathbf{m}| = 10^{-6}$. In comparison the annual tide Sa has as its principal term (equation 6.2.5)

$$|\xi'|_{\text{Sa}} = \tfrac{1}{2}(1+k_2-h_2)(A_\beta/g)a_{00}(Q^+_{20002}+Q^-_{20002})P_{20}(\sin\phi)$$
$$= (1+k_2-h_2)(A_\beta/g)a_{00}P_{20}(\sin\phi),$$

and

$$|\xi'_{\max}|_{\text{Sa}} = 0.16 \text{ cm}$$

for latitude $\phi = 90°$. This is about one-third of the annual pole tide, and its relative importance raises the question of whether this tide and the 14-month pole tide can introduce cross-coupling between wobble and l.o.d. (Dahlen 1976). That this can occur in principle is readily seen from the tide equation (6.4.7) which contains a $P_{20}(\sin\phi)$ harmonic and which can perturb the l.o.d. However most of the energy in the pole tide, at least in the present equilibrium model, is contained in the 2,1-harmonic, and the amplitude of the 2,0-harmonic will actually be less than the corresponding harmonic in the Sa tide by a factor of 3. As the ocean Sa tide is of no consequence for l.o.d. changes, neither will the pole tide be significant, and the cross-coupling is negligible: $\psi_3 \simeq 10^{-10}$ and the expected perturbations in $m_3$ with a 14-month period do not rise above the noise level of the astronomical data. Changes in l.o.d. observations cannot be used to determine to what extent the pole tide follows the equilibrium theory. Inversely, tides raised in the oceans by the change in the l.o.d. will contain 2, 1-harmonics, but their contribution to the wobble is also negligible. Finally, the zonal terms in the l.o.d. tides are small compared with the Sa and Ssa solar tides at the same frequencies.

### 6.4.3 *Some results*

Several authors have analysed l.o.d. observations for tidal perturbations at the fortnightly (Mf) and monthly (Mm) frequencies (table 6.9). Theoretical amplitudes, assuming $k_2 = 0.30$ and an equilibrium ocean response, follow from table 6.8, and any difference

Table 6.9. *Estimates of Love numbers from l.o.d. observations at the Mf and Mm frequencies*

| Author | Data | $k$(Mf) | $k$(Mm) |
|---|---|---|---|
| Pilnik (1970) | 1953–1967 | 0.313 | 0.284 |
| Guinot (1970*a*) | 1967–1969 | 0.331 | 0.265 |
| Djurovic & Melchior | 1967–1971 | 0.352 | 0.195 |
| (1972) | | 0.349 | 0.261 |
| Guinot (1974) | 1967.0–1974.0 | 0.334 | 0.295 |
| | | ±0.005 | ±0.011 |
| Djurovic (1976) | | 0.343 | 0.301 |
| | | ±0.030 | ±0.044 |

between these and observed values may be a consequence of (i) anelastic deformation of the solid Earth, (ii) non-equilibrium ocean tide response, or (iii) exchange of angular momentum between the solid Earth and the atmosphere at the tidal frequencies. Pilnik's (1970) analysis is based on 15 yr of l.o.d. data from the time service of the USSR. Guinot (1974) used BIH data since 1967. Djurovic & Melchior (1972) analysed observations from 49 observatories for the years 1967–1971. In a subsequent study Djurovic (1976) has analysed the same data with results comparable with those obtained by Guinot. All these studies show the rather curious result that $k_2$ (Mf) $> k_2$(Mm) (see also Agnew & Farrell 1978)†. Guinot's value of $k_2 = 0.33$ for Mf is comparable with the theoretically expected value for an elastic Earth and equilibrium ocean tide. Anelasticity of the mantle will also increase $k_2$, by some 2–3% if $Q$ at the tidal frequencies is $\simeq 300$, relative to the elastic values (chapter 2). At present we cannot discriminate between the oceanic and anelastic effects. Guinot's value of $k_2 = 0.30$ for Mm appears too small to be explained by ocean or anelastic contributions. Possibly zonal winds of comparable frequency contaminate the results (section 7.6).

† See Merriam (1979) for a further discussion.

# SEVEN

# SEASONAL VARIATIONS

Both latitude and l.o.d. observations show seasonal oscillations that rise well above the noise level of the astronomical spectra (figures 1.1 and 1.2). The principal seasonal oscillation in the wobble is the annual term which has generally been attributed to a geographical redistribution of mass associated with meteorological causes. Jeffreys, in 1916, first attempted a detailed quantitative evaluation of this excitation function by considering the contributions from atmospheric and oceanic motion, of precipitation, of vegetation and of polar ice. Jeffreys concluded that these factors explain the observed annual polar motion, a conclusion that is still valid today, although the quantitative comparisons between the observed and computed annual components of the pole path are still not satisfactory. These discrepancies may be a consequence of (i) inadequate data for evaluating the known excitation functions, (ii) the neglect of additional excitation functions, (iii) systematic errors in the astronomical data, or (iv) year-to-year variability in the annual excitation functions. The semi-annual term in the wobble is much smaller than the annual term, and the astronomical evidence for it is not compelling. This could be expected from the nature of the solution (4.3.12) of the polar motion for a sinusoidally varying excitation: for equal-magnitude excitation functions at the annual and semi-annual frequencies and $Q \simeq 100$, the annual pole shift will be some eight times larger than the semi-annual pole shift simply because it is closer to the Chandler resonance.

L.o.d. observations show distinct oscillations at the annual, semi-annual and biennial periods and, apart from an important tidal component at the semi-annual period, the fluctuations are almost entirely caused by the variable zonal wind circulation, the quantitative evaluation of the excitation functions agreeing well with the astronomical observations.

Several reasons exist for studying the seasonal variations in rotation, apart from a desire for an explanation of the astronomical observations: (i) the nature of the excitation of the Chandler wobble is still uncertain, and it has been suggested that this motion may be maintained by a somewhat irregular annual excitation (chapter 8); (ii) seasonal changes in l.o.d. appear to fluctuate considerably from year to year and, as the Earth's rotation appears to be quite sensitive to changes in the zonal circulation, the astronomical data may provide a measure of the strength of the global circulation patterns; (iii) a relativistic effect, referred to as the Nordtvedt effect, predicts a small annual change in the l.o.d. which may be observable if the meteorological excitation functions are known with precision (Rochester & Smylie 1974); (iv) a precise knowledge of these functions may also indicate to what extent the mantle motions are transferred to the core. In this chapter we discuss in turn the excitation functions associated with seasonal redistributions of mass in (i) the atmosphere, (ii) groundwater, and (iii) the oceans, as well as the excitation functions associated with angular momentum changes.

A closely related problem to the seasonal changes in the rotation is the periodic shift in the Earth's centre of mass with respect to astronomical and geophysical observatories on the Earth's surface. Such shifts may have two possible consequences: they may contribute to the Kimura term in the wobble equation (5.3.1), and they may lead to discrepancies between the observed and computed motions of artificial Earth satellites or of the Moon. Stolz (1976*a*, *b*) has estimated the contributions from shifting air mass, groundwater and sea level, and concludes that together they do not amount to a seasonal shift in the centre of mass of more than a few millimetres. This is below the observational level of any present or near-future measuring system and we consider it no further.

## 7.1 Atmospheric pressure

Early evaluations of the atmospheric pressure excitation function by R. Spitaler in 1901, Jeffreys in 1916, L. Rosenhead in 1929 and Byzova (1947) have all established the importance of the seasonal air mass shifts in explaining the annual polar motion, although the quantitative conclusions have remained unsatisfactory. More recently, the excitation functions have been evaluated by Munk & Hassan (1961), Kikuchi (1971), Siderenkov (1973), Wilson & Haubrich (1976*a*) and Jochmann (1976). Lamb (1972) discusses the global aspects of the seasonal pressure variations. The greatest seasonal change occurs over Central Asia, while secondary maxima occur over the North Pacific and the North Atlantic. In these last two areas, separated by about 180° in longitude, the maximum differences are in phase, occurring in summer and winter, and their contributions to the seasonal excitation functions tend to cancel each other. Also, as these secondary areas of large pressure differences lie over oceans, the ocean response will further reduce their contribution to the excitation functions. In the more important region of Central Asia, the maximum difference occurs out of phase with the two oceanic pressure anomalies and it is this feature that dominates the excitation function: the atmospheric excitation pole can be expected to oscillate approximately in a meridional plane through 90° East longitude. This is indeed observed (figure 5.13).

### 7.1.1 *Static ocean response*

In evaluating the meteorological excitation function, the interaction between air pressure and sea level must be considered. The treatment is outlined in section 6.2.3. But the more usual approach is to assume a static or inverted barometer response of the oceans to the variable air mass. The increase in the atmospheric pressure, over and above the mean pressure over the entire ocean surface, depresses the sea surface locally by approximately 1 cm for every millibar of pressure. We introduce the following notation.

$p(\phi, \lambda; t)$ = pressure at position $\phi$, $\lambda$ at time $t$,

$\tilde{p}(\phi, \lambda)$ = annual mean pressure at $\phi$, $\lambda$,

$\delta p(\phi, \lambda; t) = p(\phi, \lambda; t) - \tilde{p}(\phi, \lambda)$,

$\tilde{p}_o(t)$ = average pressure over the oceans at time $t$,

$\tilde{p}_o$ = annual mean pressure over the oceans,

$\delta\tilde{p}_o(t) = \tilde{p}_o(t) - \tilde{p}_o$.

The pressure perturbation over the ocean is defined as

$$\Delta p(\phi, \lambda; t) = \delta p(\phi, \lambda; t) - \delta\tilde{p}_o(t), \tag{7.1.1}$$

and the static inverted barometer ocean response is

$$\Delta\xi(\phi, \lambda; t) = -(1/g\rho_w)\, \Delta p(\phi, \lambda; t). \qquad (7.1.2)$$

The ocean acts so as to annul the horizontal pressure gradients on the sea floor but the ocean floor pressure will vary with time, by $\delta\tilde{p}_o(t)$, because a variable fraction of the total air mass may overlie the oceans. This model assumes that the oceans re-adjust globally to a regional pressure change, i.e. that the variable atmospheric load induces flow between the various oceans of the world. A regional compensation model may be more appropriate, especially for pressure variations of higher than annual frequency.

The analysis by Munk & MacDonald of the ocean-loading problem led to the conclusion that, for extended pressure disturbances, the ocean response was slightly larger than the inverted barometer response when the perturbations were of low frequency compared with a critical frequency $f_c$. At $f_c$ ($\sim\frac{1}{8}\,d^{-1}$) they noted a change in phase in the response and at frequencies greater than the Coriolis frequency, but less than $f_c$, the response was somewhat less than the inverted barometer response. The notion of the critical frequency is, however, a consequence of introducing spatially unbounded pressure perturbations which do not satisfy the appropriate boundary conditions. Crepon (1973) has studied a number of cases of the geostrophic adjustment of the ocean to pressure perturbations and finds that in most cases the barometric response is less than 1.

The advantages of assuming a static ocean response are (i) that it permits a rudimentary treatment of the ocean–atmosphere interaction even though the basis for this assumption is not entirely satisfactory, and (ii) that it permits the evaluation of the pressure over the oceans when data are available only for the continents. For the more rigorous approach outlined in section 6.2.3, global data are required. Evidence for the static ocean response at lower than tidal frequencies is still relatively sparse but appears to support the hypothesis. Numerous studies (see, for example, Wunsch 1972) indicate that for the island stations the inverted barometer response appears to be a reasonably good one at relatively high (greater than 1 cycle month$^{-1}$) frequencies. Along coastlines there sometimes appear to be dynamical interactions between sea level and pressure, but these are rather local effects. Wunsch's detailed study of sea level fluctuations at Bermuda indicates a static response in the period range from 40 h to 400 h. At longer periods, physical processes other than atmospheric pressure begin to dominate the sea level record and the coherency between pressure and sea level is lost, although Wunsch believes that there is still an inverted barometer effect. Lisitzin & Pattullo (1961) and Pattullo (1963) discuss the seasonal variations in sea level in the Pacific Ocean and conclude that, by and large, the observed variations can be explained by the pressure response plus steric effects. This is confirmed by the more detailed analysis by Reid & Mantyla (1976) for the North Pacific.

In these studies only a local response is considered in which sea level is modified by 1 cm for every millibar change in local pressure. This is not strictly in keeping with the above definition of the inverted barometer, which is that the ocean responds only to pressure over and above the mean pressure over the oceans. The difference in ocean response between these two definitions is $\varepsilon_\xi \simeq \delta\tilde{p}_o(t)/g\rho_w$ (equations 7.1.1 and 7.1.2). Siderenkov (1973) finds $|\delta p_o(t)| \simeq 0.82$ mbar and $\varepsilon_\xi < 1$ cm, compared with local seasonal variations in pressure of the order of 10 mbar. The presently available sea level data are inadequate to determine whether or not the ocean responds locally, regionally or globally to a pressure change at any one position.

### 7.1.2 *Excitation functions*

The excitation functions to be evaluated are given by (4.1.1). The observed quantity is not the atmospheric density, however, but the ground level pressure $p(\phi, \lambda, h; t)$ at altitude $h$ above sea level. Instead of $p$ we use the departure in pressure from its annual mean $\delta p(\phi, \lambda, h; t)$. With the reasonable assumption that the atmosphere at the seasonal frequency is in hydrostatic equilibrium

$$\delta p(\phi, \lambda, h; t) = \int_{R+h}^{\infty} \delta\rho_a g \, dr,$$

where $\rho_a(h)$ is the atmospheric density. Only the lower atmosphere will contribute significantly to this integral and we can assume that g is constant. Then, with (4.1.3),

$$\Delta I_{ij} = \frac{10^3 R^4}{g} \int_S \Pi_{ij}\, \delta p(\phi, \lambda, h; t)\, \mathrm{d}S \tag{7.1.3}$$

where the departure in pressure $\delta p$ is in millibars. The pressure should be known everywhere at the Earth's surface and, if sea level pressure charts are used, a correction for station height is necessary. Ideally the pressure should refer to the mean altitude of the area for which the value is assumed to be representative; that is, a correction should be made for the difference between station elevation and the mean altitude of the area. In regions of dense distribution, this correction is unlikely to be important unless the stations lie systematically in higher- or lower-than-average terrain, but it may become important in mountainous regions of sparse data coverage.

The seasonal variation in mean pressure over the globe $\delta\tilde{p}(t)$ can be divided into two parts: (i) $\delta\tilde{p}_o(t)$, the mean pressure over ocean areas, and (ii) $\delta\tilde{p}_L(t)$, the mean pressure over land areas. Then

$$\delta\tilde{p}(t) = (1 - a_{00})\delta p_L(t) + a_{00}\delta\tilde{p}_o(t),$$

or

$$\delta\tilde{p}_o(t) = [\delta\tilde{p}(t) - (1 - a_{00})\delta\tilde{p}_L(t)]/a_{00}, \tag{7.1.4}$$

which represents the seasonal variation of the mean pressure over the oceans as a function of the mean pressure over land, $\delta\tilde{p}_L(t)$, and the mean pressure over the globe, $\delta\tilde{p}(t)$. This, with $\delta\tilde{p}_L(\phi, \lambda; t)$, provides the necessary information for evaluating the excitation functions, provided that the ocean responds statically to air pressure.

Munk & MacDonald (1960) evaluate $\delta\tilde{p}(t)$ from the amount of water vapour in the atmosphere. Separating air pressure into two parts, the contribution from dry air and the contribution from water vapour $\delta p_M(t)$, gives, since dry air is conserved,

$$\delta\tilde{p}(t) = \delta\tilde{p}_M(t).$$

Thus the total pressure field and the static ocean response are defined by the pressure over land and the water vapour pressure. A more complete compilation of $\delta p_M(\phi, \lambda; t)$ than used by Munk & MacDonald is given by Tuller (1968) for all latitudes in the form of monthly mean maps of precipitable water.

From Tuller's data, Stolz & Larden (1979) estimated

$$\delta p_{\mathrm{M}}(t) = -0.11 \cos \odot - 0.06 \sin \odot \text{ mbar}. \qquad (7.1.5a)$$

where $\odot \equiv h$ (table 6.2).

The same authors also estimated $\delta p_{\mathrm{o}}(t)$ and $\delta p_{\mathrm{L}}(t)$ from sea level pressure maps, with corrections for land elevation applied. They find

$$\left.\begin{aligned} \delta p_{\mathrm{o}}(t) &= -0.73 \cos \odot - 0.19 \sin \odot \text{ mbar}, \\ \delta p_{\mathrm{L}}(t) &= 1.30 \cos \odot + 0.31 \sin \odot \text{ mbar}. \end{aligned}\right\} \qquad (7.1.6a)$$

With (7.1.4)

$$\delta p(t) \equiv \delta p_{\mathrm{M}}(t) = -0.11 \cos \odot - 0.04 \sin \odot \text{ mbar}. \qquad (7.1.5b)$$

Siderenkov (1973) evaluates the seasonal variation in pressure directly and finds, assuming zero phase lag with respect to January 1,

$$\delta p(t) = -0.15 \cos \odot \text{ mbar} \qquad (7.1.5c)$$

and

$$\delta p_{\mathrm{o}}(t) = -0.82 \cos \odot, \qquad \delta p_{\mathrm{L}}(t) = 1.40 \cos \odot. \qquad (7.1.6b)$$

Munk & MacDonald's comparable estimate for $\delta p_{\mathrm{M}}(t)$ is

$$\delta p_{\mathrm{M}}(t) = -0.17 \cos \odot - 0.08 \sin \odot \text{ mbar}. \qquad (7.1.5d)$$

With $\delta p_{\mathrm{o}}(t)$ defined by (7.1.4), the inertia tensor (7.1.3) becomes

$$\begin{aligned} \Delta I_{ij} &= 10^3 \frac{R^4}{g}\left[\int_{\text{Land}} \Pi_{ij} \delta p_{\mathrm{L}}(\phi, \lambda, h; t)\, \mathrm{d}S + \delta p_{\mathrm{o}}(t) \int_{\text{Ocean}} \Pi_{ij}\, \mathrm{d}S\right] \\ &= 10^3 \frac{R^4}{g} \int_{\text{Land}} \Pi_{ij}[\delta p_{\mathrm{L}}(\phi, \lambda, h; t) - \delta p_{\mathrm{o}}(t)]\, \mathrm{d}S, \end{aligned} \qquad (7.1.7)$$

and the excitation functions are

$$\left.\begin{aligned} \boldsymbol{\psi} &= (1/(C-A))(\Delta I_{13} + \mathrm{j}\, \Delta I_{23}) X_{\text{wobble}}, \\ \psi_3 &= -(\Delta I_{33}/C) X_{\text{l.o.d.}}. \end{aligned}\right\} \qquad (7.1.8)$$

### 7.1.3 *Results*

Hassan (1961) and Munk & Hassan (1961) have computed the monthly mean values of the moments and products of inertia (7.1.7) for the years 1873–1950, using observed station level pressures or sea level pressures corrected back to station level. Wilson (1975) (see also Wilson & Haubrich 1976*a*) has analysed essentially the same data as Munk & Hassan, starting with the year 1900 and with a further 20 yr of observations. Both studies use only land stations and allow for the ocean response according to (7.1.4) using Munk & MacDonald's relation for $\delta\tilde{p}_M(t)$. The distribution of stations reporting pressure is fairly uniform over the continents, with the exception of Antarctica, Greenland and Central Asia, where the coverage is sparse or non-existent. The absence of data from high latitudes is of little consequence in evaluating $\psi_1$ and $\psi_2$ because of the $\cos^2\phi$ terms in the integrals (7.1.7). Lack of data over China may be more important since a principal feature of the seasonal pressure pattern is the monsoon over Central Asia: an absence of data may lead to an underestimation of the excitation function. Wilson has also computed the excitation functions from global climatic pressure compilations prepared by Schutz & Gates (1971, 1972, 1973, 1974). These compilations give sea level pressures over the continents and oceans and have been reduced back to average altitudes. Wilson has estimated $\boldsymbol{\psi}$ from these data for different assumptions about the ocean response: (i) no response, (ii) regional inverted barometer response within the Indian, Atlantic and Pacific oceans without there being mass transfer between these oceans, and (iii) static inverted barometer response. Tables 7.1 and 7.2 summarize the various results for the annual excitation functions in the form

$$\left.\begin{aligned}\psi_1 &= A_1\cos\odot + B_1\sin\odot,\\ \psi_2 &= A_2\cos\odot + B_2\sin\odot,\\ \psi_3 &= A_3\cos\odot + B_3\sin\odot.\end{aligned}\right\} \qquad (7.1.9)$$

As expected, both zero and regional ocean response increase the excitation functions. Agreement between Wilson's two sets of excitation functions based on a static ocean response is satisfactory, suggesting that the difference between the use of station altitudes or area mean altitudes may not be very important and that the two data sets must be quite similar.

One possible source of error in the evaluation of $\boldsymbol{\psi}$ is an incorrect value for $\delta\tilde{p}_M(t)$ or $\delta\tilde{p}(t)$. If $\varepsilon_p(t)$ is the error in either of these

Table 7.1. *Meteorological excitation functions contributing to the seasonal changes in the wobble (in units of $10^{-6}$) expressed in the form (7.1.9)*

| Authors | Ocean response | Annual $A_1$ | Annual $B_1$ | Annual $A_2$ | Annual $B_2$ | Semi-annual $A_1$ | Semi-annual $B_1$ | Semi-annual $A_2$ | Semi-annual $B_2$ |
|---|---|---|---|---|---|---|---|---|---|
| Air pressure | | | | | | | | | |
| (1) Munk & Hassan | IB[c] | −1.8 | 0.2 | −12.9 | −1.0 | 0.4 | −0.8 | 1.8 | 1.4 |
| (2) Wilson (NCAR)[a] | IB | −3.1 | 0.0 | −10.5 | −0.3 | | | | |
| (3) Wilson (Rand)[b] | IB | −3.5 | −0.1 | −9.1 | 1.1 | | | | |
| (4) Siderenkov | IB | −3.8 | 0 | −18.2 | 0 | | | | |
| (5) Jochmann | IB | 2.9 | 0 | −16.8 | 0 | | | | |
| (6) Wilson | RIB[d] | −6.8 | −1.2 | −12.3 | 1.5 | | | | |
| (7) Wilson | None | −8.5 | −3.3 | −19.7 | 4.8 | | | | |
| (8) Kikuchi | None | −6.5 | −2.4 | −20.9 | −0.7 | | | | |
| (9) Siderenkov | None | −8.3 | 0 | −24.6 | 0 | | | | |
| Groundwater | | | | | | | | | |
| (10) Van Hylckama | | −0.5 | −0.2 | 4.5 | 5.3 | 0.4 | −0.4 | 0.9 | 0.8 |
| Sea level | | | | | | | | | |
| (11) Equation (7.2.5) | | 0.1 | −0.7 | 0.2 | −1.1 | −0.02 | +0.09 | −0.04 | 0.15 |
| Total | | | | | | | | | |
| (12) 1+10+11 | | −2.2 | −0.7 | −8.2 | 3.2 | 0.8 | −1.1 | 2.7 | 2.3 |
| (13) 2+10+11 | | −3.5 | −0.9 | −5.8 | 3.9 | | | | |
| (14) 4+10+11 | | −4.2 | −0.9 | −13.5 | 4.2 | | | | |
| (15) 5+10+11 | | 2.5 | −0.9 | −12.1 | 4.2 | | | | |

[a] NCAR, National Center for Atmospheric Research

[b] Rand data refer to the models of Schutz & Gates

[c] IB refers to global inverted barometer ocean response

[d] RIB refers to regional inverted barometer ocean response

Table 7.2. *Meteorological excitation functions contributing to the seasonal changes in $m_3$ (in units of $10^{-8}$) expressed in the form (7.1.9)*

| | | Annual | | Semi-annual | |
|---|---|---|---|---|---|
| Authors | | $A_3$ | $B_3$ | $A_3$ | $B_3$ |
| Air pressure | | | | | |
| (1) Munk & Hassan | IB | −0.003 | 0.003 | −0.001 | 0.000 |
| Groundwater | | | | | |
| (2) Van Hylckama | | 0.017 | 0.002 | 0.003 | −0.005 |
| Sea level | | | | | |
| (3) Equation (7.2.7) | | −0.011 | 0.064 | 0.002 | −0.009 |
| Winds | | | | | |
| (4) Mintz & Munk | 0–6 km | −0.31 | −0.20 | 0.03 | 0.01 |
| (5) Frostman *et al.* | 0–30 km | −0.26 | −0.19 | 0.14 | 0.08 |
| (6) Lambeck & Cazenave | 0–30 km | −0.344 | −0.131 | −0.051 | 0.260 |
| (7) Lambeck & Cazenave | 30–60 km | −0.024 | −0.010 | +0.050 | 0.010 |
| (8) Equation (7.4.4) | | | | −0.005 | 0.021 |
| Tides | | | | | |
| (9) Solid tide | | −0.030 | −0.001 | 0.173 | −0.065 |
| (10) Ocean tide | | −0.003 | −0.000 | 0.017 | −0.006 |
| Ocean current | | (order 0.02) | (order 0.02) | (order 0.06) | (order 0.06) |
| Total | | | | | |
| (1 − 3) + (6 − 10) | | −0.398 | −0.053 | 0.188 | 0.206 |

quantities, the error in the inertia tensor will be

$$\delta I_{ij} = \frac{10^3 R^4}{g a_{00}} \varepsilon_p(t) \int_s \Pi_{ij} \mathscr{C}(\phi, \lambda)\, \mathrm{d}S,$$

and the errors in $\boldsymbol{\psi}$ are, with $N_{21}$ given by (2.4.4),

$$\begin{pmatrix} \varepsilon_{\psi_1} \\ \varepsilon_{\psi_2} \end{pmatrix} = \frac{R^4}{3g} \frac{10^3}{C-A} \frac{\varepsilon_p(t)}{a_{00}} \begin{pmatrix} a_{21} \\ b_{21} \end{pmatrix} N_{21}^{-2} X_{\text{wobble}}$$

$$= \begin{pmatrix} 1.2 \\ 2.0 \end{pmatrix} 10^{-8} \varepsilon_p(t). \tag{7.1.10}$$

For the difference in the magnitude of Munk & MacDonald's (equation 7.1.5*d*) and Stolz & Larden's (equations 7.1.5*a, b*) estimates of $\delta\tilde{p}_{\text{M}}(t)$

$$\left.\begin{aligned} \varepsilon_{\psi_1} &\simeq 0.08 \times 10^{-8}, \\ \varepsilon_{\psi_2} &\simeq 0.14 \times 10^{-8}. \end{aligned}\right\} \tag{7.1.11}$$

These uncertainties are probably smaller than uncertainties introduced by either insufficient data or an inadequate treatment of the ocean response.

Siderenkov (1973) has evaluated $\boldsymbol{\psi}$ using monthly mean sea level pressure charts compiled by D. J. Stekhnovskiy, with mean altitude corrections applied (see also Siderenkov & Stekhnovskiy 1971). The time span of these data is 1881–1940 (N. S. Siderenkov, personal communication 1977). Siderenkov has assumed that the extreme values of the inertia tensor occur in January and July so that

$$\Delta I_{ij} = \tfrac{1}{2}[I_{ij}(\text{January}) - I_{ij}(\text{July})] \cos \odot. \tag{7.1.12}$$

Both Munk & Hassan's and Wilson's analyses indicate that this is satisfactory when the ocean responds statically to air pressure. The latter's study of the case of zero oceanic response indicates, however, that now the phase is no longer zero with respect to January 1. This is apparently a consequence of the pressure differences over the Pacific and Atlantic not being entirely out of phase with those over Asia, and the effect of this phase difference on $\boldsymbol{\psi}$ is accentuated when the ocean does not respond to the variable air pressure.

Siderenkov has evaluated the excitation functions for (i) no ocean response and (ii) static global ocean response (table 7.1). His estimate of $\psi_1$ agrees well with Wilson's values for both cases. For $\psi_2$, Siderenkov's amplitudes appear systematically larger. In view of the consistency between Wilson's two estimates based on different data, it is unlikely that the difference is a consequence of Wilson's use of station height pressure rather than average altitude pressures. Neither does the discrepancy appear to be due to the different values for $\delta\tilde{p}_M(t)$ used in the two studies since the discrepancy (7.1.11) is smaller than this. For zero ocean response, Siderenkov's values for $\psi_1$ and $\psi_2$ are in reasonable agreement with the comparable solution of Wilson. This appears to rule out any major differences in the two data sets. The discrepancies for the static ocean case are more disturbing and point to a difference in the treatment of the ocean response. If in (7.1.10) the mean pressure over the oceans, $\delta p_o(t)$, evaluated by Siderenkov (7.1.6*b*), is substituted for $\varepsilon_p(t)$ we obtain the partial excitations due to air mass over the oceans with static response of

$$\begin{pmatrix}\Delta\psi_1\\ \Delta\psi_2\end{pmatrix} = \begin{pmatrix}0.7\times10^{-8}\\ 1.1\times10^{-8}\end{pmatrix},$$

which agree well with Siderenkov's estimates of the same partial excitations and his calculations appear to be self-consistent. The two independent estimates (7.1.6*a*) and (7.1.6*b*) of $\delta p_o(t)$ are also in agreement. Wilson does not give partial excitations for the oceans and continents and we cannot compare further the two sets of results.

Kikuchi (1971) evaluated the excitation functions $\psi_1$, $\psi_2$ using monthly mean atmospheric pressure charts published by the Japanese Meteorological Research Institute. Kikuchi does not discuss the ocean response to the air mass and apparently his results are for zero ocean response. If so, they agree well with Siderenkov's and Wilson's estimates for the same excitation functions. Jochmann (1976) used 30 yr of data, 1931–1960, compiled by the World Meteorological Organization. Sea level pressures are estimated on a $5°\times5°$ grid and corrected for mean altitudes. An inverted barometer response is assumed and the seasonal variation in air mass is evaluated directly from the pressure data. The maxima and minima of the excitation functions are assumed to occur in early January and July. Jochmann's value for $\psi_2$ agrees well with that of

Siderenkov and less well with that of Wilson, but his value for $\psi_1$ differs in sign from the other solutions.

Apart from possible numerical errors, there is no apparent explanation of the differences between the solutions (2), (4) and (5) of table 7.1. A possible explanation may lie in the data sets: a displacement between the solutions of Siderenkov and Jochmann, of some 20° in longitude, for the areas of maximum pressure over Central Asia is sufficient to explain their discrepancy in $\psi_1$. Such a displacement may be a consequence of a lack of data in this critical area or it may be a consequence of shifting pressure patterns over long periods, since Siderenkov's analysis is representative of the average conditions during 1880–1940 while Jochmann's analysis is representative of the years 1931–1960. As pressure variations over the Pacific and the North Atlantic are largely compensated, this implies a shift in longitude of the area of maximum pressure difference over Central Asia from about 80° East longitude in 1910–1920 to 110° East longitude in 1945. An inspection of the limited data around Central Asia does not indicate such an important shift.

Wilson (1975) gives the time series of the excitation functions. Averages over 5 yr show considerable variation in amplitude and phase with time (figure 7.1). In the early part of this century, both $\psi_1$ and $\psi_2$ increased to a maximum at about 1930, remained approximately constant from 1930 to 1945, and decreased thereafter; $\psi_2$, in particular, changed by 25% from 1945 to 1965. These variations are very similar to the general zonal circulation trends discussed in chapter 9 and this does suggest that the zonal and longitudinal air mass shifts are related. Nowhere does $\psi_2$ reach the average value found by Siderenkov and by Jochmann.

Munk & Hassan (1961) have also determined $\psi_3$. Wilson (1975) does not, neither does Jochmann (1976). Siderenkov's results cannot be used since by applying the expression (7.1.12) only the cosine term is determined, whereas Munk & Hassan's study indicates that the sine term is dominant. Kikuchi's (1971) value is apparently valid only for zero ocean response to the air pressure. Thus, only Munk & Hassan's estimate for $\psi_3$ is valid (table 7.2). The change in $\Delta I_{33}$ is about four or five times smaller than the change in $\Delta I_{13}$ and $\Delta I_{23}$, reflecting the principal east–west rather than north–south seasonal mass transport. Compared with the total excitation

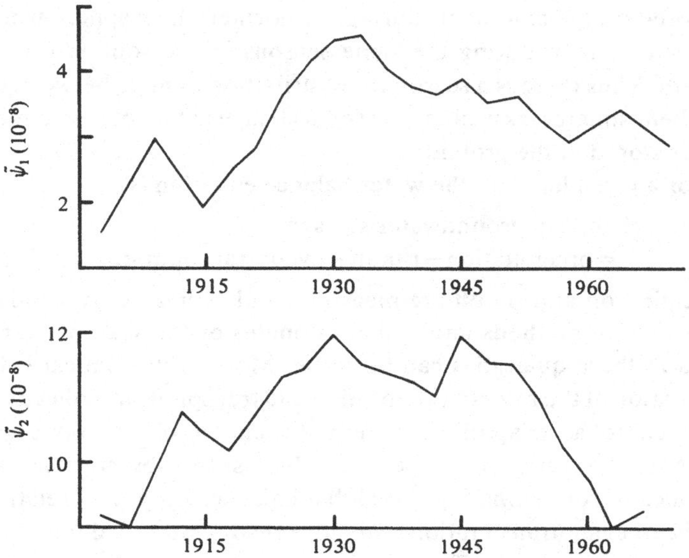


Figure 7.1. Variation in amplitude of the atmospheric $\boldsymbol{\psi}$ from 1900 to 1970 as computed by Wilson (1975). Values are averaged over 5 yr.

function required to explain the seasonal change in l.o.d., the air mass shifts contribute less than a few per cent. The variability of $\psi_3$ with time shows a similar trend to that of $\psi_1$ and $\psi_2$ with increasing amplitude from 1900 to 1930 and decreasing thereafter.

Munk & Hassan (1961) computed the semi-annual excitation function due to the pressure variations and the motion of the excitation axis also appears to be mainly in the $x_2x_3$-plane (table 7.1). This function is about 20% of the annual excitation but, as its frequency is further from the resonance frequency of 1 cycle/14 months, the semi-annual oscillation in $\mathbf{m}$ is only a few per cent of that of the annual oscillation. Astronomical observations confirm this; there is no observable semi-annual oscillation in $m_1$ and $m_2$. The semi-annual pressure contribution to $\psi_3$ is also small.

## 7.2 Groundwater

Groundwater storage includes all moisture stored on the surface in the form of snow, ice and vegetation as well as water stored in the ground. The maximum storage usually occurs during the winter and early spring. Due to the asymmetrical land distribution, more water is stored on the continents during the northern hemisphere winter and spring than during the same seasons in the southern hemisphere. Thus there is a seasonal redistribution of mass between the two hemispheres, as well as a seasonal change in the total amount of water stored in the ground.

For a given location the water balance equation is

$$\text{change in groundwater storage} = \text{precipitation} - \text{runoff} - \text{evapotranspiration}. \tag{7.2.1}$$

Precipitation and runoff are measured and estimated by standard hydrological methods and global estimates of the seasonal variations in these quantities can be made. More problematical is the estimation of the rates of evaporation and transpiration, collectively referred to as evapotranspiration. Lysimeter observations give directly the rate of evaporation but such observations are insufficient for estimating the global balance. Empirical relations between evapotranspiration and more readily observed quantities have to be resorted to. Two such relations are due to Thornthwaite and to Penman (see, for example, Lamb 1972; Budyko 1974). The former consists of a simple relation between temperature and evapotranspiration. But temperature is not a satisfactory measure of the energy available for estimating evaporation since evapotranspiration depends also on radiation, air humidity and winds. Consequently, Thornthwaite's method leads to an overestimation of the evaporation rate. Penman's method, by taking into account radiation, humidity and wind, is more precise but also less useful for global studies since these additional parameters are often unknown. Further refinements of these methods are reviewed by Federer (1975). A third method estimates the evapotranspiration by the relation

$$\text{evapotranspiration} = \text{precipitation} - \text{divergence of moisture of the atmosphere} - \text{change in precipitable water in the atmosphere}. \tag{7.2.2}$$

Substituting (7.2.2) into (7.2.1) gives

$$\text{change in groundwater storage} = -\text{runoff} + \text{divergence of moisture in the atmosphere} + \text{change in precipitable water in the atmosphere}.$$

Regional studies using this approach have been made, but there does not yet appear to have been an attempt at a global evaluation of this equation. The use of empirical relations with readily observed parameters appears to be still the only viable method for estimating the global balance.

Van Hylckama in 1956 used Thornthwaite's method to estimate the seasonal changes in evapotranspiration and the variable groundwater storage. Subsequently he scaled the evapotranspiration values so that the amplitudes agree better with results based on Penman's relation and with lysimeter readings, whenever these are available (Van Hylckama 1970). This leads to estimates for the seasonal change in total groundwater storage and for the excitation functions that are smaller than his earlier estimates (table 7.1). As for air pressure, the main contribution of groundwater is to $\psi_2$ and is of the order of 30–40% of the total annual excitation function inferred from the astronomical data. The atmospheric pressure and groundwater excitations are not in phase, air pressure leading groundwater for both the annual and semi-annual frequency by about 45°. The contribution of groundwater to $\psi_3$ is negligible (table 7.2).

Care must be taken to conserve water mass. The seasonal variation in the mean groundwater storage is

$$\Delta q'_{\rm w} = (0.02 \cos \odot + 0.60 \sin \odot) \text{ g cm}^{-2},$$

and reaches a maximum in the northern hemisphere spring. Tuller's (1968) water vapour pressure data give a variable air load of (equation 7.1.5*a*)

$$\Delta q'_{\rm w} = (-0.11 \cos \odot - 0.06 \sin \odot) \text{ g cm}^{-2}.$$

Then the total seasonal change in water is

$$\Delta q = (-0.09 \cos \odot + 0.54 \sin \odot) \text{ g cm}^{-2},$$

which must be redistributed over the oceans. Thus sea level will be uniformly modified by† $-\Delta q_{\rm w}/a_{00}$, or

$$\xi = 0.13 \cos \odot - 0.77 \sin \odot \text{ cm.} \tag{7.2.3}$$

† As the effect dealt with is small, no attempt is made to ensure that sea level remains an equipotential surface.

From (7.1.10), the additional effect on the excitation function $\psi$ is

$$\begin{pmatrix}\psi_1\\ \psi_2\end{pmatrix}=\frac{R^4}{3}\frac{\rho_w}{C-A}\begin{pmatrix}a_{21}\\ b_{21}\end{pmatrix}N_{21}^{-2}\xi X_{\text{wobble}} \tag{7.2.4}$$

$$=\begin{pmatrix}0.84\\ 1.40\end{pmatrix}(0.13\cos\odot-0.77\sin\odot)10^{-8}, \tag{7.2.5}$$

and represents about 20% of the direct groundwater effect (table 7.1).

The uncertainty of the groundwater excitation function is difficult to evaluate as it is a surface integral of a linear relation between several parameters, all of which show significant geographical variations; presumably this ensures some compensation of error. The year-to-year variability of the seasonal variations is also an unknown quantity but, in so far as precipitation and temperature show as much year-to-year fluctuation as air pressure, the groundwater excitation functions can be assumed to exhibit year-to-year changes that are of a similar magnitude to, but out of phase with, the air pressure functions.

From (4.2.2) a change in sea level $\xi(t)$ perturbs $\psi_3$ by

$$\psi_3=-\frac{8\pi R^4}{3C}\xi(t)[N_{00}^{-2}a_{00}(1+k_0')-N_{20}^{-2}a_{20}(1+k_2')]$$
$$=-0.83\times10^{-9}\xi(t), \tag{7.2.6}$$

and with (7.2.3)

$$\psi_3=(-0.011\cos\odot+0.064\sin\odot)10^{-8}. \tag{7.2.7}$$

Contributions to the semi-annual excitation function follow from Van Hylckama (1970). The results are tabulated in table 7.1. He also gives

$$\Delta q_w'=-0.03\cos 2\odot+0.07\sin 2\odot,$$
$$\Delta q_w''=0.008\cos 2\odot+0.004\sin 2\odot,$$

whence, with (7.2.4) and (7.2.6), the excitations due to variable sea level. These excitation functions amount to about 20% of the corresponding annual terms (tables 7.1 and 7.2).

## 7.3 Oceans

### 7.3.1 *Sea level fluctuations*

As indicated by the excitation functions (7.2.5), the motion of the excitation pole is quite sensitive to changes in sea level; a non-steric change of 1 cm in sea level contributes about $1.5 \times 10^{-8}$ to the excitation functions $\psi_1$, $\psi_2$. Observed seasonal changes in sea level are of the order of 20 cm and have a potentially important influence on the wobble. The global distribution of the seasonal sea level is only poorly known. In particular the non-steric parts of the ocean level fluctuations are only qualitatively understood.

Seasonal changes in sea level can be expressed as

$$\xi = \delta\xi_a + \delta\xi_s + \delta\xi_w + \delta\xi_m, \tag{7.3.1}$$

where

$\delta\xi_a$ = change in sea level due to barometric pressure changes,

$\delta\xi_s$ = steric changes in sea level due to salinity and temperature changes in a column of water,

$\delta\xi_w$ = change in sea level caused by fluctuations in wind stress,

$\delta\xi_m$ = change in sea level due to variable water mass contained in the oceans.

These contributions are not always independent. Steric changes in sea level, arising from variations in the specific volume of the water column, do not load the Earth and will not contribute to the excitation functions. Most of the contribution to $\delta\xi_s$ comes from salinity and temperature variations above the thermocline, and the maximal seasonal variations are of the order of 5–10 cm in mid-latitudes. Of this amount, the greatest contributions come from thermal changes, salinity changes contributing at most 1 or 2 cm in equatorial areas. The barometric effect, $\delta\xi_a$, is most important at high latitudes where it attains 5–10 cm. The third term, $\delta\xi_w$, in (7.3.1) accounts for changes in sea level induced by wind stress. Part of this term contributes to the steric height and will not load the Earth; the remainder will. Gill & Niiler (1973) have evaluated these two contributions for the North Pacific and North Atlantic and show that $\delta\xi_w$ is generally less than 1 cm. The last term, $\delta\xi_m$, allows for a seasonal exchange of water mass between the oceans, atmosphere and groundwater. Its amplitude is about 1 cm (equation 7.2.3). The importance of the barometric and steric factors confirm that sea level changes are mainly isostatic. This is confirmed by the studies of Reid & Mantyla (1976) and others. Thus, the principal ocean contribution to the seasonal excitation functions comes from changes in the mean atmospheric pressure over the oceans. This effect has already been included in the air pressure calculation. The secondary contribution to $\psi_i$ from $\delta\xi_m$ is given by (7.2.5) and (7.2.7). Seasonal wind effects on sea level may be of the order of 1–2 cm and could contribute up to $3\times10^{-8}$ to $\psi_1$ and $\psi_2$. At present, seasonal sea level fluctuations are not sufficiently well known to be able to directly observe the contributions from wind and changing ocean mass.

### 7.3.2 *Currents*

Large-scale ocean currents are controlled most significantly by geostrophic forces whereby the Coriolis force is balanced by horizontal pressure gradients. Only in smaller bodies of water do other forces become important. Assuming geostrophic motion, the current speeds $u_\lambda$, $u_\phi$ are

$$\left.\begin{aligned}\rho u_\lambda &= (-1/2R\ \Omega \sin\phi)\ \partial p/\partial\phi,\\ \rho u_\phi &= (1/2R\ \Omega \sin\phi\cos\phi)\ \partial p/\partial\lambda.\end{aligned}\right\} \tag{7.3.2}$$

Substituting (7.3.2) into the excitation functions (4.1.1) gives

$$\boldsymbol{\psi} = \frac{1}{\Omega^2(C-A)}\int_h\int_S\left[\frac{\partial p}{\partial\phi} - \mathrm{j}\tan\phi\,\frac{\partial p}{\partial\lambda}\right]\mathscr{C}(\phi,\lambda)\,\mathrm{e}^{\mathrm{j}\lambda}\,\mathrm{d}S\,\mathrm{d}h, \tag{7.3.3}$$

where the ocean function $\mathscr{C}(\varphi,\lambda)$ ensures that the currents are confined to the oceans. Of this function, the most important term is $a_{00}$, all other terms being less than 15% of it, and, to within this accuracy, we can replace $\mathscr{C}(\phi,\lambda)$ by $a_{00}$. Now (7.3.3) can be rewritten as (Munk & MacDonald 1960)

$$\begin{aligned}\boldsymbol{\psi} = \frac{R^2 a_{00}}{\Omega^2(C-A)}\int_h\int_\phi\int_\lambda&\left[\frac{\partial}{\partial\phi}(p\cos\phi\,\mathrm{e}^{\mathrm{j}\lambda})\right.\\ &\left.-\mathrm{j}\frac{\partial}{\partial\lambda}(p\sin\phi\,\mathrm{e}^{\mathrm{j}\lambda})\right]\mathrm{d}\phi\,\mathrm{d}\lambda\,\mathrm{d}h\\ =0.&\end{aligned} \tag{7.3.4}$$

Geostrophic currents do not perturb the excitation function $\psi$, as concluded earlier by Munk & MacDonald. Seasonal changes in the parameters controlling the ocean circulation – air pressure, surface temperature, salinity and wind stress – all introduce changes in the currents that appear to follow closely the geostrophic rule. Furthermore, as these parameters tend to be in phase over large parts of the oceans, the seasonal changes in the currents are small: Gill & Niiler (1973) estimate that the seasonal changes in velocity due to wind stress, salinity and temperature fluctuations, do not exceed a few millimetres per second. These are average values for the mid-oceans and locally the fluctuations may be more important. Any ageostrophic part of such motions will make only very minor contributions to $\psi$. This conclusion is also reached by Wilson (1975) who computed the changes in relative angular momenta of the ocean currents given by a numerical ocean circulation model of K. Takano.

Geostrophic ocean currents contribute to $\psi_3$ but no quantitative estimate is yet possible. In all probability the effect is small, since the blocking action of the continents ensures that any zonal flow is matched by a return flow either nearby along the surface or at a different depth. An exception to this is the Circum-Antarctic current. Kort (1962) gives estimates of the average volume transport of water associated with the eastward Circum-Antarctic current. Lamb (1972) tabulates similar values based mainly on a study by Treshnikov, Maksimov & Gindysh (1966) and the longitudinally averaged transport is about $1.5\times10^{14}\ \text{cm}^3\ \text{s}^{-1}$, some 20% less than found by Kort. Lamb notes that the transport across Drake Passage is only about one-half of this, $0.75\times10^{14}\ \text{cm}^3\ \text{s}^{-1}$. This implies a compensating flow, part of which occurs as a westward current within about 200 km of the Antarctic coast and part of which occurs north of the main stream. Baker *et al.* (1977), using recent current meter observations across Drake Passage, give a geostrophic velocity profile which leads to a transport of about $0.8\times10^{14}\ \text{cm}^3\ \text{s}^{-1}$. The total angular momentum of the uncompensated flow is then

$$h_3 = 2\pi R^2 \cos^2\phi \times \text{transport}.$$

The average latitude of the stream is 63° and $h_3 \simeq 0.7\times10^{32}\ \text{g cm}^2\ \text{s}^{-1}$. Apparently little is known about the seasonal variations in the position and velocity of this current. Displacements of the main stream are presumably of little importance as the total eastward transport will be controlled mainly by the topography south of Cape Horn. McKee (1971) finds pronounced sea level oscillations on each side of Drake Passage and suggests that these may be indicative of large fluctuations in transport (see also Van Loon 1972). Surface velocities of the Circum-Antarctic current are known to have a semi-annual fluctuation which relates to the pronounced semi-annual oscillation in surface atmospheric pressure at these latitudes (Van Loon 1971). Annual variations in surface winds are much smaller. Baker *et al.* confirm that this oscillation in the ocean currents occurs at all depths across Drake Passage, with amplitudes of about 50% of the velocities of their average profile. This suggests a semi-annual fluctuation in angular momentum of as much as $0.4\times10^{32}$ g cm$^2$ s$^{-1}$ or a fluctuation in $\psi_3$ of $0.06\times10^{-8}$, larger than some of the other factors included in table 7.2, but still small compared with the tropospheric wind and solid tide effect. Munk & MacDonald's conclusion, that ocean currents contribute little to $\psi_3$, does not appear to need revision.

## 7.4 Winds

The effect of winds on the Earth's rotation has received considerable attention over the last two decades due to the recognition by Munk and R. Miller that they make important contributions to the seasonal changes in the l.o.d. Recomputations by Y. Mintz and Munk confirmed this, although the then available meteorological data were inadequate to make a precise evaluation of the excitation functions. Winds play a much less important role in the discussion on the seasonal changes in wobble for, if winds follow the geostrophic rules, $\psi_1$ and $\psi_2$ vanish. This follows from (7.3.3) without the ocean function $\mathscr{C}(\phi, \lambda)$.

Wind and ocean current contributions to $\psi_i$ can be estimated by evaluating either the relative angular momenta $h_i$ or the torque $L_i$. Munk & MacDonald and Siderenkov (1968) have discussed the equivalence of the two methods. In the first approach the angular momenta are evaluated by choosing the volume $V$ of the Earth to include the atmosphere and oceans; winds and currents introduce relative motion without exerting torques on the volume. In the torque approach the volume of the Earth is the solid Earth and the winds and currents exert torques upon it. Angular momentum is exchanged between the Earth and atmosphere by the surface forces. There is now no relative motion within the volume. As Munk & MacDonald emphasize, the choice between the two methods depends on whether it is easier to measure surface stress or relative velocities. The latter is usually the more precise for little is known about the appropriate choice of the geographically variable, and possibly velocity-dependent, friction coefficient. Furthermore, the winds near the surface are quite irregular and periodic variations may be masked by this 'noise'. The biennial term, for example, is not obvious in the lower-level winds and the evaluation of the torques leads to very uncertain results. For seasonal and higher-frequency wind cycles the angular momentum approach is the most direct in view of the availability of a relatively complete description of the wind field over a 5-yr period (Lambeck & Cazenave 1973, 1974).† Lower-frequency variations can at present only be studied from surface pressure observations, suggesting the torque approach (Siderenkov 1969).

### 7.4.1 *Torque approach*

The exchange of angular momentum between the Earth and atmosphere takes place by two mechanisms: by shear at the boundary and by an excess of pressure on the windward side over the leeward side of mountain ranges. The subsequent stresses, although not entirely separable, can be considered in two parts, as surface stresses and as mountain stresses. The zonal component of the former can be written as

$$p_{\lambda,r} = -D\rho_a u_\lambda \mathbf{u}_\lambda,$$

where $D$ is a friction coefficient and $u_\lambda$ is the zonal wind; $p_{\lambda,r}$ is the stress in direction $\lambda$ on the Earth's surface whose normal is $\mathbf{e}_r$. If the shear occurs along a surface of altitude $h$ then

$$p_{\lambda,r}(h) = D(h)\rho_a(h)u_\lambda(h)\mathbf{u}_\lambda(h).$$

The torque exerted on the Earth by this stress over a latitude band of width $\Delta\phi$ is

$$\mathrm{d}L_3^s = R^3 \int_\lambda D\rho_a u_\lambda \mathbf{u}_\lambda \cos\phi \,\Delta\phi \,\mathrm{d}\lambda.$$

† Lambeck & Hopgood (1979) have extended the analysis by 10 more years.

Mountain torques arise if there is a difference in pressure between the windward and leeward side of a mountain range. If, at constant altitude, the pressure on the east side is $p_{\mathrm{E}}$ and that on the west wide is $p_{\mathrm{W}}$, the stress is

$$\Delta p_{\lambda,m} = \int_0^H \Delta p \, \mathrm{d}h,$$

where $\Delta p = p_{\mathrm{E}} - p_{\mathrm{W}}$ and $H$ is the height of the mountain. The mountain torque for a latitude band $\Delta\phi$ is

$$\mathrm{d}L_3^{\mathrm{m}} = -\sum_i \int_0^H \Delta p_i(h) R \cos\phi \, \Delta\phi \, \mathrm{d}h,$$

where the summation is over all orographic features in the band. Both torques are of similar magnitudes, with mountain torques being perhaps more important in the northern hemisphere and surface torques being more important in the southern hemisphere. Smagorinsky (1969) notes that while the mountain torques are observed to be important in numerical modelling of the circulation, the total torque is about the same whether mountains are present or not. Only the partitioning of the torques is altered. This is also shown by the model of Manabe & Terpstra (1974). Apparently the mountains influence the angular momentum budget of the atmosphere through the mountain torques, but they also change the surface torques by altering the surface wind vectors; angular momentum, in the first instance, is conserved within the atmosphere before any residual amounts are exchanged with the solid Earth. Both torques should therefore be considered in evaluating $\psi_i$. Zonal surface torques can be evaluated directly from zonal averages of the surface stresses and require a knowledge of the friction coefficient $D$ and of the surface winds. In the absence of observations for the latter, geostrophic winds computed from surface pressure values could be used (equations 7.3.2). The friction coefficient $D$ is velocity dependent and will be different over oceans and land. Mountain torques require the pressure differences across orographic features at different heights. These pressure differences may be a consequence of the mechanical disturbance of flow by the barrier or of different temperatures on the two sides, and the mountain torques do not relate directly to the wind flow. C. W. Newton

(1971*a*), for example, shows that for the Andes the torques are most important in the tropics where the winds are weak, and these torques can be ascribed to the thermal structure. Most pressure charts will only give the large-scale effects of the mountain ranges but there is no observational evidence that it is these, rather than the more detailed pressure structure of the order of 100 km, that dominate the true mountain torques, and any values for $dL_3^m$ can only be considered as indicative of the order of magnitude of these forces.

*Length-of-day.* Recent evaluations of the mountain torques have been made by Newton for the entire globe based on observed pressure fields at 1000-, 850-, 700- and 500-mbar levels. Oort & Bowman (1974) give a more detailed analysis for the northern hemisphere and the two studies are in qualitative agreement, with Newton's values somewhat larger than those of Oort & Bowman. Surface friction has been evaluated by Hellerman (1967) for the world's oceans. C. W. Newton (1971*b*) assumes that the zonal means of these ocean stresses represent the average around the globe. As winds are generally higher over the oceans than over land, whereas $D$ is less over oceans than over land, this assumption appears reasonable (Kung 1968). Table 7.3 summarizes the results of the torques exerted by the atmosphere on the Earth. These are converted to the excitation function $(\dot{\psi}_3)_L$ and compared with the excitation function $(\dot{\psi}_3)_h$ computed directly from wind observations using the angular momentum approach (Lambeck & Cazenave 1973; section 7.4.2 below). The agreement between the two is reasonable in amplitude although the phase of $(\psi_3)_L$ is quite different, reflecting the considerable uncertainty in the torque method. Siderenkov (1969) has evaluated $\psi_3$ using the torque approach, by assuming that the mountain torques are negligible compared with surface stress torques and by introducing a turbulent frictional stress in a boundary layer over the Earth's surface. Different values for the surface friction coefficient and for the turbulent viscosity coefficient have been adopted for land and sea. For the seasonal variations, Siderenkov finds an order-of-magnitude agreement with the astronomical data but the phase relationship is poor. The agreement in amplitude is a consequence of

Table 7.3. *Mountain torque $L_3^m$, surface torque $L_3^s$ and total torque $L_3$ exerted by winds on the Earth's surface in units of $10^{25}$ g cm$^2$ s$^{-2}$. $(\dot{\psi}_3)_L$ is the rate of change in the excitation function evaluated from $L_3$ (in units of $10^{-16}$ s$^{-1}$) and $(\dot{\psi}_3)_h$ is the estimate from the annual and semi-annual wind excitation functions, based on the angular momentum calculation by Lambeck & Cazenave (1973)*

| | January | April | July | October |
|---|---|---|---|---|
| $L_3^m$ | 2 | 6 | 0 | −10 |
| $L_3^s$ | −7 | 2 | 8 | −3 |
| $L_3 = L_3^m + L_3^s$ | −5 | 8 | 8 | −13 |
| $(\dot{\psi}_3)_L$ | −8.5 | 13.6 | 13.6 | −22.1 |
| $(\dot{\psi}_3)_h$ | 8.0 | −3.6 | 13.6 | −18.0 |

adopting friction and viscosity coefficients that lead to agreement with the astronomical data. This scaling is justifiable in Siderenkov's study as he is mainly interested in the longer-period fluctuations and he uses the annual term for scaling purposes only (see section 7.6.2). In view of the studies by C. W. Newton (1971*a*, *b*) and Oort & Bowman (1974), Siderenkov's neglect of the mountain torques may be more important than he estimates, for, if we add Newton's seasonal mountain torques, the agreement with the astronomical data is better, both in phase and amplitude (figure 7.2).

*Wobble.* As indicated by (7.3.4), if winds follow the geostrophic law they will not contribute to the excitation functions $\boldsymbol{\psi}$. But winds only approximate the geostrophic law and some wind contribution cannot be excluded. Available wind data are still insufficient to establish the longitude dependence of the monthly mean wind fields and the only way of obtaining an estimate of the wind effect is to use numerical atmospheric circulation models. Several such models exist (see, for example, the reviews by Arakawa 1975; Gates & Imbrie 1975) which, although they reproduce most of the principal global circulation characteristics, do possess shortcomings (see, for example, Manabe, Hahn & Holloway 1974). Wilson (1975) has attempted to evaluate the angular momenta $\mathbf{h}$ and hence $\boldsymbol{\psi}$ using a numerical model by Y. Mintz and A. Arakawa. This model gives

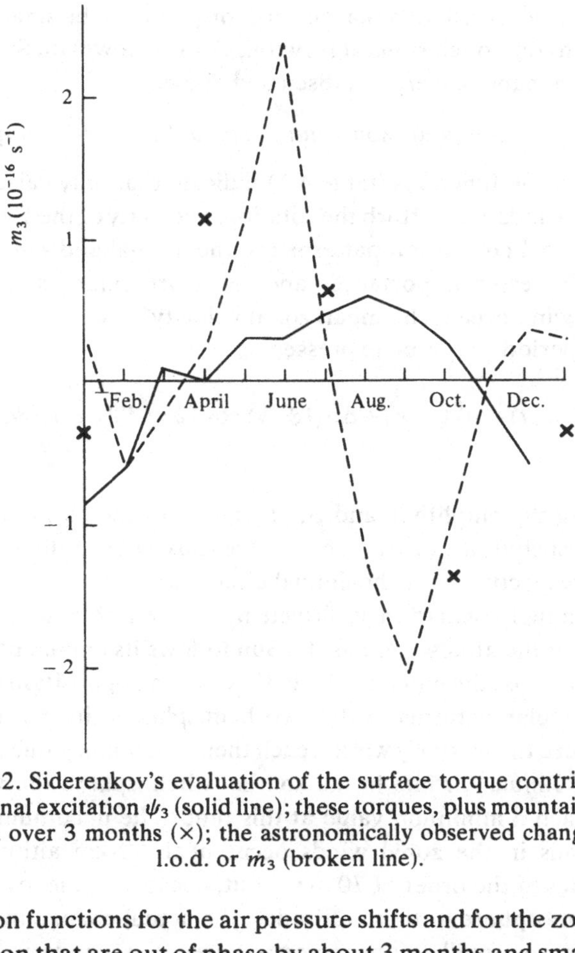


Figure 7.2. Siderenkov's evaluation of the surface torque contribution to the seasonal excitation $\psi_3$ (solid line); these torques, plus mountain torques averaged over 3 months (×); the astronomically observed changes in the l.o.d. or $\dot{m}_3$ (broken line).

excitation functions for the air pressure shifts and for the zonal wind circulation that are out of phase by about 3 months and smaller by a factor of about 2 than the estimates derived directly from observations. Thus, while the discrepancies are significant, the model calculation does give an order-of-magnitude estimate for $\boldsymbol{\psi}$: Wilson finds that the model winds contribute up to $5 \times 10^{-8}$ to $\boldsymbol{\psi}$.

To obtain a more precise estimate of the wind effect on wobble, Wilson (see also Wilson & Haubrich 1976*a*, *b*) has attempted to evaluate $\boldsymbol{\psi}$ using the torque approach. With the above-mentioned numerical model he computed the angular momenta **h** and mountain torques, found that the two lead to comparable excitation functions, and concluded that surface torques must be small. This is contrary to the conclusions of Newton, Oort & Bowman, Smagorinsky and Manabe & Terpstra discussed above.

### 7.4.2 *Angular momentum approach; length-of-day*

The excitation functions (table 4.1) indicate that only variations in the zonal winds can perturb the rotational velocity of the Earth. The two principal circulation patterns are the annual and semi-annual winds. Of lesser importance, and more irregular, is the zonal biennial wind cycle. The mean zonal velocity $u_\lambda(\phi, h; t)$ of a wind cycle of period $T$ can be expressed as

$$u_\lambda(\phi, h; t) = u_\lambda^0(\phi, h) + \Delta u_\lambda(\phi, h) \cos(2\pi/T)[t + \beta_\lambda(\phi, h)], \tag{7.4.1}$$

$\Delta u_\lambda$ being the amplitude and $\beta_\lambda$ the phase of the oscillation. The spatial distributions of $\Delta u_\lambda$ and $\beta_\lambda$ are illustrated in figure 7.3(*a*) and (*b*) respectively for the annual circulation.

The annual oscillation is driven by the variable solar energy received in the atmosphere as the Sun follows its annual path back and forth across the equator. It clearly shows an opposition in phase of the circular patterns in the two hemispheres: in the northern hemisphere the westerly winds reach their maximum value near 20° North in January, while in the southern hemisphere the westerly winds reach a minimum value at this time. The maximum annual fluctuations in the zonal winds occur near 60-km altitude with amplitudes of the order of 70 m s$^{-1}$ but, because of the low density of the atmosphere at these altitudes, the actual change in angular momentum is small. More important are the secondary maxima, occurring in mid-latitudes near the tropopause, although winds down to ground level contribute to the excitation function. Due largely to the unequal ocean–continent distribution, the circulation patterns are not completely anti-symmetric, with the consequence that periodic changes in angular momentum of the two hemispheres do not entirely cancel, the difference being exchanged with the Earth. The annual term in the l.o.d. is, therefore, essentially a measure of the angular momentum imbalance between the circulation in the two hemispheres at the annual frequency.

The semi-annual zonal wind pattern, given in figure 7.4, shows small maxima of about 5 m s$^{-1}$ near the tropopause in equatorial latitudes. Greater amplitudes are observed near the stratopause but again these do not contribute greatly to the angular momentum integral. Like the annual term, the semi-annual oscillation extends to all latitudes and down to low altitudes. Unlike the annual term the main part of the oscillation is symmetrical about the equator: there is not the partial cancellation of the angular momenta of the two hemispheres. Thus the amplitude of the semi-annual excitation is a measure of the seasonal variation of the total angular momentum of the two hemispheres at the semi-annual frequency.

Munk & MacDonald, in reviewing the problem in 1960, concluded that the annual change in l.o.d. is caused primarily by the winds although the agreement between their computed excitations and the BIH astronomical data was not very satisfactory: only a little more than 60% of the observed change was explained. For the non-tidal part of the semi-annual term their computed excitation represented only about 15% of the sought amount. These shortcomings were entirely due to the inadequate meteorological data available at that time, particularly wind data at altitudes above the 500 mbar level (6 km). A more recent evaluation by Frostman, Martin & Schwerdtfeger (1967) also does not adequately explain the observed phase and amplitude of seasonal terms (table 7.2).

The study by Lambeck & Cazenave (1973) has, however, shown conclusively the dominant nature of the meteorological influences on the seasonal oscillations using the zonal wind compilation prepared by Newell *et al.* (1974). These data cover a 5-yr period from 1957 to 1963 and extend between latitudes ±45° and up to an altitude of 10 mbar, or about 30 km. Figures 7.3 and 7.4 illustrate the annual and semi-annual harmonics in these data as determined by Newell *et al.* Substituting the variable part of the function (7.4.1) into the excitation function (4.1.2), or

$$\begin{aligned}\psi_3 &= -\frac{1}{\Omega I_{33}}\int_v \rho_{\mathrm{a}} r^3 \cos^2\phi\, u_\lambda(\phi, \lambda, r)\,\mathrm{d}h\,\mathrm{d}\phi\,\mathrm{d}\lambda \\ &= \frac{-2\pi R^3}{\Omega I_{33}}\int_{\phi=\pi/2}^{-\pi/2}\int_{r=R}^{r+h} \rho_{\mathrm{a}}(h)\cos^2\phi\, u_\lambda(\varphi, r)\,\mathrm{d}h\,\mathrm{d}\phi,\end{aligned} \tag{7.4.2}$$

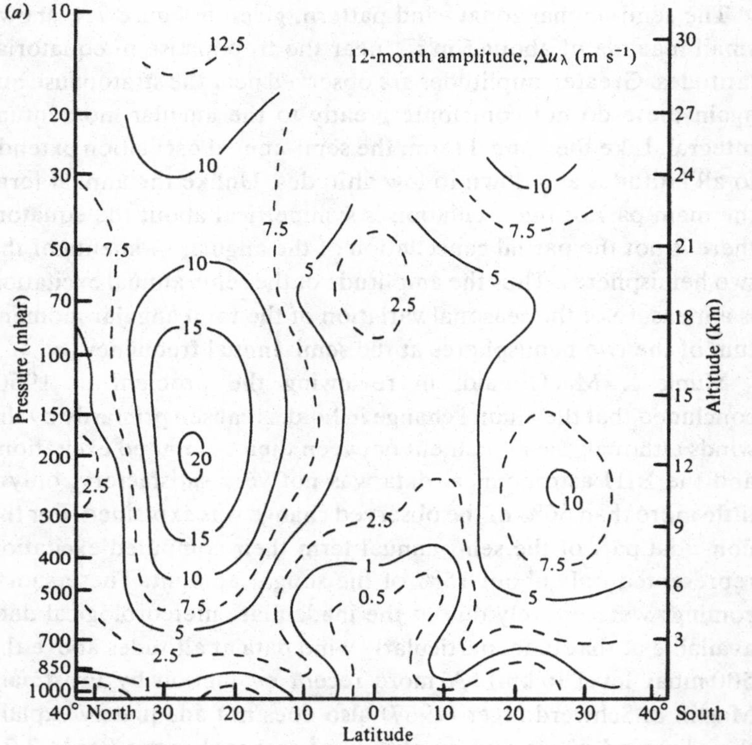


Figure 7.3. Amplitude and phase distribution ((*a*) and (*b*) respectively) of the zonal annual wind oscillation (from Newell *et al.* 1974). The phase is with respect to July 15.

gives

$$\psi_3 = \frac{-2\pi R^3}{\Omega I_{33}} \int_\phi \int_r \rho_a \, \Delta u_\lambda \cos \frac{2\pi}{T}(t+\beta_\lambda) \cos^2 \phi \, dh \, d\phi. \quad (7.4.3)$$

The results for the periods $T = 12$ months and $T = 6$ months are tabulated in table 7.2. These estimates are for the atmosphere up to 60-km altitude: the lower 30 km is based on the detailed wind data of Newell *et al.* (1974), the upper 30 km is based on wind profiles given by Belmont & Dartt (1970). Winds above 60 km contribute less than $10^{-10}$ (see Lambeck & Cazenave 1973 for a further discussion). The upper 30 km appears particularly important for the semi-annual terms and cancels in part the contribution from the

(b)

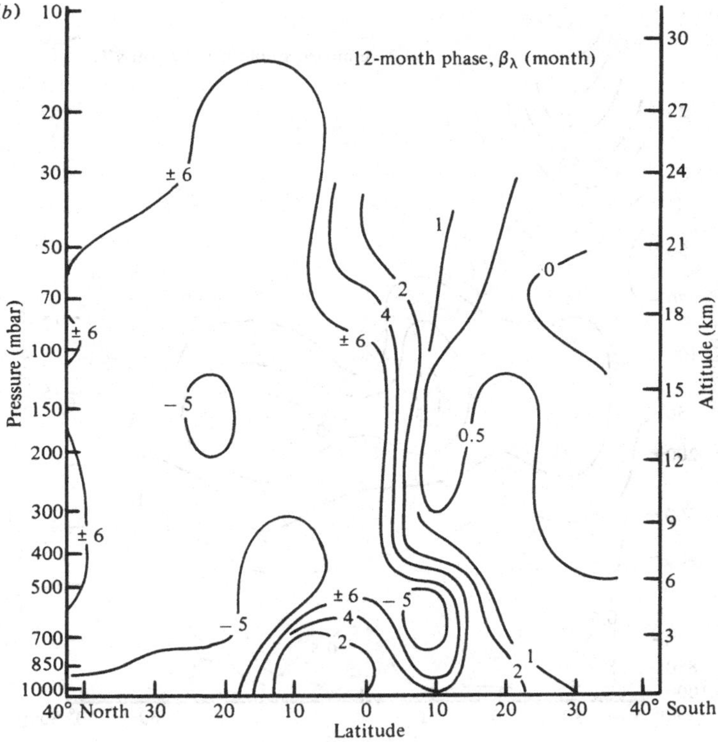


lower atmosphere. A strong semi-annual cycle has been noted in southern latitudes. Van Loon & Labitzke (1973) studied this oscillation in the lower stratosphere and found that maximum speeds of $12\ \mathrm{m\ s^{-1}}$ or more occur at latitudes 60–70° South and at altitudes 25–30 km. Their amplitude and phase curve complement Newell *et al.*'s results and indicate that the winds in higher latitudes, above 45°, between 20 km and 40 km altitude, and in both hemispheres, contribute

$$(-0.005 \cos 2\odot + 0.021 \sin 2\odot)10^{-8} \qquad (7.4.4)$$

to the total excitation. This still leaves a potentially important gap in the wind data for $|\phi| > 45°$ and $h < 20$ km and the total excitation semi-annual function may be underestimated by about 20%.

Upper atmospheric wind observations indicate the presence of a strong biennial oscillation which reaches maximum values of about

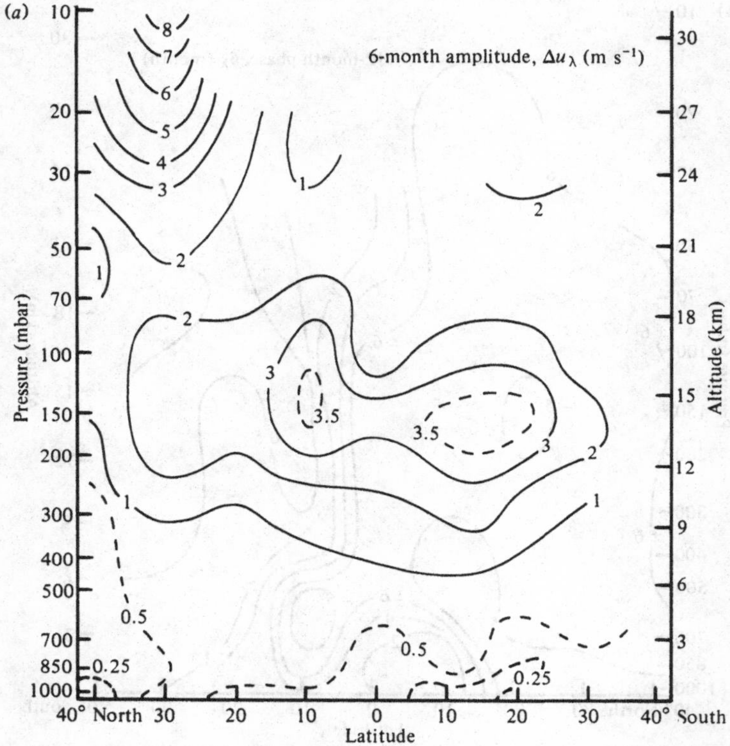


Figure 7.4. Amplitude and phase distribution ((*a*) and (*b*) respectively) of the zonal semi-annual wind oscillation (from Newell *et al.* 1974). Phase is with respect to July 15.

20 m s$^{-1}$ in the stratosphere near 25-km altitude, although it propagates from at least 60-km altitude down to the tropopause. At lower altitudes its presence is not clearly established, and surface pressure data show only a very small biennial component. Over the equator, the winds near 30 km are out of phase with those at 12 km, indicating that the biennial cycle commences at high altitudes and propagates down in altitude with time. When the lower altitudes are reached in this cycle the new cycle has already begun higher in the stratosphere (figure 7.5). The period of the biennial oscillation does not appear to be strictly 24 months but varies between about 20 months and 30 months (Dartt & Belmont 1964). Wallace & Newell (1966) have suggested that the circulation may be interrupted at

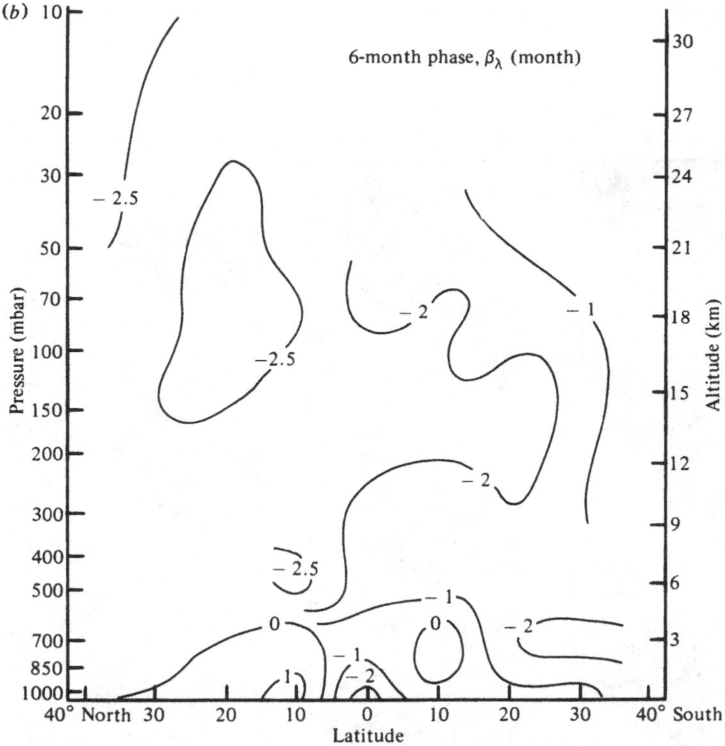


intervals and later reactivated, possibly with a change of phase. They suggest that such a breakdown may have occurred in early 1963 after the eruption of Mount Agung (see also Newell *et al.* 1974), although the data of Angell & Korshover (1970) suggest that the cycle was re-established by 1965. The data of Angell & Korshover, Dartt & Belmont (1970) and more particularly of Wallace (1973) furthermore indicate that there are important fluctuations in the altitude at which the maximum winds occur during any one cycle. These observations indicate an oscillation with non-stationary properties, and the biennial excitation function can likewise be expected to be a non-stationary function as indeed the astronomical data indicate.

Iijima & Okazaki (1966) were apparently the first to point out the existence of the biennial term in the l.o.d. fluctuations and they

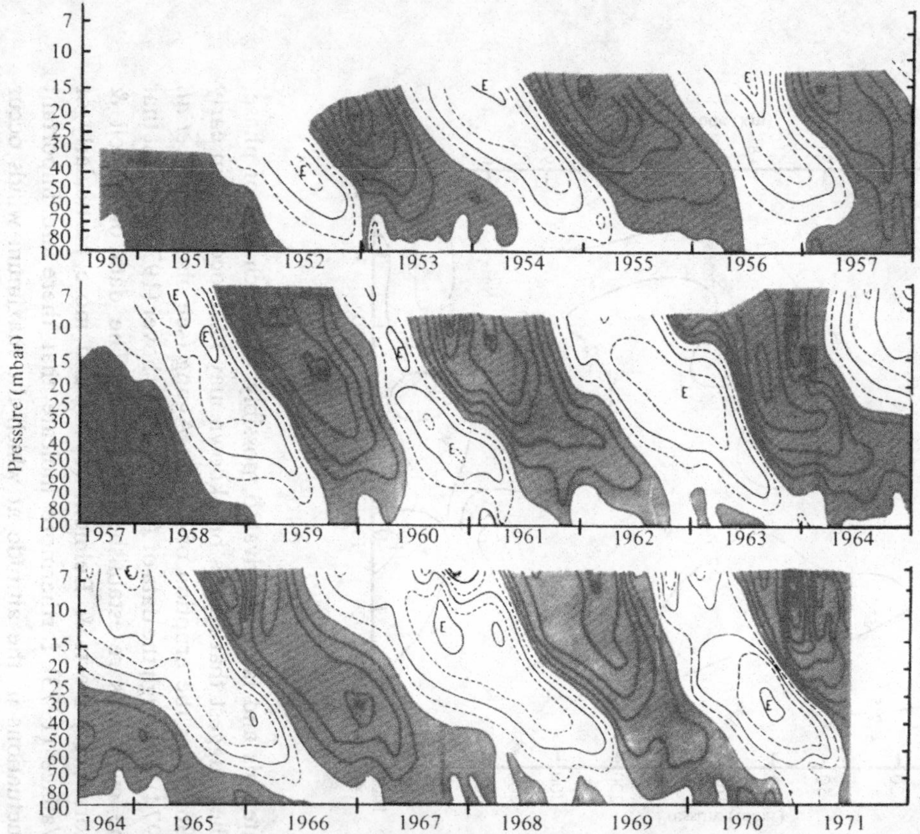


Figure 7.5. Residual zonal wind profile at $\phi = 0$ after removal of annual and semi-annual winds (after Wallace 1973). Shaded areas denote a net westerly wind, non-shaded areas a net easterly wind.

interpreted this as being caused by the quasi-biennial atmospheric circulation, although their argument was qualitative only. Lambeck & Cazenave (1973), using the profiles of Newell *et al.* (1974) illustrated in figure 7.6 in the form of (7.4.1), established quantitatively the relation between the biennial wind circulation and a rather irregular quasi-biennial oscillation in the l.o.d. Table 7.4 summarizes the results. The upper 30-km contribution, based on the data by Belmont & Dartt (1970), is small. Harmonic analyses of $m_3$ do not indicate a single significant harmonic at 24 months; instead there appear several harmonics above the noise level. This is in keeping with the non-stationary behaviour of the quasi-biennial zonal circulation and suggests that it may be more useful to compare the time series of $\psi_3$ and $m_3$ (section 7.5.2).

## 7.5 Astronomical and geophysical comparisons

### 7.5.1 *Wobble*

The total seasonal excitation function due to meteorological causes is given in table 7.1. Several estimates are given, based on the different solutions for the atmospheric contribution as this is the dominant contribution to $\boldsymbol{\psi}$. These excitation functions differ from those of Wilson (1975) and Wilson & Haubrich (1976*a*) in that the mountain torque is ignored here and Wilson has not allowed for the water conservation term (7.2.5). These differences are relatively minor. That the four totals agree in the coefficients $B_1$ and $B_2$ is entirely due to the contribution from the groundwater function, which is the same in all cases. Table 7.5 and figure 7.7 summarize the results in terms of the positive and negative frequencies for the meteorological excitation functions (4.3.12*a*) and the corresponding estimates of the motion of the rotation axis (4.3.12*b*). The four meteorological solutions refer to the last four entries of table 7.1. Also illustrated are the astronomical estimates of $\boldsymbol{\psi}$ and $\mathbf{m}$. These are based on the ILS data from 1901 to 1970, the same interval as the meteorological data used by Wilson (solution 2). Jochmann (1976) estimated $\mathbf{m}$ for the period 1931–1960, the same as for his meteorological estimate (solution 4 of table 7.5). The agreement between the observed and computed $\mathbf{m}$ is somewhat better for the

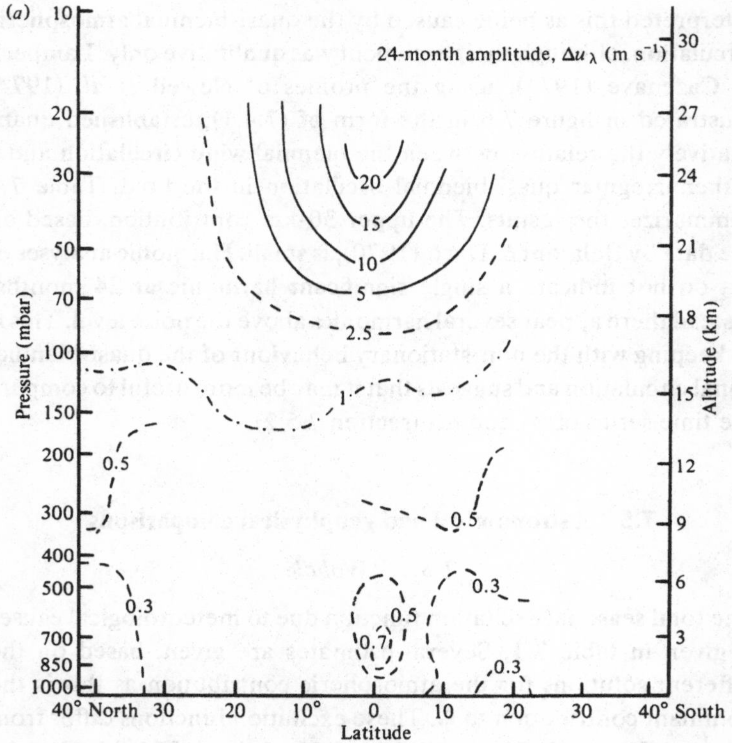


Figure 7.6. Amplitude and phase distribution ((*a*) and (*b*) respectively) of the zonal biennial wind oscillation (from Newell *et al.* 1974). The phase is with respect to July 15, of even years.

positive frequency than for the negative frequency and, overall, Jochmann's pressure excitation leads to a satisfactory agreement with the astronomical result. Siderenkov's pressure excitation (solution 3) leads to disagreement in the real part of $\mathbf{m}^+$ and this is mainly a consequence of the difference in the phase of $\psi_2$ found between his and Jochmann's solutions (table 7.1). Wilson's pressure excitation (solution 2) leads to an unsatisfactory comparison for both the real and imaginary parts of $\mathbf{m}^+$.

The role of coupling of the core motion to the wobble need not be considered for $\boldsymbol{\psi}$ as it is implicitly allowed for by using the observed Chandler wobble frequency $\sigma_0$ in the equations (4.3.12*b*) for $\mathbf{m}$. Consider, for example, a rigid shell enclosing a spherical liquid-

(b)

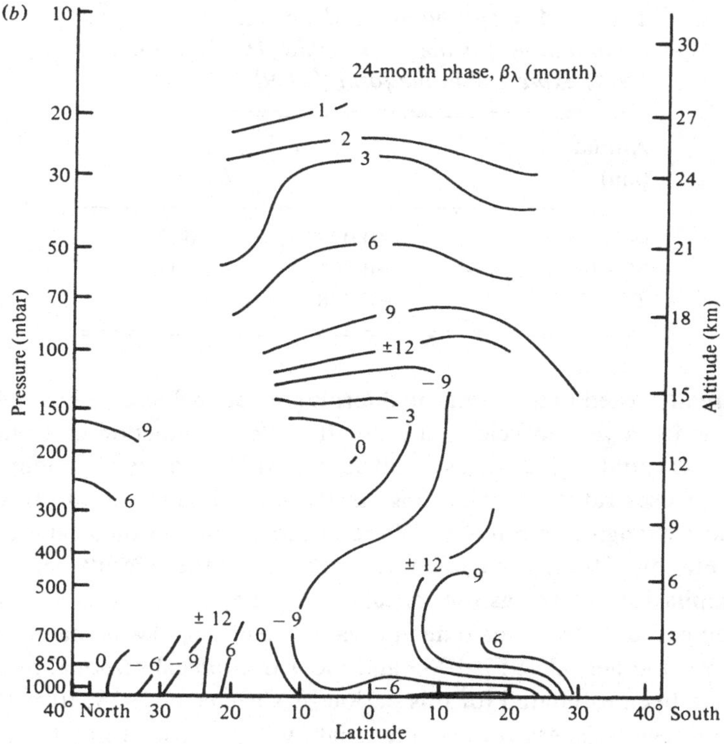


filled cavity, with an excitation function $\boldsymbol{\psi}$ acting on the shell. Equations (3.3.2) describe the motion. Their partial solution, considering for convenience only a positive frequency excitation $\psi^+$, is

$$\mathbf{m} = [\sigma_*/(\sigma - \sigma_*)]\boldsymbol{\psi}^+ \, e^{j\sigma t},$$

where $\sigma$ is the frequency of the forcing function and $\sigma_*$ is the frequency of the free nutation of the shell, or $\Omega(C-A)/A_m$. For a solid core, the solution is

$$\mathbf{m} = [\sigma_r/(\sigma - \sigma_r)]\boldsymbol{\psi}^+ \, e^{j\sigma t}.$$

### 7.5.2 *Length-of-day*

The astronomical data for the fluctuations in the l.o.d. indicate considerable variability in the amplitudes of the seasonal oscillations. Stoyko & Stoyko (1956), analysing data from 1933 to 1954,

Table 7.4. *Contribution to the biennial term in the excitation $\psi_3$ for the years 1958–1963 (in units of $10^{-8}$) expressed in the form (7.1.9)*

| Altitude range (km) | $A_3$ | $B_3$ |
|---|---|---|
| 0–30 | −0.077 | 0.029 |
| 30–60 | −0.001 | −0.003 |
| Total | −0.078 | 0.026 |

already noted that the seasonal terms appear to fluctuate in amplitude from year to year, but only after the introduction of atomic clocks could such changes be attributed with certainty to geophysical causes rather than to noise in the data. Figure 7.8 illustrates these changing amplitudes. We return to the meteorological interpretation of these changes later, but at present these results serve to emphasize that the astronomical and meteorological data must be compared for the same time intervals. The principal wind data used cover the period 1957–1963 and the astronomical seasonal terms have been estimated for this period by Cazenave (1975) from the BIH results. These results differ slightly from those of Lambeck & Cazenave (1973), in which the astronomical data corresponded to a time interval about 6 months longer than the meteorological data, and in which the biennial term was taken to be of 26-month period. The results are presented in table 7.6 in the form

$$m_3 = A\cos(\sigma t - \beta). \qquad (7.5.1)$$

Analysis of the Greenwich Observatory PZT observations for nearly the same period by Frostman, Martin & Schwerdtfeger (1967) gave a comparable result, as did the analysis of the Washington PZT data for 1956–1964 by Fliegel & Hawkins (1967). The small discrepancies in these results reflect different data, different time intervals of the data analysed, and different methods of removing lower-frequency information from the spectrum. These results can be compared with the meteorological excitation functions summarized in table 7.2 to which the tides must be added (chapter 6). The solid tide contribution is computed with $k_2 = 0.30$

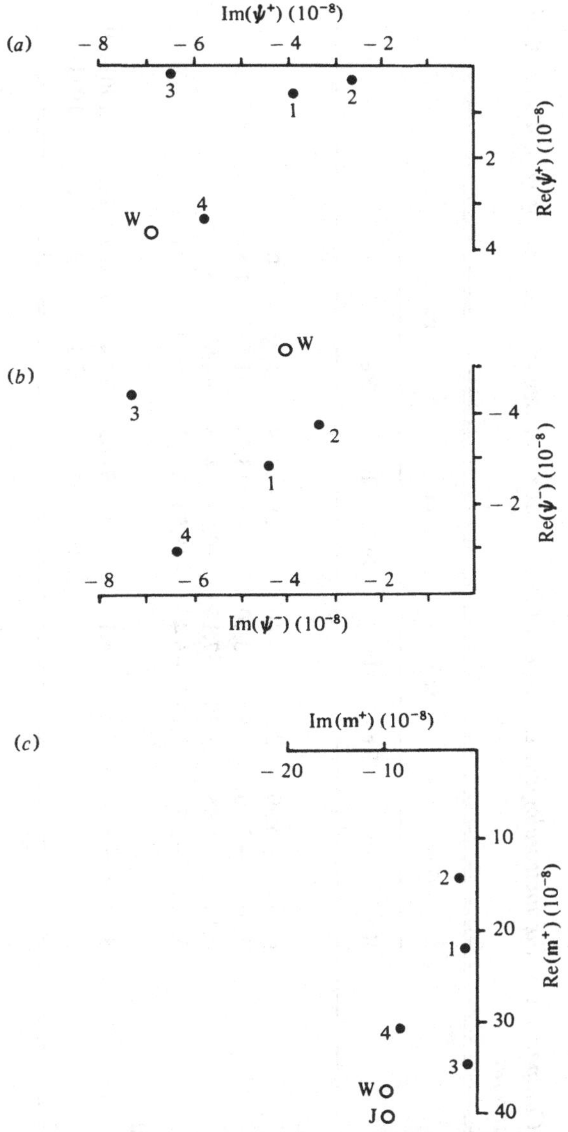


Figure 7.7. Comparison of the positive and negative annual frequencies of **ψ** ((*a*) and (*b*) respectively), and (*c*) the positive frequency of **m**. The four meteorological estimates 1–4 correspond to the solutions in table 7.5. The astronomical estimates are also given in table 7.5. The solution W refers to Wilson's estimate and J to Jochmann's solution.

Table 7.5. *Comparison of the meteorological estimates of the positive and negative rotation of* $\boldsymbol{\psi}$ *and* $\boldsymbol{m}$ *with the astronomical data*

| | $\boldsymbol{\psi}^+$ | | $\boldsymbol{\psi}^-$ | | $\mathbf{m}^+$ | | $\mathbf{m}^-$ | | | |
|---|---|---|---|---|---|---|---|---|---|---|
| | Re | Im | Re | Im | Re | Im | Re | Im | | |
| Meteorological | | | | | | | | | | |
| 1 | 0.60 | −3.90 | −2.80 | −4.40 | −1.28 | 20.94 | −1.28 | −2.01 | | |
| 2 | 0.30 | −2.60 | −3.80 | −3.30 | −1.61 | 13.96 | −1.74 | −1.51 | | |
| 3 | 0.10 | −6.45 | −4.30 | −7.15 | −0.54 | 34.63 | −1.97 | −3.27 | | |
| 4 | 3.45 | −5.75 | −0.95 | −6.45 | −18.53 | 30.87 | −0.43 | −2.95 | | |
| Astronomical | | | | | | | | | | |
| Wilson | 3.70 | −6.90 | −5.4 | −4.1 | −19.87 | 37.05 | −2.49 | −1.89 | ILS | 1901–1970 |
| Jochmann | | | | | −19.49 | 40.26 | −2.10 | −1.20 | ILS | 1931–1960 |

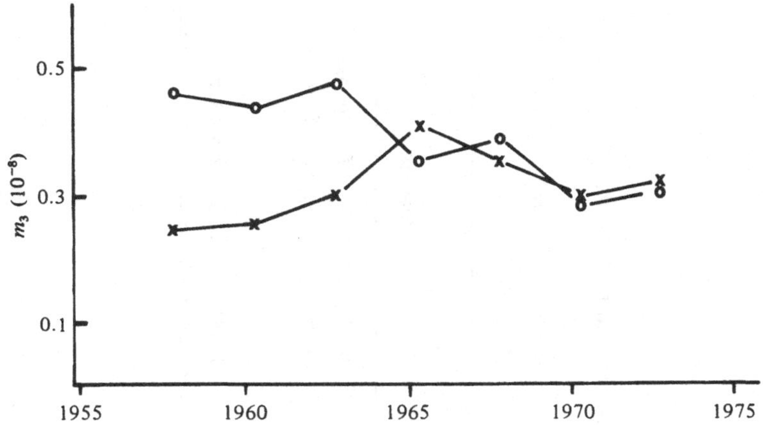


Figure 7.8. Observed astronomical variations in the amplitude of the annual (○) and semi-annual (×) changes in the l.o.d. from 1955 to 1975.

and the equilibrium theory is used to estimate the ocean tide contribution. The latter choice appears adequate since the seasonal frequencies are far from any likely resonance frequencies in the oceans and since any meteorological tides are implicitly contained in the pressure and sea level excitations. The total excitation functions differ from those of Lambeck & Cazenave in that the sea level and ocean tides are now included, but their conclusion, that the zonal wind contribution is dominant, remains valid.

The annual excitation function is somewhat smaller than the astronomical estimates and this is mainly due to the neglect of high-latitude winds. The semi-annual excitation function is also smaller than the astronomical term. This function consists of wind and tide contributions with the two being out of phase. The comparison is not entirely satisfactory and this can be attributed to the lack of high-latitude wind data. The comparison for the semi-annual frequency is complicated by a rather rapid increase in amplitude during 1962–1963, and this is reflected in the difference found in table 7.6 between the analyses of Lambeck & Cazenave (1973) and Cazenave (1975). Okazaki (1975) attributes this change to changes in the star catalogue and station longitudes introduced in 1962.0 (chapter 5). Such changes apparently do not perturb the annual terms.

Table 7.6. *Comparison of astronomical estimates of the changes in l.o.d. with the total meteorological excitation (in units of $10^{-8}$), expressed in the form (7.5.1)*

| | Annual | | Semi-annual | | Biennial | | |
|---|---|---|---|---|---|---|---|
| | $A$ | $\beta$ (degree) | $A$ | $\beta$ (degree) | $A$ | $\beta$ (degree) | |
| Lambeck & Cazenave | 0.416 | 203 | 0.346 | 24 | 0.097 | 1 | ($T = 26$ min) |
| Cazenave | 0.458 | 197 | 0.280 | 27 | 0.097 | 335 | ($T = 24$ min) |
| Frostman *et al.* | 0.413 | 203 | 0.374 | 28 | | | |
| Fliegel & Hawkins | 0.408 | 207 | 0.368 | 23 | | | |
| Excitation function | 0.402 | 188 | 0.253 | 25 | 0.085 | 315 | ($T = 24$ min) |

The incompleteness of the excitation function makes the small phase difference between $\psi_3$ and $m_3$ insignificant. Angular momentum is transferred from the atmosphere to the solid Earth by surface torques, but only about 30% of the Earth's surface interacts directly with the atmosphere and the transfer will occur mainly by the intermediary of the oceans. A lag of $m_3$ behind $\psi_3$ may therefore be expected, but the present data are inadequate for evaluating it. In evaluating $\psi_3$, the polar moment of inertia $I_{33}$ of the entire Earth has been used. This assumes that any changes in the rotation of the mantle are transmitted to the core. If, however, the core does not take part in these changes at the seasonal frequencies, $\psi_3$ will be increased in the ratio $I_{33}/I_{33}$(mantle), or by a factor of about 1.1. Also, if the mantle responds elastically to the wind surface stresses, $\psi_3$ will be further modified. At present the zonal winds are not well enough known to investigate further these modifications of the excitation functions.

The annual term in the l.o.d. is essentially a measure of the angular momentum imbalance between the circulation in the two hemispheres at the annual frequency. Variation in amplitude of the annual l.o.d. term is therefore indicative of changes in this imbalance and the astronomical data indicate that there has been an important decrease in the amplitude of the annual term since 1955 (figure 7.8). Without meteorological data the interpretation of these fluctuations is ambiguous but better astronomical data may lead to a useful measure of the year-to-year variations of the annual atmospheric wind cycle. The astronomical results show an increasing amplitude of the semi-annual term (figure 7.8), indicating a decreasing strength of the semi-annual wind oscillation. But if Okazaki (1975) is correct, the increase in amplitude around 1962.0 may merely be a consequence of the change in the astronomical reduction process in 1962.

Agreement for the biennial term is most satisfactory, both in phase and in amplitude. The astronomical observations for the years 1957–1963 suggest a period of about 26 months rather than the 24 months adopted in the analysis of the wind data. Figure 7.9 illustrates the good agreement between $\psi_3$ and $m_3$ for the interval 1958–1962. Also shown is the astronomically observed $\Lambda$ and that computed from $\psi_3$. The computed curve reproduces well the

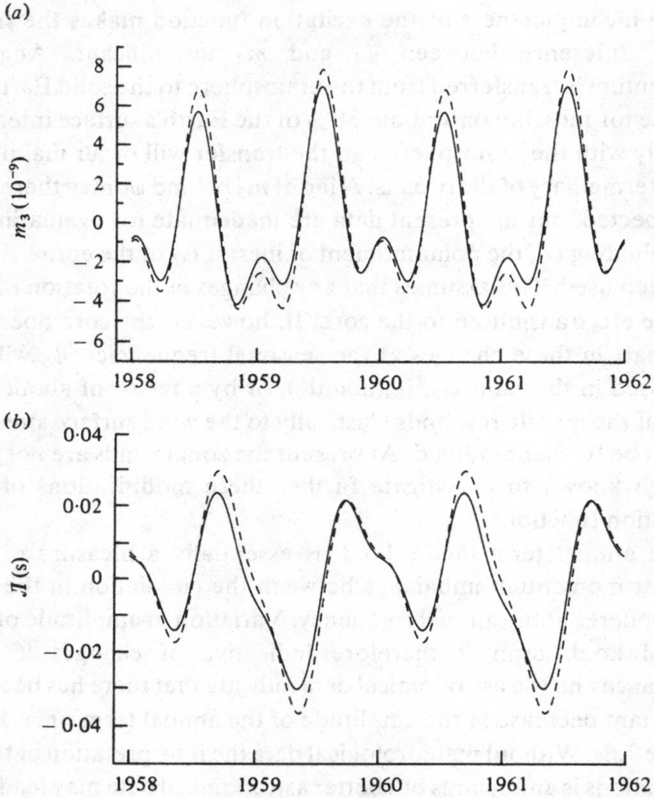


Figure 7.9. (*a*) Sum of the three seasonal excitation functions, compared with the astronomical estimate of $m_3$; (*b*) comparison of the meteorological and astronomic estimates of the seasonal amounts by which the Earth is slow. Astronomic estimates are indicated by broken curves.

slightly anomalous seasonal changes in 1959 and 1963 discussed by Fliegel & Hawkins (1967). This is of course a consequence of the biennial term, which they did not consider. Figure 7.10 illustrates the time series of $\Lambda$ for the years 1955–1972, after removal of the other seasonal terms and long-period (≃7-yr) oscillations. The $m_3$ time series is also presented schematically and its behaviour suggests the following characteristics of the biennial wind circulation (Lambeck & Cazenave 1973, 1977). Before 1960 the biennial wind pattern was well established, and propagated downwards to altitudes of as much as 12–14 km to contribute sufficiently to the

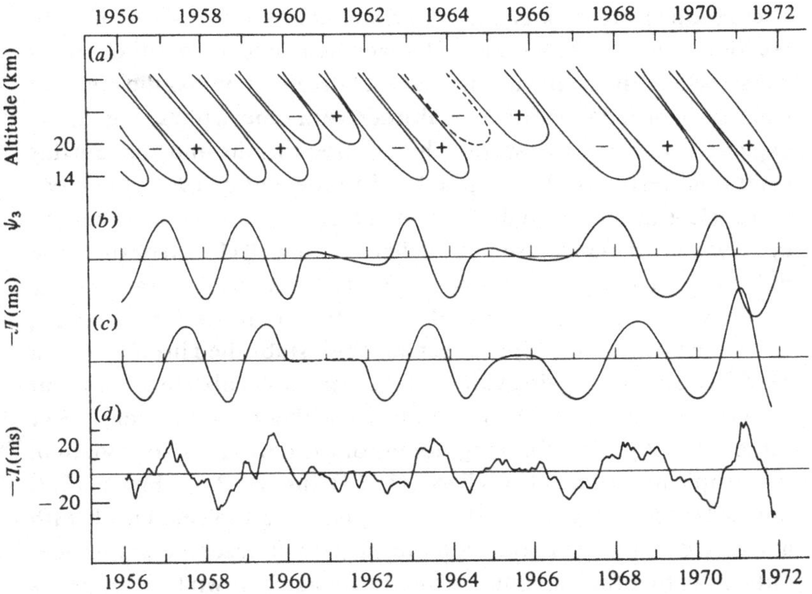


Figure 7.10. Schematic representation of the zonal biennial wind circulation near the equator as deduced from the astronomical data. Curves (*d*) and (*c*) represent −Л observed and idealized, respectively; curve (*b*) represents schematically the $m_3$ or the astronomical estimate of $\psi_3$. The upper part of the figure, (*a*), illustrates the zonal biennial wind pattern deduced from $\psi_3$. Positive cycles indicate eastward flow.

excitation function. After 1960 the cycle is not evident in the $m_3$, presumably due to the fact that the biennial wind pattern did not propagate as far down in altitude as in the previous years. This is in agreement with the equatorial wind profiles of Wallace (1966) that indicate that during 1960–1961 the westerly winds in the lower stratosphere were of lesser intensity than during the preceding period. The cycle is again evident in the astronomical data by early 1962, suggesting that the easterly winds propagated down to their previous low level at this time. The cycle then continues until the end of 1964. According to Wallace & Newell (1966), the zonal wind data exhibit a distinctly biennial component until mid-1963 only, but the astronomical data suggest that the oscillation continued for about another 12 months. The explanation of this discrepancy lies in the fact that the meteorological discussion centres around the circulation at stratospheric altitudes whereas the excitation function responds most to circulation changes near and below the tropopause, where the wind velocities are relatively small. As the biennial wind cycle propagates down in altitude with time, a phase lag can be expected between the stratospheric observations and the Earth's rotational response. From mid-1964 to the end of 1966, the biennial cycle is absent from the rotation data, suggesting that either the cycle vanished as proposed by Wallace & Newell (1966) or the cycle did not propagate as far down in altitude as usual. Stratospheric observations by Angell & Korshover (1970) indicate that the cycle, if it was absent for any time, must have re-established itself by about 1965. The cycle is again evident in the astronomical data from about the end of 1966 to early 1970, but now with a period of nearly 3 yr rather than 2 yr, and this lengthening of the period is also evident in the tropical profile of Wallace (1973) (figure 7.5). From 1970 onwards the biennial oscillation in $m_3$ has re-established itself with a period of 24 months and with the largest yet observed amplitude, indicating that the maximum winds propagated down to altitudes as low as 12 km. This is also suggested by the profiles of Wallace (1973).

This schematic wind profile represents a global pattern but it does explain the astronomical data and is not in conflict with the meteorological conclusions based largely on the wind observations at high altitudes. It does indicate that the $m_3$ record may provide a useful index of the biennial oscillation in the troposphere, a region where this oscillation is not clearly established from the meteorological record.

## 7.6 Non-seasonal meteorological excitations in l.o.d.

### 7.6.1 *High-frequency fluctuations*

Apart from the above-discussed seasonal terms, the atmospheric circulation shows considerable fluctuations at other frequencies: spectral analyses of global zonal winds reveal considerable power near 8 months (Dartt & Belmont 1964), near 40–55 d (Madden & Julian 1971), between 15 d and 25 d (Yanai & Murakami 1970; A. J. Miller 1974), between 10 d and 15 d (Wallace & Chang 1969), near 5 d (Madden & Stokes 1975), and possibly near 1–2 d. The apparent dependence of period on author is probably indicative of a continuum with the power rising above the average about the fortnightly frequency (Mitchell 1976). In addition, abrupt, aperiodic changes in circulation patterns are often observed, such as those associated with the winter stratospheric warmings. The meteorological spectrum can therefore be considered as a continuum upon which the three seasonal terms are superimposed as relatively broad spectral lines. The spectrum of the excitation function $\psi_3$ and hence of $m_3$ can also be expected to have an important continuum.

Short-duration irregular variations in the l.o.d. have been observed ever since accurate atomic time became available, and some of the more dramatic events have been thought sufficiently important to warrant separate study. Danjon (1960*b*), for example, reported an anomalous behaviour in universal time in July 1959 and associated this with a solar eruption, but more recent analyses do not suggest any particular abnormality in the rotation at this time. More recently, Guinot (1970*a*) reported a discontinuity in UT of nearly 10 ms between April 8 and April 13, 1968. Munk & MacDonald conclude that severe meteorological anomalies persisting for a few months could cause detectable variations in the l.o.d. and that there is nothing mysterious in variations of a few parts in $10^9$. Markowitz (1970) attributes the observed high-frequency (that is, higher-than-seasonal) components in the l.o.d. spectrum to a wind origin. Challinor (1971) surmises that this part of the spectrum is a consequence of the irregular behaviour of the seasonal terms. A quantitative evaluation of the high-frequency part of the excitation spectrum has been carried out by Lambeck & Cazenave (1974) who showed that indeed all irregular variations in rotation of frequencies between 0.3 cycle $\mathrm{yr}^{-1}$ and 6 cycle $\mathrm{yr}^{-1}$ are of meteorological origin and that, at frequencies higher than 6 cycle $\mathrm{yr}^{-1}$, meteorological factors also play a dominant role. In a recent study, Okazaki (1977) confirms that perturbations in l.o.d. with frequencies between 0.5 cycle $\mathrm{yr}^{-1}$ and 3 cycle $\mathrm{yr}^{-1}$ are of meteorological origin.

The global monthly mean zonal wind compilation used by Lambeck & Cazenave does not permit any evaluation of higher frequencies in the excitation, but order-of-magnitude calculations based on observed wind spectra show that the zonal winds can contribute about $10^{-11}$ to the continuum in the l.o.d. at frequencies above 6 cycle $yr^{-1}$. The contribution can possibly be more important near the tidal frequency of Mf since there are also zonal winds, possibly of tidal origin, near this frequency (A. J. Miller 1974). The *meteorological noise* in $m_3$ plus the estimated astronomical noise explain most of the very high-frequency part of the l.o.d. spectrum (figure 5.8) (Lambeck & Cazenave 1974).

The high-frequency meteorological noise has several important consequences upon the discussion of geophysical factors perturbing the Earth's rotation. It will mask other irregular short-duration excitations that are not strictly periodic. For example, it will be unlikely that changes in $m_3$ due to earthquakes will rise above this noise level. Improvements in the accuracy of the l.o.d. determination will therefore have little impact, unless the meteorological excitation can be evaluated precisely. The zonal wind analysed for the years 1958–1963 gives a month-by-month excitation function for latitudes between ±45° that is about as precise as the astronomically observed month-by-month excitation changes in l.o.d. The latter data have improved considerably since then, but there does not yet appear to have been an equal improvement in global zonal wind compilations. In particular, high-latitude southern hemisphere data are still not available. High-frequency periodic changes in l.o.d. will presumably rise above the meteorological noise if a sufficiently long period of observations is available. But excitation functions due to phenomena such as tides could be systematically biased by meteorological excitations at similar frequencies. This makes the interpretation of the rotational Love numbers uncertain, and will probably prevent the determination of reasonable phase lags between the lunar zonal potential and the rotational response in $m_3$ (chapter 6). Another problem arising from the meteorological noise concerns the identification of the mechanism responsible for the *decade* variations in l.o.d. Different possibilities are discussed in chapter 9 and a key point is the nature of the changes in $m_3$ caused by these mechanisms. Do they occur very rapidly, or over a few months or years? Clearly these details will be masked, at least in part, by the meteorological excitation.

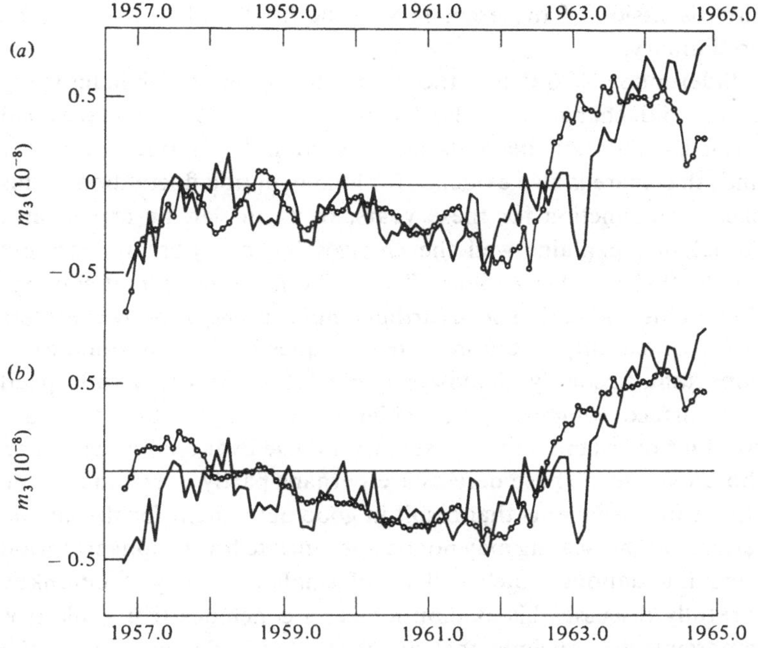


Figure 7.11. A comparison of the non-seasonal excitation function evaluated from surface wind torques with the observed changes in l.o.d. (after Siderenkov 1969). The solid line indicates the astronomical result, the open circles indicate the excitation function. Two cases are illustrated: (*a*) the case where the turbulent viscosities are 5 $m^2 s^{-1}$ for land and 1 $m^2 s^{-1}$ for sea, and the surface friction coefficients are 0.01 for land and 0.001 for sea; (*b*) the case where these coefficients are respectively 8, 0.5, 0.01 and 0.001.

### 7.6.2 *Low-frequency fluctuations*

Other than the seasonal and high-frequency oscillations, the Earth is subject to longer-period fluctuations in the l.o.d. These are generally referred to as the *decade* fluctuations and have perhaps been subject to more speculation than to geophysically sound argumentation. These changes are discussed in chapter 9 and we confine ourselves to l.o.d. changes on a time scale of some 5 yr which appear clearly in the astronomical data since 1955, and which may be a consequence of longer-period wind cycles. Meteorological evidence for such zonal winds on this time scale is minimal due to a lack of a comprehensive data set over a sufficiently long time, and any discussion of these changes in the l.o.d. will of necessity be preliminary.

Siderenkov (1969) used the surface torque approach in his study of the l.o.d. changes over the 8-yr period 1957–1965. The seasonal terms have already been discussed (section 7.4.1), but Siderenkov finds that there is also evidence for longer-period fluctuations in his excitation function for these years. In particular, he arrives at a model that explains well the changes in l.o.d. observed around 1957–1958 and 1962 (figure 7.11). The more important assumptions of his calculation need further emphasis. Siderenkov considers only surface torques but mountain torques have been found to be important. Secondly, the viscosity coefficients will vary with speed and surface roughness, but Siderenkov assumes that they are constant over land and over sea, with those over land greater than those over sea. These defects are perhaps partly overcome by the choice of coefficients that result in good agreement for the annual terms, but this scaling may not be appropriate for the longer-period wind fluctuations which will be of smaller velocity. Siderenkov carefully discusses his assumptions and concludes that while it is impossible to conclude that all of the l.o.d. changes during the interval 1957–1965 are of meteorological origin, zonal winds do make a substantial contribution. Lambeck & Cazenave (1974) and Okazaki (1977) reached a similar conclusion. Rosen, Wu & Peixoto (1976) evaluated the angular momentum of the northern hemisphere zonal circulation for a 10-yr period and confirmed these conclusions, particularly if the southern hemispheric circulation is subject to comparable changes (see figure 8 of Lambeck 1978*b*).

EIGHT

# THE CHANDLER WOBBLE

The Chandler wobble, discovered in 1891 after a long and fruitless search for a 10-month period in astronomical latitude observations, is still associated with almost as much controversy today as it was then, and many of the questions that were raised by Chandler, Newcomb, Kelvin, Volterra, Larmor, Darwin and others are still with us. These questions relate to the three essential problems associated with the Chandler wobble. (i) Can the lengthening of the period, from the 305 d predicted for a rigid Earth to the observed 434 d, be explained quantitatively? (ii) Being a free motion, the Chandler wobble will ultimately be damped out but the astronomical record of near 150 yr does not show any indication of a gradually diminishing amplitude. What maintains the motion against damping? (iii) If damping occurs, where is the rotational energy dissipated? These are also the questions that we wish to discuss once again in this chapter.

Astronomical evidence for the Chandler wobble has been discussed in chapter 5. The main characteristic is a period of about 434 sidereal days. The broad spectral peak (figure 5.14) is very suggestive of damping and, if a linearly damped oscillation is assumed, the relaxation time is of the order of 25–40 yr; the wobble amplitude would decay to $e^{-1}$ of its original value in something like 25–40 yr. The associated Chandler wobble $Q$ would be of the order of 60–100. The interpretation of the spectral peak width in terms of a decay time requires explicit knowledge of, or an implicit assumption about, the excitation mechanism. This has been frequently stressed by Jeffreys but it has not always been noted by spectral analysts.

Munk & MacDonald, in reviewing the Chandler wobble problem in 1960, concluded that, of the three problem areas, only the lengthening of the period could be explained satisfactorily. They examined several excitation mechanisms only to discard them one by one. On the other hand, they concluded that there is no shortage of energy sinks, the core, mantle and oceans all being possible contenders. Their summary, 'The situation is appallingly uncertain', was certainly apt. Since the publication of Munk & MacDonald's book, some progress has been made towards answering all three questions. Theoretical estimates of the period of the Chandler wobble have been refined by introducing more realistic core models and by a re-evaluation of the ocean contribution, with the result that the observed and computed periods agree to within a few days. Excitation of the Chandler wobble now appears to be a consequence of changes in the Earth's inertia tensor associated with both earthquakes and the atmospheric mass redistribution (section 8.4). Dissipation of the wobble energy in the core is unlikely to be important. More important may be the mantle if $Q$ is frequency dependent, as discussed in section 2.2. The most plausible sink is in the oceans, although the observational evidence for this remains marginal.

The 14-month motion of the rotation axis is referred to synonymously as the Chandler wobble, the free nutation or Eulerian precession of the Earth, but it has been recognized since the earliest theoretical studies of the subject that other free modes are possible. A number of recent studies have again focused attention on these additional modes: a nearly diurnal eigenfrequency (or several frequencies) resulting from a complex dynamical interaction between the fluid core and the mantle, and a longer-period mode or modes associated with a solid inner core. But whereas the Chandler wobble has been clearly observed, these other modes have not yet been detected (Toomre 1974; Rochester *et al.* 1974).

## 8.1 The Chandler period of the solid Earth

### 8.1.1 *A qualitative estimate*

As noted by Newcomb in 1892 (soon after Chandler's discovery), the discrepancy between the observed 435-d period and the equivalent rigid Earth period of 305 d can be attributed to the elastic yielding of the Earth subject to a variable centrifugal force. The effect of this deformation on the period has been discussed in chapter 3, using the Love number approach. However, in view of the quite different roles played by the mantle, core, and oceans, the resulting *Chandler wobble* $k_2$ is not readily interpreted and cannot be directly compared with Love number estimates based on other geophysical observations. The interpretation will be further complicated by a possible frequency dependence of the Love numbers. A more quantitative evaluation, in terms of the Earth's physical properties, is required. This has been done in several steps following Kelvin's pioneering work. S. S. Hough in 1896 and J. Larmor in 1909 investigated the lengthening of the period for a homogeneous and incompressible elastic Earth. Their results can be directly obtained from (3.4.6) with (3.4.5), using for $k_2$ the Love number calculation for the same Earth model (equation 2.1.9). Hough, in 1895, also investigated the effect of a fluid-filled core inside a rigid shell and showed that the principal dynamic effect of the liquid core is to reduce the period from its rigid body value by about 10% (equation 3.3.1). The small ellipticity of the core does not change this result significantly (equation 3.3.3).

Jeffreys, in 1949, recognized that the actual core effect on the Chandler period would be modified if the shell is not rigid, and pointed out the need for a unified approach, allowing for both the elastic deformations in the mantle and the fluid motions in the core, using realistic models for the density and elastic moduli. Such a study was first undertaken by Jeffreys & Vicente (1957*a*). Their mantle deformation theory was adapted from Takeuchi's (1950) Love number calculation, which in turn was based on one of Bullen's models for the radial variations in density and elastic moduli within the Earth. They modelled the core as a homogeneous and incompressible fluid with a central particle inner core. In a second paper, Jeffreys & Vicente (1957*b*) introduced a Roche model core in which densities vary according to pressure only. Their results for the period of the free nutation were 392 sidereal days and 395 sidereal days for the two models. Molodensky (1961) carried out a similar study and found periods of 400.9 d and 401.6 d for two different Earth models. In view of the small variations in results found by both authors for different core models, the discrepancy between the Jeffreys–Vicente and Molodensky results are larger than one would expect solely from the differences in the Earth models used. In fact, both studies, despite some important differences in the core models, yield results that are not very sensitive to the chosen model and suggest that the observations of the Chandler wobble period do not provide strong constraints on the nature of the fluid core.

The discrepancy between the computed 400-d period for the solid Earth and the observed Chandler period has been attributed to the world's oceans, and both J. Larmor in 1896 and G. H. Darwin in 1907 recognized that the small tide set up in the oceans by the centrifugal force would increase the period of the free nutation by some 40 d. This result has been confirmed by Haubrich & Munk (1959) whose more rigorous treatment leads to an increase in period of 33 d. Thus the total computed period is about 420–40 d, in general agreement with the observed value. Nevertheless, the theoretically computed value is unsatisfactory for several reasons. Both Jeffreys & Vicente and Molodensky imposed restrictions on the form of the fluid motion in the core, and it is difficult to assess the implications of this on the period of the free nutation. Also, both theories are complex and sometimes obscure, and, apart from a desire to understand this problem better, there is a geophysical interest in refining the computations. The available results suggest that the Chandler period is not very model dependent and that its use as a constraint in the inversion of geophysical data for the Earth's internal structure may be quite limited. Possibly a more important use will be in the study of the Earth's anelastic response, where a precise comparison of the computed elastic response with the observed response may give some constraints on the Earth's $Q$ at low frequencies. Two recent results may make this possible: M. L. Smith (1977) has computed the elastic response of the Earth to the variable centrifugal force using normal mode theory, and Dahlen (1976) has re-evaluated the ocean response using a refined equilibrium tide theory.

### 8.1.2 *Yielding of the mantle*

For convenience, the equations of motion (3.2.6) are expressed with respect to axes chosen so that the $h_i$ vanish at all times. Thus

$$\left.\begin{aligned} \dot{m}_1 + m_2\sigma_r &= -(1/A)(\Delta\dot{I}_{13} - \Omega\,\Delta I_{23}) = \sigma_r\psi_2, \\ \dot{m}_2 - m_1\sigma_r &= -(1/A)(\Delta\dot{I}_{23} + \Omega\,\Delta I_{13}) = \sigma_r\psi_1, \\ \dot{m}_3 &= -\Delta\dot{I}_{33}/C = \dot{\psi}_3. \end{aligned}\right\} \qquad (8.1.1)$$

The inertia tensor $I_{ij}$ is given by (3.1.4), or

$$I_{ij}(t)=\int_V \rho(x)[x_k(t)x_k(t)\delta_{ij}-x_i(t)x_j(t)]\,\mathrm{d}V.$$

Suppose that the body is deformed by an applied body force whose gravitational potential is $U'$. If $d_i(x,t)$ is the infinitesimal Lagrangian particle motion, the mass point originally at its equilibrium position $x_i$ is moved to $x_i+d_i(x,t)$ during the deformation and the change in the inertia tensor $\Delta I_{ij}(t)$ is

$$\Delta I_{ij}(t)=\int_V \rho(x)[2x_kd_k(x,t)\delta_{ij}-x_id_j(x,t)-d_i(x,t)x_j]\,\mathrm{d}V. \qquad (8.1.2)$$

The force under consideration is the centrifugal force whose potential is given by (3.4.1) and, for small $m_i$, has the form

$$U'=U'_km_k, \qquad k=1,2,3. \qquad (8.1.3)$$

The essential problem is to evaluate the deformation $d_i(x,t)$ for realistic Earth models. Gilbert (1971) has emphasized that the response of a linear system can be expressed in terms of the normal modes of that system and that statistical or dynamical theories of the body tides, Love number calculations and similar problems can be expressed in terms of the Earth's normal mode eigenfunctions. This approach is particularly suited for two relevant applications: the determination of the period of the Chandler wobble and the investigation of the hypothesis that earthquakes excite this motion (Dahlen 1976; M. L. Smith 1977).

Following Gilbert, the deformation of an elastic body subject to a periodic perturbing force $\mathbf{f}=\mathbf{f}_*\cos\sigma_*t$ is given by

$$\mathbf{d}(\mathbf{x},t)=\sum_n \mathbf{s}_n(\mathbf{x},t)\mathcal{T}\frac{\cos\sigma_*t}{\sigma_n^2-\sigma_*^2}, \qquad (8.1.4a)$$

with

$$\mathcal{T}=-\int_V \rho(\mathbf{x})\mathbf{s}_n^{\ddagger}(\mathbf{x},t)\cdot\mathbf{f}_*(\mathbf{x})\,\mathrm{d}V, \qquad (8.1.4b)$$

where $\mathbf{s}_n(\mathbf{x},t)$ $(\equiv\mathbf{s}_n(\mathbf{x})\,\mathrm{e}^{\mathrm{j}\sigma_nt})$ are the normal mode eigenfunctions of the Earth and $^{\ddagger}$ denotes a complex conjugate. The $\mathbf{s}_n$ are computed for the particular Earth model under investigation and can be verified from studies of the Earth's free elastic oscillations. The form of the potential (8.1.3) indicates that $d_i(x, t)$ is a linear function of $m_i$, or

$$d_i = \alpha_{ik}(x, t)m_k, \tag{8.1.5}$$

and from (8.1.2)

$$\Delta I_{ij}(x, t) = D_{ijk}(x, t)m_k(t),$$

in which the $D_{ijk}(x, t)$, a rather complex third-order tensor, is a function of time and of the physical parameters describing the Earth model. In the static limit $t = \infty$, $\mathbf{s}_n = \mathbf{s}_n(\mathbf{x}, \infty)$ and $\Delta I_{ij}(x, t) = D_{ijk}(x)m_k(t)$. The excitation functions on the right-hand side of (8.1.1) become

$$\sigma_r\psi_1 = (\Omega/A)D_{13k}m_k + (D_{23k}/A)\dot{m}_k,$$
$$\sigma_r\psi_2 = (\Omega/A)D_{23k}m_k - (D_{13k}/A)\dot{m}_k,$$
$$\dot{\psi}_3 = -(1/C)D_{33k}\dot{m}_k.$$

For convenience, we also assume that we can decouple wobble from spin, and this can be justified *a posteriori* in that changes in the Earth's deformation tensor induced by $m_3$ tend to be very small (section 3.4). The equations of motion become

$$\left(1+\frac{D_{131}}{A}\right)\dot{m}_1 + \frac{D_{132}}{A}\dot{m}_2 - \frac{D_{231}}{A}\Omega m_1 + \frac{\Omega}{A}(C - A - D_{232})m_2 = 0,$$
$$\frac{D_{231}}{A}\dot{m}_1 + \left(1+\frac{D_{232}}{A}\right)\dot{m}_2 - \frac{\Omega}{A}(C - A - D_{131})m_1 + \frac{D_{132}}{A}\Omega m_2 = 0, \tag{8.1.6}$$

and their solution is assumed to be of the form

$$m_1 = m_0 \cos \sigma_* t,$$
$$m_2 = m_0 \sin \sigma_* t,$$

since for $A = B$ the motion is circular (equation 3.1.8). Substituting this form into (8.1.6) and ignoring products and squares of small quantities gives

$$\sigma_* = \Omega(C - A)/A - \tfrac{1}{2}(D_{131} + D_{232})\Omega/A, \tag{8.1.7}$$

for the frequency of the free nutation. For the rigid Earth $D_{ijk}$ vanish and the frequency is $\Omega(C - A)/A$, as found by Euler. Dahlen (1976) discusses the solution for the static limit in greater detail.

An important approximation made is the replacement of $D(\mathbf{x}, t)$ by its static limit $D(\mathbf{x}, \infty)$. For a solid, predominantly elastic, Earth, this will in general be a valid process since the elastic modes have frequencies $\sigma_n$ that are high compared with $\sigma_*$. But with the core and oceans, this situation may be much changed if free modes of longer periods occur, in particular if $\sigma_n$ approaches $\sigma_*$. Pekeris & Accad (1972) encountered some of the complexities that such modes introduce in Love number calculations (see also Dahlen 1974), and similar modifications in the rotational deformation tensor $D_{ijk}(x, t)$ may occur. A more general approach is required.

### 8.1.3 *The core and the wobble period*

As already noted, a principal effect of the fluid core is to reduce the Chandler period by some 10%. Section 3.3 discusses briefly the solution for a rotating spheroid with a fluid-filled ellipsoidal cavity when the fluid is of constant density and incompressible. Factors to be included in a more complete theory include the modification of the fluid pressures on the mantle by the elastic deformation of the latter, self-gravitation, radial variation of density and rigidity in the core, and the presence of a solid inner core. M. L. Smith (1977) has considered all these effects by generalizing the normal mode approach (8.1.4) to include the fluid core with its free modes and without resorting to the static limit approximation. Smith's treatment places fewer restrictions on the nature of the permissible core motion than do the models by Jeffreys and Molodensky. The solid inner core is introduced and various degrees of density stratification have been considered. This generalized approach has enabled Smith to investigate, in addition to the Chandler wobble, other free core models, in particular the nearly diurnal wobble.

His conclusions concerning the Chandler period are as follows. (i) The fluid motion in the core remains essentially of the form assumed by Kelvin and Poincaré during a cycle of the Chandler wobble. That is, the core remains mostly unaffected by the rigid rotation component of the mantle past it. (ii) The distribution of elastic properties in the mantle is relatively unimportant; the period is quite independent of the fine detail of the radial structure. (iii) The core structure is also not very critical in determining the period, a conclusion already evident from earlier studies by Jeffreys & Vicente (1957*a*, *b*) and Molodensky (1961). For a core model in which the Adams–Williamson law holds, i.e. $N^2(r) = 0$ (equation

Table 8.1. *Summary of contributions to the theoretical period of the Chandler wobble*

| | Model 1 ($N^2=0$) | | Model 2 ($N^2=3.8\times10^{-7}$) | |
|---|---|---|---|---|
| Mantle and core (M. L. Smith 1977) | 403.6 | | 405.2 | |
| Mantle anelasticity | | | | |
| $Q_m=300$ Constant-$Q$ model | | 3.9 | | |
| $Q_m=600$ Constant-$Q$ model | | 1.8 | | |
| Frequency-dependent-$Q$ model | | 7.6 | | |
| Ocean equilibrium theory | | 27.4 | | |
| Non-equilibrium ocean tide | | | | |
| $Q_w=20$; $\varepsilon=15°$ | | 27.6 | | |
| $Q_w=40$; $\varepsilon=8°$ | | 27.5 | | |
| $Q_w=80$; $\varepsilon=4°$ | | 27.4 | | |
| Total period | $Q_m=300$ | $Q_m=600$ | $Q_m=300$ | $Q_m=600$ |
| $Q_w=20$ | 435.1 | 433.0 | 436.7 | 434.6 |
| $Q_w=40$ | 435.0 | 432.9 | 436.6 | 434.5 |
| $Q_w=80$ | 434.9 | 432.8 | 436.5 | 434.4 |
| $Q_w=\infty$ | 434.9 | 432.8 | 436.5 | 434.4 |
| Observed period (Jeffreys 1968*a*) | | $434.3\pm2.2$ | | |

2.3.1), Smith finds a Chandler period of 403.6 d for the solid Earth (model 1, table 8.1). Model 2 in table 8.1 is based on a quite strongly stably stratified fluid core for which $N^2=3.8\times10^{-7}$, not an entirely plausible value (figure 2.3). Only for such large values does the Chandler period differ from that of model 1 by more than 2 d.

In a complete treatment of the problem, dissipative processes in the core and mantle and electromagnetic interactions should also be included. These do not, however, appear to be important and they are investigated separately below (see also Sasao, Okamoto & Sakai 1977; Sasao, Okubo & Saito 1979).

## 8.2 Dissipation in the solid Earth

### 8.2.1 *Energetics of the Chandler wobble*

The kinetic energy $E_k$, relative to axes fixed in space, is

$$E_k = E_k^r + \int_M \sum \omega_i h_i \, dM + \tfrac{1}{2} \sum_{jk} \omega_i I_{jk} \omega_k,$$

where

$$E_k^r = \tfrac{1}{2} \int_M u_i u_i \, dM$$

is the kinetic energy due to velocity $u_i$ of particles relative to body-fixed axes, rotating with angular velocity $\omega_i$. The relative angular momenta $h_i$ and the $I_{jk}$ both refer to the body in its deformed state. The elastic energy $E_e$ is

$$E_e = \tfrac{1}{2} \int T_{ij} e_{ij} \, dV,$$

where $T_{ij}$ are the stresses and $e_{ij}$ the corresponding strains. For elastic deformation, the appropriate stress–strain relation is given in section 2.1.1 or

$$T_{ij} = \tfrac{1}{3} K e_{mm} \delta_{ik} + 2\mu e'_{ij},$$

where $e'_{ij}$ is the deviatoric stress tensor. Hence,

$$E_e = \int (\tfrac{1}{6} K e_{mm} e_{nn} + \mu e'_{ij} e'_{ij}) \, dV.$$

Analogously to the preceding problem of the elastic yielding of the mantle, the strains are proportional to the $\omega_i$ (equation 8.1.5) and the elastic energy can be expressed as

$$E_e = \beta_{E_e}(\omega_1^2 + \omega_2^2)$$

where the factor $\beta_{E_e}$ is a function of the elastic moduli, the functions $y_i$ defined in section 2.1, and the magnitude of the perturbing potential. The gravitational energy, $E_g$, due to perturbations in the gravity field associated with the deformation, is also proportional to $(\omega_1^2 + \omega_2^2)$ and completes the expression for the total energy of a rotating, deformed body. Complete expressions for $E_k^r$, $E_e$ and $E_g$ are given by Kovach & Anderson (1967) (see also Kaula (1963, 1964) for similar problems). Merriam (1976) has evaluated these energy integrals for an Earth model subject to tidal and centrifugal forces. To evaluate the $Q$ of the Chandler wobble, we require the total energy that is dissipated during the decay of the wobble and the peak elastic energy stored in the motion (equation 2.2.1). Of the total energy, the main term is that part of the kinetic energy associated with the change in the rigid body rotation. In the initial state

$$2E_k^i \simeq A(\omega_1^2+\omega_2^2)+C\omega_3^2,$$

and in the final state, after the wobble is completely damped,

$$2E_k^f \simeq C\omega^2.$$

As angular momentum is conserved

$$A^2(\omega_1^2+\omega_2^2)+C^2\omega_3^2 \simeq C^2\omega^2$$

and the change in energy is

$$\begin{aligned} E = E_k^i - E_k^f &= \tfrac{1}{2}A(1-A/C)(\omega_1^2+\omega_2^2) \\ &= \tfrac{1}{2}AH\Omega^2(m_1^2+m_2^2). \end{aligned} \tag{8.2.1}$$

In equating the angular momenta in the two states, the contributions due to the deformation have been ignored. This is adequate for the present. To $E$ must be added the change in energy associated with the deformation. The more detailed calculation of Merriam & Lambeck gives

$$\begin{aligned} E &\simeq 0.9AH\Omega^2(m_1^2+m_2^2) = 0.9AH\Omega^2 m_0\,\mathrm{e}^{-2\alpha t} \\ &\simeq 1.2\times10^{34}\,m_0^2\ \text{erg}. \end{aligned} \tag{8.2.2}$$

For small damping, the rate at which this energy is dissipated is

$$\mathrm{d}E/\mathrm{d}t = 1.8\alpha AH\Omega^2 m_0^2\,\mathrm{e}^{-2\alpha t}$$

and the loss of energy during one cycle of the wobble is

$$\begin{aligned} \Delta E &= \oint \frac{\mathrm{d}}{\mathrm{d}t}E\,\mathrm{d}t = \frac{2\pi}{\sigma_0}\frac{\mathrm{d}E}{\mathrm{d}t} \\ &= 3.6\pi\alpha\sigma_0^{-1}AH\Omega^2 m_0^2. \end{aligned}$$

The $Q$, as defined in (2.2.1), is

$$Q^{-1} = \frac{1}{2\pi}\frac{\Delta E}{E_e} = 1.8\frac{AH\alpha}{\sigma_0\beta_{E_e}} = \frac{2\alpha}{\sigma_0}f_c, \tag{8.2.3}$$

(see Merriam & Lambeck 1979 for details) where

$$f_c = 0.9AH/\beta_{E_e} \simeq 9.$$

As pointed out by Stacey (1977), this definition is not that usually employed by astronomers to estimate the wobble decay time $\alpha^{-1}$. Instead, the wobble $Q$, $Q_w$, is usually defined as

$$1/Q_w = (1/2\pi)\,\Delta E/(E_k^i - E_k^f)\delta = 2\alpha/\sigma_0,$$

where $\Delta E$ is usually evaluated from (8.2.1) rather than (8.2.2). Comparing this with (8.2.3) gives

$$1/Q = f_c/Q_w \simeq 9/Q_w.$$

It is $Q$, and not $Q_w$, that must be compared with seismic estimates. For a nominal value of $Q_w \simeq 100$, $Q \simeq 11$. For $Q_w \simeq 600$, as proposed by Graber (1976) from analyses of short periods of latitude data, $Q \simeq 70$.

The rate at which the energy is dissipated follows from (8.2.2) as

$$dE/dT = 3.2 \times 10^{26} m_0^2 Q^{-1} \text{ erg s}^{-1}. \qquad (8.2.4a)$$

With $m_0 = 10^{-6}$ and $Q^{-1} = 11$

$$dE/dt \approx 2.1 \times 10^{13} \text{ erg s}^{-1}. \qquad (8.2.4b)$$

Despite, or perhaps because of, the smallness of the dissipation rate (equation 8.2.2) when compared with about $3\text{–}4 \times 10^{19}$ erg s$^{-1}$ for the energy dissipated in the semi-diurnal tides, the energy sink has not been conclusively identified. Oceans, mantle and core have all been proposed, dismissed, and reinstated at least once and there is still no unanimous view on the subject. Ocean dissipation, by shallow sea friction, was proposed by Jeffreys in 1920 but he subsequently dismissed this source as being inadequate. At first glance this is surprising in view of the importance of the oceanic dissipation at the higher tidal frequencies. Jeffreys' argument, also summarized by Munk & MacDonald, depends heavily on the assumption that the dissipation occurs by bottom friction in a few shallow seas but, as discussed in chapter 10, this may not be so. Munk & MacDonald conclude that the oceans cannot be ruled out and more recently Wunsch (1974*a*) confirmed the apparent importance of the oceans as the Chandler wobble energy sink. Imperfections in the mantle elasticity have generally been ruled out, since the seismic free oscillations of the Earth suggest a mantle $Q$ that is considerably higher than the comparable $Q_w$ deduced from the Chandler wobble. Jeffreys, in the sixth edition of *The Earth*, argues that the mantle may be an important energy sink after all, using as arguments (i) the apparent inadequacies of the core and oceans as sinks, (ii) the evidence for tidal dissipation in planets and satellites devoid of oceans and atmospheres, (iii) a creep law for mantle materials proposed by C. Lomnitz, although the applicability of this law to the mantle is considered questionable (see, for example, Ritsema 1972). Dissipation in the core by viscous damping has been proposed by Bondi & Gold (1955) but Munk & MacDonald suggest that the core viscosity would have to be $10^9$ P and higher for this mechanism to be effective; recent geophysical evidence points to a considerably lower value (chapter 2). Turbulent friction in the core also appears to be inadequate for dissipating the energy (Toomre 1966; Rochester 1970). We reexamine these mechanisms below. Electromagnetic damping, suggested as a possible mechanism by Jeffreys in 1956, is also ruled out as a possible sink of Chandler wobble energy (Rochester & Smylie 1965; section 9.2.4).

### 8.2.2 *Mantle dissipation*

Evidence for, and mechanisms of, dissipation in the mantle have been briefly discussed in chapter 2 and it appears that the dissipation can be characterized by a broad absorption band, encompassing the seismic frequencies. Whether or not this broad absorption band also encompasses the Chandler frequency is unknown at present. At the tidal periods of 12 h, the $Q$ of the solid Earth appears to be of the order of 100 or more (Lambeck 1977); only a lower limit can be established due to the inadequacy of the ocean tide correction. If the broad band continues beyond about 14 months, the dissipation of the pole tide energy cannot occur in the mantle unless the Chandler $Q$ is much higher than is generally thought. For if we take $Q = 200$, $Q_w$ must exceed 1500. Alternatively, with $Q_w \simeq 100$, $Q \simeq 11$, much lower than any comparable seismic $Q$ – but not outrageous if the recent frequency-dependent-$Q$ model proposed by Anderson & Minster (1979) is well founded.

The effect of dissipation on rotation can be modelled in the same way as in the tide problem (chapter 6). Now the rotational deformation is given by a degree-2 order-1 bulge that lags the potential of the centrifugal force by an angle $\varepsilon$. Thus if $\tilde{m}_1$, $\tilde{m}_2$ represent the position of a fictitious rotation axis that lags the actual axis by $\varepsilon$,

$$\begin{pmatrix} \tilde{m}_1 \\ \tilde{m}_2 \end{pmatrix} = \begin{pmatrix} \cos\varepsilon & \sin\varepsilon \\ -\sin\varepsilon & \cos\varepsilon \end{pmatrix} \begin{pmatrix} m_1 \\ m_2 \end{pmatrix}.$$

The potential of the anelastically deformed Earth follows from (3.4.1$c$). For small lag angles, $\cos\varepsilon \simeq 1$, $\sin\varepsilon \simeq Q_w^{-1}$ and

$$\begin{pmatrix} C_{21}^* \\ S_{21}^* \end{pmatrix} = -\frac{R^3}{3}\frac{\Omega^2}{GM}k_2\begin{pmatrix} 1 & Q_w^{-1} \\ -Q_w^{-1} & 1 \end{pmatrix}\begin{pmatrix} m_1 \\ m_2 \end{pmatrix}.$$

The excitation functions (3.4.3) become

$$\boldsymbol{\psi} = (k_2/k_o)(1 - \mathrm{j}/Q)\mathbf{m},$$

leading to equations of motion

$$\dot{\mathbf{m}}/\sigma_r - \mathrm{j}[(1 - k_2/k_o + \mathrm{j}k_2/k_o)Q_w]\mathbf{m} = 0,$$

whose solution is

$$\mathbf{m}(t) = \mathbf{m}_0\, \mathrm{e}^{\mathrm{j}\boldsymbol{\sigma}_0 t}, \tag{8.2.5$a$}$$

with

$$\boldsymbol{\sigma}_0 = \sigma_0[1 + \mathrm{j}(1/Q)k_2/(k_o - k_2)]; \tag{8.2.5$b$}$$

or

$$\mathbf{m}(t) = \mathbf{m}_0\, \mathrm{e}^{-\alpha t}\, \mathrm{e}^{\mathrm{j}\sigma_0 t} \tag{8.2.6$a$}$$

with

$$\alpha = [k_2/(k_o - k_2)]\sigma_0/Q_w \simeq \sigma_0/2Q_w. \tag{8.2.6$b$}$$

The difference between (8.2.3) and (8.2.6$b$) is due to the approximate treatment of the effect of deformation in the former.

The decay of the wobble can therefore be described simply by replacing the wobble frequency $\sigma_0$ by a complex frequency

$$\boldsymbol{\sigma}_0 = \sigma_0 + \mathrm{j}\alpha,$$

where $\sigma_0$ is the frequency in the absence of damping. The introduction of the complex frequency is equivalent to the introduction of a complex Love number

$$\mathbf{k}_2 = k_2(1 + \mathrm{j}\kappa),$$

with

$$\kappa = [(k_o - k_2)/k_2]\alpha\sigma_0 = Q^{-1} \equiv \varepsilon.$$

For $Q_w \simeq 30$, anelasticity modifies the period by only 1 part in 7000 or less than 0.1 d.

However, the assumption that a pure imaginary part of the elastic moduli provides an adequate account of anelastic behaviour is at variance with the interpretation given in chapter 2 of the observation that the $Q$ of the Earth is nearly frequency independent over a wide range of seismic frequencies. A more appropriate expression for the shear modulus may be that for the standard linear solid in which the real part contains a frequency-dependent contribution (equation 2.2.4). No rigorous calculation of this effect has yet been attempted but an order-of-magnitude estimate follows from (3.4.5) with the frequency-dependent $k_2$ of (2.2.10) (see also Dahlen 1979). That is, the change in period $P_e$ is

$$\mathrm{d}P_e/P_e = [k_2/(k_o - k_2)]\,\mathrm{d}k_2/k_2.$$

Here $P_e$ ($\simeq$403 d) represents the period of the elastic Earth without oceans and $k_2$ is the corresponding Love number

$$k_2 = k_o(1 - P_r/P_e),$$

where $P_r = 305$ d. If the absorption at seismic frequencies is also pertinent for the Chandler wobble then, for $Q \simeq 600$, $\mathrm{d}k_2/k_2 \simeq 1.4\%$, and $\mathrm{d}P_e \simeq 1.8$ d. For $Q \simeq 300$, $\mathrm{d}k_2/k_2 \simeq 3.0\%$, and $\mathrm{d}P_e \simeq 3.9$ d. If the $Q$-proportional-to-(frequency)$^{1/3}$ law is adopted, then, with the results of table 2.3, $\mathrm{d}P_e \simeq 8$ d. This is considerably more important than the modification of the period by the imaginary part of the elastic moduli or of $k_2$.

### 8.2.3 *Core dissipation*

The effect of viscosity on the fluid motion in a spherical rotating cavity has been studied by, amongst others, Stewartson & Roberts (1963), Roberts & Stewartson (1965) and Busse (1968), for the problem of a precessing shell. It appears that, apart from a thin boundary layer, the flow in the core will be essentially the same as the Kelvin–Poincaré flow for the inviscid case. If, within this boundary or Ekman layer, a quasi-geostrophic state exists, the Coriolis force is balanced by a viscous shear force and this determines the thickness $\Delta$ of the layer (see Greenspan 1968). Balancing these two forces,

$$|2\boldsymbol{\omega} \wedge \mathbf{u}_t| \simeq |\nu \nabla^2 \mathbf{u}_t|,$$

or

$$2\Omega u \cos\theta \simeq 2\nu u_t/\Delta^2, \tag{8.2.7}$$

where $\mathbf{u}_t$ is the relative velocity of the fluid past the boundary just below the Ekman layer and $\nu$ is the kinematic viscosity. Thus

$$\Delta \simeq (\nu/\Omega)^{1/2}. \tag{8.2.8}$$

Magnetohydrodynamic effects may complicate the structure of the Ekman layer but Hide (1970) shows that $\Delta$ is not sensibly modified. For $\nu \simeq 10^8$ cm$^2$ s$^{-1}$, $\Delta \simeq 12$ km; for $\nu \simeq 10^0$ cm$^2$ s$^{-1}$, $\Delta \simeq 1$ m.

The order of magnitude of the tangential shear stress exerted on the boundary by the fluid flow is

$$O\{\mathcal{S}_t\} \simeq \rho\nu\, \partial u_t/\partial r \simeq \rho\nu u_t/\Delta \simeq \rho\Omega u_t\Delta,$$

and the total torque acting on the shell is of magnitude

$$O\{L_0\} = \int \mathcal{S}_t \wedge \mathbf{r}\, ds = \rho\Omega\Delta \int \mathbf{u}_t \wedge \mathbf{r}\, ds.$$

The magnitude of $\mathbf{u}_t$ is $|\dot{m}R_b|$ and

$$O\{L_0\} = \tfrac{4}{3}\pi R_b^4 \rho(\Omega\nu)^{1/2}\dot{m} = L_0'|\dot{m}|. \qquad (8.2.9a)$$

The component $L_3$ is proportional to $\dot{m}_3$ and is negligible. With $\dot{m} = m_0\sigma_0 \simeq 2\times 10^{-13}$,

$$O\{L_0\} \simeq 10^{21}\nu^{1/2} \text{ g cm}^2 \text{ s}^{-2}. \qquad (8.2.9b)$$

The torque $\mathbf{L}$ acts in a direction opposite to the motion of the shell shifting in consequence of the wobble. That is,

$$\mathbf{L} = (L_0'/\Omega^2)\boldsymbol{\omega} \wedge \dot{\boldsymbol{\omega}},$$

$$= L_0' \begin{pmatrix} m_2\dot{m}_3 - \dot{m}_2 \\ \dot{m}_1 - m_1\dot{m}_3 \\ m_1\dot{m}_2 - m_2\dot{m}_1 \end{pmatrix} = L_0' \begin{pmatrix} -\dot{m}_2 \\ \dot{m}_1 \\ 0 \end{pmatrix},$$

and the excitation functions are

$$\begin{pmatrix} \psi_1 \\ \psi_2 \end{pmatrix} = \frac{L_0'}{\Omega^2(C-A)} \begin{pmatrix} -\dot{m}_1 \\ -\dot{m}_2 \end{pmatrix}.$$

The equations of motion become

$$\mathrm{j}\dot{\mathbf{m}}(1/\sigma_0 - \mathrm{j}\lambda) + \mathbf{m} = 0, \qquad (8.2.10)$$

with

$$\lambda = L_0'/\Omega^2(C-A) \simeq 0.34\nu^{1/2},$$

where $\sigma_0 = \sigma_r(1 - k_2/k_o)$ is the frequency in the absence of the torque. In the presence of the torque the frequency is further modified and damping occurs. That is, the solution of (8.2.10) is

$$m_0\, \mathrm{e}^{-\alpha t}\, \mathrm{e}^{\mathrm{j}\sigma_0' t},$$

and represents a damped motion with relaxation factor

$$\alpha = \lambda\sigma_0^2/(1+\lambda^2\sigma_0^2),$$

and whose period is modified from $\sigma_0$ to $\sigma_0'$ according to

$$\sigma_0' = \sigma_0/(1-\lambda^2\sigma_0^2).$$

The effect of core viscosity is to increase the period of the wobble. This is what we would expect since, as the coupling becomes tighter, more and more core mass partakes in the wobble until, in the limit, the entire core follows the shell. For all plausible values of core viscosity $\lambda^2\sigma_0^2 \ll 1$ and

$$\alpha \simeq \lambda\sigma_0^2 \simeq 10^{-14}\nu^{1/2}\,\mathrm{s}^{-1}.$$

Evidence for the core viscosity has been reviewed in section 2.3.2. For $\nu = 10^0\,\mathrm{cm}^2\,\mathrm{s}^{-1}$, $\alpha^{-1} \simeq 3\times10^6$ yr while for $\nu \simeq 10^8\,\mathrm{cm}^2\,\mathrm{s}^{-1}$, $\alpha^{-1} \simeq 300$ yr, still long compared with an observed relaxation time of about 30–60 yr. The kinematic viscosity must be of the order of $10^{10}\,\mathrm{cm}^2\,\mathrm{s}^{-1}$ if the core provides the sink for the Chandler energy. The effect of core viscosity on the period of the free nutation is entirely negligible, even for $\nu \simeq 10^{10}\,\mathrm{cm}^2\,\mathrm{s}^{-1}$. The uncertainty in the knowledge of $\nu$ is of no importance in so far as coupling of the wobble is concerned.

Elastic yielding of the mantle in response to the core motions is unlikely to modify significantly the above results (Bondi & Gold 1955). Nonlinear terms in the boundary layer cause the motion in the core to depart from that prescribed by the Kelvin–Poincaré flow, but Busse (1968) estimates that these are extremely small for the Earth's core and unlikely to change the above result. Density gradients in the core and the presence of the inner core are also unlikely to be important here, since Smith's (1977) analysis suggests that the flow in the outer core is not seriously modified by these factors. Unless the core viscosity is as high as $10^{10}\,\mathrm{cm}^2\,\mathrm{s}^{-1}$, laminar viscous friction is totally inadequate for damping the Chandler wobble.

The flow stirred by the relative motion along the core–mantle boundary may be turbulent rather than laminar, due to the roughness of the boundary or to instabilities in the laminar boundary layer. The condition for turbulence is that the Reynolds number

$$Re \simeq u_t l/\nu \gg 1,$$

where $l$ is the characteristic scale of the turbulence. If, following Rochester (1970), we take $u_t \simeq 0.1\,\mathrm{cm\,s}^{-1}$ (inferred from the geomagnetic westward drift), $l \simeq 10^3$ km (of the order of magnetic anomalies), then $\nu \ll 10^7$ cm$^2$ s$^{-1}$ for turbulence to occur. The Reynolds stress arising from eddy friction is of the order

$$\mathscr{S}_t \simeq \alpha\rho u_t^2,$$

where $\alpha$ is typically 0.1–0.01 or even smaller (Toomre 1966). Then with $u_t = 0.1$ cm s$^{-1}$, $\mathscr{S}_t \simeq 0.1$–$0.01$ dyn cm$^{-2}$. Integrating over the boundary will give an upper limit to the torque on the base of the mantle of $10^{16}$–$10^{17}$ g cm$^2$ s$^{-2}$ compared with about $10^{21}\nu^{1/2}$ g cm$^2$ s$^{-2}$ for the magnitude of the viscous laminar torque (8.2.9). Clearly turbulent friction is an even less effective mechanism than laminar friction. This argument does assume that typical core velocities are similar to the drift rate of the anomalies in the magnetic field. The actual velocities may be higher if their wavelengths and periods are sufficiently short for the associated magnetic field not to penetrate to the surface. But these core velocities would have to reach values of tens of centimetres per second in order for turbulent viscous coupling to become important.

## 8.3 The pole tide

The variable rotation of the Earth sets up a small tide in the oceans, referred to by Munk & MacDonald (1960) as the pole tide, which is normally considered as an ocean curiosity of no real importance. For geophysics, however, the pole tide is of some interest; it significantly modifies the period of the wobble from what it would be if the Earth were without oceans, and it may provide the sink for the damping of this wobble. The actual tide set up in the oceans by the centrifugal force can be expressed in a general way by equations (6.2.1) for the lunar or solar tide. That is,

$$\xi(\phi, \lambda; t) = \sum_u \sum_v \sum_{+}^{-} D_{uv}^{\pm} \cos(\sigma t \pm v\lambda - \varepsilon_{uv}^{\pm}) P_{uv}(\sin\phi), \quad (8.3.1)$$

in which the amplitude will depend on the magnitude of the wobble. Only the coefficients of degree 2 and order 1 in this expansion contribute to the excitation function and to the wobble but, in view of the small amplitude of the observed pole tide, usually not more than a few millimetres, its global structure is unknown. It is not possible to estimate the coefficients $D^{\pm}_{21}$ and $\varepsilon^{\pm}_{21}$ from the few available observations. It is not even known whether the pole tide follows an equilibrium theory or not, and while this may not be very important in discussing its effect on the period, it is a key question in the discussion of the dissipation.

### 8.3.1 *Evidence for the pole tide*

Tide records, read hourly, are precise to about 1 cm and, if the measuring error is random, monthly mean tidal amplitudes should be better than a few tenths of a millimetre, although Rossiter (1962) suggests that the monthly means can be considered accurate only to 3 mm. More severe than observational errors will be the consequence of background meteorological and oceanic noise, and all analyses of low-frequency tides suggest that the noise level at the Chandler frequency may approach 1 cm for records of 25-yr duration and longer (see, for example, the analysis by Miller & Wunsch 1973). Thus the search for the pole tide, and in particular the estimation of the important 2, 1-harmonic, becomes a marginal and ambiguous undertaking. In spite of this, a number of attempts have been made to separate this tide from the background noise. Haubrich & Munk (1959) carried out the first detailed and objective search for the pole tide using spectral analyses. In their study of 16 stations, six of which lie along the Dutch coast, they found evidence for the pole tide for the Netherlands group, Swinemunde and Marseille, marginal evidence for Bombay and no evidence at all for Brest, Buenos Aires, Baltimore, Portland, Seattle, Wazima and Hososima. Comparing the observed pole tide amplitudes for the Netherlands group with the amplitudes predicted by an equilibrium theory, Haubrich & Munk found that the former were significantly greater than their equilibrium values. Such enhancement has also been observed in the Baltic Sea by Maksimov & Karklin (1965). Tide records for Canadian waters indicate the presence of a pole tide with apparently some local enhancement occurring along the eastern seaboard (Holland & Murty 1970). More recently, Miller & Wunsch (1973) (see also S. P. Miller 1973) have analysed tide records of 66 stations for which observations exist for 25 yr and longer. They found some evidence for a worldwide pole tide such as at stations at Honolulu, Ketchikan, St John and Bangkok, but there are too few ports with reliable and lengthy records to give an indication of the structure of the global tide. Certainly the amplitude and phase information for the stations facing the open seas, away from areas of possible local magnification, does not permit the 2, 1-harmonics to be estimated. Currie (1975) has analysed 126 station records using the maximum-entropy method with similar inconclusive results: the evidence for the pole tide is worldwide but its global geometry remains hidden in the noise. This conclusion must also be drawn from the study by Hosoyama, Naito & Sato (1976), who searched for a latitude dependence of the pole tide phase and amplitude. They do find a hint of a lag of the tide behind the equilibrium tide for high northern latitudes and a lead in lower northern latitudes. But (i) the uncertainties of the lags are large ($\approx 50°$), (ii) the result is dominated by the Baltic and North Sea stations, and (iii) for many of their stations, Currie's (1975) study indicates a tide signal that lies below the noise level. This appears to be particularly so for the stations responsible for the phase leads. Unfortunately Hosoyama *et al.* do not tabulate their results for each station, and a more detailed evaluation of their lag angles is not possible. Despite the more powerful analysis techniques available since the time of Haubrich & Munk's study, the conclusion remains much the same: description of the pole tide by an equilibrium theory cannot be confirmed or refuted from the available amplitude observations alone.

Miller & Wunsch (1973), Currie (1975) and Hosoyama *et al.* (1976) all confirm the earlier observations by Haubrich & Munk (1959) and by Maksimov & Karklin (1965) that the pole tide in the North, Baltic and Bothnia seas is unusually strong; at Delfzijl and other Dutch stations the tide is about five times its equilibrium value. In particular Miller & Wunsch's study indicates an increasing amplitude of the pole tide with longitude along the Dutch and Danish coasts (figure 8.1), suggesting the presence of a relatively strong boundary current and hence strong dissipation. The much smaller slope along the Baltic coastline is indicative of smaller pole tide currents and of less dissipation. Thus, while there is little indication on the nature of the global pole tide, there is observational evidence for relatively important tides in some local areas.

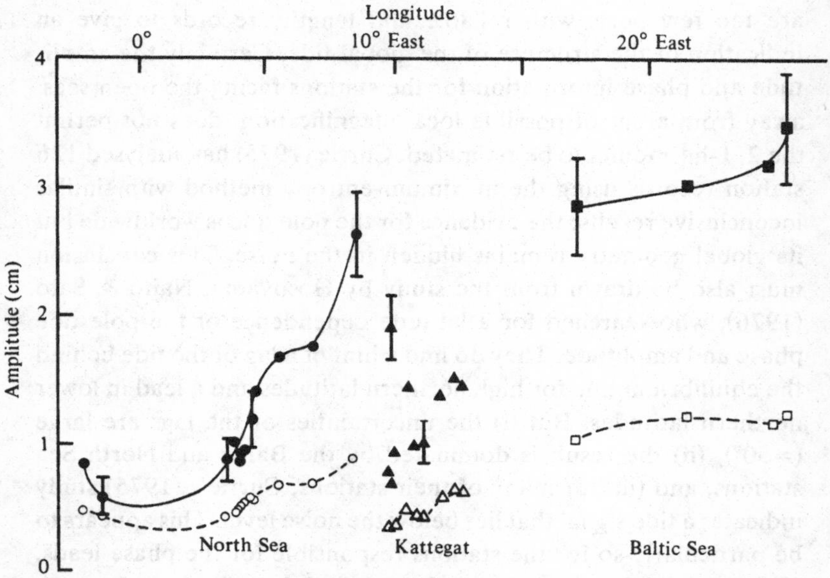


Figure 8.1. The observed amplitude of the pole tide for stations in the North and Baltic seas (solid symbols). The noise level is indicated by the open symbols (from Miller & Wunsch 1973).

Published information on the lag of the pole tide with respect to the equilibrium tide is even more sparse than the amplitude information. Holland & Murty (1970) indicate that the observed tide in western Canadian waters lags the equilibrium tide by approximately 20°, while Miller & Wunsch (1973) find that for the North European stations the variation in phase with longitude is roughly the same as predicted by the equilibrium theory, but that the observed tide lags behind the latter by some 30° or 40° (see also Wunsch 1974*a*). Jessen (1964) analysed sea level records of nine stations and found phase lags that range from 130° to −35°. The average lag, taking the three Baltic–North Sea stations as one, is 50°. His lag for the Baltic–North Sea stations is, however, about three times larger than found by either Miller & Wunsch (1973) or Hosoyama *et al.* (1976). In view of this limited observational evidence, it is tempting to appeal to theoretical considerations to determine to what extent the pole tide follows an equilibrium theory, but the theoretical justification is also inconclusive (section 6.2.1).

### 8.3.2 *The equilibrium pole tide and period of Chandler wobble*

The relevant part of the equilibrium ocean tide raised by the variable centrifugal force is given by (6.4.8) and the excitation functions are (section 4.2)

$$\left.\begin{matrix}\psi_1\\ \psi_2\end{matrix}\right\} = \frac{3}{10}\frac{\rho_w}{\bar{\rho}}\frac{1}{k_o}(1+k_2-h_2)X_{\text{wobble}}\left\{\begin{matrix}m_1A_1+m_2A_2\\ m_1B_1+m_2B_2\end{matrix}\right.$$

$$=\psi_0''\frac{\sigma_r}{\sigma_e}\left\{\begin{matrix}m_1A_1+m_2A_2\\ m_1B_1+m_2B_2\end{matrix}\right., \tag{8.3.2}$$

with $\psi_0''\simeq 0.030$. This excitation allows for the yielding of the crust under the variable ocean load and for the first-order effect of this yielding on the tide itself; $\sigma_e$ is the frequency of the wobble for an elastic Earth in the absence of the oceans. The equations of motion (3.2.6) are

$$\left.\begin{aligned}\frac{1}{\sigma_e}\frac{dm_1}{dt}-m_1\frac{\sigma_r}{\sigma_e}\psi_0''B_1+m_2\left(1-\frac{\sigma_r}{\sigma_e}\psi_0''B_2\right)=0,\\ \frac{1}{\sigma_e}\frac{dm_2}{dt}-m_1\left(1-\frac{\sigma_r}{\sigma_e}\psi_0''A_1\right)+m_2\frac{\sigma_r}{\sigma_e}\psi_0''A_2=0,\end{aligned}\right\} \tag{8.3.3}$$

and their solution has the form

$$\left.\begin{aligned}m_1&=M_1\cos\sigma_0 t,\\ m_2&=M_2\sin(\sigma_0 t+\beta).\end{aligned}\right\} \tag{8.3.4}$$

Substitution of (8.3.4) back into (8.3.3) gives

$$\sigma_0^2=\sigma_e^2\left[1-\left(\frac{\sigma_r}{\sigma_e}\psi_0''\right)^2(B_1A_2-B_2A_1)-\frac{\sigma_r}{\sigma_e}\psi_0''(A_1+B_2)\right]. \tag{8.3.5a}$$

Also

$$\left.\begin{aligned}\beta&=\tan^{-1}[(\sigma_r/\sigma_e)\psi_0''B_1],\\ M_1/M_2&=(\cos 2\beta)^{1/2}.\end{aligned}\right\} \tag{8.3.5b}$$

From (6.4.8*b*) and table 4.2

$$\left.\begin{aligned}\psi_0''^2(B_1A_2-B_2A_1)&=1.98\times10^{-3},\\ \psi_0''(A_1+B_2)&=9.00\times10^{-2},\end{aligned}\right\} \tag{8.3.6}$$

and, with $\sigma_r=2\pi/305$, $\sigma_e=2\pi/403.6$,

$$\sigma_0=0.936\sigma_e,\qquad \beta\simeq 3°,\qquad M_1/M_2\simeq 1.$$

The motion remains essentially circular. The principal effect of the oceans on the wobble is to decrease the frequency of the wobble. If we adopt M. L. Smith's (1977) value for $\sigma_e=2\pi/(403.6$ sidereal days) then $\sigma_0 = (2\pi/431.0)$ and the oceans increase the period by 27.4 d. Munk & MacDonald estimate an increase in period of about 33 d but their analysis is defective for a number of reasons, most important being their neglect of the modification of the ocean tide by the solid Earth yielding. Dahlen (1976) has given a more refined development of the equilibrium tide by including the complete loading effect; that is, by solving (6.4.7) for the harmonic terms $\xi_n$. He finds a lengthening of the period of 27.6 d.

### 8.3.3 *Non-equilibrium tide and dissipation*†

The equilibrium theory and the limited observational evidence indicate that the Chandler wobble does set up a small oscillation in the oceans, and any friction at the ocean–crust interface, along internal surfaces, or elsewhere in the ocean, will oppose the motions of the solid Earth; the wobble will be gradually damped out at a rate that is determined by the lags in the tide expansion (8.3.1). The problem is similar to that of dissipation of energy in the shorter-period tides raised by the Moon and Sun, in which dissipation is fully described by the same harmonics in the ocean tide as in the potential that actually raises the tide (chapter 10). Variations in the tide topography described by other harmonics do not contribute to the dissipation. This is also true for the pole tide, and dissipation can be evaluated by modelling the tide as the 2, 1-harmonic in the equilibrium tide but lagging the potential of the centrifugal force by an amount $\varepsilon$.

For small $\varepsilon$ the wobble equations follow from (8.3.3) and page 206 as

$$\left.\begin{aligned} \frac{1}{\sigma_e}\frac{\mathrm{d}m_1}{\mathrm{d}t} + (m_2 - \varepsilon m_1)\left(1 - \frac{\sigma_r}{\sigma_e}\psi_0'' B_2\right) = 0 \\ \frac{1}{\sigma_e}\frac{\mathrm{d}m_2}{\mathrm{d}t} - (m_1 + \varepsilon m_2)\left(1 - \frac{\sigma_r}{\sigma_e}\psi_0'' A_1\right) = 0 \end{aligned}\right\}, \qquad (8.3.7)$$

where we have assumed, on the basis of the previous equilibrium solution, that the wobble remains very nearly circular. The solution of (8.3.7) is of the form

$$\mathbf{m} = \mathbf{m}_0\,\mathrm{e}^{\sigma_0 t}$$

† See also Dickman (1979).

where the complex wobble frequency is given by

$$\boldsymbol{\sigma}_0 \simeq \sigma_e\left[1-\frac{1}{2}\frac{\sigma_r}{\sigma_e}\psi_0''(A_1+B_2)+\mathrm{j}\varepsilon\right]. \tag{8.3.8a}$$

The first two terms represent the frequency for the elastic Earth plus equilibrium ocean tide. The departure of this tide from equilibrium modifies the real part of the frequency from $\sigma_0$ to

$$\sigma_0\left[1-\varepsilon^2\left(\frac{\sigma_e}{\sigma_0}\right)^2\right]^{1/2}$$

or

$$\sigma_0[1-1.17\,\varepsilon^2]^{1/2} \tag{8.3.8b}$$

The relaxation time is

$$\tau=\alpha^{-1}=\varepsilon/\sigma_e$$

and with (8.2.6*b*)

$$Q_w^{-1}=2(\sigma_e/\sigma_0)\,\varepsilon. \tag{8.3.9}$$

In this discussion the non-equilibrium pole tide has been modelled as a bulge equal in amplitude to the equilibrium tide and lagging the centrifugal force potential. If the amplitude of this bulge exceeded the equilibrium value by a factor $\delta$

$$\boldsymbol{\sigma}_0 \simeq \sigma_e\left\{1-\delta\left[\frac{1}{2}\frac{\sigma_r}{\sigma_e}\psi_0''(A_1+B_2)+\mathrm{j}\varepsilon\right]\right\}, \tag{8.3.8c}$$

and the consequences on period could be significant if globally $\delta$ was much different from unity. As the Chandler wobble period is far from the periods of the slowest gravity models in the oceans (see, for example, Platzman 1975) it is improbable that $\delta$ will be very different from unity apart from possible local magnification.

The lag angle $\varepsilon$ must be considered as an equivalent lag angle representing the lag of the combined solid and ocean deformation. It is similar to the equivalent lag angles introduced in section 10.2 for the lunar tides. To relate $\varepsilon$ to the lag angles in the ocean tide we can equate the centrifugal force potentials due to the combined elastic deformation and the ocean tide expansion (8.3.1) to that due to motion of the fictitious axis $\tilde{\mathbf{m}}$ (page 206). The result is

$$D_{21}^{-} \sin \varepsilon_{21}^{-} = -\frac{5}{12\pi} \frac{R\Omega^2}{G} \frac{k_2}{1+k_2'} m_0 \varepsilon$$

$$D_{21}^{+} = 0.$$

The amplitude of the $D_{21}^{-}$ term is given by the equilibrium theory as

$$D_{21}^{-} \simeq 0.25 \text{ cm}$$

and the lag angle of the 2,1 harmonic in the ocean tide is

$$\varepsilon_{21}^{-} \simeq 1.16 \times 10^7 \, m_0 \varepsilon.$$

The very limited observation of the pole tide suggests that the global lag may be of the order 10–30°. This, with $m_0 \simeq 10^{-6}$, leads to $\varepsilon \simeq 1$–$2.5°$, $Q_w = 30$–$10$ and to an increase in the Chandler period of only from 0.1 d to 0.5 d. For $Q_w \simeq 100$ as suggested from the wobble analysis, $\varepsilon \simeq 0.3°$ and the change in period is negligible.

As for the $M_2$ tide, much of the discussion on the dissipation of the pole tide has centred on the rate of bottom friction in shallow seas. Wunsch (1974*a*) has constructed a simple dynamical model of the pole tide for the North Sea that consists of a rectangular basin open at, and deepening to, the north. This model qualitatively describes the increasing amplitudes observed along the Dutch and Danish coastlines and the requirements for this to occur are apparently (i) a broad opening of the sea to the open ocean, (ii) strong bottom slope in the shallow water, and (iii) meridional walls to support the eastern boundary currents. Wunsch suggests that other seas such as the Sea of Okhotz, Yellow Sea and Bering Sea, traditionally important sources for the dissipation of the short-period tides, may behave like the North Sea. Wunsch models the Baltic by a circular, flat-bottomed, shallow basin with a narrow opening, and finds that this qualitatively describes the observed rise and fall with an essentially constant amplitude and phase over the entire basin. Other shallow enclosed seas such as the Hudson Bay may behave in this manner, while open shallow seas such as the Patagonia Shelf would not be expected to have a pronounced pole tide. Wunsch's North Sea model leads to a dissipation of $8 \times 10^{11}$ erg s$^{-1}$, although he emphasizes that this value is extremely model dependent and that the actual dissipation could be an order of magnitude less. Extrapolation to all of the world's shallow seas yields a total dissipation of the order of $4\times10^{13}$ erg s$^{-1}$ and possibly as low as $4\times10^{12}$ erg s$^{-1}$. These estimates may be further reduced by perhaps a factor of 2 or 3 since the three conditions iterated above for the establishment of strong tide currents are not met in all the shallow seas. These values must be compared with the astronomical estimate (8.2.4) or $2\times10^{13}$ erg s$^{-1}$ for $Q_w\simeq100$ or $\simeq7\times10^{13}$ erg s$^{-1}$ for $Q_w\simeq30$.

Dissipation in shallow seas as important as the calculation by Wunsch suggests, implies that the pole tide currents in a few shallow seas would need to exercise a strong influence on the global tidal lag. For the short-period tides this appears improbable and it is unlikely that the shallow sea dissipation is very important. We must look for other mechanisms that give a more uniform dissipation, either along the coastlines or throughout the oceans. Some such mechanisms are discussed in chapter 10 in relation to the short-period tides but at present, while a number of possibilities exist, their quantitative evaluation is still lacking.

### 8.3.4 *The Chandler wobble period*

The various contributions to the Chandler wobble period are summarized in table 8.1. The total theoretical period, assuming that dissipation occurs in the oceans, ranges between 432.8 d and 436.7 d for the various ocean, mantle and core model combinations, compared with Jeffreys' (1968*a*) observed period of $434.3\pm2.2$ sidereal days. The observed period is compatible with a neutrally stratified core, a mantle $Q$ of about 300 and a wobble $Q_w$ of 80 with the dissipation occurring in the oceans. The required lag angles of the pole tide are small, of the order of 5°, and the present sea-level data are inadequate to detect this globally. The observed period is not inconsistent with dissipation in the mantle rather than oceans, provided that $Q$ follows the (frequency)$^\alpha$ rule. Then the wobble period, for $N^2=0$, is $(403.6+7.6+27.4)$ or 438.6 d. Jeffreys' observed wobble period gives a Chandler wobble $k_2$ of $0.280\pm0.003$, in general agreement with other estimates of $k_2$. Oceans, however, reduce this value to $k_2=0.230$, significantly different from that computed from seismic data using static theory. This is due mainly to the dynamic core contribution, with mantle anelasticity accounting for a few per cent of this difference.

Numerous authors have suggested that the Chandler wobble period fluctuates with time such that, when the wobble amplitude is larger than average, the period is longer than average (see, for example, Proverbio, Carta & Mazzoleni 1972). The only part of the Earth that could cause an amplitude-dependent period is a non-equilibrium ocean, since the wobble period is a function of the phase lag (equation 8.3.8). Thus the larger the amplitude, the greater the pole tide currents, and the greater the dissipation and lag – assuming that non-linear ocean bottom friction is a dominant mechanism. Hence the expected relation between wobble period and amplitude is similar to what Proverbio *et al.*'s analysis suggests, although the predicted variation is small.

An increase in the non-equilibrium amplitude of the pole tide (equation 8.3.8*c*) with increasing wobble amplitude could give rise to the observed effect although it is not at all clear how this can be treated independently of a change in lag angle: an increase in the ratio of the non-equilibrium to equilibrium tide will also result in an increase in lag angle.

## 8.4 Excitation of the Chandler wobble

### 8.4.1 *Seismic excitation*

The question of the influence of earthquakes on the Earth's rotation has a long history. Kelvin in 1876 speculated on what the effects of sudden changes in the solid Earth's mass distribution would have on the as yet unobserved polar motion. J. Milne in 1893, soon after Chandler's discovery, suggested that there may be some relation between polar motion and seismic activity, while J. Larmor estimated the effect of tectonic activity on the secular polar motion. Later, G. Cecchine, in 1928, also suggested that there may be a correlation between seismic activity and the amplitude of the Chandler wobble, an observation that appeared to be based essentially on the rapid increase in the wobble amplitude observed during the years 1906–1908, just after several years of intense seismic activity. More recently Stoyko & Stoyko (1969), Myerson (1970),

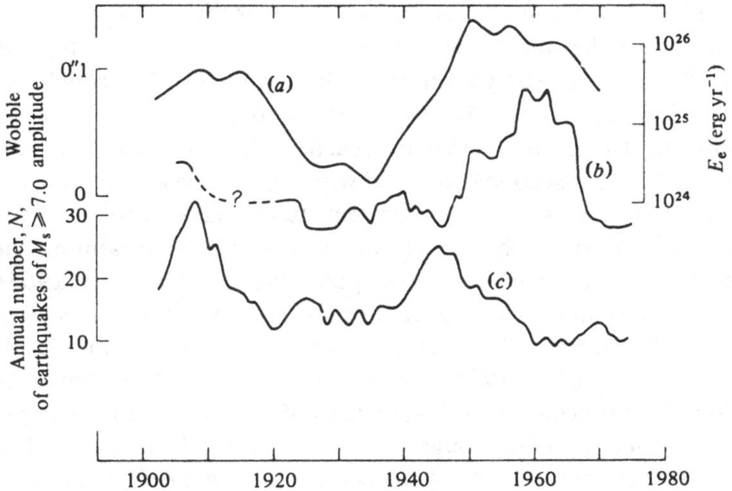


Figure 8.2. Comparison of (*a*) the amplitude of the Chandler wobble, (*b*) the elastic energy, $E_e$, released by earthquakes (5-yr running averages), and (*c*) the annual number, *N*, of earthquakes of $M_s \geq 7.0$ (5-yr running averages). From Kanamori (1977*a*).

Whitten (1971) and Anderson (1974) have stressed this similarity between the amplitude of the polar motion and earthquake activity, either in terms of the number of earthquakes or in terms of the seismic energy released. Figure 8.2 illustrates the comparisons based on the recent data by Kanamori (1977*a*). These qualitative comparisons do not establish the truth of the seismic excitation hypothesis and, as Myerson and Anderson emphasize, these similarities could indicate that both seismic activity *and* the variable polar motion have a common source mechanism.

We wish to investigate the relation between the Earth's rotation and earthquake activity. Specifically we wish to test the hypothesis, here referred to as the seismic excitation hypothesis, that large earthquakes excite the Chandler wobble. Since the annual amount of seismic energy released is of the order of $10^{26}$ erg yr$^{-1}$, while the amount of energy required to maintain the Chandler wobble is given by (8.2.4*a*) as $10^{22} Q^{-1}$ erg yr$^{-1}$, or less than $10^{21}$ erg yr$^{-1}$, this hypothesis is *a priori* not unreasonable.

Munk & MacDonald investigated the hypothesis quantitatively, by assuming that the displacements associated with an earthquake could be represented by a simple block fault model. They concluded that earthquakes could not excite the wobble: a block of 30-km depth and 100 km by 100 km in area, lifted by 1 m, shifted the pole by only $10^{-5}$ arc second compared with the observed wobble of the order of $10^{-1}$ arc second. Subsequent studies have, however, ruled out the validity of this simple block model as representing the displacement fields of large earthquakes. Work by Chinnery (1965), Press (1965) and others indicated that the dislocation fields associated with large earthquakes were more extensive than previously recognized. Press's (1965) results, in particular, show that if an offset of 10 m occurs on a dip-slip fault of length 600 km and depth 30 km, corresponding roughly to the March 27, 1964 Alaskan earthquake, the horizontal displacement at 1500 km from the fault is about 5 mm. Clearly, a large amount of movement of material accompanies major earthquakes.

The recognition of the importance of the displacements far from the seismic event led Mansinha & Smylie (1967) to re-open the question of whether or not large earthquakes can excite the Chandler wobble. Using a homogeneous half-space approximation, they computed the displacement field and mapped it on to a sphere to estimate the changes in the Earth's inertia tensor and the corresponding changes in the direction of the rotation axis. Finding an excitation at least $10^3$ greater than that obtained by Munk & MacDonald, they concluded that earthquakes can indeed be important in exciting the wobble. One year later they presented evidence for correlations between the changes in the motion of the instantaneous pole path and major earthquakes (Smylie & Mansinha 1968). Assuming that the change in $\psi$ can be expressed by the step function, (4.3.1) and figure 4.1 give the subsequent motions of the inertia and rotation axes. Plotted as a function of time, the pole path appears as a series of circular arcs with breaks in the curvature occurring at the time of the earthquakes. By fitting such arcs to the observed polar motion data, Smylie & Mansinha tried to establish the occurrence of the principal breaks in the pole path, and to correlate these with known events. A more rigorous examination of the astronomical data by Haubrich (1970) showed that the significance of this correlation is marginal at best (but see also the rebuttal by Mansinha & Smylie 1970): a small modification in the analysis procedure produces large changes in the result and, in particular, the largest breaks in the pole path do not correlate at all with known earthquakes. This lack of clear correlation is not surprising since the expected changes in the direction of the principal axis are of the same order as or smaller than the noise in the astronomical data. It is now generally recognized that the present astronomical data cannot be used in this way to test the seismic excitation hypothesis because (i) of noise in the data, and (ii) the surface wave magnitudes are inadequate indicators of the seismic moment (see below) and the whole large-earthquake time series has to be considered. Only during the last 10 yr has the pole path been determined with some reliability by the BIH but no sufficiently large earthquakes have occurred in this interval.

A more convincing test of the hypothesis is to compute the seismic excitation function and to determine whether or not this is sufficient to maintain the Chandler wobble. For this, a more complete model is required for the dislocation field than that used by Mansinha & Smylie. Ben Menahem & Israel (1970) extended the dislocation theory to a homogeneous non-gravitating sphere and, later, both Smylie & Mansinha (1971) and Dahlen (1971*b*) extended the theory to more realistic Earth models by allowing for the liquid core, for self-gravitation, and for radial variations in the elastic moduli and density. All three studies concluded that seismic activity could play an important role in exciting the Chandler wobble; Smylie & Mansinha found nearly an order-of-magnitude increase in the pole displacement compared with their 1967 results. An essential difference in the developments by Smylie & Mansinha and by Dahlen lies in the treatment of the liquid core and of the core–mantle boundary conditions. Considerable controversy has surrounded this aspect of the theory (see, for example, Dahlen 1971*a*, 1974; Pekeris & Accad 1972; Saito 1974; Wunsch 1974*b*, 1975*b*; Chinnery 1975; Crossley & Gubbins 1975; Dahlen & Fels 1978). This has tended to confuse the discussion on the seismic excitation hypothesis even though it does not appear to have important consequences for the computed excitation functions.†

† Mansinha, Smylie & Chapman (1979) have reviewed this problem in some detail.

Table 8.2. *Comparison of estimates of the shift of the inertia axis due to the 1960 Chilean and 1964 Alaskan earthquakes for the parameters given in table 8.3; (1) is the main shock and (2) is the fore-shock of the Chilean event*

| Event | | Wobble excitation: Magnitude (0″.01) | Wobble excitation: Direction (degree) | Author |
|---|---|---|---|---|
| Chile 1960 | (1) | 2.12 | 114 | M. L. Smith (1977) |
| | (2) | 2.80 | 118 | M. L. Smith (1977) |
| | (1) | 2.56 | 109 | O'Connell & Dziewonski (1976) |
| | (1) | 2.2 | 101 | Mansinha *et al.* (1979) |
| Alaska 1964 | | 0.72 | 201 | M. L. Smith (1977) |
| | | 0.73 | 202 | Dahlen (1973) |
| | | 1.11 | 203 | O'Connell & Dziewonski (1976) |

Some important errors, however, did lead Dahlen (1973) to review his earlier results and he now concludes that earthquakes cannot excite the Chandler wobble, a result substantiated by the independent study of Israel, Ben Menahem & Singh (1973). The various arguments for and against the seismic hypothesis up until 1973 are briefly reviewed in Kaula *et al.* (1973). Most recently O'Connell & Dziewonski (1976) again reinstate the hypothesis, although Kanamori (1976*a*, 1977*a*) argues that their evidence is inadequate. Finally, Mansinha, Smylie & Chapman (1979) corrected a shortcoming in their formalism for mapping the displacement field on to a sphere, and their results for the deformation caused by individual earthquakes now agree with those of Dahlen and Israel *et al.*

What is the reason for these fluctuating conclusions? In the early computations of the dislocation fields, widely different models have been used, but the independent studies of Dahlen (1973), Israel *et al.* (1973), O'Connell & Dziewonski (1976) and M. L. Smith (1977) for realistic Earth models are in agreement, in so far as the computation of the dislocation field is concerned (table 8.2). These

Table 8.3. *Source parameters for the 1960 Chilean and 1964 Alaskan earthquakes. The Chilean parameters are from Kanamori & Cipar (1974) and correspond to the two separate events of their two-event source model. (1) is the main shock and (2) the fore-shock. The Alaskan parameters are from Kanamori (1970)*

| Event | | Depth (km) | $\phi_o$ (degree) | $\lambda_o$ (degree) | Moment (dyn cm) | Strike (degree) | Dip (degree) | Slip (degree) |
|---|---|---|---|---|---|---|---|---|
| Chile 1960 | (1) | 25 | −38.5 | 285.5 | $2.7\times10^{30}$ | 170 | 10 | 80 |
| | (2) | 50 | −38.5 | 285.5 | $3.5\times10^{30}$ | 170 | 10 | 80 |
| Alaska 1964 | | 50 | 61 | 213 | $7.5\times10^{29}$ | 114 | 160 | 270 |

comparisons also suggest that any numerical errors in the computations are unimportant, while any conceptual errors in the theory, in particular in the treatment of the core–mantle boundary, also appear to be relatively unimportant. The real basis for the disagreement appears to lie in how the moments of earthquakes are estimated, how the cumulative excitation function is deduced from the inadequate historic seismic data, whether significant mass shifts occur immediately prior to or after the earthquake, and how the computed quantities are compared with the astronomical observations.

If the earthquake, occurring at time $t_s$, modifies the inertia tensor by $H(t-t_s)(\Delta I_{13}+\Delta I_{23})$, the solution of the equations of motion for $t>t_s$ follows as

$$\mathbf{m}(t)=\mathbf{m}_0\,\mathrm{e}^{\mathrm{j}\sigma_0 t}+\left[\frac{\Delta\mathbf{I}}{C-A}-\mathrm{e}^{\mathrm{j}\sigma_0(t-t_0)}\left(1+\frac{\sigma_0}{\Omega}\right)\frac{\Delta\mathbf{I}}{C-A}\right]X_{\text{wobble}}.$$

The second term represents the shift in the mean position of the rotation axis, and the last term represents the wobble. With $\Omega\gg\sigma_0$ and $X_{\text{wobble}}/(C-A)=\Omega/\sigma_0 A$,

$$\mathbf{m}(t)\simeq\mathbf{m}_0\,\mathrm{e}^{\mathrm{j}\sigma_0 t}+\frac{\Omega}{\sigma_0 A}\Delta\mathbf{I}-\mathrm{e}^{\mathrm{j}\sigma_0(t-t_0)}\frac{\Omega+\sigma_0}{\sigma_0 A}\Delta\mathbf{I}.$$

Thus, in the absence of other perturbing forces and of damping, the instantaneous rotation pole follows a circular path at an angular frequency $\sigma_0$ until, at the instant of the earthquake, a shift occurs in the position of the mean pole and the instantaneous axis begins to rotate about this new position. No instantaneous change in the position of the rotation pole occurs; there is only a change in the path it follows in time (figure 4.1). A succession of earthquakes, associated with sufficiently large changes in the excitation function, could then maintain the Chandler wobble and explain at the same time the secular drift observed in the mean pole. That is

$$m(t) = m_0\, e^{j\sigma_0 t} + \frac{\Omega}{\sigma_0 A} \sum_{t_s} \Delta\mathbf{I}_{t_s} - e^{j\sigma_0 t} \frac{\Omega}{\sigma_0 A} \sum_{t_s} \Delta\mathbf{I}_{t_s}\, e^{-j\sigma_0 t_s}, \tag{8.4.1}$$

where the summation is carried out over all events occurring at times $t_s$. The shift in the position of the mean pole of rotation is given by

$$S = \frac{\Omega}{\sigma_0 A} \sum_{t_s} \Delta\mathbf{I}_{t_s}, \tag{8.4.2}$$

and the modification of the Chandler wobble is

$$-\frac{\Omega}{\sigma_0 A} \sum_{t_s} \Delta\mathbf{I}_{t_s}\, e^{j\sigma_0 (t - t_s)}. \tag{8.4.3}$$

This formalism has been followed by most authors, and its validity rests on the assumption that the earthquake displacement field modifies the inertia tensor in a step-wise manner. But the problem may be more complicated. An earthquake represents a sudden release of elastic strain that has been accumulating slowly within the Earth prior to the seismic event. This build up of strain may give rise to a slow change in the inertia tensor, and the mean pole of rotation will migrate slowly during the interval between successive earthquakes. A more representative form of the inertia tensor variation may be the ramp function (4.3.8). If the time interval in which the strain builds up is long compared with the period of the Chandler wobble, the amplitude is not significantly modified, and when the strain is eventually released the mean pole will revert back towards its original position, and over a long time period there will not be any large-scale polar wandering as predicted by (8.4.2).

Calculations of the static displacement field resulting from an earthquake have been performed for realistic Earth models by

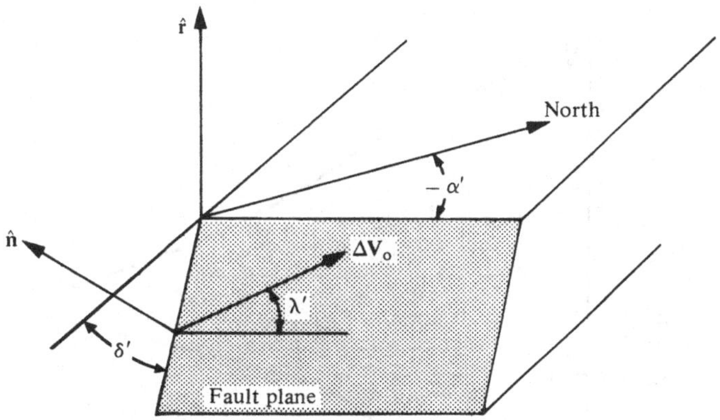


Figure 8.3. Definition of the fault plane parameters $\alpha'$, $\delta'$, $\lambda'$; $\mathbf{\Delta v}_o$ is the direction of motion, $\hat{\mathbf{n}}$ is the normal to the fault plane, $\hat{\mathbf{r}}$ is the radial direction.

Smylie & Mansinha (1971), Dahlen (1971*b*, 1973) and Israel *et al.* (1973), using elastic dislocation theory. Expressions for $\Delta I_{ij}$ can be written in the form (Dahlen 1971*b*)

$$\left.\begin{aligned} \Delta I_{13} &= M_o \sum_{i=1}^{3} \Gamma_i(h) g_i(\phi_o, \lambda_o, \delta', \alpha', \lambda'), \\ \Delta I_{23} &= M_o \sum_{i=1}^{3} \Gamma_i(h) h_i(\phi_o, \lambda_o, \delta', \alpha', \lambda'). \end{aligned}\right\} \tag{8.4.5}$$

$M_o$ refers to the moment of the earthquake, defined as

$$M_o = \mu(r)[A\ \Delta v_o], \tag{8.4.6}$$

where $A$ is the area of the fault plane and $\Delta v_o$ is the average displacement on the fault; $\mu(r)$ is the rigidity of the material in the earthquake zone. The $\Gamma_i(h)$ are functions of the depth of the earthquake and depend on the radial variations of elastic moduli and density. The $g_i$ and $h_i$ are functions of the latitude $\phi_o$ and longitude $\lambda_o$ of the source, and of the parameters defining the orientation of the fault plane and the direction of motion; specifically, the parameters used are (figure 8.3)

$\alpha'$ = strike azimuth, measured anti-clockwise from North,
$\delta'$ = dip with respect to the Earth's surface,
$\lambda'$ = slip angle measured in the fault plane, anti-clockwise from horizontal.

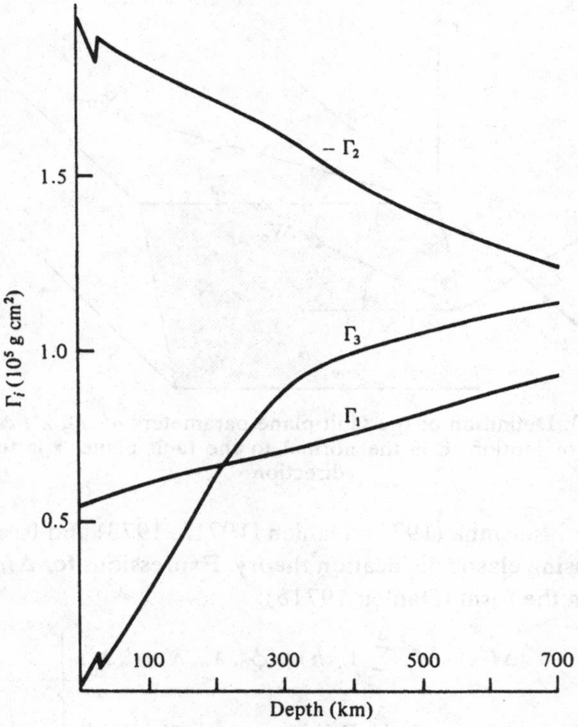


Figure 8.4. Functions $\Gamma_i(h)$ of (8.4.5) (from Dahlen 1973).

Dahlen (1973) has computed the $\Gamma_i(h)$ (figure 8.4) and these are in agreement with comparable functions computed by Israel *et al.* (1973). The $g_i$ and $h_i$ are

$$\left.\begin{aligned} g_1 &= a_1 \sin 2\phi_o \cos\lambda_o - a_2 \cos\phi_o \sin\lambda_o, \\ g_2 &= a_3 \sin 2\phi_o \cos\lambda_o, \\ g_3 &= a_4 \cos 2\phi_o \cos\lambda_o + a_5 \sin\phi_o \sin\lambda_o, \\ h_1 &= a_1 \sin 2\phi_o \sin\lambda_o + a_2 \cos\phi_o \cos\lambda_o, \\ h_2 &= a_3 \sin 2\phi_o \sin\lambda_o, \\ h_3 &= a_4 \cos 2\phi_o \sin\lambda_o - a_5 \sin\phi_o \cos\lambda_o, \end{aligned}\right\} \qquad (8.4.7)$$

with

$$\left.\begin{aligned} a_1 &= \sin 2\alpha' \sin \delta' \cos \lambda' + \tfrac{1}{2} \cos 2\alpha' \sin 2\delta' \sin \lambda', \\ a_2 &= \sin 2\alpha' \sin 2\delta' \sin \lambda' - 2 \cos 2\alpha' \sin \delta' \cos \lambda', \\ a_3 &= -\sin 2\delta' \sin \lambda', \\ a_4 &= \cos \alpha' \cos \delta' \cos \lambda' - \sin \alpha' \cos 2\delta' \sin \lambda', \\ a_5 &= \sin \alpha' \cos \delta' \cos \lambda' + \cos \alpha' \cos 2\delta' \sin \lambda'. \end{aligned}\right\} \qquad (8.4.8)$$

It is quite evident from these expressions that the efficiency with which an earthquake of given moment can excite the polar wobble depends very much on both the location of the source and the source parameters. Large, deep-focus events with intermediate dip angle and occurring at mid-latitudes will in general contribute most to $\Delta I_{ij}$.

O'Connell & Dziewonski (1976) evaluated the changes in the inertia tensor due to earthquakes using the development of Gilbert (1971) in which the displacements due to a body force **f** are expressed in terms of the normal modes of the Earth. This is similar to (8.1.4) except that the body force **f**, caused by the change in stress before and after the earthquake, is represented by a step function. In the static limit ($t = \infty$) the displacement vector is (Gilbert 1971)

$$\mathbf{d}(x, t) = \sum_n \mathbf{s}_n(x) \mathcal{T} \frac{1}{\omega_n^2}$$

where $\mathcal{T}$ is defined by (8.1.4*b*). Once **f** is evaluated from the stress drop associated with the event, the change in inertia tensor follows from (8.1.2). Only modes of degree 0 and 2 contribute to $\Delta I_{ij}$ and, of these, O'Connell & Dziewonski find that the fundamental mode and the first overtones contribute most; the spherical modes ${}_0S_0$, ${}_0S_2$ and ${}_1S_2$ account for more than 95% of the total amount. All of these modes have been observed, and the eigenfrequencies and excitation are in agreement with realistic models of the Earth's structure and of seismic source mechanisms. Shifts in the mean pole for the 1960 Chilean and 1964 Alaskan earthquakes are summarized in table 8.2 for the source parameters given in table 8.3. The results by Dahlen (1973), O'Connell & Dziewonski and M. L. Smith (1977) are comparable and provide a check on both theory and numerical errors.

### 8.4.2 *Seismic moments and fault parameters*

If the moments and fault parameters are known for the major earthquakes that have occurred during this century, it becomes a relatively simple matter to estimate the cumulative excitation function and to test the seismic excitation hypothesis. But reliable parameters are available for only a few recent events and much of the controversy surrounding the seismic hypothesis centres around the validity of the estimates of seismic moments for past earthquakes.

Seismic moments, defined by (8.4.6), are difficult to infer with precision from individual events. The extent of the fault and the total slip can best be estimated from observations of the ground deformations in the vicinity of the earthquake, but only for some quite recent large events is reliable information available. Seismic methods are more widely used, in which the moment is determined from the amplitudes of long-period waves. Typically the spectrum of seismic waves contains a relatively flat portion at low frequencies and it is the amplitude of this part of the spectrum that relates to the seismic moment (see, for example, Brune 1968). Long-period waves are relatively little affected by structural complexities, both at the source and throughout the mantle, and the seismic moment is usually well determined, particularly for large earthquakes, provided that sufficiently long-period information is available. This condition is satisfied only for recent events. For past earthquakes, the readily available observed quantity is the seismic magnitude $M_s$ and is defined in terms of seismic waves of periods of about 20 s. The relation between the moment $M_o$ and magnitude $M_s$ is complex. Under certain conditions, $M_s$ is proportional to the seismic moment (see, for example, Ben Menahem & Harkrider 1964) but, particularly for large earthquakes, $M_s$ does not necessarily provide a true measure of the deformation. Above a certain frequency, referred to as the corner frequency, the spectral amplitudes decrease rapidly. This frequency is a function of the earthquake size, and moves to lower frequencies when the source size increases. If the corner frequency falls below that of the 20-s waves used in defining $M_s$, the observed magnitudes will change only very slowly with the size of the event. A second difficulty in relating $M_o$ and $M_s$ occurs when an earthquake propagates for much longer than 20 s:

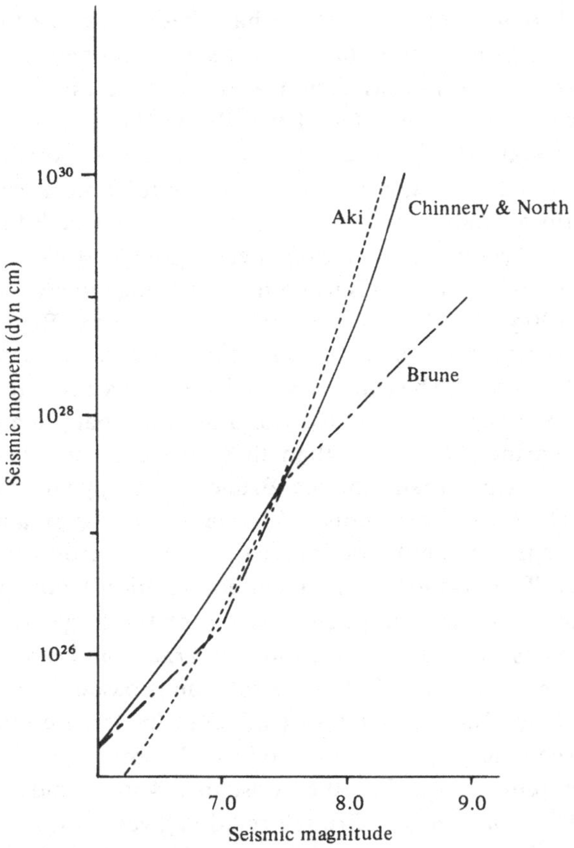


Figure 8.5. Moment–magnitude relations due to Brune (1968), Aki (1972) and Chinnery & North (1975).

measured $M_s$ values will be underestimated and this effect becomes increasingly more important the larger the earthquake. For these reasons spectral amplitude should be measured at periods of about 100 s rather than 20 s (Brune & King 1967), and in a study of some 20 events Brune & Engen (1969) find differences in magnitude of 0.5, depending on whether $M_s$ is estimated from 20-s or 100-s data. But for many past events only 20-s magnitudes are available from the existing catalogues (see also Kanamori & Anderson 1975*b*). Kanamori (1977*a*, *b*) has revised the magnitudes of many of the large past earthquakes.

Several semi-empirical relations have been developed in recent years to relate the magnitude $M_s$ to the seismic moment $M_o$. Three have been used in the discussion of the seismic excitation hypothesis: by Brune (Brune & King 1967; Brune 1968; Brune & Engen 1969), by Aki (1967, 1972) and by Chinnery & North (1975). Figure 8.5 illustrates these results. Brune's relation is based on a study of the excitation of 100-s surface waves, and leads to average cumulative slip rates that are in general agreement with slip rates estimated by other methods when the averaging interval is long, at least 70–100 yr (Davies & Brune 1971). According to Brune (1968) this relation is most reliable for earthquakes occurring along transform faults, although he considers that it could give estimates for $M_o$ that may be uncertain by a factor of 5 or more, particularly if it is used for earthquakes other than those on transform faults. For magnitudes larger than 7.5, the Brune curve appears to underestimate the seismic moments. Aki's curve is based on a model in which the spectral amplitudes decrease inversely proportionally to frequency. The Aki moment–magnitude relation tends to predict somewhat better the moments of some of the larger events but overestimates the moments for the two largest, and most studied, recent events. Chinnery & North's relation is based on a compilation of 87 published estimates of $M_o$ and, for large earthquakes, falls between the Aki and Brune models. The three relations lead to quite different results. For the Alaskan 8.4-magnitude event of March 1964, for example, Brune's relation gives a seismic moment of $2.6 \times 10^{28}$ dyn cm, that of Chinnery & North $1.0 \times 10^{30}$ dyn cm, and that of Aki $4.0 \times 10^{30}$ dyn cm. Values deduced from ground observations and from long-period surface waves range from $4.0 \times 10^{29}$ dyn cm (Plafker 1969) to $7.5 \times 10^{29}$ dyn cm (Kanamori 1970).

Fault parameters, like seismic moments, are known approximately for only some of the large earthquakes. P-wave first-motion studies allow one to determine the orientation of the fault plane and the direction of slip, if a sufficient number of observations from distant, well-distributed stations are available, but this condition is seldom satisfied. The data are more often appallingly crude. McKenzie & Parker (1967) have shown that, for the North Pacific, such studies give solutions that are in general agreement with the directions predicted by the plate tectonics hypothesis. This suggests

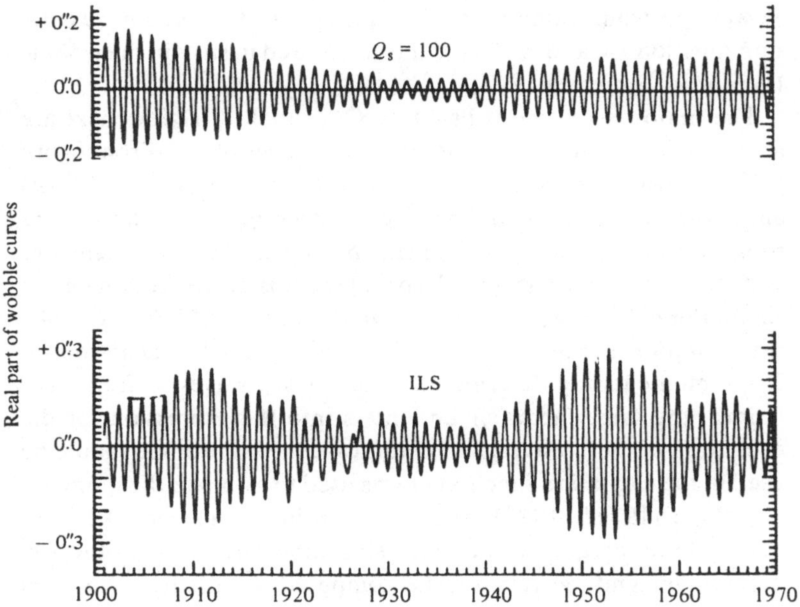


Figure 8.6. Changes of amplitude of the Chandler wobble from 1900 to 1970 due to the seismic excitation function of O'Connell & Dziewonski for $Q_w = 100$, compared with the astronomical results (lower curve).

that, if the relative plate motions are known, the fault strike, azimuth and the horizontal projection of the slip vector can be deduced, at least for shallow events. Specifically, as most large earthquakes all occur on plate boundaries, nominal parameters can be attributed to most of the events with the aid of the motions predicted by the plate tectonics hypothesis.

O'Connell & Dziewonski (1976) evaluated $\psi(t)$ from past events using the Chinnery–North moment–magnitude relation and fault plane and slip parameters estimated from the movements indicated by the plate tectonics hypothesis. The polar motion $m(t)$ is then estimated from (8.4.1) with an appropriate dissipation term. With $Q_w \simeq 100$ their simulated Chandler amplitude variations show some similarity to the observed Chandler amplitude variations (figure 8.6) and the magnitudes are comparable, indicating that seismic events may play an important role in exciting the Chandler wobble.

Power spectrum estimates of the time series of the seismic excitation function are about 25% of those required to sustain the wobble (Wilson & Haubrich 1977).

The majority of events listed by O'Connell & Dziewonski are assumed to be of shallow depth with dip angles of 20°. From figure 8.4 the most important of the $\Gamma_i(h)$ in (8.4.5) are $\Gamma_1(h)$ and $\Gamma_2(h)$ and, with (8.4.7) and (8.4.8), the changes in inertia tensor are roughly proportional to sin $\delta'$ or sin $2\delta'$. Hence, if the dip angles are systematically underestimated, so will the seismic excitation be too small. For an average value of $\delta' \simeq 40$ rather than 20°, for example, the excitation is about twice that given by O'Connell & Dziewonski. If the moments are determined from surface waves or free oscillations, the same factor sin $2\delta'$ enters into the expression for the excitation of these waves, and in this case the same angle should be used in the excitation function as was used to estimate the moment.

Dahlen (1971*b*, 1973) grouped the principal earthquakes into one of three categories and then, using either the Brune or the Aki moment–magnitude relation, he computed the moment for each event and deduced the pole shift:

$$|S| = (M_o)_s < \Gamma_i^2(h)(g_i^2 + h_i^2) >^{1/2}.$$

Attributing random phase factors exp (j $\sigma_0 t_s$) to each pole shift, the power spectrum $\mathcal{S}_\psi(\sigma)$ of the seismic excitation function was estimated assuming that it is the outcome of a two-dimensional random walk. From 1000 simulated two-dimensional random walks, Dahlen (1973) estimated $\mathcal{S}_\psi(\sigma) \simeq 4.6 \times 10^{-14}$ rad$^2$/(cycle yr$^{-1}$) using Aki's moment–magnitude relation. This yields a theoretical total Chandler power of $88 \times 10^{-14}$ rad$^2$ for $Q_w = 100$. As Dahlen points out, the main contribution to this power comes from a very few large seismic events whose moments may be overestimated by the Aki rule. By constraining the largest moments not to exceed $10^{30}$ dyn cm, the theoretical Chandler power is reduced to $3.7 \times 10^{-14}$ rad$^2$, while Brune's relation, always for a $Q_w$ of 100, gives $4.6 \times 10^{-16}$ rad$^2$. These values should be compared with the observed Chandler power of about $74 \times 10^{-14}$ rad$^2$ (page 103). Clearly the choice of the moment–magnitude relation is a crucial factor in deciding whether or not the seismic excitation hypothesis is acceptable.

From Dahlen's (1973) results, it is clear that if Aki's moment–magnitude curve is accepted as valid, earthquakes can indeed excite the Chandler wobble even if $Q_w$ is relatively low. O'Connell & Dziewonski come to a similar conclusion using Chinnery & North's moment–magnitude relation, although they find that the computed amplitudes tend to be too small by a factor of about 2 if $Q_w$ is 100. Since the Chinnery & North relation predicts somewhat lower moments of large events than does the Aki curve, these two studies yield compatible conclusions. The question is whether these moment–magnitude relations overestimate the moments of the largest events. Dahlen believes that they do. He argues that, as the largest observed seismic moment measured is $7.5 \times 10^{29}$ dyn cm, it is unlikely that any events of moment much greater than $10^{30}$ dyn cm have occurred in the past, but Aki's relation yields 18 events with moments in excess of $10^{30}$ dyn cm and these dominate the computed Chandler wobble power. Thus Dahlen prefers the results obtained with his modified Aki rule in which no moment exceeds $10^{30}$ dyn cm, and he concludes that this leads to an excitation function that is inadequate to maintain the Chandler wobble. In Chinnery & North's compilation of seismic moments, one event has a seismic moment greater than $10^{30}$ dyn cm: the Chilean earthquake, for which Kanamori & Cipar's (1974) revised estimate gives a moment of $2.7 \times 10^{30}$ dyn cm. The next-highest measured moment is that of the 1964 Alaskan event for which Kanamori (1970) and Abe (1970) estimate $7.5 \times 10^{29}$ dyn cm from studies of surface wave and free-oscillation amplitudes. It is perhaps curious that these two largest moments are for the most recent and most studied of large earthquakes, and it may not be unreasonable to suppose, as does Anderson (1975), that moments for some of the past large seismic events may be underestimated. McGarr (1976) has attempted to place an upper limit on the largest moments by computing the average annual amount that is required to accommodate the observed relative plate motions. He concludes that an upper limit of the order of $3 \times 10^{30}$ dyn cm can be placed on earthquakes: that events with large moments are unlikely to occur. But, as he points out, the plate velocities that he uses in this calculation are averages for at least the last 100 000 yr or longer and it may not be very meaningful to compare them with a seismic frequency–moment relation that is based on observations over about 60 yr only.

Kanamori (1976*a*) argues that many of the seismic moments used by O'Connell & Dziewonski are overestimated for two reasons. First, he points out that the magnitudes used by O'Connell & Dziewonski, taken from Duda's catalogue (Duda 1965), are not compatible with the Richter and Gutenberg scale used by Chinnery & North, and that the former overestimate the magnitudes on average by about 0.3 for the events used. This results in an overestimation of the moments by a factor of about 5. The same criticism can be made of Dahlen's study. Secondly, Kanamori argues that Chinnery & North's relation seriously overestimates the moments for several large events for which seismic moments are known. In fact, Chinnery & North's relation agrees with observed moments for the Kamchatka (1952), Chile (1960) and Alaska (1964) events, but overestimates the Assam (1950), Tokachi-Oki (1952) and Nankaido (1946) events (using the Richter–Gutenberg magnitude scale) by a factor of about 10. Kanamori believes that many of the other events are also overestimated and he concludes that earthquakes are inadequate for exciting the Chandler wobble. Kanamori (1977*b*) has summarized recent re-evaluations of the seismic surface wave magnitudes $M_s$ and seismic moments carried out by him and co-workers, by Okal (1976, 1977) and by Chen & Molnar (1977). He has considered all large events since 1900 and his list is relatively complete after 1920. The new values are based on (i) direct observations of long-period waves, (ii) an empirical relation between 100-s magnitudes and moments, and (iii) an empirical relation between the moment and the surface of the aftershock area. Figure 8.7 illustrates the $M_o$–$M_s$ relation. Clearly it is far more complicated than suggested by the simple moment–magnitude relations discussed earlier. Figure 8.7 also illustrates the uneven temporal distribution of the large earthquakes: 6 events with $M_o \geq 10^{29}$ dyn cm occurred in the 15 yr from 1950 to 1965 while only one occurred in the preceding 50 yr.

An observation by Kanamori & Cipar (1974) suggests that large aseismic motions may accompany some earthquakes. They observe for the 1960 Chilean event a slow deformation in the epicentral area prior to the major shock, and estimate that this precursory dis-

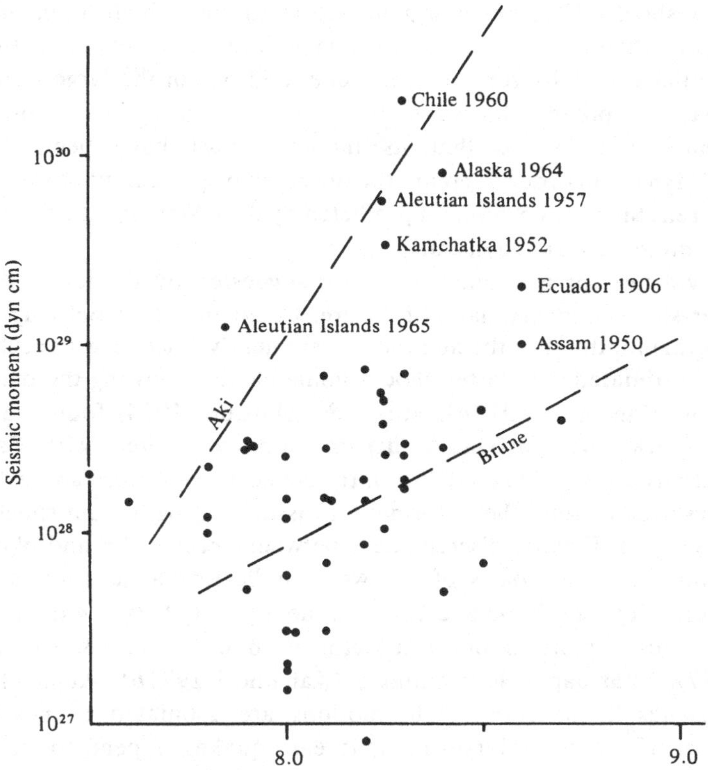


Figure 8.7. Seismic surface wave magnitude–seismic moment relations for large earthquakes since 1900 (data from Kanamori 1977*a*). The Aki and Brune relations are indicated.

placement has as large a moment as that associated with the main shock. Kanamori & Anderson (1975*a*) estimate that the total amount is of the order of $5 \times 10^{30}$ dyn cm. Evidence for the aseismic or *silent* events is very limited for other earthquakes: Dziewonski & Gilbert (1974) have reported stress release just before two deep earthquakes, and Thatcher (1974) finds aseismic slip associated with the 1906 San Francisco earthquake, giving a greater moment than that associated with the earthquake itself (see also Fukao & Furumoto 1975; Kanamori 1976*b*).

The so-called *Tsunami earthquakes*, discussed by Kanamori (1972), are also indicative of significant stress release just before the main shock. These events generate abnormally high Tsunamis despite their often small seismic magnitudes. An event in the Aleutians in 1946, for example, generated one of the largest and most widespread tidal waves in the Pacific during this century. Kanamori estimated that its moment must have been $5 \times 10^{28}$ dyn cm in order to create the observed displacement of water, whereas the seismic moment predicted by the Chinnery–North rule is a modest $2 \times 10^{27}$ dyn cm.

Evidence for aseismic slip is also suggested by studies of the after-shock area: estimates of the area based on after-shocks, up to several months after the main shock, are nearly twice as large as the area estimated from after-shocks immediately following the main event (Kanamori 1977*a*). Stuart & Johnston (1974) found that pre-shock and post-shock deformations associated with San Andreas events of magnitudes in the range 3 to 4.4 were about 10 times greater than the deformations associated with main shocks themselves. Finally, discrepancies between seismic slip and plate motions in many parts of the world indicate that aseismic slip occurs. Between India and Eurasia, the disparity between the two estimates of plate motion is a factor of about 3 (Chen & Molnar 1977). Near Japan it is about 5 (Kanamori 1977*b*). Along the Marianas Trench, the plate motions are estimated as nearly 10 cm $yr^{-1}$ but no large-moment earthquakes appear to have occurred at all during the last few centuries. If these aseismic slips occur over time intervals that are short compared with the Chandler wobble period, they could provide an important excitation mechanism. This has been stressed previously by M. A. Chinnery and F. A. Dahlen, in the discussion summarized in Kaula *et al.* (1973) and, more recently, by O'Connell & Dziewonski (1976). This also suggests that a useful measure, for comparison with the wobble, is the annual number, $N$, of earthquakes above a certain magnitude. This is shown in figure 8.2 together with the elastic energy $E_e$ released by large earthquakes. Both $N$ and $E_e$ show trends that are comparable with the wobble amplitude fluctuations.

### 8.4.3 *Atmospheric excitation*

The suggestion that changes in atmospheric mass distribution may not be strictly periodic and hence may excite the Chandler wobble was apparently first made by V. Volterra in 1895. Jeffreys in 1940, P. Rudnick in 1956, and Munk & MacDonald have also discussed this possibility without coming to any firm conclusions about the efficacy of the mechanism. Munk & Hassan (1961) concluded, from an evaluation of the atmospheric excitation function, that the atmosphere plays only a minor role in generating the Chandler wobble. More recently Wilson (1975) (see also Wilson & Haubrich 1976*a*) re-opened the discussion and concluded that the atmospheric role may be more important than previously suggested by Munk & Hassan.

The test of the atmospheric excitation hypothesis, that the temporal variability in the mass distribution of the atmosphere is sufficiently large to excite the wobble and to maintain it against dissipation, is relatively straightforward. The excitation function, computed from ground level pressure, has been discussed in chapter 7, and it suffices to compare the power in its spectrum at the Chandler frequency with the power required to maintain the observed wobble, but, as arises so frequently in discussions of the Earth's rotation, data are neither sufficiently reliable nor complete to permit an unambiguous interpretation. This is demonstrated by the different conclusions drawn by Munk & Hassan (1961) and by Wilson from analyses of largely the same data set.

The meteorological data set and the month-by-month excitation function $\boldsymbol{\psi}_A$ used by Wilson (1975) are the same as discussed in chapter 7 in relation with the annual wobble. The excitation function includes the mountain torques although the validity of this contribution is questionable (section 7.4.1). To test the atmospheric excitation hypothesis one can either (i) infer the polar motion $\mathbf{m}$ from the excitation function and compare it with the observed motion $\mathbf{m}$, or (ii) infer the excitation $\hat{\boldsymbol{\psi}}$. Munk & Hassan (1961) adopted the first procedure; Wilson adopted the second. Wilson uses the ILS/IPMS polar motion record for the years 1900–1970 and, after elimination of the annual term, $\hat{\boldsymbol{\psi}}$ is inferred using the relation (4.3.13*b*) for a discrete data set. Wilson finds no evidence for a significant overall correlation between $\hat{\boldsymbol{\psi}}$ and $\boldsymbol{\psi}_A$ but, near the Chandler frequency, there is a suggestion of correlation between the two (figure 8.8). This spectrum is for $Q_w = 100$ and $\sigma_0 = 0.8415 \times 2\pi \text{ yr}^{-1}$. Near the Chandler frequency the two functions are also in phase, and Wilson concludes that the coincidence of

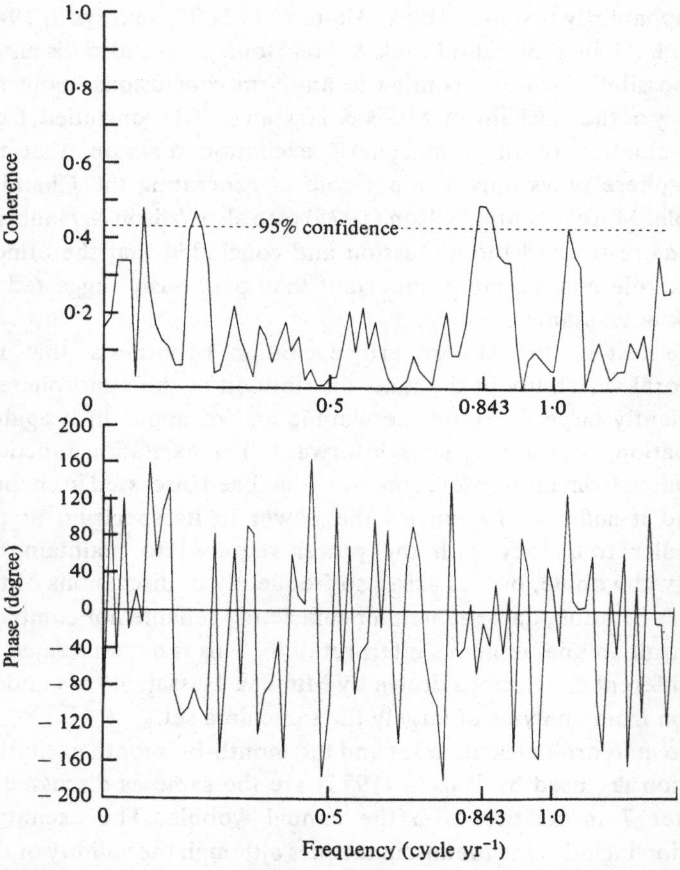


Figure 8.8. Coherence and phase between the atmospheric excitation function $\psi_A$ and the astronomically deduced excitation $\psi$ in the frequency range 0–1.2 cycle yr$^{-1}$ (from Wilson & Haubrich 1976*a*).

maximum coherence with minimum phase indicates that $\psi_A$ and $\hat{\psi}$ are correlated with 97% probability.

If this correlation is accepted, it does not establish the reality of the atmospheric excitation hypothesis: conceivably such a correlation could be a consequence of a pole tide in the atmosphere, and I. V. Maksimov has long maintained that such a tide exists (see Lamb 1972). Analyses of pressure data in the Baltic region and elsewhere indicate a marked 14-month oscillation (Cazenave,

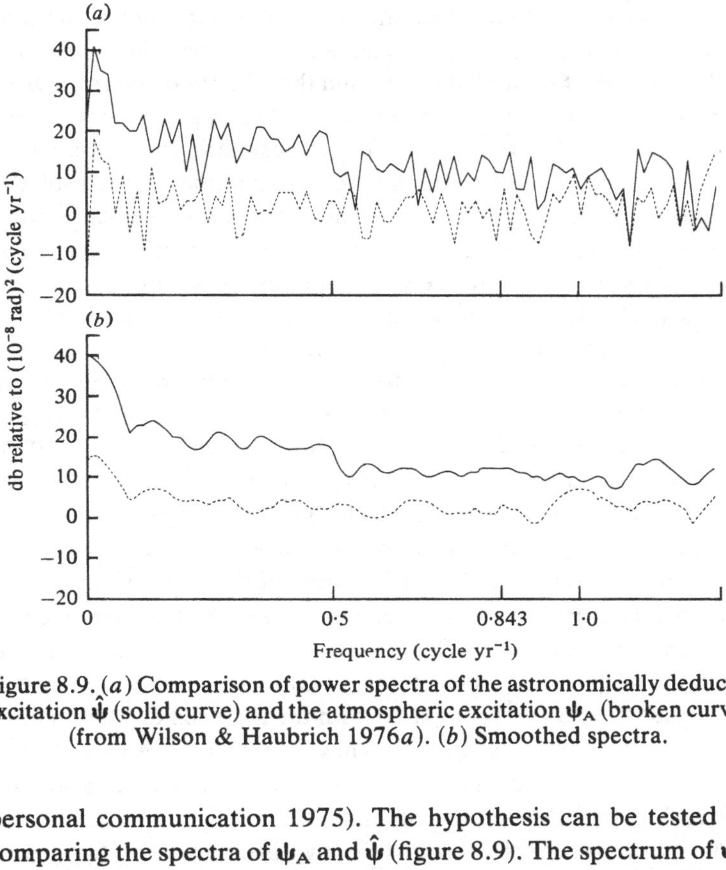


Figure 8.9. (*a*) Comparison of power spectra of the astronomically deduced excitation $\hat{\psi}$ (solid curve) and the atmospheric excitation $\psi_A$ (broken curve) (from Wilson & Haubrich 1976*a*). (*b*) Smoothed spectra.

personal communication 1975). The hypothesis can be tested by comparing the spectra of $\psi_A$ and $\hat{\psi}$ (figure 8.9). The spectrum of $\psi_A$ is approximately flat over all frequencies, down to the Nyquist frequency of 6 cycle yr$^{-1}$, with a value of about $2.0\times 10^{-16}$ rad$^2$/(cycle yr$^{-1}$). About the Chandler frequency, between 0.67 cycle yr$^{-1}$ and 1.0 cycle yr$^{-1}$, Wilson estimates an average power of $2.2\times10^{-16}$ rad$^2$/(cycle yr$^{-1}$), compared with $1.5\times 10^{-16}$ rad$^2$/(cycle yr$^{-1}$) found by Munk & Hassan. Part of this difference can be explained by the inclusion of the mountain torque contribution in Wilson's $\psi_A$, part may be due to the additional 20 yr of data, and a further part may be a consequence of an improved estimation process. The average power of $\hat{\psi}$ estimated by Wilson over the same frequency range is of the order of $14\times 10^{-16}$ rad$^2$/(cycle yr$^{-1}$) for the same $Q_w$ and $\sigma_0$ as above. This is about seven times greater than that contained in the meteorological excitation function. Munk & Hassan found that the latter was inadequate by a factor of about 10 and that the atmosphere plays an unimportant role in exciting the wobble. Wilson's comparison does not greatly improve the case for the atmospheric excitation hypothesis. He argues, however, that the astronomic estimate of the excitation function may be too high due to noise in the polar motion observations, and he suggests that the noise contribution may be as much as $5\times10^{-16}$ rad$^2$/(cycle yr$^{-1}$). But noise in the meteorological data will also lead to an overestimation of the spectrum of $\boldsymbol{\psi}_A$. Secondly, Wilson argues that the atmospheric excitation function may be underestimated due to an inadequate coverage of the pressure data over central Asia. Comparison of the annual pressure excitation function with that computed by Siderenkov (1973) indicates a $\psi_2$-component that may be too large by a factor of nearly 2. Thirdly, other meteorological and hydrological excitation functions may have some year-to-year variability and contribute to the Chandler excitation. If either or both of these omissions contribute to the excitation power spectrum at the Chandler frequency, in the same proportion to Wilson's $\boldsymbol{\psi}_A$ as they appear to do at the annual frequency, the total excitation will be $5–8\times10^{-16}$ rad$^2$/(cycle yr$^{-1}$) compared with the required excitation of at least $9\times10^{-16}$ rad$^2$/(cycle yr$^{-1}$). These *corrective terms* can be no more than orders of magnitude, and it would be unwise to draw any conclusions that depend heavily upon them.

A comparison of the form of the two excitation spectra (figure 8.9) also suggests that there is more to the wobble excitation than just the atmosphere, for while the atmospheric spectrum is white, the astronomical spectrum is definitely pink. This is indicative of step function changes with a variable and relatively long time interval between successive steps (section 4.3). Year-to-year variability in the seasonal functions may make a substantial contribution to the Chandler wobble excitation but it alone does not appear to be adequate to sustain the wobble.

### 8.4.4 *Conclusions*

The nature of the mechanism maintaining the Chandler wobble remains elusive. Seismic excitation alone appears inadequate, but

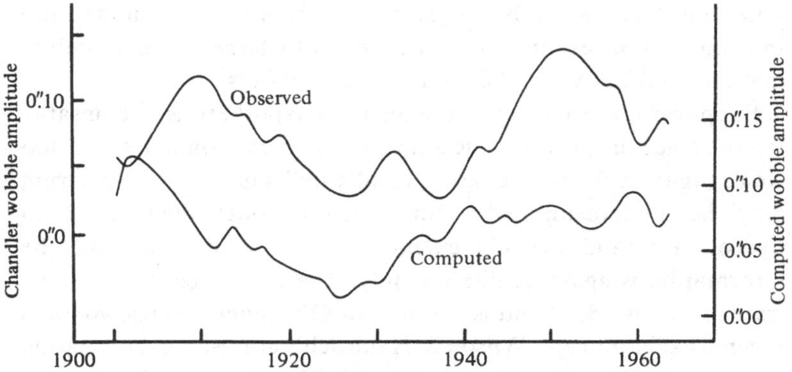


Figure 8.10. Fluctuations in the Chandler wobble amplitude as observed and as computed from the seismic and atmospheric data. Note that the vertical scales of the two curves have been displaced.

aseismic deformation may be important. Atmospheric excitation is also insufficient. Probably the most likely outcome is that a combination of seismic, aseismic and meteorological factors is responsible. Evidence in support of the seismic excitation hypothesis is inconclusive. Most of it is qualitative only and includes: (i) the similarity between the Chandler wobble amplitude fluctuations and some indices of seismic activity (figure 8.2), (ii) the ability to reproduce, in part, this amplitude fluctuation by O'Connell & Dziewonski's synthetic curve (figure 8.6), (iii) the suggestion that the astronomically deduced excitation spectrum (figure 8.9) increases slightly in power with decreasing frequency, in agreement with the expected spectrum (4.3.2) for successive step changes occurring at relatively long intervals. The use of Kanamori's revised moments, fault plane parameters $\lambda'$, $\delta'$, $\alpha'$ estimated from the plate tectonics motions, and the formalism (8.4.5) does not significantly improve the comparison. In particular, it has not been possible to reproduce the rapid change in phase observed in the pole path (figure 5.15) with the available excitation functions. A growing body of evidence indicates that aseismic deformation may be important, but it is not possible to test this hypothesis in detail since there is little understanding so far as to which past earthquakes have been associated with large aseismic deformations. An attempt to invert the wobble data for seismic moments – by assuming that the times and $\lambda'$, $\delta'$, $\alpha'$ of the largest events ($M_s \geq 8.0$) are known, and that only these events are associated with large aseismic deformation – has not yielded comprehensible results.

Support for the atmospheric excitation hypothesis is also unsatisfactory since the power in the atmospheric excitation function is too small (figure 8.9). Siderenkov's results indicate that this spectrum may be underestimated. Other non-seasonal fluctuations in hydrological and oceanic excitations may contribute, but the information is again qualitative only. The total excitation is illustrated in figure 8.10 and is the sum of O'Connell & Dziewonski's seismic excitation and Wilson & Haubrich's atmospheric excitation with the latter increased by a factor of 2. The agreement is not much better than that for the seismic excitation alone (see also Wilson & Haubrich 1977). Removal of the atmospheric contribution, before inverting the wobble path for the seismic moments, has not led to better results than before. It is not yet possible to infer parameters defining large seismic and aseismic events from past Chandler wobble data.

NINE

# DECADE FLUCTUATIONS

In this chapter I discuss the fluctuations in the Earth's rotation in the frequency range from about 0.2 cycle $yr^{-1}$ to very low frequencies. These are often referred to loosely as the *decade fluctuations*, although they cover a much longer-period range from, say, 5 yr to Newcomb's *Great Empirical Term* of 300 yr. The astronomical evidence for these fluctuations in both wobble and l.o.d. has been discussed in chapter 5, and it is in the l.o.d. that they are most pronounced and well above the observational noise level. Most of the discussion will, in consequence, concern the l.o.d. decade variations. Observed since the early nineteenth century, these changes are illustrated in figure 5.3 in the form of $m_3$ and $\dot{m}_3$. The characteristic time constant of the changes is 10–20 yr. Since the introduction of atomic time in 1955, the improved observations are indicative of a shorter time constant, about 5 yr. The wobble observations suggest a secular drift of the pole upon which an approximately 20-yr oscillation appears to be superimposed. But the reality of this term is open to question and may be a characteristic of the observing process itself, rather than of some geophysical excitation (see section 5.3).

The geophysical origin of the decade fluctuations has been subject to considerable debate and most discussions have centred around the possible role of the core. This may be attributed to a lack of adequate mechanisms originating in the more observable parts of the Earth, or to the inability to test in a satisfactory manner the various core mechanism hypotheses. But it is a fact that the core is the only sufficiently mobile part of the Earth with sufficient mass to modify the rotation by the observed amounts on the decade time scale. These core problems are collectively referred to as core–mantle coupling and deal with the degree to which motions in the core and motions of the mantle as a whole are related. They concern the extent to which mantle rotation is transferred to the core or, vice versa, the extent to which motions in the fluid core are transmitted to the mantle. Long-period changes in the distribution of water, ice and air are quite inadequate to explain the totality of the decade l.o.d. changes, although they may contribute partially. A third origin, a solar torque originating in the electromagnetic interaction between the Earth's magnetic field and solar particles, is totally inadequate.

## 9.1 Core–mantle coupling

The coupling of core and mantle motions may have several astronomical consequences, some of which have already been invoked in previous chapters. The dependence of the Chandler wobble period on the degree to which the core follows the mantle motion has been discussed in chapter 8. A second example is given by an external torque operating on the mantle; for example, a tidal torque (chapter 10) or wind torques (chapter 7). The rotational response to this torque will also depend on the degree to which the core follows the mantle. These are examples of passive coupling. A third example is the consequence of fluctuations in the flow field inside the fluid core due to, for example, magneto-hydrodynamic processes, for if the mantle is partly coupled to these motions, changes in its rotation will result. This is an example of direct coupling.

Several mechanisms have been proposed for coupling the motions of the core and mantle. *Pressure* or *inertial coupling*, a function of the ellipticity of the core–mantle boundary, received considerable attention in the nineteenth century by Hopkins, Kelvin, Hough and others. A rotation of the mantle about an axis inclined to the axis of figure brings into play an asymmetric distribution of pressure forces along the core–mantle boundary which entrain the core to follow the mantle rotation. This type of coupling has been discussed in relation to the Chandler wobble in chapter 8. It does not play an important role in any discussion on the decade fluctuations in l.o.d., unless the rotation axis of the core is very much misaligned with the rotation axis of the mantle. *Topographic coupling*, introduced by Hide (1969), is a variant of inertial coupling in which the boundary is assumed to be irregular. This *topography*† will modify both the flow of fluid past the boundary and the pressure distribution on the mantle. *Viscous coupling*, or viscous friction, can also transfer angular momentum from the core to the mantle or vice versa. Bondi & Lyttleton (1948) considered laminar viscous boundary layer friction and Toomre (1966) discussed turbulent boundary layer friction. Both mechanisms have also been discussed in relation to the Chandler wobble and do not appear to be important. *Electromagnetic coupling* aroused considerable interest in the early 1950s by the development of geomagnetic dynamo theories by E. C. Bullard and W. M. Elsasser and by Brouwer's re-analysis of the l.o.d. fluctuations on the *decade* time scale.

Rochester (1970, 1974) reviewed the various coupling mechanisms and their astronomical consequences. The strength of the interactions is given locally by a tensor $\mathcal{S}$ describing the stress on the liquid just below the core–mantle boundary. The net torque that this stress exerts on the mantle is

$$\mathbf{L} = \int_S \mathbf{R}_b \wedge \mathcal{S} \cdot \mathbf{e}_r \mathrm{d}S, \tag{9.1.1}$$

where $\mathbf{e}_r$ is the normal to the core–mantle boundary $S$ at $\mathbf{R}_b$. The sum of inertial $\mathbf{L}_I$, topographic $\mathbf{L}_T$, viscous $\mathbf{L}_V$ and electromagnetic $\mathbf{L}_E$ torques determines the mantle's rotational response. That is, according to equations (3.2.6), (3.2.7) and (4.2.6),

$$\mathrm{j}\dot{\mathbf{m}}/\sigma_0 + \mathbf{m} = [\mathrm{j}\mathbf{L}/\Omega^2(C_m - A_m)]X_{\mathrm{wobble}}, \tag{9.1.2a}$$

and

$$\dot{m}_3 = (1/\Omega C_m)L_3, \tag{9.1.2b}$$

where $A_m$ and $C_m$ are the moments of inertia of the mantle. The problem is to estimate the stress tensor $\mathcal{S}$ acting on the core–mantle boundary when both the forces and the physical properties of this region are very inadequately known.

### 9.1.1 *Electromagnetic core–mantle coupling*

Magneto-hydrodynamic theory predicts that in the Earth's fluid core any fluid motion will drag along with it the magnetic field, if the electrical conductivity of the core is high enough (see, for example,

† Hide has referred to the study of the core–mantle topography as core-phrenology.

Elsasser 1950; Roberts 1971). If the core is tightly coupled to the mantle there will be no drift of the magnetic field, whereas, if there is no coupling, the secular tidal decelerations of the mantle result in an eastward drift of the magnetic field. The observed westward drift is then considered as evidence for some coupling between the core and mantle (Bullard *et al.* 1950). Changes in the rate of the westward drift indicate a change in the differential rotation between the core and mantle, and imply an exchange of angular momentum between the core and mantle and a modification of the mantle's rotation. By a process of elimination, Munk & Revelle (1952) concluded that Brouwer's decade changes in the l.o.d. are probably a consequence of variations in the angular momentum of the core, and that electromagnetic forces are the most likely coupling mechanism. Vestine (1953) presented evidence for a correlation between the westward drift and the l.o.d., but such evidence is only marginal due to the uncertainty in the magnetic data observed at the Earth's surface (section 9.1.3). Theories of core–mantle coupling have been discussed quantitatively by Bullard *et al.* (1950), Takeuchi & Elsasser (1954), Elsasser & Takeuchi (1955), Rochester (1960, 1968), Kakuta (1961, 1965), Roden (1963), Rochester & Smylie (1965), Roberts (1972) and Yukutake (1972).

*Coupling model.* Bullard's model of the core (Bullard *et al.* 1950) is a convenient starting point for assessing the validity of the coupling mechanism. In this model, thermal convection, driven by radioactive heating, causes a radial flux of matter between the inner and outer parts of the core, outwards in some regions, inwards in others. In a rotating core such a motion must also be accompanied with a radial variation in angular velocity in order to conserve angular momentum, and material in the outer part of the core, $R_a < r < R_b$, will rotate more slowly than material in the inner part of the core $r < R_a$. Rochester (1960) refers to the sphere $r = R_a$ as the Bullard discontinuity; $\omega_1$ denotes the angular velocity of the inner core $r < R_a$, $\omega_2$ that of the outer core $R_a < r < R_b$ and $\omega_3$ that of the mantle. In the model $\omega_2 < \omega_3$ and $\omega_1 > \omega_2$. Within the core, a poloidal dipole field is arbitrarily supposed to exist and the relative motions of two highly conductive parts of the core in the presence of this field produce electrical currents and further magnetic fields.

Specifically, two toroidal quadrupole fields are produced which are symmetric about the rotation axis. First, a toroidal field $\mathbf{B}_{\mathrm{T}}^{(a)}$ is produced by the differential rotation at the discontinuity in the flow field at $r = R_a$. This field diffuses out of the interior of the core and leaks into the lower mantle ($r < R_c$), whose electrical conductivity $\sigma_{\mathrm{m}}$ is assumed to be small but finite. The upper mantle $R_c < r < R$ is assumed to have zero conductivity. The toroidal field does not penetrate beyond $r = R_c$; it cannot be observed at the Earth's surface. A second toroidal field $\mathbf{B}_{\mathrm{T}}^{(b)}$, is generated at the core–mantle boundary by the differential motion of the fluid past the mantle. The details of the generation of these fields need not be specified. It is sufficient that they exist, and reach and penetrate the core–mantle boundary. Figure 9.1 shows schematically the distribution of these two fields in the core and mantle; from Lenz's law they are of opposite sign since the relative slip at the two surfaces $R_a$ and $R_b$ is opposite. The currents $\mathbf{J}$ producing the toroidal fields penetrate the mantle and interact with the stationary dipole field $\mathbf{B}_{\mathrm{P}}$. The resulting Lorentz force $\mathbf{J} \wedge \mathbf{B}_{\mathrm{P}}$ exerts a torque

$$\mathbf{L} = \int_{V_{\mathrm{m}}} \mathbf{r} \wedge (\mathbf{J} \wedge \mathbf{B}_{\mathrm{P}})\, \mathrm{d}V, \qquad (9.1.3a)$$

on the mantle. The integral is over the conducting part of the mantle. The component $L_3$ about the rotation axis reduces to

$$L_3 = \frac{1}{4\pi} \int_{V_{\mathrm{m}}} r \cos\phi \, (\mathrm{curl}\, \mathbf{B}_{\mathrm{T}} \wedge \mathbf{B}_{\mathrm{P}})_\lambda \, \mathrm{d}V$$

$$= -\frac{1}{4\pi} \int_{S(R_b)} r \cos\phi \, B_{r,\mathrm{P}} B_{\lambda,\mathrm{T}} \, \mathrm{d}S, \qquad (9.1.3b)$$

(Rochester 1960, 1962). In (9.1.3$b$)

$$B_{\lambda,\mathrm{T}} = B_{\lambda,\mathrm{T}}^{(a)} + B_{\lambda,\mathrm{T}}^{(b)}.$$

The integral is taken over the core–mantle interface $R = R_b$.

At the core boundary the radial component of $\mathbf{B}_{\mathrm{P}}$ is

$$-2g_{10}(R/R_b)^3 \sin\phi, \qquad (9.1.4)$$

where $g_{10}$ is the dipole term in the magnetic potential observed at the Earth's surface. The torque $L_3^{(a)}$ created by $\mathbf{B}_{\mathrm{T}}^{(a)}$ tends to accelerate the mantle and to rotate it with the inner part of the core $r < R_a$, while the toroidal field $\mathbf{B}_{\mathrm{T}}^{(b)}$, being in the opposite sense,

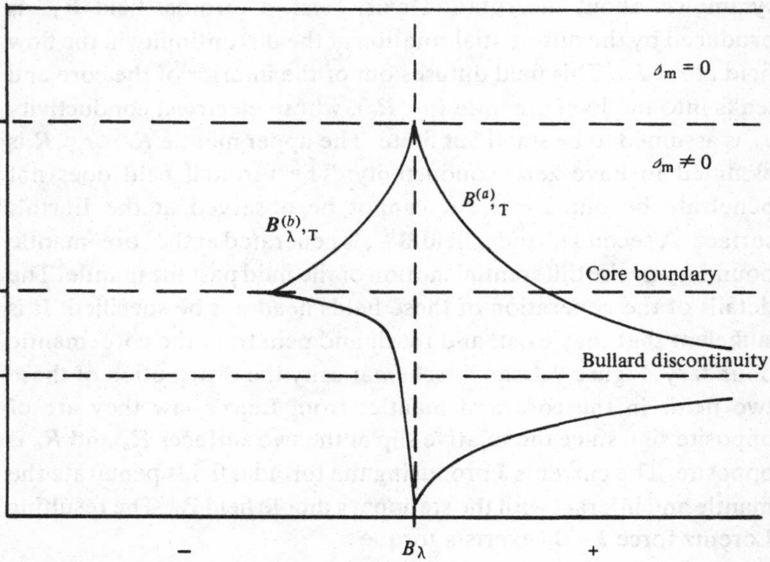


Figure 9.1. Steady-state toroidal fields in the core and mantle (after Rochester 1960).

creates a Lorentz torque $L_3^{(b)}$ that tends to decelerate the mantle and to rotate it with the outer part of the core $R_a < r < R_b$. The two toroidal fields are (Bullard *et al.* 1950; Rochester 1960; Rikitake 1966; Le Mouël 1976*b*)

$$B^{(a)}_{\lambda,\mathrm{T}} = \tfrac{2}{3}\mathcal{M}\left(\frac{R_b}{r}\right)^3 \frac{1-r^5/R_c^5}{1-R_b^5/R_c^5}\frac{\mathrm{d}P_{20}}{\mathrm{d}\phi}, \tag{9.1.5a}$$

$$B^{(b)}_{\lambda,\mathrm{T}_2} = -\tfrac{4}{3}\pi\sigma_{\mathrm{m}}(\omega_3-\omega_2)R_b^2 g_{10}\left(\frac{R}{r}\right)^3 \frac{1-r^5/R_c^5}{1+\frac{3}{2}R_b^5/R_c^5}\frac{\mathrm{d}P_{20}}{\mathrm{d}\phi}, \tag{9.1.5b}$$

in which $\mathcal{M}$ is the maximum value attained by the toroidal field $\mathbf{B}_{\mathrm{T}}^{(a)}$ at the core–mantle boundary. Writing the sum of (9.1.5*a*) and (9.1.5*b*) as

$$B_{\lambda,\mathrm{T}} = B^{(a)}_{\lambda,\mathrm{T}} + B^{(b)}_{\lambda,\mathrm{T}} = \Upsilon(r)r^2\,\mathrm{d}P_2/\mathrm{d}\phi, \tag{9.1.5c}$$

and with (9.1.4), the torque (9.1.3*b*) reduces to

$$L_3 = -\tfrac{4}{5}R^3R_b^2 g_{10}\Upsilon(r=R_b). \tag{9.1.6}$$

In the equilibrium state the net torque on the mantle vanishes. This requires that $\Upsilon(r=R_b)=0$, or, with (9.1.5) at $r=R_b$,

$$\omega_3-\omega_2=\frac{\mathcal{M}}{2\pi\sigma_m R_b^2 g_{10}}\left(\frac{R_b}{R}\right)^3\frac{1+\frac{3}{2}R_b^5/R_c^5}{1-R_b^5/R_c^5}. \tag{9.1.7}$$

With $(\omega_3-\omega_2)$ known from the westward drift, $\mathcal{M}$ is determined.

Consider a disturbance in the velocity field of the outer core. The strength of the magnetic field carried along by the fluid as well as the toroidal fields is modified; induced currents flow in the mantle so as to oppose the penetration of the field change. As the change diffuses out of the core, the mantle is subject to a modification of the Lorentz torques provided by the poloidal–toroidal field interactions. The net torque now is such that the resulting disturbance in $\omega_3$ reduces the production of excessive toroidal fields at the boundary and brings the core and mantle into the equilibrium (9.1.7) again. Consider a step function change in $\mathcal{M}$ at time $t_0$:

$$\Delta\mathcal{M}(t)=\Delta\mathcal{M}^0 H(t-t_0).$$

Provided that the diffusion of the toroidal fields into the mantle is rapid, the Lorentz torque results in a change in rotation according to (equation 9.1.2$b$)

$$\begin{aligned} C_m \mathrm{d}^2\omega_3/\mathrm{d}t^2 = \dot{L}_3 &= -\tfrac{4}{5}R^3R_b^2 g_{10}\dot{\Upsilon}(r=R_b)\\ &= \tfrac{8}{15}R^3 g_{10}[\Delta\dot{\mathcal{M}}(t)-\beta(\dot{\omega}_3-\dot{\omega}_2)], \end{aligned} \tag{9.1.8a}$$

with

$$\beta=2\pi\sigma_m g_{10}R_b^2\left(\frac{R}{R_b}\right)^3\frac{1-R_b^5/R_c^5}{1+\frac{3}{2}R_b^5/R_c^5}. \tag{9.1.8b}$$

Angular momentum of the Earth is conserved and any changes in rotation relate by

$$C_1\dot{\omega}_1+C_2\dot{\omega}_2+C_m\dot{\omega}_3=0,$$

where $C_1$ is the moment of inertia of the inner part of the core $r<R_a$ rotating with velocity $\omega_1$, and $C_2$ is the moment of inertia of the outer region of the core $R_a<r<R_b$ rotating with velocity $\omega_2$. Assuming that most of the inertia of the core is contained in the outer part,

$$\dot{\omega}_2\simeq-(C_m/C_c)\dot{\omega}_3$$

with $C_c=C_1+C_2$. There is no experimental evidence to suggest how the core may be divided into the two parts. Substituting into

(9.1.8*a*) gives

$$\frac{d^2\omega_3}{dt^2}+\frac{8}{15}\beta R^3 g_{10}\frac{C}{C_c C_m}\frac{d\omega_3}{dt}=\frac{8}{15}\frac{R^3}{C_m}g_{10}\Delta\mathcal{M}^0\delta(t-t_0) \tag{9.1.9a}$$

and

$$\Delta\omega_3(t>t_0)=\Delta\mathcal{M}^0[1-e^{-\alpha(t-t_0)}]C_c/C\beta. \tag{9.1.9b}$$

The angular velocity immediately after the initial disturbance decays with a time constant $\tau=\alpha^{-1}$:

$$\tau=\left[\frac{16}{15}\pi\sigma_m(g_{10})^2R_b^5\left(\frac{R}{R_b}\right)^6\frac{C}{C_cC_m}\frac{(1-R_b^5/R_c^5)}{(1+\frac{3}{5}R_b^5/R_c^5)}\right]^{-1}. \tag{9.1.10}$$

This is the time constant given by Bullard *et al.* (1950) (see also Rochester 1960; Rikitake 1966).

A change in the westward drift of $\Delta(\omega_3-\omega_2)$ will change the mantle velocity by (equations 9.1.7 and 9.1.9*b*)

$$\Delta\omega_3=(C_c/C)\Delta(\omega_3-\omega_2).$$

The average westward drift is of the order of 0°.2 $yr^{-1}$. Fluctuations of as much as 50% about this value have been reported but their significance is uncertain (section 9.1.3., figure 9.3). For a 20% change in the westward drift $\Delta\omega_3\simeq3.5\times10^{-13}$ and $\Delta m_3\simeq5\times10^{-8}$. This is of the same magnitude as the change observed in $m_3$ from 1900 to 1930. From (9.1.10) the time constant is about 20 yr for $\sigma_m\simeq2\times10^{-9}$ e.m.u. (or $2\times10^2$ $ohm^{-1}$ $m^{-1}$), $g_{10}=0.3$ e.m.u., $R_c=5.5\times10^8$ cm.

A potentially serious defect with the above model is the assumption that the magnetic fields diffuse instantaneously through the mantle. The time for a magnetic field to diffuse through a layer $R_b<r<R_c$ of material of conductivity $\sigma_m$ is given approximately as $4\pi\sigma_m(R_c-R_b)^2$. With $\sigma_m\simeq10^{-9}$ e.m.u., $R_c-R_b\simeq2000$ km, some 15 yr are required for the magnetic field to diffuse from the core–mantle boundary through the mantle. As $\sigma_m$ decreases with increasing radius this suggests that only the lowermost part of the mantle effectively contributes to the coupling, and this is also suggested by Roden's (1963) calculations and by Yukutake's (1972) model. Thus the boundary $r=R_c$ should have little influence on the time constant and Rochester (1960) shows that, when diffusion is taken

into account, a more appropriate form of the torque and time constant is

$$\left.\begin{aligned} L_3(t) &= \frac{8}{15} R^3 g_{10} \Delta \mathcal{M}^0 - \frac{16}{15} \pi \sigma_m (g_{10})^2 R_b^5 \left(\frac{R}{R_b}\right)^6 (\omega_3 - \omega_2) \\ \tau &= \frac{15}{16} \frac{C_c C_m}{C} \frac{1}{\pi \sigma_m R_b^5 (g_{10})^2} \left(\frac{R_b}{R}\right)^6 \end{aligned}\right\} \qquad (9.1.11)$$

This decreases the time constant by about 30% from that given by (9.1.10).

A further limitation of Bullard's model is the treatment of the non-dipole field; playing only a passive role, it is swept along with the outer layers of the core and takes no part in the coupling. Rochester (1960) shows that the non-dipole field will result in additional toroidal fields penetrating the core boundary and these interact with the dipole and non-dipole parts of the mantle field. This increases the effectiveness of the coupling by a factor of about 2 or more if the non-dipole field at the core–mantle interface is much greater than the dipole field. More importantly, the inclusion of the non-dipole field reduces the time constant, and Rochester's model yields $\tau \simeq 12$ yr for $\sigma_m = 10^{-9}$ e.m.u. Roberts (1972) has extended Rochester's theory to allow for arbitrary slow motions of the core surface past the mantle, rather than a simple solid body rotation assumed in the Bullard model. Perhaps a more important limitation is that the models of Rochester, Roden and Roberts all ignore the possibility that currents may occur in the mantle with time scales so short that they do not penetrate the totality of the mantle. Hide (1966) in his study of the free hydromagnetic core modes suggests that the spectrum of the magnetic fluctuations at the core boundary could contain components with periods of a few years or less and these would enhance the strength of coupling.

Yukutake (1972) considers the coupling associated with a Bullard–Rochester model in which the poloidal dipole field varies in magnitude with time. Additional toroidal fields are formed due to the interaction of the fluctuating dipole field with (i) the differential rotation of the two layers of the core and (ii) the differential rotation between the outer core and mantle. Once these fields are evaluated the additional torques and changes in rotation follow from equations similar to (9.1.6), (9.1.8) and (9.1.9). These torques further enhance the coupling at the decade and longer periods. Archeomagnetic observations suggest that the dipole field may have changed by about 50% over 8000 yr and Yukutake concludes that such a change can explain a non-tidal secular acceleration of the Earth of the order $3\times10^{-22}$ rad s$^{-2}$ as well as cause periodic variations with characteristic time scales of about 65 yr and 400 yr (Yukutake 1973; Watanabe & Yukutake 1975).

The electromagnetic torque (9.1.6) assumes that the leakage into the mantle and the consequent flux redistribution take place over the entire core–mantle boundary. If $l$ is the scale of the perturbed field, the torques will be reduced by a factor $\pi l^2/4\pi R_b^2$, or by a factor of about 50 if $l = 1000$ km. In order for the mantle to respond to such a local perturbation either $\sigma_m$ or the magnitude of the perturbation must be larger than previously assumed. The regional pattern of secular variations has been observed to alter appreciably within a few decades. An example is the strong secular variation focus that developed below South Africa late in the nineteenth century due to the westward drift of a non-dipole anomaly. This has been described by Bullard (1948): the field strength at Capetown prior to 1895 was decreasing at a rate of about 35 $\gamma$ yr$^{-1}$ and at this time the rate increased suddenly to 100 $\gamma$ yr$^{-1}$. It has been suggested by A. F. Moore that this particular anomaly is related to the rapid change in the l.o.d. that occurred a few years later (see Roberts 1972).

### 9.1.2 *Mantle conductivity*

The value of $\tau$ depends critically on the value of the lower-mantle conductivity $\sigma_m$ and reliable values for this parameter are scarce. Price (1970), Le Mouël (1976*a*) and Stacey (1977) review some of the available evidence. Three independent observations provide estimates of $\sigma_m$. A fluctuating electromagnetic field of external origin induces currents in the mantle and creates an additional field at the surface of the Earth. The characteristics of this field depend upon the conductivity distribution in the mantle, as well as upon the frequency of the fluctuations in the external field. Mathematically there is no unique solution to the problem of finding the electric current distribution in the Earth that produces a given field at the surface, and any attempt at estimating $\sigma_m$ will have a similar non-uniqueness. The depth of penetration of these currents depends upon the conductivity distribution and upon the frequency of the time variations of the field, and to obtain a depth profile of $\sigma_m$ requires the study of currents over a wide range of frequencies. Banks & Bullard (1966) obtain $\sigma_m \simeq 2 \times 10^{-11}$ e.m.u. at a depth of 1200 km from a study of the annual fluctuations in the field (see also Banks 1969, 1972). Yukutake (1965) found $6 \times 10^{-10}$ e.m.u. at 1600 km from a study of the 11-yr cycle associated with solar activity. The most careful analysis of this long-period fluctuation in the magnetic field appears to be the study by Courtillot & Le Mouël (1976) who conclude that the presently available data do not provide strong constraints on the range of lower-mantle conductivity models. Also $\sigma_m$ has been estimated from studies of the secular variation spectrum. In this case both the source field and the conductivity distribution are unknown and the result depends upon the validity of the assumed source field. Runcorn (1955) assumed that this field has a white noise spectrum and that the diminution of the spectrum observed at the Earth's surface determines the conductivity. He obtained in this manner $\sigma_m(R_c - R_b)^2 \simeq 10^9$ and for $R_c - R_b \simeq 10^3$ km, $\sigma_m \simeq 10^{-7}$ e.m.u. Runcorn considers this to be an upper limit to permissible values of $\sigma_m$ (but see Smylie 1965). This argument has been revived by Stacey *et al.* (1979) who also conclude that, at the base of the mantle, $\sigma_m \simeq 10^{-7}$ e.m.u. A second argument for estimating $\sigma_m$ from the attenuation of the field by the mantle is also due to Runcorn (1955). In the geomagnetic literature frequent reference is found to *secular variation impulses*, sudden changes in the rate of secular variation that are established within a few years. Attributed to an internal origin, their time constants give an estimate of $\sigma_m$ and Runcorn finds that $\sigma_m(R_c - R_b)^2 \simeq 10^{-7}$ or, for $R_c - R_b \simeq 10^3$ km, $\sigma_m \simeq 10^{-9}$ e.m.u. Similar arguments have been used by McDonald (1957) and Currie (1968) who both conclude that, near the core, $\sigma_m \simeq 2 \times 10^{-9}$ e.m.u. These estimates are, however, open to question. Recently Alldredge (1975) showed that the secular variation impulses correlate with sunspot activity and that they are therefore of external origin, while Courtillot & Le Mouël (1976) have shown that these impulses are merely manifestations of the 11-yr solar cycle. Currie's estimate for $\sigma_m$ is based on what appears to be a rather arbitrary assumption that the observed secular variations in the period range from 4 yr to 33 yr are of internal origin. Alldredge (1977) assumed that 13 yr was the shortest period that effectively penetrates the mantle and this leads to an order-of-magnitude increase in conductivity. Further, by requiring his conductivity profile to merge with the values given by Banks for the upper mantle, the conductivity at the base of the mantle is raised to $10^{-6}$ e.m.u. Courtillot & Le Mouël, however, conclude that most of the power in the secular variation spectrum between periods 2 yr and 20 yr relates to the solar cycle and its first two harmonics.† Clearly the above estimates of $\sigma_m$ for the lower mantle must be taken with extreme caution.

Kolomiytseva (1972), using the morphology of a 60-yr anomaly observed over much of Europe, estimates $\sigma_m$ to be much higher than the above estimates. Evidence for this secular variation anomaly is discussed in Golovkov & Kolomiytseva (1971) and its amplitude and phase variations are explained by the screening effect of a conducting mantle (Golovkov *et al.* 1971). Kolomiytseva finds $\sigma_m \simeq 8 \times 10^{-9}$ e.m.u. at 2000-km depth and, perhaps more important, concludes that the observations are at variance with models that assume a sharp conductivity contrast across the core–mantle boundary. A number of difficulties exist with analyses of this kind. One is that the result is dependent on the assumed source. Kolomiytseva assumes that this is a ring current near the core boundary, but the dependence of $\sigma_m$ on the choice of model needs further study. Closely related is the assumption that the observed anomaly can be isolated from the global secular variation, but probably it is part of a global re-ordering of the magnetic field. This also warrants further study. Finally, in terms of mantle properties, a value of $\sigma_m$ close to that of the core seems implausible, and possibly this value may be a consequence of a screening of the secular variation by the outermost part of the core. A recent review by Berdichevsky *et al.* (1977) of Soviet experimental results for $\sigma_m$ does

† R. Hide (personal communication July 1978) also rejects such high values for $\sigma_m$ on different grounds. Recently Hide (1978) proposed a new method for locating the core–mantle boundary from external magnetic observations and obtains a result in surprisingly good agreement, to within 2%, of the seismically observed value. Large conductivity values for the lower mantle would invalidate Hide's argument and render the agreement fortuitous.

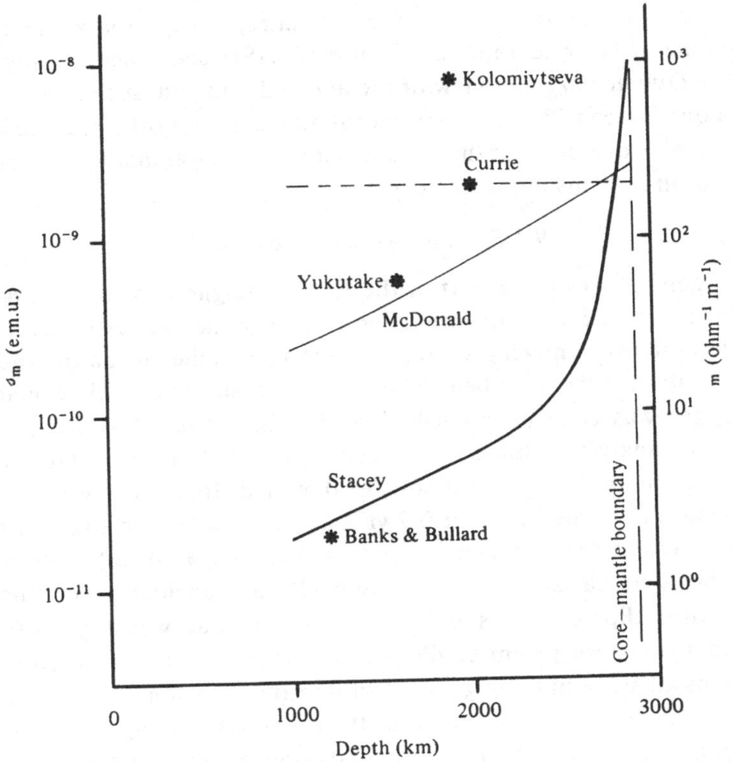


Figure 9.2. Estimates of the electrical conductivity of the lower mantle.

not discuss these high values for the lower mantle. Another high-conductivity estimate sometimes mentioned in the literature is by Braginski & Nikolaichik (1973). Less often mentioned is that this value is based on the assumption that core–mantle coupling occurs and that l.o.d. changes correlate with the geomagnetic secular variation. While this approach may ultimately yield the surest estimate of $\sigma_m$, we cannot use it to substantiate the electromagnetic coupling hypothesis. Values for $\sigma_m$ of the order of $10^{-9}$ e.m.u. are compatible with theoretical estimates for magnetism-rich olivine under pressures and temperatures that are representative of the lower mantle (Tozer 1959; Shankland 1975). Figure 9.2 illustrates the various estimates for the lower-mantle $\sigma_m$. Stacey (1977) proposes a model in which the lower-mantle conductivities are generally lower than the above estimates, except for a highly conducting layer near the core (figure 9.2). (See also Stacey, Conley & McQueen 1979.) In view of the above discussion, such a model can only be considered as one speculation amongst others; there is no reliable evidence for the conductivity of the lower mantle and the situation remains unsatisfactory.

### 9.1.3 *Observational evidence*

The general westward drift of the Earth's magnetic field was first noticed in the seventeenth century, magnetic observations at several stations showing a progressive change in the inclination and declination of the main field. The worldwide pattern of such secular changes was clearly established by Vestine *et al.* (1947) and a detailed analysis of these results by Bullard *et al.* (1950) led to the conclusion that the non-dipole field was drifting in a westerly direction at a rate of about $0^\circ\!.2\ \mathrm{yr}^{-1}$ at all latitudes and that most harmonics in the geomagnetic potential participated in this drift, although at variable rates. Vestine (1953) concluded that the eccentric dipole was subject to a substantial westerly drift. Numerous more recent studies of the magnetic field at different epochs all agree in finding a westward drift of the non-dipole field, but the estimates of the rate of drift vary greatly and appear to be strongly dependent upon the methods used to analyse the observations.

Of interest in studies of the Earth's variable rotation are possible changes in the drift rate, for, if indeed the drifting of the field implies a differential motion between the core and mantle, such changes should be accompanied by changes in the l.o.d. Evidence for such changes in the drift rate and for their correlation with l.o.d. changes is limited for several reasons. (i) The temporal variations in the worldwide magnetic field observed at the surface are known with an inadequate precision and, as we require the field at the core–mantle boundary, this limitation is much magnified when the field is extrapolated downwards. At the surface the time variations are known to about 20% (Lowes 1974), but in the downward extrapolation to the core – assuming insignificant electric currents flowing in the mantle – each harmonic is multiplied by $(R/R_b)^{n+2}$ and the time variation of the field becomes extremely uncertain;

that part due to harmonics $n < 6$ will be known only to about 50% accuracy. Clearly, to extrapolate the surface field to the core, to compute the flow field there, and to evaluate the angular momentum variations of the core (see, for example, Ball, Kahle & Vestine 1969) will lead to very uncertain results. (ii) There is also a question as to what part of the magnetic field drift should best correlate with the l.o.d. changes. The toroidal field in the core may be 100–200 times larger than the poloidal field according to many of the earlier dynamo theories, yet the former is reduced to zero at the surface of the Earth due to the very low conductivity of the upper mantle. More recent dynamo models (see, for example, Busse 1975) and power considerations (Stacey 1977) suggest that the two fields may be of comparable strength. Possible fluctuations in the toroidal field can only be deduced from observations of changes in the poloidal field and from model-dependent relations between the two. This will give little more than order-of-magnitude estimates at best. The absence of any clear correlation between l.o.d. and magnetic fluctuations cannot be used to infer the inadequacy of electromagnetic core–mantle coupling. (iii) Finally, perhaps to add further confusion to the subject of core–mantle coupling, Hide (1966) proposed an explanation of the westward drift that does not involve movement of core material past the mantle. Hide considers the hydrodynamic oscillations in the presence of a strong magnetic field inside a spherical liquid-filled shell, and concludes that slow waves exist whose oscillation periods are comparable with the geomagnetic secular variations and which move westward relative to the core material (see also Acheson & Hide 1973; R. Hide, personal communication August 1977). If these waves do contribute significantly to the secular variations, even a rough estimate of the core flow becomes impossible.

Despite these limitations, considerable research has been carried out since 1950 to determine whether or not a correlation exists between l.o.d. and the poloidal field. Vestine (1953) concludes that there is little chance of finding real evidence for fluctuations in the field prior to 1900 and questions whether the changes observed around 1900 (figure 9.3) are real. Dicke (1966), using data back to 1820, compiled by Vestine, concluded that there is strong evidence for perturbations having occurred in the drift of some of the

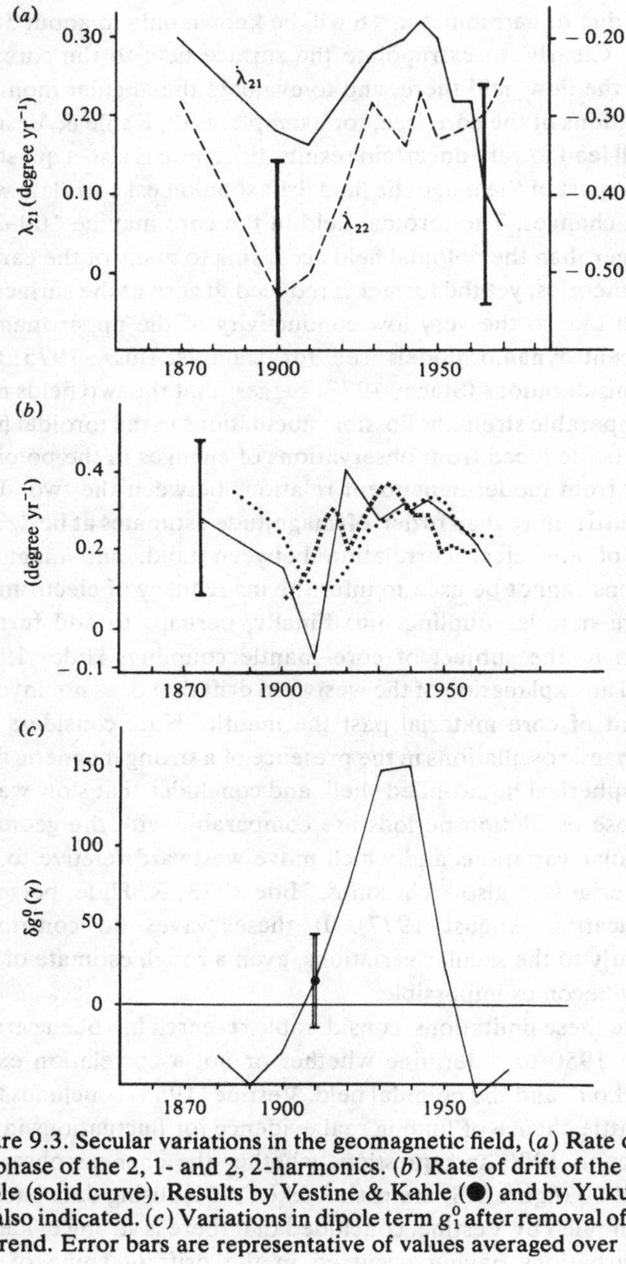


Figure 9.3. Secular variations in the geomagnetic field, (*a*) Rate of drift of the phase of the 2, 1- and 2, 2-harmonics. (*b*) Rate of drift of the eccentric dipole (solid curve). Results by Vestine & Kahle (●) and by Yukutake (×) are also indicated. (*c*) Variations in dipole term $g_1^0$ after removal of a secular trend. Error bars are representative of values averaged over 10 yr.

harmonics around 1860–1880 and again around 1900. But more recent analyses of the older data do not confirm the reality of such changes. Neither do they corroborate Vestine & Kahle's (1968) conclusion that significant changes have occurred in the rate of the westward drift of the eccentric dipole from 1900 to 1968.

Numerous analyses of the magnetic field exist for different epochs. McDonald & Gunst (1967) review some of the earlier solutions, and Barraclough (1976) reviews results for the secular variations of the field. Differences between the low-degree gauss coefficients $g_{mn}$, $h_{mn}$ in different models are often large, even for very recent models (see, for example, Lowes 1974), considerably larger than the internal precision estimates would indicate. We consider the following models of the geomagnetic potential in the subsequent comparisons:

(i) Vestine *et al.*'s (1947) analysis (see also McDonald & Gunst 1967) for the gauss coefficients at 10-yr intervals from 1905 to 1945. No information on the precision of these coefficients is available.

(ii) Malin's (1969) estimates of the gauss coefficients from 1942.5 to 1962.5 at 5-yr intervals. His precision estimates appear to be commensurable with the spread in values obtained by other authors for models during this time interval. For the 5-yr means of the harmonics of degree 1 to 3 the error estimates are of the order 40–50 $\gamma$. For want of further information, we assume that Vestine's analysis is of similar accuracy and for his 10-yr mean gauss coefficients we assume uncertainty estimates of $50/\sqrt{2}\gamma$. Malin & Clark (1974) give further models for 1962.5 and 1967.5 and Barraclough *et al.* (1975) for 1975.0. Presumably these latter models are compatible in data and reduction methods with Malin's models for the epochs from 1942.5 to 1962.5, although they are probably of greater precision.

(iii) Barraclough's (1974) analysis of the available data for epochs between 1600 and 1910. He estimated the field at 50-yr intervals from 1600 to 1850 and for 1890 and 1910. The available observations do not warrant the estimation of models at more closely spaced epochs. We only consider his results for 1850, 1890 and 1910.

Figure 9.3(*a*) illustrates the variations in the phase of second-degree harmonics defined by $m\lambda_{mn} = \tan^{-1}(h_{mn}/g_{mn})$ and based upon the above models. These changes are suggestive of perturbations having occurred in the drift rate around 1900 and again near 1940. In particular the changes in $\lambda_{21}$ are similar to the known variations in the l.o.d.: prior to about 1910 the drift rate is decreasing, increasing again from 1910 to about 1945 and thereafter steadily decreasing. But the years 1910 and 1945 also correspond to changes in the analyses: the models prior to 1910 are from Barraclough, those from 1910 to 1940 are based on Vestine's study, and the post-1940 data are from Malin and colleagues. It is possible that these changes may simply be a consequence of differences in data and in methods of data analysis. Furthermore, Malin's precision estimates of the gauss coefficients lead to uncertainties in the 5-yr and 10-yr mean drift rates that exceed the fluctuations themselves. We cannot conclude that the variations in the rate of change of the phase of $\lambda_{21}$ and $\lambda_{22}$ are statistically significant. Likewise any accelerations in the phases of the third-degree harmonics are insignificant. Dicke's (1966) conclusions cannot be maintained.

Changes in the westward longitudinal drift of the eccentric dipole are illustrated in figure 9.3(*b*) and they exhibit similar trends to $\lambda_{21}$. It can be readily demonstrated that, with the present field parameters, this drift of the dipole is almost entirely determined by the drift of $\lambda_{21}$; the two results cannot be considered independent. Vestine & Kahle's (1968) curve of the eccentric dipole drift based on the same coefficients from 1905 to 1945 represents a very smoothed version of figure 9.3(*b*), but the magnitude of the fluctuations remains smaller than the uncertainties. Yukutake (1973) has also computed the eccentric dipole drift towards the west, with results that are quite different from those of Vestine & Kahle. The fluctuations $\delta g_{10}$ in the dipole coefficient $g_{10}$ are illustrated in figure 9.3(*c*), after the removal of the mean secular change of $-19.6\ \gamma\ \mathrm{yr}^{-1}$ over the 125-yr period from 1850 to 1975. Here the situation is somewhat better, with the uncertainty estimates being in general inferior to $|\delta g_{10}|$. No meaningful results can be established prior to 1850. There is a suggestion of a minimum in $\delta g_{10}$ around 1890–1900 followed by a maximum around 1940, and possibly a second minimum around 1960–1970. Yukutake (1973) finds a similar fluctuation in the dipole field for the years 1890–1960. The variation in the dipole strength from 1900 to 1940 is of the order of 0.7%. If this induces a comparable change in the toroidal field leaking into the mantle, the corresponding change in $m_3$ is of the order of $1.5\times10^{-8}$ for $\sigma_m \simeq 10^{-9}$ e.m.u. Thus a not entirely unreasonable lower-mantle conductivity of $3\times10^{-9}$ e.m.u. is required to explain the observed change of $m_3 \simeq 5\times10^{-8}$ in the l.o.d. from 1900 to 1940.

Golovkov & Kolomiytseva (1971) studied the approximately 60-yr variation in the field over Europe, and their results suggest some correlation between l.o.d. and the secular geomagnetic variations. Records over about 100 yr at a number of stations showed considerable similarity in the changes in the vertical component of the field (figure 9.4), changes that are also similar to the l.o.d. curve: minima around 1900 followed by maxima near 1930–1940, and decreasing from 1940 to the present. Kolomiytseva (1972) estimated that the source of this anomaly has a length scale of about 2000 km and the mantle conductivity will have to be of the order of $1–2\times10^{-8}$ e.m.u. if this anomalous change in the magnetic field is also responsible for the l.o.d. change. It is difficult to assess the reliability of the older vertical component data, but the fact that a number of different stations – each with its own instrumental problems – give very similar results does give some confidence in the general trends sketched in figure 9.4.

### 9.1.4 *Polar motion*

The above discussion on electromagnetic coupling and l.o.d. changes can be extended to the excitation and damping of the wobble components since, should the electromagnetic torques have components about an axis in the plane of the equator, the direction of the instantaneous rotation axis could be modified. This problem has been discussed by Rochester & Smylie (1965) and by Rochester (1968).

The equations of motion for the mantle and core subjected to torques have been discussed in section 3.3. For the mantle they are, including damping (equations 3.3.2),

$$\mathbf{j}\dot{\mathbf{m}}/\sigma_* + \mathbf{m} = \mathbf{j}\mathbf{L}/\Omega^2(C_m - A_m) - \alpha\dot{\mathbf{m}}/\sigma_*$$

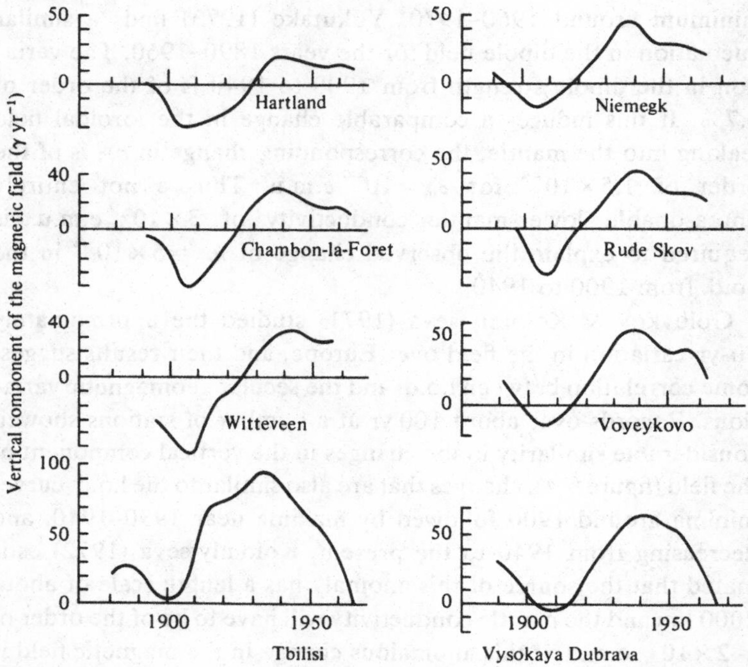


Figure 9.4. Secular variation in the vertical component of the magnetic field at eight different stations in Europe (after Golovkov & Kolomiytseva 1971). Joseph Stalin, after he left the Church, was employed at the Tbilisi geophysical observatory at the time of this rapid change in the secular variations! From Stalin's biographies covering this period, it is clear that he did not devote much time to his recording duties at the observatory and one can only wonder as to how he influenced subsequent geophysical discussions.

with $\sigma_* = \Omega(C_m - A_m)/A_m$, and

$$\mathrm{j}\dot{\mathbf{n}} + \Omega(\mathbf{m} - \mathbf{n}) = -\mathrm{j}\mathbf{L}/A_c\Omega.$$

The core is assumed to rotate as a rigid body and its motion $\mathbf{n}$ is with respect to mantle-fixed axes. The torque $\mathbf{L}$ includes both the direct perturbing torque and the passive response torque caused by the generation of the toroidal field as the mantle moves past the core. The direct torque will be most efficient at exciting the wobble if it has the same frequency as the resonance wobble of the mantle, and it can be expressed by

$$\mathbf{L}^{(a)} = \delta\mathbf{L}^{(a)}\, e^{\mathrm{j}\sigma_* t} H(t). \qquad (9.1.12a)$$

The restoring torque, analogous to (9.1.5$b$), follows from Rochester (1968), with adequate precision,

$$\mathbf{L}^{(b)} = \frac{12}{15}\pi\frac{\sigma_{\mathrm{m}}}{C_{\mathrm{c}}}\Omega(\mathbf{m}-\mathbf{n})R_b^5\left(\frac{R}{R_b}\right)^5(g_{10})^2\frac{1-R_b^5/R_c^5}{1+\frac{3}{2}R_b^5/R_c^5}. \quad (9.1.12b)$$

Rochester & Smylie's solution of the equations of motion gives the following rotational response to these torques:

$$\Delta\mathbf{m} = (1/A_{\mathrm{m}}\Omega\alpha)\delta\mathbf{L}^{(a)}\,\mathrm{e}^{\mathrm{j}\sigma_* t}(1-e^{-t\alpha}), \quad (9.1.13)$$

where the time constant of the damping $\tau = \alpha^{-1}$ is

$$\tau^{-1} = \frac{12\pi}{15}\frac{\sigma_{\mathrm{m}}}{C_{\mathrm{m}}}R_b^5\left(\frac{R}{R_b}\right)^5(g_{10})^2\frac{\sigma_*}{\Omega}\frac{1-R_b^5/R_c^5}{1+\frac{3}{2}R_b^5/R_c^5}. \quad (9.1.14)$$

For $\sigma_{\mathrm{m}} \simeq 3\times10^{-9}$ e.m.u., $\tau \simeq 10^5$ yr, compared with an observed relaxation time of the order of 10–100 yr. Electromagnetic damping of the wobble is entirely insufficient unless the lower-mantle conductivity approaches that of the core. The postulated perturbing torque (9.1.12$a$) is of relatively short period and by computing $L^{(a)}$ at different epochs using time-averaged gauss coefficients, one obtains what can be considered as an equilibrium value $L^{(a)}(0)$. Rochester & Smylie obtain $L^{(a)}(0) \simeq 10^{24}$ dyn cm and this represents about 20% of the zonal components of the torque. Assuming that fluctuations of 20% occur in $L^{(a)}$

$$\Delta\mathbf{m}(t) \simeq (3.4\times10^{-18}/\alpha)\,\mathrm{e}^{\mathrm{j}\sigma_* t}(1-\mathrm{e}^{-t\alpha}) \quad (9.1.15)$$

and, with $\alpha \simeq 30\ \mathrm{yr}^{-1}$,

$$\Delta\mathbf{m}(t) \simeq 0''.68\times10^{-3}.$$

This compares with an observed wobble of about 0''.15. Electromagnetic core–mantle coupling provides a quite inadequate means of transferring angular momentum from the core to the mantle in order to sustain the Chandler wobble. Runcorn (1968) argued that the Chandler wobble is maintained by electromagnetic torques but he gives no details and his conclusion is clearly at variance with the results of Rochester & Smylie. More recently, Runcorn (1970) proposed that high-frequency, localized geomagnetic secular variations exert impulsive torques on the mantle and that these would perturb both the l.o.d. and the wobble. Again Runcorn gives no details or even order-of-magnitude estimates of these torques, but the estimate given in section 9.1.1 for such localized impulses indicates that they must be of very large amplitude or that the lower mantle must have a very high electrical conductivity. Sasao *et al.* (1979) point out that such torques would also give rise to a nearly diurnal nutation as large as the Chandler wobble!

Rochester (1968) discussed in a more rigorous manner the astronomical consequences of the electromagnetic torques and concludes that there is no significant indirect *cross-coupling* between changes in $\mathbf{m}$ on the one hand and $m_3$ on the other, and that the two components can be treated independently: there will be no significant transfer of angular momentum about the instantaneous rotation axis to an axis in the equatorial plane. This is for a linear coupling mechanisms. Stacey (1970) proposes a non-linear model which would permit such a transfer from the precessional motion to the wobble, but the details of this mechanism have not been worked out yet.

### 9.1.5 *Topographic coupling*

If the core–mantle interface is sufficiently rough to penetrate the viscous hydromagnetic boundary layer, pressure forces of the fluid moving along the interface may be sufficiently large to couple the mantle to the core motion. Consider a rectangular *bump* on the interface of length $l$ and height $h$. Assume that underneath the boundary layer a quasi-geostrophic state exists, in which the horizontal pressure forces are in approximate balance with the horizontal component of the Coriolis acceleration of the fluid. That is,

$$\Delta p/l \simeq |\mathbf{e}_t \cdot 2\rho\hat{\mathbf{\Omega}} \wedge \mathbf{u}| \simeq 2\rho\Omega u \sin\phi.$$

The tangential shear stress is

$$\mathcal{S}_t \simeq D_T(\Delta p/l)(h-\Delta) = 2\rho\Omega u(h-\Delta)D_T \sin\phi, \quad (9.1.16)$$

where $D_T$ is a dimensionless drag coefficient and $\Delta$ the thickness of the boundary layer (section 8.2.3). For probable values of core viscosity, $\Delta$ will be small, of the order of a few metres. We see below that $h$ must be of the order of a kilometre for topographic coupling to be effective so that we shall be justified in neglecting $\Delta$ in (9.1.16). Only if the viscosity is of the order of $10^6\ \text{cm}^2\ \text{s}^{-1}$ will $\Delta$ become important.

To estimate the total torque on the mantle one requires (i) the form of the core topography and (ii) the flow field past the boundary. Both are inaccessible to direct observations, at least at present, and even their power spectra are unknown. One can only hazard an order-of-magnitude estimate. From (9.1.1) and (9.1.16)

$$L_3 = \frac{R_b}{4}\int_{S_c} \cos\phi \mathcal{S}_t \, dS,$$
$$\simeq (2/\pi) R_b^3 D_T \Omega \rho \{\langle u^2\rangle\langle h^2\rangle\}^{1/2},$$

where $\langle u^2\rangle$ and $\langle h^2\rangle$ are the mean square values of the velocity and the topography. Hide (1977) discusses possible values for the drag coefficient $D_T$ and stresses the possible dependence of $D_T$ on $h$ and on the magnetic field in the core (see also Moffatt 1978). In particular, a large topography does not necessarily imply strong topographic coupling, and, if $h$ exceeds a critical value, $D_T$ will actually decrease with increasing $h$. Hide concludes that probably $D_T \simeq 1$ but that further work is required to substantiate this choice.† Then

$$L_3 \simeq 2\times 10^{22}\{\langle u^2\rangle\langle h^2\rangle\}^{1/2},$$

and the resulting change in the speed of rotation is

$$\dot{m}_3 = \dot{\omega}_3/\Omega = L_3/C_m\Omega \simeq 4\times 10^{-19}\{\langle u^2\rangle\langle h^2\rangle\}^{1/2}.$$

The drift of the non-dipole field is of the order 0°.2 $\text{yr}^{-1}$ corresponding to $\langle u^2\rangle^{1/2} \simeq 0.04 \text{ cm s}^{-1}$. Consider a change in the flow field of 10%. The corresponding change in the rotational acceleration is $\Delta\dot{m}_3 \simeq 1.5\times 10^{-21}\langle h^2\rangle^{1/2}$ compared with observed changes of the order of $2\times 10^{-16}$ (figure 5.3). Thus the topography must be at least of the order of 1.5 km if topographic coupling is to be effective.

The notion that the core–mantle boundary may not be a smooth surface has been discussed by Elsasser & Takeuchi (1955) (see also Garland 1957), who argue that diffusion and differentiation processes over geological time will lead to a rough boundary surface. Convection, if it is mantle wide, may also contribute to the deformation. Seismic observations cannot yet discern any core topography, but impose an upper limit of perhaps 5 km. Hide &

† According to R. Hide (personal communication March 1978) some recent work by I. N. James of the Meteorological Office confirms that the choice $D_T \simeq 1$ is not unreasonable.

Horai (1968) interpret the low-degree harmonics in the gravity field as arising from such a topography and deduce amplitudes of the order of 4–5 km. Apart from an arbitrary decision as to which harmonics do and which do not contribute, this argument does not consider the consequences of such topography. Stresses required to maintain it are high, of the order 2000 bar or more, and dynamical forces, associated with lower-mantle convection, will be required. But once this is permitted, lateral density anomalies in the mantle also contribute to the observed field and gravity alone does not permit us to separate the two sources. Using a statistical argument, Lambeck (1976) suggests that the topography probably does not exceed 150–200 m. If convection is mantle wide, however, these limits could be exceeded if the gravitational effect of the mass excess in the descending limb of the convection cycle were compensated by a mass deficit due to a downward depression of the core–mantle boundary.

## 9.2 Oceanic and atmospheric contributions

### 9.2.1 *Sea level*

A source for changes in the Earth's inertia tensor, and hence rotation, is the exchange of mass between the world's oceans on the one hand and the glaciers and ice sheets on the other. Munk & MacDonald review the work prior to 1960. Cazenave (1975) and Lambeck & Cazenave (1976) rediscuss the evidence for this transfer and its consequences on the l.o.d.

The load due to an ocean layer of depth $\xi_O(t)$ is

$$q_O(t) = \rho_w \mathscr{C}(\phi, \lambda)\xi_O(t),$$

and the load due to an ice layer of depth $\xi_I(t)$ is

$$q_I(t) = \rho_I \mathscr{C}_I(\phi, \lambda)\xi_I(t).$$

$\mathscr{C}(\phi, \lambda)$ is the ocean function (equation 4.2.10) and $\mathscr{C}_I(\phi, \lambda)$ is a similarly defined land–ice function; $\mathscr{C}_I = 1$ where there is ice on land and 0 elsewhere. The ice load is considered to be of two parts, that due to the Antarctic ice cap, defined by $\mathscr{C}_I^A(\phi, \lambda)$, and that due to the Greenland ice sheet, defined by $\mathscr{C}_I^G(\phi, \lambda)$. Both functions are expanded in spherical harmonics and table 9.1 gives the coefficients $a_{ij}^A$, $b_{ij}^A$, $a_{ij}^G$, $b_{ij}^G$ relevant to the present discussion. Any temporal

Table 9.1. *Selected coefficients in the spherical harmonic expansion of the ocean and ice functions.* $\mathscr{C}_{\mathrm{I}}^{\mathrm{A}}$ *represents the Antarctic ice function,* $\mathscr{C}_{\mathrm{I}}^{\mathrm{G}}$ *the Greenland ice function*

| | $\mathscr{C}(\phi, \lambda)$ | $\mathscr{C}_{\mathrm{I}}^{\mathrm{A}}(\phi, \lambda)$ | $\mathscr{C}_{\mathrm{I}}^{\mathrm{G}}(\phi, \lambda)$ |
|---|---|---|---|
| $a_{00}$ | 0.697 | 0.039 | 0.0068 |
| $a_{21}$ | −0.051 | −0.0014 | 0.0058 |
| $b_{21}$ | −0.065 | −0.0043 | −0.0054 |
| $a_{20}$ | −0.134 | 0.170 | 0.030 |

change in the ice thickness of either hemisphere is assumed constant over the ice sheet and is denoted by $\dot{\xi}_{\mathrm{I}}^{\mathrm{A}}$ or $\dot{\xi}_{\mathrm{I}}^{\mathrm{G}}$. Changes in sea level are denoted by $\dot{\xi}_{\mathrm{o}}$. Long-term variations in the water vapour content of the atmosphere are assumed negligible as are changes in groundwater storage. The contribution from mountain glaciers is ignored; their total surface area is only about 3% of that of Antarctica and their mass is only about 1% of the total ice mass (Lliboutry 1965). As the variation in the ice budget of these glaciers may be more important than that of the polar ice sheets, we return to this later. With these assumptions, conservation of mass requires that

$$\rho_{\mathrm{w}} a_{00} \dot{\xi}_{\mathrm{o}} + \rho_{\mathrm{I}} (a_{00}^{\mathrm{G}} \dot{\xi}_{\mathrm{I}}^{\mathrm{G}} + a_{00}^{\mathrm{A}} \dot{\xi}_{\mathrm{I}}^{\mathrm{A}}) = 0. \tag{9.2.1}$$

We ignore the gravitational attraction of the ice mass upon the surrounding sea so that the sea surface will not be strictly an equipotential surface (Farrell & Clark 1976). The excitation functions associated with these surface loads allow from (4.2.1):

$$\left.\begin{aligned} \begin{Bmatrix} \psi_1^{\mathrm{o}} \\ \psi_2^{\mathrm{o}} \end{Bmatrix} &= -\frac{4\pi}{5} X_{\text{wobble}} \frac{R^4}{C-A} \frac{\rho_{\mathrm{I}}}{a_{00}} \begin{Bmatrix} \alpha_1 \\ \alpha_2 \end{Bmatrix} \\ \psi_3^{\mathrm{o}} &= \frac{8\pi}{15} X_{\text{l.o.d.}} \frac{R^4}{C} \frac{\rho_I}{a_{00}} \alpha_3, \end{aligned}\right\} \tag{9.2.2}$$

with

$$\begin{aligned} \alpha_1 &= \dot{\xi}_{\mathrm{I}}^{\mathrm{A}} (a_{21} a_{00}^{\mathrm{A}} - a_{21}^{\mathrm{A}} a_{00}) + \dot{\xi}_{\mathrm{I}}^{\mathrm{G}} (a_{21} a_{00}^{\mathrm{G}} - a_{21}^{\mathrm{G}} a_{00}), \\ \alpha_2 &= \dot{\xi}_{\mathrm{I}}^{\mathrm{A}} (b_{21} a_{00}^{\mathrm{A}} - b_{21}^{\mathrm{A}} a_{00}) + \dot{\xi}_{\mathrm{I}}^{\mathrm{G}} (b_{21} a_{00}^{\mathrm{G}} - b_{21}^{\mathrm{G}} a_{00}), \\ \alpha_3 &= \dot{\xi}_{\mathrm{I}}^{\mathrm{A}} (a_{00} a_{20}^{\mathrm{A}} - a_{00}^{\mathrm{A}} a_{20}) + \dot{\xi}_{\mathrm{I}}^{\mathrm{G}} (a_{00} a_{20}^{\mathrm{G}} - a_{00}^{\mathrm{G}} a_{20}). \end{aligned}$$

The factors $X_{\text{wobble}}$ and $X_{\text{l.o.d.}}$ include the Earth's elastic response to the time-variable surface load and, as the decade variations in rotation are short compared with a characteristic time of the viscous response of the Earth of about 2000 yr (Cathles 1975), an elastic response appears adequate. This will not be true for a secular change in sea level. The problem is to evaluate the excitation functions from observations of sea level and the ablation and accumulation of the ice sheets. Sea level fluctuations contribute to $\psi_i$ only if they result in a horizontal redistribution of water. Long-term variations, greater than a few years duration, can result from various factors, including

(i) variations in the amount of ice stored in the polar ice caps and mountain glaciers;
(ii) variations in the annual mean atmospheric pressure and the wind stresses on the ocean surface;
(iii) variations in the annual mean water temperature and other oceanographic factors such as salinity and ocean currents;
(iv) long-period tides;
(v) changes in the holding capacity of the ocean basins, due to subsidence or uplift of the sea floor;
(vi) subsidence or uplift of the coastlines, resulting in apparent changes in mean sea level.

If sea level observations are geographically uniformly distributed, tides will not contribute to the apparent mean sea level fluctuations. Other than tectonic changes on a geological time scale, (v) and (vi) are important now due to the post-glacial adjustment of the Earth. Melting of the Late Quaternary ice caps has resulted in a very significant redistribution of mass, to which the Earth's rotation would long ago have adjusted were it not for the backsurge flow in the asthenosphere which is still very active today (see the collection of papers edited by Andrews 1974). Temperature variations can have important consequences on the sea level curve, but these will not significantly modify the inertia tensor since the mass displacements are essentially vertical and small. Changes in Atlantic mean surface temperatures of about 0°.05 $\text{yr}^{-1}$ have been recorded (Wahl & Bryson 1975) but these appear to be restricted to the upper 100–200 m of water. The accompanied sea level changes are of the order of 1–2 mm $yr^{-1}$. Such changes will not be uniform over the ocean surface, due to geographical variations in the long-term climatic changes, and the temperature effect on the global sea level curve should be less than the above values. Changes in atmospheric pressure of as much as 10 mbar over periods of about 30 yr have been recorded (see, for example, the pressure indices of Lamb & Johnson 1966), corresponding to a change in sea level of 10 cm if the ocean response is that of an inverted barometer. These pressure variations are related to the north–south migration of the main circulation streams and their contribution to the global sea level will be reduced if the sea level observations are globally distributed. A complete discussion requires a unified treatment of air pressure, sea level and solid Earth interactions. This is not attempted here because the oceanic and atmospheric excitation functions are thought to be quite small compared with the observed astronomical changes (Lambeck & Cazenave 1976). Also, the atmospheric excitation function (see below) is not corrected for the sea level response either, and the two effects will largely cancel. Wind stress may make important contributions to the sea level observations in shallow seas, but their effects on the global sea level should vanish. Annual mean values of salinity or density are poorly known and variations in their global mean values are presumed to be small.

Several recent analyses of sea level variations have been made. For European waters the most complete study is that by Rossiter (1967), for North American waters it is that by Hicks (1973). Lisitzin (1974) gives a recent review. These studies confirm that sea level has been increasing by about 1 mm $yr^{-1}$ but the results are heavily influenced by the post-glacial rebound that occurs in both regions. This rebound can be either positive or negative, depending on the position of the fore-bulge produced by the elastic upward bending of the lithosphere subject to the ice load (see, for example, Walcott 1970). Fairbridge & Krebs (1962) have updated Gutenberg's (1941) analysis and give the global sea level curve for the years 1863–1963 (figure 9.5). Their analysis is based on the tide-gauge records for the Pacific, Indian and Atlantic oceans, but the data for the first two oceans and for the South Atlantic are sparse, and their world curve remains dominated by the North Atlantic data; sea level fluctuations associated with tidal and meteorological

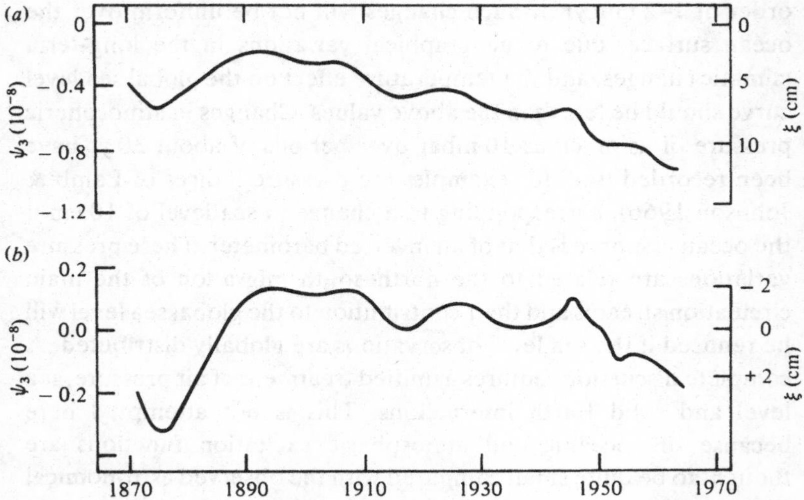


Figure 9.5. (*a*) Secular variation in mean sea level after Fairbridge & Krebs (1962) and oceanic excitation function. (*b*) The residual sea level curve and oceanic excitation function after removal of a secular trend.

phenomena may, therefore, contribute to this curve whereas they would not modify the inertia tensor of the Earth.

A second approach to estimating changes in sea level is through the analysis of the ice and snow budgets of the polar regions. Since Munk & MacDonald's discussion, several new studies of the ice storage have been published but there has not been any significant improvement in the estimates of the resulting change in sea level. The problem is one of estimating the small difference between the uncertain rates of accumulation and ablation. The various estimates of the ice and snow budgets for Greenland and Antarctica do not even agree in whether there is a net loss or a net gain, although both ice sheets appear to have been quite stable and surrounded by receding outlet glaciers (see, for example, Lliboutry 1965). For Greenland, which contains 9% of the total land snow and ice, Bauer (1966) estimates a net loss of $0.12 \times 10^{18}$ cm$^3$ yr$^{-1}$. H. Bader (Lliboutry 1965) places an upper limit of a net gain of $0.12 \times 10^{18}$ cm$^3$ yr$^{-1}$. Bauer's value leads to a rise in sea level of 0.3 mm yr$^{-1}$ while Bader's estimate results in a fall in sea level by the same amount. Loewe (1964) suggests that the Greenland ice budget may be in a balanced state. Antarctica, containing about 90% of the total land ice, leads to similar contradictory values. Estimates quoted by Lliboutry (1965) are $0.5 \times 10^{18}\,\mathrm{cm^3\,yr^{-1}}$ by Dolgouchine, $-0.4 \times 10^{18}\,\mathrm{cm^3\,yr^{-1}}$ by Buitnisky, and $0.1 \times 10^{18}\,\mathrm{cm^3\,yr^{-1}}$ by Averyanov. These values result in changes in sea level that range from $+1.4\,\mathrm{mm\,yr^{-1}}$ to $-1.1\,\mathrm{mm\,yr^{-1}}$. A more recent estimate by Loewe (1967) indicates a net gain of about $0.24 \times 10^{18}\,\mathrm{cm^3\,yr^{-1}}$ and also illustrates the uncertainty of the estimation process. Accumulation is estimated to be $1.9 \times 10^{18}\,\mathrm{cm^3\,yr^{-1}}$. Losses by runoff of meltwater, evaporation, and the discharge of the ice sheet, glaciers and ice shelves total $1.5 \times 10^{18}\,\mathrm{cm^3\,yr^{-1}}$, giving a net gain of about $0.4 \times 10^{18}\,\mathrm{cm^3\,yr^{-1}}$. But a potentially important source of melting occurs underneath the ice shelves and floating ice tongues. Estimates of this are poor but Loewe assumes that it may be of the order of $0.2 \times 10^{18}\,\mathrm{cm^3\,yr^{-1}}$. This represents one-half of the imbalance. Loewe suggests that the various estimates are sufficiently imprecise that an overall balance may exist. A further problem is that, as far as the sea level response is concerned, it is only the change of land ice volume that enters into the discussion, but there appears to be no way of separating this from the ice budget of the ensemble of land and surrounding sea ice.

Assuming that the observed sea level variations are entirely a consequence of the melting of land ice, the excitation functions (9.2.2) depend on whether the melting is confined to one hemisphere or whether it occurs in both hemispheres. Table 9.2 summarizes some results for different situations. If melting occurs at the same rate over the two ice sheets, then the change in ice thickness required to explain a rise in sea level of $1\,\mathrm{mm\,yr^{-1}}$ is $1.7\,\mathrm{cm\,yr^{-1}}$ and the total net ice loss is $0.4 \times 10^{18}\,\mathrm{cm^3\,yr^{-1}}$. This is within the range of values discussed above. The pole responds at a rate of about $0''.001\,\mathrm{yr^{-1}}$ in a direction 30° West. The corresponding change in rotation, $m_3$, is $10^{-9}\,\mathrm{yr^{-1}}$, much smaller than the observed decade fluctuations. If melting occurs only in Greenland the change in ice thickness is $11\,\mathrm{cm\,yr^{-1}}$ and the pole responds at a rate of about $0''.003\,\mathrm{yr^{-1}}$ in a direction 40° West. This agrees roughly with what is observed. If melting occurs only in Antarctica, the decrease in ice thickness is of the order of $2\,\mathrm{cm\,yr^{-1}}$ and the pole shift is much smaller than before. The change in l.o.d. is of the same magnitude as

Table 9.2. *Magnitude $|\psi|$ and direction of the secular pole shift and changes in $\psi_3$ due to an exchange of mass between the ice sheets and the oceans*

| | $\dot{\xi}_o$ (cm $yr^{-1}$) | $\dot{\xi}_I$ (cm $yr^{-1}$) | $|\psi|$ (arc second $yr^{-1}$) | Direction | $|\psi_3^o|$ ($10^{-10}$) |
|---|---|---|---|---|---|
| Equal melting both ice sheets | 0.1 | 1.7 | 0.0006 | 30° West | 0.9 |
| Greenland only | 0.1 | 11 | 0.0030 | 40° West | 0.9 |
| Antarctic only | 0.1 | 2 | 0.0001 | 25° West | 0.9 |
| Mountain glaciers only | 0.1 | 33 | 0.0004 | 45° East | |

before. While the pole is quite sensitive to the ice budget of Greenland, it is quite insensitive to the Antarctic ice budget. This is a consequence of the rather symmetrical longitudinal ice distribution of the latter ice sheet. For the pole position, it is the geometry of the areas where melting occurs that is important, while for l.o.d. it is the area over which the melted ice is redistributed.

If, as some studies of the ice budgets in the two hemispheres suggest, both Antarctica and Greenland are in near balance, the rise in sea level of 1 mm $yr^{-1}$ cannot be attributed to these water sources. Possibly it is a consequence of mountain glaciers, for, while they contain only about 1% of the total land ice, they are much more variable than the polar ice and may make a significant contribution to any change in sea level. Observations of European and other glaciers suggest that there has been a global recession for at least the last 100 yr (for a review see Lliboutry 1965). Finsterwalder (1954) estimated that the secular loss of eight typical glaciers in the Eastern Alps for the years 1856–1950 was of the order of 500 mm $yr^{-1}$/(unit of surface area). Superimposed upon this are fluctuations of the order of 200 mm $yr^{-1}$ following the observed temperature fluctuations on a time scale of 20–40 yr. According to Manley (1954), the Alpine glacier loss–gain variations follow closely those observed in Iceland and Norway (see also Le Roy Ladurie 1973). Assuming that these rates are typical of all European glaciers of total surface area $4 \times 10^5$ km$^2$, the expected rise in sea level is of the order of 0.5 mm yr$^{-1}$, half the observed amount! The regional trends for Europe are also reflected by glaciers in North America and Asia. If similar loss rates occurred everywhere, the rise in sea level would be of the order of 1 mm yr$^{-1}$, equal to the observed rate. This would represent an upper limit, since not all glaciers will exhibit the same loss rates, nor will the loss–gain rates always be in phase, but it does point to a potentially important role of loss–gain rates of non-polar glaciers in eustatic sea level variations. For an order-of-magnitude estimation, we assume that the principal mountain glaciers occur at 45° North and at longitudes +15° for Europe, +90° for Asia and −120° for North America and that their surface areas are each $4 \times 10^5$ km$^2$. Then $a_{00}^{\mathrm{M}} \simeq 2.3 \times 10^{-3}$, $a_{21}^{\mathrm{M}} \simeq 7 \times 10^{-5}$, $b_{21}^{\mathrm{M}} \simeq 6 \times 10^{-5}$. High-latitude and equatorial ice fields will not contribute significantly to these coefficients. The mean rate of melting follows from (9.2.1) or

$$\dot{\xi}_{\mathrm{M}} = -(\rho_{\mathrm{w}}/\rho_{\mathrm{I}})(a_{00}/a_{00}^{\mathrm{M}})\dot{\xi}_{\mathrm{o}}$$

and $O\{\psi\} \simeq 0''.0004$ yr$^{-1}$ for $\dot{\xi}_{\mathrm{o}} \simeq 1$ mm yr$^{-1}$. Though smaller than the observed amount, non-polar glaciers may contribute more to the secular polar motion that does the Antarctic ice sheet.

Figure 9.5 gives the contribution to $\psi_3$ from the sea level fluctuations given by Fairbridge & Krebs (1962), assuming that these variations are eustatic in origin. Curve (*a*) represents the total change and is mainly secular in nature; the superimposed fluctuations (curve (*b*)), if real, represent only a few per cent of that required to explain the *decade* fluctuations in the l.o.d.

### 9.2.2 *Atmospheric contributions*

Variations in the mass distribution of the atmosphere and of its angular momentum may contribute partially to the excitation functions. Lambeck & Cazenave (1976) conclude that the total contribution may be of the order of 10–15% for the decade and longer fluctuations in the l.o.d., while Siderenkov (1973) concludes that fluctuations in angular momentum may contribute significantly to the l.o.d. changes of periods up to about 5 yr (section 7.6.2). We consider possible effects on $\psi_3$ only. Any long-period contributions from relative angular velocity variations, $\psi_3^{\mathrm{A}}(h)$, can be evaluated only from ground pressure and the geostrophic approximation (equations 7.3.2). This has been done by Lambeck & Cazenave for the years 1890–1970. They find $|\psi_3^A(h)| \simeq 6\times10^{-9}$. Most of this appears to be associated with fluctuations in the northern hemispheric circulation (figure 9.6) and represents some 10% of the changes in $m_3$ for the corresponding period. Upper tropospheric winds may contribute further to $\psi_3^A(h)$ but their magnitudes are unknown. The 5-yr wind data set used by Lambeck & Cazenave (1974) indicates that most of the power in the low-frequency part of the excitation spectrum comes from winds in the lower troposphere below about 5 km. $\psi_3^A(\Delta I)$, computed from the ground level pressure for the years 1880–1970, is of the order $10^{-9}$ (figure 9.6). Any yielding of the solid Earth under this surface load reduces the excitation function, but this modification has been ignored since $\psi_3^A(\Delta I)$ is already quite small compared with $\psi_3^A(h)$. The ocean response to the pressure variations should also be considered. This has not been done either, since the additional excitation from observed sea level variations, containing in part this response, has already been computed. The trends in $\psi_3^A(\Delta I)$ (figure 9.6) follow those of $\psi_3^A(h)$ but they amount to only about 30% of $\psi_3^A(h)$. Thus the total atmospheric excitation function is dominated by the northern hemisphere surface winds. The functions $\psi_3^A(h)$, and their sum, all appear to lag $m_3$ by about 15 yr. Only from about 1890 onwards can the total excitation $\psi_3^A + \psi_3^o$ be considered as more than an order-of-magnitude indicator of trends in this function. The changes corresponding to the years 1905–1935 and 1870–1900 represent about 10% of $m_3$. An increase in the computed meteorological excitation function by the required order of magnitude is possible only if there have been very important changes in the upper atmospheric circulation over periods of about 10 yr and longer, changes that are proportionally more important than those observed at low altitudes. There do not appear to be any observational data available pertaining to such long-period changes. If the geostrophic winds are representative of the circulation in more than the first 3 km of atmosphere as supposed above, the total excitation function can be increased by a factor of about 2 giving a total excitation of the order of 20% of $m_3$. Two observations of limited reliability indicate that this extrapolation may be invalid.

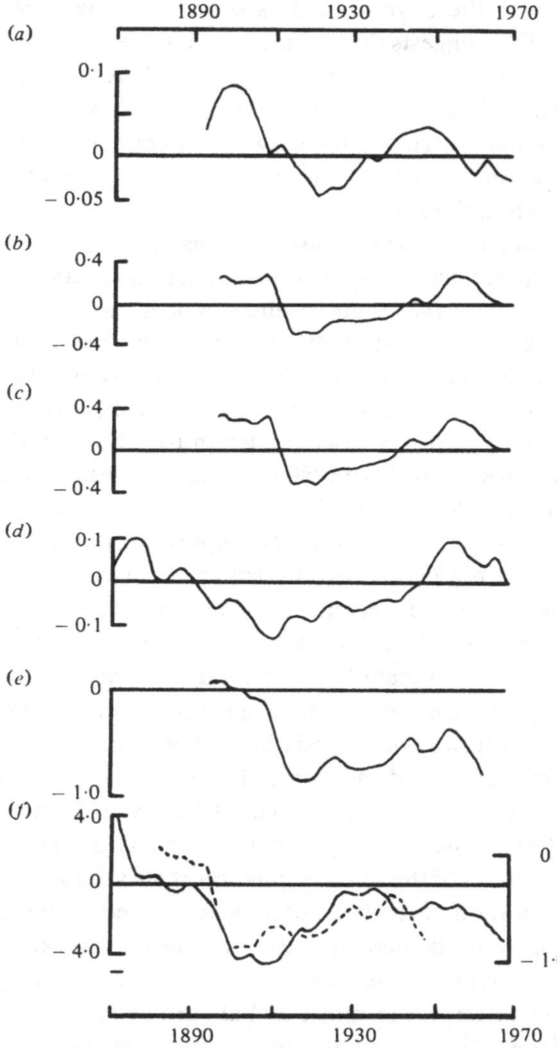


Figure 9.6. Meteorological and oceanic excitation functions contributing to the decade fluctuations in $m_3$: (*a*) $\psi_3^A(h)$ for geostrophic winds in the southern hemisphere; (*b*) $\psi_3^A(h)$ for the northern hemisphere; (*c*) total geostrophic winds; (*d*) $\psi_3^A(\Delta I)$ for the variable atmospheric inertia tensor; (*e*) the total excitation function $(a)+(b)+(d)+$ residual oceanic excitation function from figure 9.5; (*f*) the observed $m_3$ (solid line) and the total $\psi_3$ (broken line) translated by 15 yr. Note the different scales for these two functions.

The first is that the 5-yr wind data set utilized by Lambeck & Cazenave (1974) suggests that most of the low-frequency contribution to $\psi_3(h)$ comes from the lower troposphere. Secondly, the phase lag of about 15 yr between $\psi_3$ and $m_3$, if real, is troubling, as it is difficult to imagine such a lag between upper and lower atmospheric circulation at low frequencies. The ocean contribution could be overestimated, since the variation in sea level will include effects that do not change the inertia tensor, and, as $\psi_3$ (ocean) is out of the phase with $\psi_3$ (atmosphere), the total excitation function could be increased slightly. Other indirect effects appear to be small. Evidence of long-term global fluctuations in groundwater storage is lacking but, as the annual variation contributes only about $3 \times 10^{-10}$ to the total seasonal excitation in $m_3$ (table 7.2), any long-term effects are not likely to exceed a few parts in $10^{10}$. Long-period body tides do not exceed $10^{-9}$. Uncertainties in their estimated effects due to the ocean tide are unlikely to exceed 10–20% of this. Ocean currents also appear unlikely to be important as their seasonal effects on $m_3$ do not exceed a few parts in $10^{10}$ (section 7.3.2).

While the combined oceanic–atmospheric excitation function quite fails to explain the l.o.d. changes, it does appear to follow a trend that is similar to that observed in the l.o.d. Similar trends are seen in a large number of meteorological indices collected during the last two centuries which describe, albeit imperfectly, climatic trends in various parts of the world. These trends may be characterized by two types of atmospheric circulation, alternating typically every 20–40 yr. The first type (type I) is characterized by an increasing intensity of the zonal circulation at all latitudes and with a poleward migration of the belts of maximum wind intensities. The circulation is accompanied by a decrease in the overall range of surface-air temperatures between the equator and the poles, and by an overall increase in the mean global surface-air temperatures. Ocean-surface temperatures also tend to increase at high latitudes. The type-II circulation is characterized by a weakening of the zonal circulation, by a migration of the main streams to lower latitudes, and by an overall decrease in surface temperature. For both types of circulation the migration in latitude and the changing intensities are global phenomena, occurring at all longitudes and in both hemispheres, although the trends in different regions are not always in phase. Both easterly and westerly winds increase with the type-I circulation and decrease during the type-II circulation. Lamb (1972) discusses in detail the available observational evidence for these two main types of circulation during the last 150 yr. Lambeck & Cazenave (1976) have summarized some of these indices, which indicate the following global pattern. The years 1900–1930 are typical of the type-I circulation, being characterized by an increasing strength of the zonal circulation at mid-latitudes that reaches a maximum for the years 1930–1935. There is a poleward shift by a few degrees of all the pressure-extrema belts and a global surface warming. A secondary maximum occurred as late as 1950. Subsequent years are of the type-II circulation, with an increasing weakening of the circulation becoming persistent and global by the 1960s. Before 1900 the indices are less reliable but the following global pattern emerges (Lamb 1972). The years 1870–1895 are of type II, with the zonal circulation being weak. The years 1840–1870 are an interval of increasingly vigorous zonal circulation (type I), while the period 1820–1840 is characteristic of the type-II circulation. Indices of the Arctic Sea ice and of the high-latitude tree growth rates suggest that this last period may have commenced as early as about 1790 and that it was preceded by type-I circulation from about 1700 onwards. These dates are quite approximate, and it does not appear possible to establish relative amplitudes of the intensity of the global circulation patterns for the different intervals. The general trend of circulation changes is sketched schematically in figure 9.7. Comparing this with $\dot{m}_3$ suggests that periods of increasing strength of the zonal circulation are accompanied by periods of a rotational acceleration while periods of decreasing circulation are accompanied by a rotational deceleration. Such similarities are seen particularly between $m_3$ and global surface temperature variations since 1900 (figure 9.8).

Cause and effects cannot be distinguished from this observation alone and three alternative hypotheses are possible: (*a*) the atmospheric circulation causes long-period changes in l.o.d., as it does for periods of less than 2–3 yr; (*b*) the fluctuations in l.o.d. and climatic change are both the consequence of a third phenomenon; or (*c*) the fluctuation in l.o.d. causes the observed variations in the circulation. If the correlations of the indices and of the atmospheric–oceanic

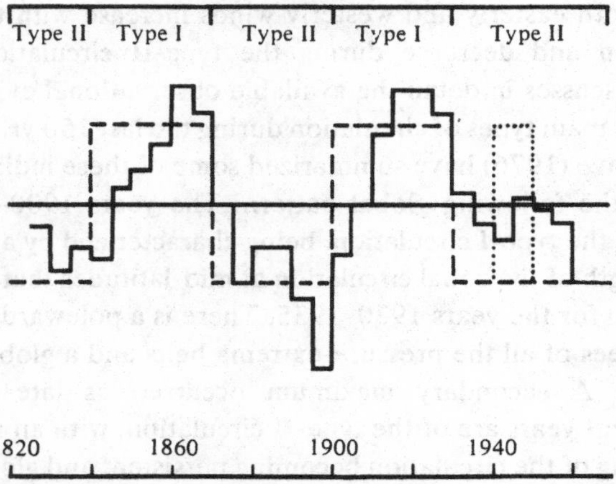


Figure 9.7. Schematic representation of the circulation pattern (broken line) in terms of the type-I and type-II circulation defined in the text. The dotted line indicates a secondary period of type-I circulation. The astronomically observed $\dot{m}_3$, running mean values over 15 yr, are defined by the solid line. The type-I circulation corresponds to positive $\dot{m}_3$ and the type-II circulation corresponds to negative $\dot{m}_3$.

excitation functions with $m_3$ are real, the second hypothesis (*b*) is the most convincing even if little quantitative information appears to be available to substantiate it. One phenomenon that has been evoked to explain climatic changes is volcanic activity. Volcanic dust injected into the upper atmosphere will persist for several years and, by modifying the Earth's albedo, may cause important meteorological changes (Lamb 1970). In particular, periods of intense activity such as occurrèd in the 1880s have been used to explain the changing climate. Anderson (1974) has suggested that this activity may also be responsible for the important change in l.o.d. at the turn of the century although quantitative arguments are lacking. If the effect of volcanic dust cannot change the atmospheric excitation function by the required amount, the following scenario is suggested. Volcanic activity affects climate, leading to the observed, but insufficient, excitation function. Associated with the volcanic activity are earthquakes resulting in a changing inertia tensor of the solid Earth, which would explain the missing 90% of the sought excitation. Stoyko & Stoyko (1969) have already suggested such a correlation between the l.o.d. variations and seismic activity,

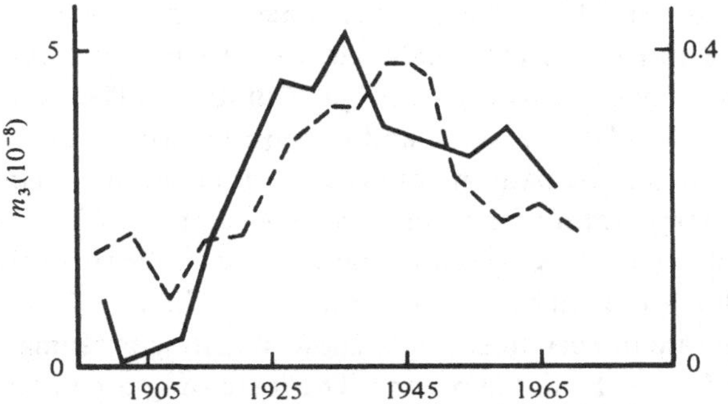


Figure 9.8. Variations in $m_3$ (solid line) and in global surface temperature $T$ (broken line) from 1900 to 1970.

as have more recent and more detailed studies by Anderson (1974, 1975), Press & Briggs (1975), and M. A. Chinnery (personal communication 1975).† But there are two potential weaknesses in these arguments. First, the seismic energy released appears to be inadequate to change the l.o.d. by the required amounts, but aseismic, lithospheric motions could be more effective (section 9.3). Secondly, the observed phase lag between $m_3$ and $\psi_3$ is not explained unless the volcanic activity lags seismic activity by as much as 15 yr. A second phenomenon sometimes evoked to explain climatic changes is the variation in solar activity but the evidence for this is inconclusive (for a review see Lamb 1972). Also, the possible indirect effect of a variable solar-wind torque acting on the Earth appears to be completely inadequate to explain the missing excitation.

† This suggestion is hardly original. The Reverend Dr Stephen Hales in an essay entitled 'Some Considerations on the Causes of Earthquakes', read before the Royal Society in April 1750, discussed likely causes of recent earthquakes in England.

'As the late Earthquakes in London and some other parts of England have roused the attention of mankind, to consider the causes of them, both in a religious view and natural view: And as in a religious view they have been considered by the Bishop of London. . . so I shall here give a short account of what seems to me to be a probably natural cause of them.' Hales goes on to note that 'In the history of Earthquakes it is observed that they generally begin in calm weather, with a black cloud' and 'signs of plenty of inflammable sulpherous matter in the air: which cloud would probably be dispersed like a fog, were there a wind: which dispersion would prevent the Earthquake, which is probably caused by the explosive lightning of this sulpherous cloud' (Hales 1750). Earthquake prevention appears to be simple!

As discussed above, sea level fluctuations appear to be inadequate to explain the secular drift of the pole of rotation, if these fluctuations are of eustatic origin. But they will be a combination of closely related eustatic and meteorological factors. During the type-I circulation pattern, for example, there is a global increase in surface temperatures and this may be accompanied by a general increase in sea level. At the same time the Antarctic snow accumulation index compiled by J. Fletcher (Lamb 1972, p. 337) shows an increasing rate of accumulation of snow, resulting presumably in a removal of water from the oceans. The increasing ocean temperatures at high latitudes will increase ablation by melting at the bottom of the floating ice, but this will not modify the sea level curve. It is not clear if losses by other processes will be increased. The point is that the sea level curve may represent a quite delicate balance between eustatic and climatic factors, and as such we cannot exclude the possibility that the secular pole shift has its origin in sea level changes, particularly if the Greenland ice sheet is subject to temporal fluctuations in the ice budget. The fact that the drift is in the predicted direction is suggestive of it. More speculative is Markowitz's (1968) secular pole shift curve (figure 5.12) that suggests that the drift rate may be variable, that it was less than the average from 1900 to 1910, 1927 to 1938 and again from 1952 to 1966, and that it was greater than average from 1910 to 1927 and from 1938 to 1952. Such a trend is similar to that observed in the l.o.d. (for example, figure 9.7).

## 9.3 Earthquakes

From Dahlen's (1971*b*) work we can determine an expression similar to (8.4.5) for $\Delta I_{33}$; or

$$\Delta I_{33} = M_o \sum_{i=1}^{3} \Gamma_i(h) j_i(\phi, \lambda_o, \delta', \alpha', \lambda')$$

with

$$j_1 = 3a_1 \cos^2 \phi_o,$$
$$j_2 = a_3(\tfrac{3}{4} \sin^2 \phi_o - \tfrac{1}{4}),$$
$$j_3 = -\tfrac{3}{2} a_4 \sin 2\phi.$$

The $a_i$ are given by (8.4.7) and the $\Gamma_i(h)$ by figure 8.4. For the 1960 Chilean earthquake, with the parameters given in table 8.3, the change in $m_3$ is of the order of $2\times10^{-10}$, corresponding to a change in l.o.d. of 0.015 ms. For the 1964 Alaskan event, $\Delta$(l.o.d.)$\simeq$0.003 ms. Aseismic deformation may be more important than in the case of the Chandler wobble excitation since the contribution is now independent of the time interval over which the deformation occurs: in the Chandler wobble problem, aseismic deformation contributes to the wobble amplitude only if the duration is short compared with the wobble period. From Kanamori's (1977*b*) study, aseismic deformation in many parts of the world may be an order of magnitude larger than the seismic deformation but this is still inadequate to explain changes in $m_3$ of $10^{-8}$. Earthquakes do not appear to make an important contribution to the decade changes in l.o.d.

## 9.4 Conclusions

Electromagnetic coupling of the mantle to fluid motions in the core appears to be the most plausible explanation of the decade variations in the l.o.d. This mechanism can quantitatively explain the observed changes in $m_3$, of the order of $5\times10^{-8}$ in about 20 yr, and it gives rise to a spectrum of $\dot{m}_3$ that is similar to the observed spectrum. Important unknowns in the mechanism are the lower-mantle electrical conductivity $\sigma_m$ and the core motions. Values for $\sigma_m$ of $2$–$3\times10^{-9}$ e.m.u. ($2$–$3\times10^2$ ohm$^{-1}$ m$^{-1}$) are required, and this cannot be excluded from the presently available evidence. The simplified theory outlined above indicates that the time constant of the coupling mechanism is of the order of 8–10 yr, comparable with that suggested by the decade fluctuations in l.o.d. Present models do not take into account the electromagnetic eddy currents in the mantle with time constants so short that they do not penetrate through to the surface, and, if present, they could enhance the coupling. Observed changes in l.o.d. with time constants of the order of 5 yr are more difficult to attribute to this mechanism unless the conductivity of the lower mantle is considerably increased. For example, changes in l.o.d. of the order of $1.5\times10^{-8}$ in about 3 yr as occurred from 1962 to 1965 (figure 5.5) require $\sigma_m\simeq10^{-8}$ e.m.u. or important eddy currents in the mantle. These conclusions are not very different from those of Rochester (1970, 1974). Observational evidence for fluctuations in the surface magnetic field that could support the electromagnetic coupling mechanism is limited. Hydromagnetic core oscillations may be responsible for part of the secular variations but, even in their absence, only the poloidal part of the field is observed: the toroidal part responsible for the coupling can only be deduced from the poloidal part and is a function of the dynamo model adopted. Of the fluctuations in the poloidal field discussed in the literature, only those of the dipole part, $g_{10}$, appear to rise above the noise level of the data. They show some similarity to the l.o.d. changes. Regional temporal variations in the magnetic field also show some similarity to the l.o.d. fluctuations since 1900. But the evidence remains circumstantial.

Other mechanisms for coupling core motions to the mantle appear to be inadequate. All recent estimates of the core viscosity are sufficiently low to rule out any form of viscous coupling. Inertial coupling can possibly be effective only if the core and mantle axes are very misaligned. Topographic coupling is more plausible, if the lower mantle can support an irregular core–mantle interface of amplitudes of about 1–2 km. Gravitational considerations appear to exclude this unless the topography is of relatively short wavelength. To evaluate realistically this coupling mechanism, both the fluid velocity field and the topography are required, but at present neither is known. Munk & MacDonald (1960) discussed the possible consequences of an interaction between the solar wind and the geomagnetic field on the Earth's rotation. From the preliminary information then available on the nature and physical properties of the magnetosphere and solar wind, they concluded that these interactions may contribute to the non-secular variations in the l.o.d. More recent studies by Schatzman (1966) and Coleman (1971) rule this out. Likewise, the mass exchange between oceans and the redistribution of mass in the atmosphere are inadequate. The earthquake excitation function also appears to be insufficient.

While the secular and long-period variations in l.o.d. are quite clearly established, evidence for such changes in the position of rotation axis is more debatable. The secular drift in a direction of about 70° West longitude appears reasonably well established. The rate of about 0″.003 $yr^{-1}$ is an order of magnitude greater than that predicted by current models of plate tectonics but as four of the five ILS stations lie in seismically active areas, close to plate margins, it is not clear whether, over only 70 yr, the station motions can be compared with plate motions deduced from paleomagnetic observations. Secular changes in sea level, if of eustatic origin, are inadequate to explain the magnitude of the pole drift but predict a motion in the direction roughly between 30° West and 40° West longitude, close to that observed.† Climatic factors may influence sea level but in an opposite way to the eustatic changes, and as such the importance of sea level in causing the secular drift cannot be ruled out.

† Recently Nakiboglu & Lambeck (1979) have shown that this observed drift can be accounted for by the late Pleistocene deglaciation, concomitant sea level change and mantle rebound if the Newtonian viscosity of the mantle averages about $10^{23}$ P.

TEN

# TIDAL DISSIPATION

## 10.1 Introduction

Tidal dissipation and its consequences on the lunar orbit and Earth's rotation have become a classic problem, yet there is probably no other subject in geophysics that has had as long a history of frustration and still attracts very considerable attention from geophysicists, astronomers and oceanographers. That this is so is as much a reflection of the fascination of the subject as an indication of a problem of some importance in understanding the origin and dynamical evolution of the Moon. In his Harold Jeffreys lecture, entitled 'Once Again – Tidal Friction', Walter Munk introduced the subject by saying that in 1920 it appeared Jeffreys had solved the problem of tidal dissipation but that we have gone backwards ever since (Munk 1968). Now, some 10 yr later, we have gone full circle, for once again there is agreement between observations and theory. Future new developments may mean that we have to go through the cycle of agreement and disagreement once more, before we can finally conclude that the subject is closed. But if these new results, such as those that may come from lunar laser-ranging analysis, disagree with our present knowledge we can always use Jeffreys' dictum '[The analysis] covers only a short interval of time and will probably be improved' (Jeffreys 1973).†

Recent improvements in the question of tidal dissipation and lunar orbit evolution include:

(i) revised estimates of the recent (since the seventeenth century) astronomical data of observations of the Moon, Sun and Mercury, including improvements in the planetary ephemeris;
(ii) re-evaluation of the ancient astronomical records, in particular the solar eclipse observations, the enlargement of the reliable

† This may not be necessary. Williams, Sinclair & Yoder (1979) find, from 8 yr of laser data, $\dot{n}_{☾} \simeq 24''$ century$^{-2}$, a value in general agreement with other estimates.

data set and the extension of this data set further back in time, to about 1400 BC;

(iii) improvements in the ocean tide models by numerical modelling;

(iv) recognition that the dissipation in the oceans is fully described by second-degree harmonics in the ocean tide expansion, making the computation of dissipation, by evaluating the rate at which work is done on the ocean surface, the most precise and direct method.

A number of new developments may result in further improvements in the near future:

(i) the widespread use of sea floor tide gauges in strategic positions on the sea floor will lead to further improvements in the ocean tide models;

(ii) use of tidal parameters perturbing the orbits of close Earth satellites can be applied directly to the lunar problem;

(iii) the lunar laser-ranging data will provide a check, or improve upon, the astronomical estimate of the present rate of the lunar acceleration.

The importance in studying the problem of tidal dissipation lies in its relevance to the study of the origin of the Moon and perhaps, by extrapolation, to the evolution of other planetary satellite systems. The problem of integrating the present lunar orbit into the past under the influence of tidal dissipation is well known and goes back to G. H. Darwin in the late nineteenth century. More recent studies are by Gerstenkorn (1955, 1969), Slichter (1963), MacDonald (1964), Kaula (1964), Goldreich (1966) and Singer (1968). The consequences of the backward extrapolation are also well known; if the present rate of dissipation is representative of the past, the Moon will have been within the Roche limit of the Earth about $1.5 \times 10^9$ yr ago. Neither the Earth's nor the Moon's surface shows evidence for such a geologically recent catastrophic event that a lunar sojourn within the Roche limit is generally supposed to imply. The youngest rocks found on the Moon were crystallized at least $3 \times 10^9$ yr ago while stromatolites indicate that lunar tides existed on the Earth $2.5 \times 10^9$ yr ago and earlier. Recent discussions on the constraints imposed by tidal dissipation on theories of the lunar origin are given by Kaula (1971) and Kaula & Harris (1975).

The future evolution of the lunar orbit is perhaps of less immediate consequence, for when the l.o.d. has become equal to the lunar month and the Moon has begun its long spiralling motion back to the Earth, several thousand million years will have passed (Jeffreys 1929). But such an evolution is of interest for other planets whose satellites may already have passed through this stage as has been proposed by McCord (1968), and others, in an attempt to explain the absence of large satellites around Venus and Mercury.

Clearly, in any extrapolation, either into the past or into the future, the manner in which the tidal energy is dissipated is a crucial element in the theory. As Munk (1968) states 'Those who have been heavily involved in calculating past orbits have a vested interest in dissipation by bodily tides rather than ocean tides, for the solid Earth is less ephemeral than the ocean basins', but unless the present estimates for the accelerations are vastly in error, only a variable energy sink can solve the time-scale problem and the only energy sink that can vary significantly with time is the ocean. All the evidence now points to the fact that the oceans are a much more important energy sink. The actual mechanics of the dissipation remain unclear nevertheless. If, as generally supposed since the work of G. I. Taylor and Jeffreys, dissipation is by friction in shallow seas, then any extrapolation into the past becomes very uncertain indeed since we know that important changes in the ocean configuration have occurred in the past. The recognition that the amount of dissipation is contained in only some of the second-degree harmonics of the ocean tide points to a more uniform dissipating mechanism, unless the shallow seas have an inordinate influence on the phase of the global ocean tide represented by these harmonics. As discussed below, this is unlikely; ocean dissipation may not be as ephemeral as Munk suggests and the extrapolation into the past may still be valid. In particular, the modelling of the tide as a lagged ellipsoid remains a valid means of representing the ocean tide throughout the past history of the oceans since (see equation (10.2.5) below) the ratio of Earth radius to Earth–Moon distance will always be small for any plausible $Q$ history and only the $s = 2$ harmonics will have significant effects on the lunar orbit.

Only the lag angle may have changed but possibly not by great amounts. Thus, even if the time scale of the orbital evolution cannot be established, extrapolation of the variations in inclination or eccentricity of the orbit as a function of the semi-major axis remains valid. This has not always been recognized. Munk (1966, 1968), for example, argues that MacDonald's (1964) results must be revised if shallow seas were important in the past. Goldreich (1966, p. 428) also argues that possible intense local dissipation in the past may vitiate his results, while Hipkin (1975) criticizes the use of the lagged tidal bulge as representing the ocean tide in studies of the tidal perturbations in satellite motions (see also Goldreich & Peale 1968, p. 291; Alfvén & Arrhenius 1969). Gold & Soter (1969) in the similar spin–orbit coupling problem of Venus recognized that no matter what the form of the tide, in this case the atmospheric tide, only second-degree terms contribute to the tidal torques. MacDonald (1964, p. 530) and Gold, in a lecture at the University of Paris in 1975, have also stressed the importance of these terms in the Earth–Moon problem.

A closely related study of some geophysical significance is the secular acceleration of the Earth's spin. The exchange of angular momentum between Earth and Moon results in a tidal deceleration of the Earth that is actually greater than that observed for the last 3000 yr. Hence there is apparently some mechanism that accelerates the Earth. One proposal (Dicke 1966) is that a secular change in the gravitational constant is responsible for part of this acceleration. A second proposal (Dicke 1969; O'Connell 1971) is that it is a consequence of the isostatic post-glacial rebound. A third proposal (Yukutake 1972) is that it is caused by core–mantle electromagnetic coupling. Other proposals invoke the growth of the core or variations with time in the depth of the principal mantle transition zones. Clearly a precise determination of the acceleration of the Earth's spin may provide important constraints on cosmological theories and geophysical models. However, evidence for the non-tidal acceleration remains weak due to it being a relatively small difference between two larger quantities that are both known to within about 10% only. Muller (1976) and Lambeck (1977) in fact conclude that the present astronomical evidence does not require the existence of such an acceleration.

## 10.2 The problem

The effect of the Earth's tidal deformation on the lunar motion and on the Earth's tidal acceleration is well known and discussed in a general manner in most geophysics textbooks (for example, Jeffreys 1970; Stacey 1977). Lambeck (1977) has discussed the problem in greater detail. The potential outside the Earth created by the solid tidal potential is given by (6.1.12) and to a high degree of approximation can be represented by a series of second-degree spherical harmonics. The delay $\Delta t$ in the Earth's non-elastic response means that the tidal bulge will be displaced with respect to the Earth–Moon axis by a small amount $\varepsilon_{lmpq}$ given by (6.1.13*b*) (figure 6.1) and the attraction by the tide-generating potential $U_n$ (equation (6.1.10) on this bulge creates a torque

$$\mathbf{L}=\int_{V_\mathrm{E}} \rho\mathbf{x}\wedge\nabla U_n \,\mathrm{d}V_\mathrm{E},$$

whose value, integrated over the volume $V_\mathrm{E}$ of the Earth and averaged over one cycle of the lunar motion, will not vanish. In consequence, the Earth's spin is modified. The equal but opposite torque of the tidal bulge of the Moon

$$\mathbf{L}=-\int_{V_☾} \rho\mathbf{x}\wedge\nabla\,\Delta U_n \,\mathrm{d}V_☾,$$

adds to the latter's orbital momentum an amount equal to that lost by the Earth and modifies the Moon's motion. Once the above torques are evaluated, the changes in the rotation of the Earth follow from Euler's equations (3.1.1). A more convenient method of estimating the accelerations is to evaluate directly the axial component of the angular momentum in the system. That associated with the orbital motion of the Earth and Moon about their centre of mass is given by

$$H_☾=\frac{Mm_☾}{M+m_☾}r_☾^2\dot{f}_☾\simeq\frac{Mm_☾}{M+m_☾}a_☾^2(1-e_☾^2)^{1/2}n_☾, \qquad (10.2.1)$$

where $f_☾$ is the true anomaly of the lunar motion. The component parallel to the rotation axis is $H_☾\cos I_☾$. The angular momentum of the Earth's spin $\Omega$ is

$$H=C\Omega,$$

and that of the Moon's spin is $C_{☾}\Omega_{☾}$. This last contribution is negligibly small since

$$C_{☾}/C = m_{☾}R_{☾}^2/MR^2 < 10^{-3}$$

and

$$\Omega_{☾}/\Omega \simeq 1/27.$$

If $C$, $M$ and $m_{☾}$ do not vary with time, the conservation of angular momentum in the Earth–Moon system,

$$H_{☾} \cos I_{☾} + C\Omega = \text{constant}, \tag{10.2.2a}$$

requires that

$$\frac{1}{C}\frac{Mm_{☾}}{M+m_{☾}}\frac{\mathrm{d}}{\mathrm{d}t}[a_{☾}^2 n_{☾}(1-e_{☾}^2)^{1/2}\cos I_{☾}] + \dot{\Omega}_{\mathrm{T}} = 0,$$

or

$$\dot{\Omega}_{\mathrm{T}} = \frac{1}{C}\frac{Mm_{☾}}{M+m_{☾}} n_{☾}a_{☾}^2\left(\tfrac{1}{3}\cos I_{☾}\frac{\dot{n}_{☾}}{n_{☾}} + e_{☾}\cos I_{☾}\dot{e}_{☾} + \sin I_{☾}\frac{\mathrm{d}I_{☾}}{\mathrm{d}t}\right). \tag{10.2.2b}$$

In these expressions $I_{☾}$ represents the inclination of the lunar orbit on the equator averaged over one period of the lunar ascending node on the ecliptic (Kaula 1964). These expressions also assume that there is no tidal effect on the other elements of the lunar orbit (MacDonald 1964; Lambeck 1977). The $\dot{\Omega}_{\mathrm{T}}$ represents the Earth's tidal acceleration due to the transfer of angular momentum from the Earth's spin to the lunar orbit. The lunar torque on the Earth follows from

$$L_{☾} = \dot{H} = C\dot{\Omega}_{\mathrm{T}}. \tag{10.2.3}$$

The $\dot{a}_{☾}$, $\dot{e}_{☾}$ and $\dot{I}_{☾}$ follow upon substituting the tidal potential into the Lagrangian equations of motion (6.3.2). We are concerned here with the action of a tidal potential raised by the Moon at the fictitious position $\tilde{\mathbf{r}}$ on the real Moon at $\mathbf{r}$. This potential follows from (6.3.1) where the asterisks refer to elements of the tide-raising body and the other elements now refer to the actual position of the Moon. To evaluate the Lagrangian equations of motion, the partial derivatives of this potential with respect to these elements are required. Once evaluated, the elements of the tide-raising body and the tidally perturbed body can be equated. Furthermore, we are only concerned with secular changes in the lunar orbit, in those changes for which $\dot{v}^*_{lmpq} - \dot{v}_{lmjg} = 0$. This occurs when $p = j$ and $q = g$. Then the secular changes in the elements $a_☾$, $e_☾$, $I_☾$ are, dropping the subscripts ☾,

$$\left.\begin{aligned}
\dot{a}_{lmpq} &= 2K_{lm}[F_{lmp}(I)]^2[G_{lpq}(e)]^2(l-2p+q)\sin\varepsilon_{lmpq},\\
\dot{e}_{lmpq} &= K_{lm}[(1-e^2)^{1/2}/ae][F_{lmp}(I)]^2[G_{lpq}(e)]^2\\
&\quad\times[(1-e^2)^{1/2}(l-2p+q)-(l-2p)]\sin\varepsilon_{lmpq},\\
\dot{I}_{lmpq} &= K_{lm}\frac{(l-2p)\cos I - m}{a(1-e^2)^{1/2}\sin I}\\
&\quad\times[F_{lmp}(I)]^2[G_{lpq}(e)]^2\sin\varepsilon_{lmpq},
\end{aligned}\right\}. \quad (10.2.4a)$$

with

$$K_{lm} = \frac{Gm_☾ k_l}{[G(M+m_☾)a]^{1/2}}\left(\frac{R}{a}\right)^{2l+1}\frac{(l-m)!}{(l+m)!}(2-\delta_{0m}),$$

and

$$\varepsilon_{lmpq} = (l-2p)\omega + (l-2p+q)M^* + m(\Phi-\theta).$$

In a conservative system the elements $a$, $e$, $I$ will not exhibit secular variations since they define the energy (equation 10.2.7) and angular momentum (equation 10.2.1) of the orbital motion of the Moon. The three angular elements $\Phi$, $\omega$ and $M$ will undergo secular changes due to the Earth and solar attractions, in addition to tidal effects that are unimportant compared with the effect on $a$. Also, with Kepler's law $n^2a^3 = G(M+m_☾)$,

$$\dot{n}_{lmpq} = -\tfrac{3}{2}(n_{lmpq}/a_{lmpq})\dot{a}_{lmpq}. \quad (10.2.4b)$$

Table 10.1 summarizes the principal secular contributions to $\dot{a}$, $\dot{e}$ and $\dot{I}$ for a constant lag angle $\varepsilon_{lmpq}$. The principal contribution to $\dot{a}$ comes from the $M_2$ tide ($lmpq = 2200$); for $\dot{e}$ the $N_2$ tide (2201)† is most important; for $\dot{I}$ three tides, $M_2$, $O_1$ (2100) and $K_1$ (2110), contribute about equally with the last two almost cancelling each other.

† The term radial tide or 'push-pull' tide is sometimes introduced to denote the tidal terms proportional to $e$.

Table 10.1. *Percentage contribution to the secular changes in $a_☾$, $e_☾$ and $I_☾$ of the lunar orbit by the principal tidal frequencies*

| Tide | *lmpq* | $\dot{a}_☾$ | $\dot{e}_☾$ | $\dot{I}_☾$ |
|---|---|---|---|---|
| $M_2$ | 2200 | 80.3 | −7.5 | 81.2 |
| $N_2$ | 2201 | 4.5 | 91.0 | 3.0 |
| $L_2$ | 220−1 | | −1.9 | |
| $2N_2$ | 2202 | | 3.2 | |
| $K_2$ | 2210 | | | 7.7 |
| $O_1$ | 2100 | 14.3 | | −70.3 |
| $Q_1$ | 2101 | 0.8 | 16.2 | −2.6 |
| $K_1$ | 2110 | | | 79.3 |
| Others | | 0.1 | −1.0 | 1.7 |

The development sketched above is particularly relevant if the dissipation occurs in the solid Earth, where the tidal bulge is harmonic in the same degree and order as the tide-raising potential. It has often been supposed that severe local dissipation in the oceans may make this model a very poor fit to reality. As shown by Lambeck (1975*a*), this is not the case and this is also implied by Kaula's (1969) equations. The Love number $k_2$ and the phase lag $\varepsilon_{lmpq}$ entering into the potential (6.3.1) are tidal effective parameters in that they reflect the total response of the solid Earth *and* its fluid layers to the gravitational attraction. These parameters are, therefore, not immediately comparable with other estimates of the Love numbers, either of observational or of theoretical origin. In particular, due to the resonance frequencies of some oceans being near the frequencies of some of the forcing functions (see, for example, Platzman 1975), the tidal effective $k_2$ and phase lag $\varepsilon_{lmpq}$ must be considered frequency dependent. This suggests that it may be more appropriate to develop the ocean tide effect independently of the solid tide. This has been done by Lambeck *et al.* (1974) for the ocean tide perturbations in the orbits of close satellites and applied to the lunar orbit by Lambeck (1975*a*, 1977). The procedure is the same as for the solid tide but the potential (6.2.2) is used instead of (6.3.1). The results for *a*, *e* and *I* due to the tide-raising potential $U_\beta$

are

$$\left.\begin{aligned}
\dot{a}_{\beta,stuv} &= K'_{\beta,stuv}(s-2u+v)\begin{bmatrix}\sin\\ \cos\end{bmatrix}_{s-t\text{ odd}}^{s-t\text{ even}} \overset{+}{\varepsilon}_{\beta,st},\\
\dot{e}_{\beta,stuv} &= K'_{\beta,stuv}\frac{(1-e^2)^{1/2}}{ae}[(1-e^2)^{1/2}(s-2u+v)\\
&\quad -(s-2u)]\begin{bmatrix}\sin\\ \cos\end{bmatrix}_{s-t\text{ odd}}^{s-t\text{ even}} \overset{+}{\varepsilon}_{\beta,st},\\
\dot{I}_{\beta,stuv} &= K'_{\beta,stuv}\frac{[(s-2u)\cos I - t]}{a\sin I(1-e^2)^{1/2}}\\
&\quad \times\begin{bmatrix}\sin\\ \cos\end{bmatrix}_{s-t\text{ odd}}^{s-t\text{ even}} \overset{+}{\varepsilon}_{\beta,st},
\end{aligned}\right\} \qquad (10.2.5)$$

with

$$K'_{\beta,stuv} = \frac{3GMF_{stu}(I)G_{suv}(e)}{R[G(M+m)a]^{1/2}}\,\frac{1+k'_s}{2s+1}\,\frac{\rho_w}{\bar{\rho}}\left(\frac{R}{a}\right)^2 \overset{+}{D}_{\beta,st}.$$

Existing developments and computer programs of the evolution of the lunar orbit mostly use the solid tide development as discussed above, and for this reason it may be useful to determine equivalent Love numbers and phase lags that include the ocean effects. This is readily possible for the lunar orbit since the terms of degree $s = 2l$ in the ocean potential (6.2.2) are of no consequence. The equivalent phase lags are obtained by equating the secular perturbations in a given element due to the solid tide with those due to the ocean tide for the same frequency. The result is

$$\begin{aligned}
k_l \sin \varepsilon_{lmpq} &= \frac{3M}{m}\frac{\rho_w}{\bar{\rho}}\frac{(1+k'_l)}{(2l+1)}\left(\frac{a}{R}\right)^{l+1}\frac{1}{R}\overset{+}{D}_{\beta,lm}\\
&\quad \times \frac{(l+m)!}{(l-m)!(2-\delta_{0m})}\,\frac{1}{F_{lmp}(I)G_{lpq}(e)}\begin{bmatrix}\sin\\ \cos\end{bmatrix}_{l-m\text{ odd}}^{l-m\text{ even}} \overset{+}{\varepsilon}_{\beta,lm}.
\end{aligned} \qquad (10.2.6)$$

The coefficients $\overset{+}{D}_{\beta,lm}$, $\overset{+}{\varepsilon}_{\beta,lm}$ will change as the lunar orbit evolves with time and, apart from physical changes in the sea floor configuration or in the volume of the oceans, it is the quantity

$$\left(\frac{a}{R}\right)^{l+1}\frac{\overset{+}{D}_{\beta,lm}}{F_{lmp}(I)G_{lpq}(e)}\begin{bmatrix}\sin\\ \cos\end{bmatrix}_{l-m\text{ odd}}^{l-m\text{ even}} \overset{+}{\varepsilon}_{\beta,lm}$$

that remains constant (Lambeck 1977). The equivalent phase

Table 10.2. *Equivalent phase lags for the solid Earth–ocean system*

| Ocean model results | $\varepsilon_{lmpq}$ |
|---|---|
| $M_2$ | 6.4 |
| $S_2$ | 4.4[c] |
| $N_2$ | 6.6 |
| $K_2$ (lunar) | 4.6 |
| $K_2$ (solar) | 4.6 |
| $L_2$ | 5.5 |
| $2N_2$ | 6.6 |
| $T_2$ | 4.8 |
| $K_1$ (lunar) | 1.8 |
| $K_1$ (solar) | 1.8 |
| $O_1$ | 1.7 |
| $P_1$ | 1.5 |
| $Q_1$ | 1.7 |
| Atmospheric tides | |
| $S_2$[a] | −0.1 |
| $S_2$[b] | −0.3 |
| Satellite results | |
| $M_2$ | 5.7 |
| $S_2$ | 4.6[c] |

[a] To be used with the ocean model results
[b] To be used with the satellite results
[c] Equivalent phase angle for combined ocean and atmosphere

angles, based on the ocean models discussed in chapter 6, are given in table 10.2.

A complete analysis also requires a development of (i) the tide raised by the Earth on the Moon and (ii) the tide raised on the Earth by the Sun. The external potential of the first tide follows from (6.3.1) by interchanging Earth and Moon parameters. This gives the potential per unit mass of the Earth. To estimate the effect of this on the Earth–Moon motion, it must first be multiplied by $M/m_{☾}$ before substituting it into the Lagrange equations of motion. Complete expressions are given by Kaula (1964). For a given $2mpq$ the ratio of perturbations in $a$, $e$, $I$ due to the Moon and Earth tides is given by

$$\frac{\text{Moon tide effect}}{\text{Earth tide effect}}=\left(\frac{M}{m_☾}\right)^2\left(\frac{R_☾}{R}\right)^5\frac{(k_2 \sin \varepsilon_{2mpq})_☾}{k_2 \sin \varepsilon_{2mpq}}\simeq 0.67\frac{(\sin \varepsilon_{2mpq})_☾}{\sin \varepsilon_{2mpq}}$$

with $(k_2)_☾ \simeq 0.02$. Data from the Apollo seismic network indicate an upper-mantle shear-wave $Q$ higher than several thousand. Below 600-km depth $Q$ is of the order of 300 with a central-region $Q$ of about 100 (Toksöz *et al.* 1974).† For a lower limit of $Q \simeq 150$, $\varepsilon_{lmpq} < 1°$, compared with an observed tidal effective lag of about 5° for the Earth. Hence

$$(\text{Moon tide effect})/(\text{Earth tide effect}) \lesssim 0.15.$$

The Moon tide equivalent to $M_2$ on the Earth ($lmpq = 2200$) is a permanent deformation and does not contribute to the dissipation. Other tides raised on the Moon do not contribute significantly to changes in the semi-major axis of the lunar orbit. The contribution of Moon tides to $\dot{I}$ is also very small but the contributions to the eccentricity of the lunar orbit may be relatively important since the principal effect on eccentricity results from the ellipticity of the lunar orbit or from the $N_2$ component.

Solar tides raised on the Earth give imperceptible perturbations in the Earth's orbital motion and these can be estimated from the expressions for the Moon tide perturbations upon substituting the Sun for the Earth, and the Earth for the Moon; that is, by substituting solar parameters for the lunar constants in (10.2.4) or (10.2.5) and multiplying by $m_\odot/M$. The Earth's orbit has undergone little tidal evolution, even when integrated over the age of the solar system, but the Sun's torque on the solar tide raised on the planet does modify the Earth's spin by a significant amount and must be included in discussions of the spin history and energy dissipation. Thus a term $H_\odot \cos I_\odot$ must be added to (10.2.2*a*).

The rate at which tidal energy is dissipated in the Earth–Moon system can be calculated in several equivalent ways. These include the computation of (i) the time average of the rate of work done by the Moon on the Earth, (ii) the rate of work done by the Earth on

† Yoder (1979), however, finds a $Q$ of about 25 at the libration frequencies! This provisional value must be considered with considerable caution but if correct it would support Anderson & Minster's argument that $Q \propto \omega^{1/3}$.

the Moon, and (iii) the energy balance between the spin and orbital motions. This last method, involving directly the quantities $\dot{\Omega}_T$ and $\dot{n}_☾$, does not require a knowledge of the energy sink and follows directly from the astronomical data.

The rotational energy associated with the Earth's spin is

$$E_1 = \tfrac{1}{2}C\Omega^2,$$

and the rate of change of rotational energy is

$$\mathrm{d}E_1/\mathrm{d}t = (\mathrm{d}/\mathrm{d}t)(\tfrac{1}{2}C\Omega^2) = C\Omega\dot{\Omega}_T,$$

where $\dot{\Omega}_T$ is the total tidal acceleration of the Earth, $(\dot{\Omega}_T)_☾ + (\dot{\Omega}_T)_\odot$. The energy associated with the orbital motion is

$$E_2 = \tfrac{1}{2}a_☾^2 n_☾^2[m_☾ M/(m_☾ + M)] - GMm_☾/a_☾ = -GMm_☾/2a_☾$$

and includes the potential and kinetic energies. The change in this energy state is

$$\mathrm{d}E_2/\mathrm{d}t = -\tfrac{1}{3}m_☾ n_☾ a_☾^2 \dot{n}_☾$$

and represents about $n_☾/\Omega$ or 4% of the total spin energy. The total rate of energy dissipation is†

$$\frac{\mathrm{d}E}{\mathrm{d}t} = \frac{\mathrm{d}E_1}{\mathrm{d}t} + \frac{\mathrm{d}E_2}{\mathrm{d}t} = C\Omega\dot{\Omega}_T - \tfrac{1}{3}m_☾ n_☾ a_☾^2 \dot{n}_☾, \qquad (10.2.7a)$$

$$= (5.86\dot{\Omega}_T - 9.58\dot{n}_☾)10^{40}\ \mathrm{erg\ s^{-1}}. \qquad (10.2.7b)$$

$\dot{\Omega}_T$ relates to the lunar acceleration by (10.2.2*b*). Thus to determine $\dot{\Omega}_T$ the $\dot{e}_☾$ and $\dot{I}_☾$ are also required, as are the contributions from the solar tides. Of these various contributions, that related to $\dot{n}_☾$ represents about 75% of the total, and the remainder can be calculated with adequate precision once the energy sink has been established. Table 10.3 summarizes the various quantities based on a nominal acceleration $\dot{n}_☾$ of $-1.5 \times 10^{-23}$ rad s$^{-2}$. Of this amount, about 80% is due to the $M_2$ tide (table 10.1). The change in semi-major axis follows from (10.2.4*a*) and with (10.2.4*b*) this determines the phase lag $\varepsilon_{lmpq}$. Once $\varepsilon_{lmpq}$ is known $\dot{e}_☾$ and $I_☾$ follow from the second and third equations of (10.2.4*a*). The tidal acceleration of the Earth follows from (10.2.2*b*). Solar contributions are estimated as indicated above, once the corresponding

† In the energy balance one should include the kinetic, elastic and gravitational energies associated with changes in the deformation. These are small compared with $E_1$.

Table 10.3. *Estimates of tidal accelerations, the observed secular acceleration of the Earth, lunar and solar torques and the rate of energy dissipation based on nominal values of* $\dot{n}_{☾} \equiv 30''$ *century*$^{-2}$ *and* $\dot{\Omega} \equiv 1100''$ *century*$^{-2}$

| | | |
|---|---|---|
| $\dot{n}_{☾}$ | $-1.49 \times 10^{-23}$ rad s$^{-2}$ $\equiv -30''$ century$^{-2}$ | Observed |
| $\dot{a}_{☾}$ | $1.44 \times 10^{-7}$ cm s$^{-1}$ $\equiv 4.5$ cm yr$^{-1}$ | Equation 10.2.4*a* |
| $\dot{e}_{☾}$ | $5.7 \times 10^{-19}$ s$^{-1}$ | Equation 10.2.4*a* |
| $\dot{I}_{☾}$ | $-4.2 \times 10^{-19}$ s$^{-1}$ | |
| $(\dot{\Omega}_T)_{☾}$ | $-6.19 \times 10^{-22}$ rad s$^{-2}$ | Equation 10.2.2*b* |
| $(\dot{\Omega}_T)_{\odot}$ | $-1.34 \times 10^{-22}$ rad s$^{-2}$ | |
| $\dot{\Omega}_o$ | $-5.5 \times 10^{-22}$ rad s$^{-2}$ $\equiv 1100''$ century$^{-2}$ | Observed |
| $L_{☾}$ | $4.98 \times 10^{23}$ dyn cm $\equiv 4.98 \times 10^{16}$ N m | Equation 10.2.3 |
| $L_{\odot}$ | $1.08 \times 10^{23}$ dyn cm $\equiv 1.08 \times 10^{16}$ N m | Equation 10.2.3 |
| $dE/dt$ | $4.5 \times 10^{19}$ erg s$^{-1}$ $\equiv 4.5 \times 10^{12}$ W | Equation 10.2.7*b* |

phase lags are known. We use the equivalent phase angles of table 10.2. The estimate of the rate of energy dissipation, about $4 \times 10^{19}$ erg s$^{-1}$, compares with an average rate of $3 \times 10^{18}$ erg s$^{-1}$ released by earthquakes and a geothermal flux of $3 \times 10^{20}$ erg s$^{-1}$.

There are three approaches to estimating the above-mentioned quantities. The first is from the analysis of astronomical observations of the Sun's and Moon's motion. The second method is to evaluate the tidal energy dissipation in the world's oceans as first attempted by Jeffreys (1920) and Heiskanen (1921). This method has generally been considered to be much less precise than the first approach, but the most recent calculations by Lambeck (1975*a*, 1977) indicate that this is no longer the case: the tidal dissipation calculations are as precise as the present astronomical data. The third method is to apply the results for tidal parameters estimated from close Earth satellites directly to the lunar problem, since the parameters that cause short-period perturbations in the satellite orbits also describe the secular evolution of the lunar orbit (Lambeck 1975*a*; Lambeck & Cazenave 1977). This method has several advantages: (i) no assumptions need be made as to where dissipation occurs in the Earth; (ii) it enables a separation of the amount of dissipation that occurs in the Earth from that occurring in the Moon, should the latter be significant, by comparing satellite results with astronomical accelerations; and (iii) it enables the accelerations to be estimated separately for each tidal frequency of both lunar and solar tides. Results obtained from the satellite analysis, although only preliminary, are in essential agreement with those obtained by the other two methods.

## 10.3 Astronomical evidence

Astronomers traditionally observe the motion of stars, planets and satellites either with respect to each other or with respect to their observatory's meridian, and compare these measurements with theories of planetary motion. The observed positions are referenced to the universal time (UT) kept by the Earth whereas the computed positions assume a uniform time scale, ephemeris time (ET). Moreover, the computed positions are usually derived from a gravitational theory describing the motions in a conservative system. Resulting discrepancies between the observed and computed positions, apart from observational and computational errors, could then arise from (i) non-uniform rotation of the Earth resulting in a non-linear relation between UT and ET, and (ii) accelerations of the planets and satellites due to dissipative forces.

Only tidal forces lead to significant dissipation in the present solar system. Of the bodies observed with the purpose of establishing the differences between UT and ET – that is, Sun, Moon, Mercury and, to a lesser extent Venus – only the Moon has a non-Newtonian acceleration (chapter 5). For the Sun, the discrepancies in longitude are expressed as (equation 5.1.5$b$)

$$\Delta\lambda_{\odot} = a_{\odot} + b_{\odot}T + \tfrac{1}{2}c_{\odot}T^2 + \beta(T). \qquad (10.3.1)$$

For a planet of mean motion $n_{\mathrm{p}}$

$$\Delta\lambda_{\mathrm{p}} = (n_{\mathrm{p}}/n_{\odot})\,\Delta\lambda_{\odot}, \qquad (10.3.2)$$

but for the Moon (equation 5.1.8)

$$\Delta\lambda_{☾} = a_{☾} + b_{☾}T + \tfrac{1}{2}c_{☾}T^2 + (n_{☾}/n_{\odot})\beta(T). \qquad (10.3.3)$$

Observations of $\Delta\lambda_{\odot}$ or $\Delta\lambda_{\mathrm{p}}$ and $\Delta\lambda_{☾}$ establish both the non-uniform rotation $\dot{\omega}_3(T)$ of the Earth according to (equation 5.1.7)

$$\dot{\omega}_3(T) = -(\Omega/n_{\odot})[c_{\odot} + \ddot{\beta}(T)], \qquad (10.3.4)$$

and the lunar tidal acceleration according to (equation 5.1.12)

$$\dot{n}_{☾} = c_{☾} - (n_{☾}/n_{\odot})c_{\odot}. \quad (10.3.5)$$

The success in estimating these accelerations from the observed planetary, solar, and lunar longitude discrepancies depends on a number of factors, including: (i) completeness of the planetary and lunar gravitational theories, (ii) correctness of the assumption that only the Earth and Moon undergo tidal accelerations, (iii) reliability of the data, and (iv) ability to separate the secular accelerations in $\omega_3$ from the long-period fluctuations $\beta(T)$.

The correctness and completeness of the orbital theories have been consistently debated since Halley first suggested in 1695 that the Moon may be accelerated in longitude. The circumstances and the contributions to this discussion by Euler, Lagrange, Laplace, Adams, Delaunay and others are well known (Berry 1961; Munk & MacDonald 1960; Muller 1975). Delaunay's analytical theory has been verified by De Prit, Henrard & Rom (1971) using algebraic manipulation programs on a computer. Brown's lunar theory, used in many of the recent discussions, has also been verified analytically with the aid of a computer (Woolard 1953) and by numerical integration (Garthwaite, Holdridge & Mulholland 1970). It appears that these theories are adequate for the present purposes, with the possible exception that the integration constants may not be known with sufficient precision (Kovalevsky 1977). Evidence in the astronomical record for the tidal accelerations span a time interval of more than 3000 yr and the discussion of the data is conveniently divided into three parts: (i) the telescope data referenced to atomic time from 1955 to the present, (ii) the pre-atomic-time telescope observations spanning an interval of nearly three centuries, and (iii) the ancient and medieval observations of eclipses, conjunctions and oscillations, back into the thirteenth century BC. The significance and reliability of these data are discussed below.

Successful separation of $\dot{\Omega}(T)$ and $\dot{n}_{☾}$ will depend very much upon the time span covered by the available data, and the discussion in chapter 5 suggests that at least 300 yr of data are required if $\beta(T)$ can be considered the consequence of random events accelerating the Earth. The separation becomes barely possible with the telescope observations (Morrison 1973). Ancient astronomical observations extending back several millennia should result in estimates of the secular accelerations that are free from contamination by the irregular fluctuations $\beta(T)$, but, in any event, the precision of these observations will in general be less than the amplitude of $\beta(T)$ (Muller 1975).

### 10.3.1 *Pre-atomic time telescope observations*

The nominal value for the lunar acceleration, $\dot{n}_{☾} = -1.12 \times 10^{-23}$ rad s$^{-2}$, is that attributed to Spencer Jones (1939) (see also Clemence 1948) and is based on telescope observations made since 1680 of occultations of stars by the Moon, of longitudes of the Sun, and of transits of Mercury and Venus across the Sun's face. Spencer Jones's results have been verified by K. P. Williams in 1940 and by Clemence in 1943. His error estimate of $\dot{n}_{☾}$ is $\pm 0.04 \times 10^{-23}$ rad s$^{-2}$ but Morrison (1972) argues that $\pm 0.35 \times 10^{-23}$ rad s$^{-2}$ may be a more realistic value. The most recent discussion of the Mercury transit data is by Morrison & Ward (1975) who have used all available data from 1677 to 1973. They consider that the discrepancies between the observed (UT) and computed (ET) times of transit are a consequence of (i) the irregular nature of UT and (ii) errors in the constants defining Mercury's motion. As the transit observations occur always at one of the same two positions in the orbit, any errors in the adopted elements can give rise to constant or secular errors in the longitude discrepancy. A combination of these discrepancies with those in the lunar longitude (Morrison 1973) yields two sets of equations of the type (10.3.2) with (10.3.1) and (10.3.3). Their solution determines $\dot{n}_{☾}$, $\dot{\Omega}$, $\beta(T)$ as well as any corrections to the orbital elements of Mercury. A high degree of correlation between the Earth's rotation and these corrections, however, results in unreliable estimates for $\dot{\Omega}$ and $\beta(T)$. Morrison & Ward's estimate for $\dot{n}_{☾}$, based on data from 1677 to 1973, gives $(-1.29 \pm 0.1)10^{-23}$ rad s$^{-2}$, essentially in agreement with the Spencer Jones–Clemence value, the difference being due to (i) more Mercury transit data, (ii) revised lunar data, and (iii) adjusting the orbital constants. It has been argued (see, for example, Van Flandern 1975) that Spencer Jones's value could be erroneous as it depends upon early observations that may be subject to large systematic errors. To test this assertion, Morrison & Ward have also determined $\dot{n}_☾$ from observations made after 1788 and obtain $\dot{n}_☾ = (-1.39 \pm 0.1)10^{-23}$ rad s$^{-2}$, a result not significantly different from the earlier value based on the complete data set. There is no reason for rejecting the earlier data. The solutions by Spencer Jones, Clemence, Morrison and Ward all solve for only a subset of parameters: Morrison and Ward, for example, do not solve for the orbital elements of the Sun and Moon. Oesterwinter & Cohen (1972) find, however, that the lunar integration constants are quite sensitive to small changes in the orbital elements of the other planets and they stress the need for a general re-adjustment of the elements of all the planets and the Moon. They have attempted just this, using planetary and lunar observations from 1911 to 1969, and amongst many other parameters find $\dot{n}_☾ = (-1.9 \pm 0.40)10^{-23}$ rad s$^{-2}$. Further analyses, extending the optical data back to about 1750 and including radar observations of Mars, Mercury and Venus, are anticipated. Table 10.4 summarizes the various results.

### 10.3.2 *Post-atomic time telescope observations*

Since 1955 it has been possible to compare the Moon's motion directly against an atomic time scale that, apart from possible non-Newtonian gravitational effects, should maintain a well-defined linear relationship with ephemeris time. These observations are independent of the Earth's rotation and, as atomic time is several orders of magnitude more precise than ephemeris time, $\dot{n}_☾$ can be estimated from quite short intervals of data. A drawback of the reduced time span, however, is that any short-period deficiencies in the lunar theory will not average to zero. Several results have been published in recent years (see table 10.4): Van Flandern (1970) analysed some 7000 occultation observations from 1755 to 1969 and obtained $\dot{n}_☾ = (-2.6 \pm 0.8)10^{-23}$ rad s$^{-2}$. Later, with a further 5 yr of observations, using only photo-electric timing rather than visual timing, Van Flandern (1975) obtained $\dot{n}_☾ = (-3.2 \pm 0.9)10^{-23}$ rad s$^{-2}$ but Muller (1975) remarks that upon correcting some observations, Van Flandern reduces this value to $-1.8 \times 10^{-23}$ rad s$^{-2}$. Morrison (1973), using about 40 000 occultations in the period 1955–1972, found $(-2.1 \pm 0.3)10^{-23}$ rad s$^{-2}$. It is probably premature to draw conclusions from these values as

Table 10.4. *Summary of estimates of the acceleration $\dot{n}_☾$ and $\dot{\Omega}$ from telescope observations*

| Authors | $-\dot{n}_☾$ ($10^{-23}$ rad $s^{-2}$) | $-\dot{\Omega}$ ($10^{-22}$ rad $s^{-2}$) |
|---|---|---|
| Spencer Jones[a] | $1.12 \pm 0.35$[b] | |
| Munk & MacDonald | | 4.90[f] |
| Morrison & Ward (1975) | $1.29 \pm 0.10$[c] | |
| | $1.39 \pm 0.10$[d] | |
| Van Flandern (1970) | $2.58 \pm 0.80$ | |
| Van Flandern (1975) | $3.22 \pm 0.89$[e] | |
| Morrison (1973) | $2.08 \pm 0.30$ | 4.12[g] |
| Oesterwinter & Cohen (1972) | $1.89 \pm 0.20$ | |

[a] See Clemence (1948)
[b] Standard deviation estimated by Morrison (1972)
[c] Mercury transits from 1677 to 1973
[d] Mercury transits from 1789 to 1973
[e] According to Muller (1975) this value should be reduced to $1.78 \pm 0.89$
[f] Based on telescope observations analysed by Brouwer (1952)
[g] From telescope observations from 1663 to 1972. The values for $\dot{n}_☾$ and $\dot{\Omega}$ are highly correlated.

their time span is short. All values do lie above the telescope results based on nearly 300 yr of data and, if this difference is real, this may be a consequence of (i) deficiencies in the lunar theory that over a 15–20 yr time span give the appearance of a secular effect, (ii) secular changes in the length of the atomic second relative to the ephemeris second, and (iii) secular changes in the gravitational constant. Perhaps the only categorical statement that can be made is that it is not due to a change in the tidal acceleration over the last few centuries.

### 10.3.3 *Ancient astronomical observations*

Invaluable information on the secular accelerations of the Earth and Moon is found in the literature and chronicles of ancient civilizations. These records fall into two broad categories, (i) the large or total eclipses that do not require scientific observations and (ii) the conjunctions, occultations, positions and magnitudes that all require specialized observations. The former – often not recorded for their astronomical interest but rather for historical, literary or other reasons – frequently pose special problems of interpretation. The latter observations pose fewer problems of interpretation but assume that the observing techniques used in antiquity are known. Some special problems may arise when the early astronomer, concerned with problems other than tidal accelerations, may have been rather selective in his choice of data, as was, apparently, Ptolemy (Britten 1967; R. R. Newton 1976). These records usually specify the location from which certain phenomena were observed or the local solar time and data at which they occurred. Computed times and locations, assuming purely gravitational motions, will differ from the observed values due to the accelerations $\dot{n}_{☾}$ and $\dot{\Omega}$. Table 10.5 summarizes the types of ancient astronomical observations that have been analysed for the secular accelerations. Also given are the linear relations between $\dot{n}_{☾}$ and $\dot{\Omega}$ for each type of observation, in the form

$$X_i\dot{n}_{☾} + Y_i\dot{\Omega} = f(O-C),$$

where $f(O-C)$ is a function of the observed less computed positions or times of the recorded events.

(i) *Planetary conjunctions and occultations.* The secular acceleration of the Earth puts the observer's meridian ahead by the amount $\frac{1}{2}\dot{\Omega}T^2$ after a time $T$ (equation 5.1.4*b*, and neglecting the linear part). A fixed star therefore transits at a time $T-\frac{1}{2}\dot{\Omega}R^2/\Omega$ instead of at time $T$. A planet, moving with angular velocity $n_{\mathrm{p}}$ with respect to the stars, will be behind its computed position by an amount $-\frac{1}{2}\dot{\Omega}T^2 n_{\mathrm{p}}/\Omega$, and observations of the time of an occultation provide a direct measure of $\dot{\Omega}$. The time of the observation is recorded in local solar time, as the time elapsed since sunrise or sunset, for example. Numerous descriptions of occultations or conjunctions of stars by planets in antiquity exist but the useful records are limited to the inner planets, whose mean motions are relatively rapid. The most recent and complete discussion of the sources of such observation is by R. R. Newton (1976) who investigated Babylonian, Hellenic and Islamic sources from the sixth century BC to the

Table 10.5. *Observation equations of the form* $X_i\dot{n}_☾ + Y_i\dot{\Omega} = f(O-C)$ *for the various types of ancient observations*

| Type of observation | $X_i$ | $Y_i$ | Reference |
|---|---|---|---|
| (i) Time of occultation of star by planet | 0 | 1 | Table 10.6 |
| (ii) Time of solar equinox | 0 | 1 | Table 10.6 |
| (iii) Time of occultation of star by Moon | −27.4 | 1 | Equation 10.3.6*b* |
| (iv) Time of lunar or solar eclipse[a] | −29.6 | 1 | Equations 10.3.8 |
| (v) Maximum lunar eclipse magnitude | 1 | 0 | Equation 10.3.11*a* |
| (vi) Maximum solar eclipse magnitude | −29.8[b] | 1 | Equation 10.3.11*b* |
| (vii) Place of total solar eclipse | −29.0[b] | 1 | Figure 10.1 |

[a] Average value of R. R. Newton's (1970) condition equations
[b] These coefficients are dependent upon the position of the observer, and average values only are given

eleventh century AD. Newton has explored the Islamic records in particular detail but the increased precision of this extensive series of observations is largely offset by their comparatively recent epoch. One of the objectives set out by Newton in this study was to determine whether or not the planets undergo any real acceleration of their own but, in view of a total lack of physical mechanisms for such accelerations, it appears more useful to assume that the apparent accelerations of the planets relate to that of the Sun by (equation 10.3.2)

$$\Delta\lambda_p = (n_p/n_\odot)\,\Delta\lambda_\odot.$$

Then each planet provides an estimate of the apparent solar acceleration. The results from the planetary observations are unsatisfactory (table 10.6), due to rather large errors in the position and time measurements and the small amount of usable observations. Newton suggests that unpublished Babylonian and Islamic literature may contain further astronomical references and that

Table 10.6. *Summary of secular accelerations estimated from the solar and planetary observations*

| Data source | Data types | Epoch | $-\dot{\Omega}$ ($10^{-22}$ rad s$^{-2}$) from Newton (1970) | $-\dot{\Omega}$ ($10^{-22}$ rad s$^{-2}$) from Muller (1975) |
|---|---|---|---|---|
| Babylonian | Planetary conjunctions | 196 BC | $1.84 \pm 2.04$ | |
| Hellenic | Solar observations | 141 BC | $5.41 \pm 1.29$ | $6.30 \pm 0.65$ |
| Islamic | Solar observations | AD 925 | $5.59 \pm 1.39$ | $5.76 \pm 0.55$ (for epoch near AD 840) |
| Islamic | Solar and planetary data | AD 940 | $3.48 \pm 4.17$ | |

these may eventually lead to improved results. Certainly the Chinese astronomical records contain many references to occultations and conjunctions (see, for example, Ho 1966).

(ii) *Solar observations.* Observations of the times of the Sun's equinox passage give discrepancies between the observed and computed times of $-\frac{1}{2}\dot{\Omega}T^2/\Omega$. Principal sources of such observations are the records by Hipparchus in the second century BC (see, for example, Fotheringham 1918) and the Islamic records of Ibn Iounis of the ninth century AD (R. R. Newton 1970). Of the Hellenic solar data only the equinox observations of Hipparchus, as recorded by Ptolemy, are useful. Apparently the time of equinox passage is determined from observations of the solar declination by interpolating for the instant of zero declination. Times of the solstices are likewise determined from observing the solar declination but now, as the rate of change of declination is very small, these observations are not very sensitive. Muller (1975) has analysed the same data and finds comparable results, any difference being a consequence of different reliabilities attached to the data and of different estimation processes. Table 10.6 summarizes the Hellenic results by Newton and Muller. They are more satisfactory than the planetary data. The Islamic solar data, including equinox and the Hakemite longitude observations, are also more significant than the Islamic planetary data. Muller's analysis of the solar data agrees with that by Newton although the latter used more data.

(iii) *Lunar observations.* The Moon, with its own acceleration, will be ahead of its computed position in time $T$ by an amount $\frac{1}{2}\dot{n}_{☾}T^2$ and its total displacement in longitude with respect to the Earth is

$$\tfrac{1}{2}[\dot{n}_{☾} - n_{☾}\dot{\Omega}/\Omega]T^2. \qquad (10.3.6a)$$

Thus the Moon has an apparent acceleration of $\dot{n}'_{☾} = \dot{n}_{☾} - n_{☾}\dot{\Omega}/\Omega$. Comparisons of times of observed lunar conjunctions or occultations with computed positions give a linear relation between the two accelerations $\dot{n}_{☾}$ and $\dot{\Omega}$. These observations are few in number and the most famous of such observations are those recorded by Ptolemy in the second century AD. They have been discussed by Newcomb in 1875, Fotheringham (1915) and R. R. Newton (1970). Newton obtains a linear relation corresponding to (10.3.6*a*) of

$$-27.4\dot{n}_{☾} + \dot{\Omega} = (-11.9 \pm 2.6)10^{-23} \text{ rad s}^{-2}. \qquad (10.3.6b)$$

(iv) *Timed eclipses.* After a time $T$, sunrise or sunset will be late by $\frac{1}{2}\dot{\Omega}T^2/\Omega$ and the Moon is displaced in longitude by $\frac{1}{2}\dot{n}_{☾}T^2$. An eclipse will therefore occur late by

$$\tfrac{1}{2}[\dot{n}_{☾}/(n_{☾} - n_{\odot}) - \dot{\Omega}/\Omega]T^2, \qquad (10.3.7)$$

when measured with respect to local solar time. For nominal values of $\dot{n}_{☾} = -1.5 \times 10^{-23}$ rad s$^{-2}$ and $\dot{\Omega} = -5.5 \times 10^{-22}$ rad s$^{-2}$,

$$\Delta T \simeq 7.5 \times 10^{-19} T^2 \text{ s}.$$

After 20 centuries, $|\Delta T| \simeq 3 \times 10^3$ s and a precise determination of the time of the eclipse may provide a useful estimate of the linear relation (10.3.7). R. R. Newton (1970) has considered many such observations, but for most the times of contact are recorded to within the nearest third or half of an hour only. The mean of the Babylonian, Hellenic and Islamic lunar eclipse observations from 720 BC to AD 1001 gives the condition equation

$$-29.6\dot{n}_{☾} + \dot{\Omega} = (1.18 \pm 0.78)10^{-23} \text{ rad s}^{-2}. \qquad (10.3.8a)$$

The timed solar eclipse observations considered to be of possible interest by Newton are the eclipse of Theon of AD 364 and some Islamic observations in the ninth and tenth centuries AD. From Newton's results the mean observation equation is

$$-29.6\dot{n}_{☾}+\dot{\Omega}=(8.7\pm0.7)10^{-23}\ \text{rad s}^{-2}. \qquad (10.3.8b)$$

For two Babylonian observations discussed by Fotheringham (1935) and Stephenson (1972), but overlooked by Newton, the times of contact are estimated to be precise to within a few minutes. For the solar eclipse of 322 BC the following condition equation can be deduced from Stephenson's result:

$$-29.97\dot{n}_{☾}+\dot{\Omega}=-9.42\times10^{-23}\ \text{rad s}^{-2}. \qquad (10.3.8c)$$

For the lunar eclipse of 425 BC the observation equation is

$$-29.76\dot{n}_{☾}+\dot{\Omega}=-11.65\times10^{-23}\ \text{rad s}^{-2}. \qquad (10.3.8d)$$

The precision of these equations is high but their accuracy is difficult to assess, for, as Stephenson stresses, the accuracy depends not only on the precision of the time keeping but also on the determination of the instant of first contact, and, unless the Babylonians could predict eclipses with precision and the astronomers were awaiting the event, the actual times of contact may be quite uncertain. Stephenson is in the process of analysing a large number of Babylonian timed contacts recorded with a precision of ±4 min, but errors in the clock drifts may be much larger than this. Stephenson and L. V. Morrison are also in the process of analysing more recent and more precisely timed contacts and these results may contribute to the establishment of the $\beta(T)$ curve further back into the seventeenth century than the present telescope observations.

(v) *Magnitudes of eclipses.* The magnitude of a lunar eclipse is defined as the fraction of the Moon's diameter lying in the Earth's shadow and provides a measure of the angular separation of the centre of the Moon at a given instant of time. Usually the maximum magnitude $\mu_m$ attained during the eclipse and defined as

$$\mu_m=(R'-R'_E-D)/2R'$$

is recorded. $R'$ is the semi-diameter of the Moon and $R'_E$ is the radius of the Earth's shadow cone at the distance of the Moon, both as seen from the Earth. $D$ is the minimum distance between the centres of the Moon and axis of the shadow zone. An acceleration $\dot{n}_{☾}$ modifies the Moon's distance from the node by $\frac{1}{2}\dot{n}_{☾}T^2$ and the time of maximum magnitude is changed approximately by

$$\tfrac{1}{2}\dot{n}_{☾}T^2/(n_{☾}-n_{\odot}).$$

During this interval the shadow zone has moved through an angle

$$-\tfrac{1}{2}\dot{n}_{☾}T^2[n_{\odot}/(n_{☾}-n_{\odot})],$$

and the discrepancy in magnitude due to an error in acceleration $\delta\dot{n}_{☾}$ is

$$\begin{aligned}(\mathrm{d}\mu_{\mathrm{m}})_{☾} &= -\mathrm{d}D/2R'_{☾} \simeq -(\mathrm{d}\lambda_{\odot}/2R'_{☾})\tan I\\ &= (1/4R')[n_{\odot}/(n_{☾}-n_{\odot})]\delta\dot{n}_{☾}T^2\tan I. \qquad (10.3.9)\end{aligned}$$

Lunar magnitude observations therefore provide a direct estimate of the Moon's tidal acceleration and do not depend on the Earth's rotation. This is a consequence of the magnitude of the eclipse being the same for all observers capable of seeing the eclipse at all.

Solar eclipse magnitudes provide a more complex relation since the Moon's shadow on the Earth's surface is a small spot and the observed magnitude is very much position dependent. The change in magnitude due to the lunar acceleration is given by an expression similar to (10.3.9) but, due to the Earth's rotation, the shadow spot will be displaced on the Earth's surface by an amount $\frac{1}{2}\dot{\Omega}T^2$ and the magnitude will be further modified. For an observer on the equator and the Moon in the ecliptic, this further modification is given by replacing $\delta\dot{n}_{☾}$ in (10.3.9) by $\delta\dot{\Omega}$. The total discrepancy in the maximum magnitude becomes

$$(\mathrm{d}\mu_{\mathrm{m}})_{\odot} \simeq \frac{1}{4R'}\tan I\frac{n_{\odot}}{n_{☾}-n_{\odot}}(\delta\dot{n}_{☾}-\delta\dot{\Omega})T^2. \qquad (10.3.10)$$

The actual form of (10.3.10) depends on the position of the observer and in consequence the coefficients of $\delta\dot{n}_{☾}$ and $\delta\dot{\Omega}$ vary from one eclipse to the next. Records of both lunar and solar eclipse magnitudes have been analysed by R. R. Newton (1970) for Babylonian, Hellenic, Chinese and Islamic periods. He concludes that the precisions of these records are respectively 0.05, 0.024, 0.04, 0.012, based mainly on the assumption that the observations are accurate to the figures given in the original records. The Babylonian observations pose a special problem in that it is not clear whether the magnitudes are defined as the fraction of the diameter or of the area of the body being eclipsed. Substituting the above precision estimates into (10.3.9) suggests that the magnitude observations are quite sensitive to $\dot{n}_{☾}$ but the results obtained by Newton indicate the opposite. Apparently the magnitudes are much less precise than the above estimates. This is borne out by two Islamic observations of a partial solar eclipse of AD 901. The two observers are sufficiently near to each other that their observed magnitudes should not differ by more than 0.01 or 0.02, yet they actually differ by more than 0.10 (R. R. Newton 1970). Newton's analysis of the Chinese and Islamic lunar eclipse magnitudes leads to insignificant results, whereas the Babylonian and Hellenistic results are marginally significant. As we do not expect $\dot{n}_{☾}$ to have varied with time during the last few millennia (section 10.5.1), we take the mean of these two results. The condition equation is

$$0.47\dot{n}_{☾} = (-1.45 \pm 0.40)10^{-23} \text{ rad s}^{-2}. \qquad (10.3.11a)$$

Observations of magnitudes of solar eclipses have also been evaluated by R. R. Newton (1970). The mean of these observations gives the relation

$$-29.8\dot{n}_{☾} + \dot{\Omega} = (4.1 \pm 0.8)10^{-23} \text{ rad s}^{-2}. \qquad (10.3.11b)$$

Normally it would not be advantageous to compute a mean condition equation since the coefficients of $\dot{n}_{☾}$ vary considerably from one observation to the next, but Newton notes that these observations are subject to large biases: neither the lunar nor the solar magnitude observations contribute very much to Newton's solutions for the accelerations.

(vi) *Place of solar eclipse.* The path of totality of a solar eclipse is quite narrow, of the order of 100 km, and a comparison of the observed position of a total eclipse on the Earth's surface with the predicted position provides a precise measure of the amount that the Moon has gained on the Sun in time $T$. With (10.3.7) the eclipse occurs late by $\Delta T$ during which its path will have rotated through an angle

$$\tfrac{1}{2}\Omega[\dot{n}_{☾}/(n_{☾} - n_{\odot}) - \dot{\Omega}/\Omega]T^2. \qquad (10.3.12a)$$

With $\dot{n}_{☾} = -1.5 \times 10^{-23}$ rad s$^{-2}$, $\dot{\Omega} = -5.5 \times 10^{-22}$ rad s$^{-2}$, after 20 centuries the displacement of the path of totality is 14° or about

1400 km. Thus, the mere knowledge that an observer at a known position was in the zone of totality provides a precise measure of the linear relation between the two accelerations. For any given event the relation of the form (10.3.12*a*) depends upon the position of the observer and the general form is

$$a_i \dot{n}_☾ - \dot{\Omega} = f_i(O - C). \tag{10.3.12b}$$

It has long been recognized that the large or total solar eclipses are the most useful of ancient observations because of the narrow path width and because the dramatic nature of the event makes it a much more unmistakable phenomenon than a partial eclipse. These points have been emphasized most recently by Stephenson and Muller. Nevertheless records of total eclipse observations present severe problems in interpretation. Fotheringham in 1920 and W. De Sitter in 1927 discussed a large number of eclipse narratives. Munk & MacDonald have reviewed their work. Most of these discussions centred on the Greek and Roman classical records. More recently, Curott (1966) has discussed some Chinese records and Stephenson (1972) has explored the Far Eastern chronicles in greater detail. The latter's efforts have led to a longer record of eclipses which are further discussed by Muller & Stephenson (1975) and by Muller (1975). In these studies, as well as that by R. R. Newton (1970), there has been a critical re-evaluation of the records analysed by Fotheringham, and a corresponding drift away from the classical records to more reliable historical documents has occurred. The hazardous nature of interpreting the eclipse narratives is well illustrated by the following examples:

(*a*) 'On the day of the new Moon, in the month of Hiyar, the Sun was put to shame, and went down in the daytime, with Mars in attendance.' This tablet was found in the city of Ugarit and is discussed by Sawyer & Stephenson (1970). The questions to be considered are: is this a record of a total eclipse?; if it is, when and where was it observed? Stephenson (1972) concludes that this eclipse occurred in 1375 BC but that there is some uncertainty whether or not it was in fact total (see also Muller & Stephenson 1975). Newton has not discussed this record. This tablet is the oldest viable record of a solar eclipse known at the present time. Sawyer & Stephenson's *rediscovery* of this text makes it hopeful that further early records may come to light in the future, especially from still untranslated Sumerian tablets. Sawyer & Stephenson's approach was to first select an area for which a sufficiently representative block of literature survived, in this case ancient Ugarit; next they computed the times of principal solar eclipses that may have been visible there during the period covered by the texts; and then they searched the literature for possible astronomical references written after these predicted times.

(*b*) 'On the next day [after the escape] there occurred such an eclipse of the Sun that utter darkness set in and the stars were seen everywhere; wherefore Agathocles' men believing that the prodigy portended misfortune for them, fell into even greater anxiety about the future.' This is the eclipse of Agathocles, and the record is from Diodorus and was written some three centuries later. The date of the event is well established as 310 BC, as Diodorus states that Hieromnenon was archon in Athens at the time. The record does have other problems. Did the eclipse really occur during Agathocles' escape from Sicily or did the writer evoke it to lend further excitement to his story, perhaps assimilating it with an eclipse that occurred at some other place and time? Secondly, where was the eclipse observed? Did Agathocles sail to the north, or to the south, from Sicily? Muller gives this record zero weight.

(*c*) 'Once when it [i.e. the Sun] was observed to be completely eclipsed at the Hellespont, at Alexandria it was eclipsed with the exception of a fifth of its diameter.' This is the famous eclipse of Hipparchus, and may be one of the most discussed observations in the history of astronomy. The eclipse is clearly said to be total and the place of observation is approximately known. The unknown quantity is the date. Hipparchus's original record is lost and it is not clear whether he observed it himself during his lifetime or whether he used previous records to illustrate his parallax calculations. Stephenson and Muller both stress that this record could refer to any suitable eclipse between the founding of Alexandria in 332 BC and the death of Hipparchus circa 120 BC. Fotheringham concludes that the most likely date is 129 BC; Muller gives it zero weight.

Fotheringham discussed many other records but he considers this one, plus the Plutarch eclipse of AD 71 and the Eponym Canon of 762 BC, as the most reliable. It is constructive to consider the views of more recent writers. Newton and Stephenson discard the Plutarch record as being unreliable, holding the opinion that the circumstances surrounding it are too vague to establish its reality and, if it is real, its place of observation. Muller also doubts that this record is real and assigns a low weight to it. The Eponym Canon is a chronicle of important or striking events that occurred in all parts of ancient Assyria and includes a reference to an eclipse dated at 762 BC. The place of observation must be considered as unknown. Also, the record does not explicitly state that the eclipse was total. Newton considers this a reliable record. Stephenson and Muller are more cautious and the latter excludes it altogether from his solution for the accelerations. Newton considers that the *Thucydides* eclipse of 431 BC merits the most confidence of all the classical observations even though he is unsure of the place of observation. Stephenson questions the veracity of the records of this event, and Muller concludes that it is unsafe to use. *Phlegon*'s eclipse of AD 29 is discarded by all these authors, either because it is considered a *magical* eclipse (Newton) or because the place of observation is unknown (Muller). Dicke (1966), following Fotheringham, does accept that this record is reliable. Most other classical eclipse records are subject to similar divergences of opinion.

The main differences in the studies of R. R. Newton (1970), Stephenson (1972) and Muller (1975) are that the latter two authors use only eclipses that are total, that are annular or whose totality is specifically denied so that the magnitude is known precisely. R. R. Newton (1970, 1972) has made a detailed study of solar eclipses during some twenty centuries up to the thirteenth century AD. Most of these eclipses are partial, however; large magnitudes are implied only. From medieval chronicles from the fifth to thirteenth century, R. R. Newton (1972) discussed nearly 400 records but Stephenson (1972), covering similar source material, finds references to only about 20 eclipses that can be interpreted as total or annular. Newton adopts a statistical approach to estimating the accelerations. Stephenson and Muller, however, point out that such a method may lead to serious biases in the estimates. Many of the medieval eclipses have been recorded independently in town and monastery records. For example, G. Celoria discussed 48 descriptions of an eclipse in AD 1239. F. K. Ginzel also lists many eclipses that have been described in at least 20 independent records. Muller and Stephenson stress that the use of partial eclipses may lead to serious *population biases* in that a least-squares estimation process tends to place the narrow central path through the middle of the observatories, whereas the real path may be considerably displaced from this. This would not be very serious if the observers were quite uniformly distributed along both sides of the path of totality but, with the tendency for monasteries and towns during the Middle Ages to cluster in certain areas of Europe, this condition is seldom met. To avoid this, Stephenson and Muller use a much smaller data set than does Newton, but they argue convincingly that in this manner they avoid the population bias that is probably responsible for R. R. Newton's (1972) aberrant result for $\dot{n}_{☾}$ (see table 10.7). Other important differences between the solutions of Newton on the one hand, and of Muller and Stephenson on the other, centre around the reliability estimates associated with the various records. Important differences of opinion exist some of which have been mentioned above. Also Newton considers the Chinese records to be generally less reliable than do Stephenson and Muller. Some 37 solar eclipses, covering a time span from 1375 BC to AD 1567 have been used in the final solutions by Muller & Stephenson (1975) and Muller (1975, 1976). The corresponding observation equations of the form (10.3.12*b*) differ sufficiently in the coefficients $a_i$ to permit a satisfactory separation of the two accelerations.

The linear relations between $\dot{n}_{☾}$ and $\dot{\Omega}$ are summarized in figure 10.1 for all observations except for the total solar eclipse results. They include: (i) Muller's (1975) solution for the equinox observations (table 10.6); (ii) lunar occultations (equation 10.3.6*b*); (iii) timed eclipses (equations 10.3.8*a–d*); and (iv) eclipse magnitudes (10.3.11*a, b*). The standard deviations associated with each solution are indicated, as is the weighted least-squares solution of all these data except for the two precisely timed Babylonian eclipses (10.3.8*c, d*). From Muller's observation equations, solutions can also be be determined based on (i) only the large solar eclipse data and (ii) these solar eclipses *plus* equinox observations. These agree well with each other, with the precise Babylonian times of contact, and with Newton's lunar occultation result. Agreement with the other data sets is less satisfactory. The uncertainty estimates of the

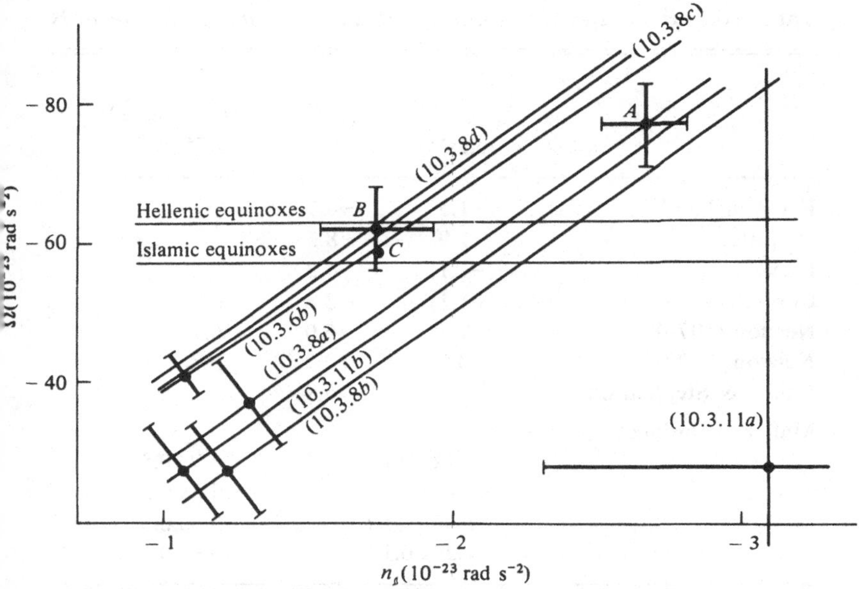


Figure 10.1. Summary of linear relations between $\dot{n}_{☾}$ and $\dot{\Omega}$: (10.3.6*b*) lunar occultations; (10.3.8*a*) timed lunar eclipses; (10.3.8*b*) timed solar eclipses; (10.3.8*c*) timed solar eclipse of 322 BC; (10.3.8*d*) timed lunar eclipse of 425 BC; (10.3.11*a*) lunar eclipse magnitudes; (10.3.11*b*) solar eclipse magnitudes. Point *A* indicates the least-squares solution of these plus the equinox data. Point *B* is Muller's solution for large solar eclipses only, and point *C* is Muller's solution for large solar eclipses and the equinox observations.

timed contacts and lunar occultations are, however, such that, when they are combined with the large eclipse and equinox data, they do not contribute much to the overall solution.

Table 10.7 summarizes a variety of results. Fotheringham's solution, as given by Munk & MacDonald, is determined essentially from three eclipses: Plutarch, Hipparchus and the Eponym Canon, three accounts that must be viewed with little confidence for the reasons discussed above. De Sitter's analysis, also quoted by Munk & MacDonald, covers similar data but with the addition of a Babylonian lunar eclipse. Dicke (1966) and Curott (1966) adopt Spencer Jones's value for $\dot{n}_{☾}$ and use the eclipse observations to determine the non-tidal acceleration of the Earth. These solutions are of interest only in that they represent the pitfalls inherent in the

Table 10.7. *Estimates of $\dot{n}_☾$ and $\dot{\Omega}$ from ancient astronomical records*

| | $\dot{n}_☾$ ($10^{-23}$ rad $s^{-2}$) | $\dot{\Omega}$ ($10^{-22}$ rad $s^{-2}$) | $\dot{\Omega}$ (corrected)[e] ($10^{-22}$ rad $s^{-2}$) |
|---|---|---|---|
| Fotheringham[a] | −1.5 | −6.7 | |
| De Sitter[a] | −1.9 | −8.3 | |
| Dicke[b] | (−1.1) | −4.3 | |
| Curott[b] | (−1.1) | −3.8 | −6.9 |
| Newton (1970) | 2.1[c] | −6.0 | −6.7 |
| Newton (1972) | 3.9 | −10.2 | |
| Muller & Stephenson | 1.9 | −6.9 | |
| Muller's solutions[d] | | | |
| | −1.72±0.10 | | −5.9±0.5 |
| | −1.49±0.10 | | −5.5±0.5 |
| | −1.40±0.10 | | −5.5±0.5 |
| | −1.36±0.10 | | −6.6±0.5 |

[a] As quoted by Munk & MacDonald
[b] With adopted values of −1.1 for $\dot{n}_☾$
[c] Mean of values determined for epochs 200 BC and AD 1000
[d] See text for explanation of these four sets of values
[e] Values corrected by Muller and Stephenson

analysis of ancient eclipse records. Of the five eclipses used in Dicke's solution, for example, two cannot be dated, the places of two others cannot be located and the fifth is the eclipse of Hipparchus! As Stephenson points out, the fact that Dicke finds close agreement between the results is in part due to the fact that the two earliest eclipses have been dated from the astronomical observations themselves! R. R. Newton (1970) found two values for $\dot{n}_☾$, one centred at epoch 200 BC and the other centred at epoch AD 1000, from a detailed discussion of ancient and medieval records. In a second exhaustive study of the medieval records, including a large quantity of partial eclipses, R. R. Newton (1972) finds a value that is considerably larger than his previous values and argues that an important change in the lunar acceleration occurred near the year AD 700. This is improbable (see section 10.5.1), and

Muller & Stephenson (1975) argue that Newton's results are strongly biased by his use of partial eclipses (see also Muller 1975). They find no evidence for such a change. There is also considerable discrepancy between the various estimates of $\dot{\Omega}$. If this quantity is constant then $\mathcal{J}(T)$ is a quadratic function of $T$ and $\dot{\Omega}$ follows from $\Omega C_{\odot}/n_{\odot}$. Newton, Curott and others have analysed $\mathcal{J}(T)$ by assuming that its linear part vanishes at the epoch AD 1900, whereas Muller and Stephenson show that it actually vanishes near AD 1770. This oversight introduces a significant error in $\dot{\Omega}$, and the last column of table 10.7 gives the values as corrected by Muller & Stephenson (1975).

Muller's solutions for $\dot{n}_{☾}$ and $\dot{\Omega}$ are given in the second part of table 10.7. Line 1 is the solution based on the eclipse and equinox observations only and is illustrated in figure 10.1. Muller considers that the potentially most important source of systematic errors comes from the uncertainty with which the rate of the lunar node, $\dot{\Omega}_{☾}$, is known. The motion of the node is caused by the Sun's, Earth's and other planets' attraction and depends on the mass distributions in the Earth and Moon. Until relatively recently, the latter were not sufficiently well known to enable $\dot{\Omega}_{☾}$ to be computed theoretically and its value is usually based on observations. Eckert (1965) reviews the unsatisfactory status of the comparison of observed and theoretical values of $\dot{\Omega}_{☾}$.

Muller (1975) estimates that an error $\delta\dot{\Omega}_{☾}$ in $\dot{\Omega}_{☾}$ will reduce the lunar tidal acceleration by about $-3.0\times10^{-10}\delta\dot{\Omega}_{☾}$. Three recent results indicate that $\delta\dot{\Omega}_{☾}$ may be important. (i) Martin & Van Flandern (1970) have analysed the lunar occultation data between 1780 and the present and find $\delta\dot{\Omega}_{☾}=(+6.6\pm0.7)10^{-15}$ rad s$^{-2}$. (ii) Muller (1975) has solved simultaneously for the accelerations and $\delta\dot{\Omega}_{☾}$ using the eclipse and equinox results, Morrison & Ward's (1975) lunar occultation and Mercury transit results, and Van Flandern's (1975) corrected estimate of $\dot{n}_{☾}$. His result is $\delta\dot{\Omega}_{☾}=(+10.8\pm5.2)10^{-15}$ rad s$^{-2}$. (iii) P. M. Muller, X. Newhall, T. C. Van Flandern & J. G. Williams (personal communication 1978) have integrated the equations of motion of the Sun–Moon and planets back into time for more than 250 yr and find a theoretical correction of $(6.7\pm0.3)10^{-15}$ rad s$^{-1}$. Line 2 of table 10.7 gives the solar eclipse estimate of $\dot{n}_{☾}$, after a correction of $6.7\times10^{15}$ rad s$^{-2}$ has been applied to the node (from Muller 1976). Line 3 gives a combined solution based on (i) the solar eclipses, (ii) the equinox data, (iii) Morrison & Ward's (1975) telescope solution (section 10.3.2), (iv) Van Flandern's corrected solution based on the atomic time comparison (section 10.3.1) and (v) the above node correction. In addition, Muller has also considered the consequences on the accelerations of a secular change in the gravitational constant. This gives the result in line 4. The cosmologies of Hoyle & Navlikar, Brans & Dicke, Peebles & Dicke and Dirac all give very similar results for $\dot{n}_☾$.

### 10.3.4 *Summary of results*

We adopt the last two solutions of table 10.7 for the accelerations of the Earth and Moon, depending on whether or not a secular change in the gravitational constant is permitted. For $\dot{n}_☾$ there is no significant difference and we adopt

$$\begin{aligned}\dot{n}_☾ &= (-1.35 \pm 0.10)10^{-23}\ \text{rad s}^{-2}\\ &\equiv (-28 \pm 2)''\text{century}^{-2}.\end{aligned} \qquad (10.3.13)$$

This value represents essentially the mean of the recent telescope observations and the ancient eclipse records and, as it is improbable that $\dot{n}_☾$ has varied significantly over the last three or four millennia, these two data sources should provide comparable estimates. Future improvements in the determination of $\dot{n}_☾$ may result from (i) an improved lunar ephemeris, based in part on improved constants of integration and in part on an improved theory for some of the planetary perturbations, (ii) additional eclipse records that extend the record further back into time, and (iii) precise laser ranging observations.

Muller's estimates for $\dot{\Omega}$ are somewhat larger than that estimated by Morrison (1973) from the telescope observations (table 10.4) and this may be a consequence of long periodic changes in l.o.d. that do not average to zero over the time span of the telescope data. Whereas a secular change in the gravitational constant does not modify $\dot{n}_☾$ significantly, it does have an important consequence on $\dot{\Omega}$, and further improvements in the determination will depend in part on independent evidence for $\dot{G}$. We adopt for the present the value

$$\dot{\Omega} = (-5.5 \pm 0.5)10^{-22} \text{ rad s}^{-2}$$
$$\equiv (1100 \pm 100)'' \text{ century}^{-2}. \qquad (10.3.14)$$

## 10.4 Ocean tide dissipation

The astronomical estimate of the amount of tidal energy dissipated in the Earth–Moon system is given by (10.2.7) and (10.2.2). With the accelerations (10.3.13 and 10.3.14),

$$\mathrm{d}E/\mathrm{d}t \simeq 4 \times 10^{19} \text{ erg s}^{-1}.$$

This amount excludes the small contributions from $\mathrm{d}e/\mathrm{d}t$ and $\mathrm{d}i/\mathrm{d}t$. The uncertainty is of the order of 10%. More problematical is the nature of the energy sink: where is the energy dissipated and by what mechanism? Recent studies confirm the dominance of the oceans as the sink and, by implication, the dissipation of tidal energy in the mantle, in the core and in the Moon is small (Lambeck 1977). The mechanism by which the energy is dissipated in the oceans remains, however, unclear.

### 10.4.1 *Energy dissipation integrals*

Laplace's tidal equations, excluding dissipation, are (see, for example, Lamb 1932; see also page 125)

$$\frac{\partial u_\lambda}{\partial t} - 2\Omega \sin\phi \, u_\phi + \frac{g}{R\cos\phi}\frac{\partial \xi'}{\partial \lambda} = \frac{1}{R\cos\phi}\frac{\partial \mathcal{U}}{\partial \lambda}, \qquad (10.4.1a)$$

$$\frac{\partial u_\phi}{\partial t} + 2\Omega \sin\phi \, u_\lambda + \frac{g}{R}\frac{\partial \xi'}{\partial \phi} = \frac{1}{R}\frac{\partial \mathcal{U}}{\partial \phi}, \qquad (10.4.1b)$$

where $u_\lambda$, $u_\phi$ are the longitudinal and latitudinal velocities and represent depth averages. $\mathcal{U}$ is the total ocean-tide-producing potential. It includes the potential $U_2$ of the direct lunar or solar attraction, the potential $k_2U_2$ of the solid tide, and the potential of the self-attraction of the ocean tide and of the Earth's deformation under this load. That is,

$$\mathcal{U} = (1+k_2)U_2 + \sum_n 3\frac{(1+k_n')}{2n+1}\frac{\rho_w}{\bar{\rho}}\xi_n. \qquad (10.4.2)$$

In (10.4.1) $\xi'$ is the geocentric tide while $\xi = \sum_n \xi_n$ is the tide with respect to the sea floor. Thus

$$\xi' = \xi + \xi^*$$

where

$$\xi^* = \frac{h_2 U_2}{g} + \sum_n \frac{3h'_n}{2n+1} \frac{\rho_w}{\bar{\rho}} \xi_n \tag{10.4.3}$$

is the deformation of the ocean floor due to the solid tide and the ocean load. The equation of continuity, equating the vertical tide with the flux out of a columnar space of height equal to the ocean depth $D$, is

$$\frac{d\xi}{dt} + \frac{1}{R \cos\phi}\left[\frac{\partial}{\partial\lambda}(u_\lambda D) + \frac{\partial}{\partial\phi}(u_\phi D \cos\phi)\right] = 0. \tag{10.4.4}$$

Applying the operation

$$(\text{equation } 10.4.1a) \times u_\lambda D + (\text{equation } 10.4.1b) \times u_\phi D$$

to the tide equations, and substituting into the continuity equation (10.4.4), gives

$$\begin{aligned}
&\rho_w D\left(u_\lambda \frac{\partial u_\lambda}{\partial t} + u_\phi \frac{\partial u_\phi}{\partial t}\right) + \rho_w g\left[(\xi + \xi^*)\frac{d\xi}{dt} + (\xi + D)\frac{d\xi^*}{dt}\right] \\
&\quad + \frac{\rho_w}{R \cos\phi}\left\{\frac{\partial}{\partial\lambda}[gDu_\lambda(\xi + \xi^*)] + \frac{\partial}{\partial\phi}[gDu_\phi \cos\phi(\xi + \xi^*)]\right\} \\
&= \frac{\rho_w}{R \cos\phi}\left[\frac{\partial}{\partial\lambda}(Du_\lambda \mathcal{U}) + \frac{\partial}{\partial\phi}(Du_\phi \cos\phi\, \mathcal{U})\right] \\
&\quad + \rho_w \mathcal{U}\frac{d\xi}{dt} + \rho_w g(\xi + D)\frac{d\xi^*}{dt}.
\end{aligned} \tag{10.4.5}$$

This equation represents the energy balance per unit of sea floor. The first term on the left-hand side, rewritten as

$$\tfrac{1}{2}\rho_w D(\partial/\partial t)(u_\lambda^2 + u_\phi^2) \equiv dE_k/dt,$$

represents the rate of change of the kinetic energy of the tidal currents per unit of ocean surface. The second term can be rearranged as

$$\tfrac{1}{2}\rho_w g(\partial/\partial t)[\xi^2 + 2(D + \xi)\xi^*] \equiv dE_p/dt,$$

and represents the rate of change of potential energy per unit of surface. The third term on the left-hand side can be written as

$$\rho_w g \nabla_s \cdot [D\mathbf{u}(\xi + \xi^*)] \equiv (d/dt)\nabla_s \cdot (\text{energy flux}),$$

where the operator $\nabla_s \cdot$ is the divergence with components along the horizontal surface, and $\mathbf{u} = (u_\lambda, u_\phi)$. This term represents the rate of divergence of the energy flux carried along by the currents. The right-hand side of (10.4.5) represents the total work done on the unit ocean surface by both the potential $\mathcal{U}$ and the sea floor.

Equation (10.4.5) can also be written as

$$(\mathrm{d}/\mathrm{d}t)(E_k + E_p) + (\mathrm{d}/\mathrm{d}t)\nabla_s \cdot (\text{energy flux}) = \mathrm{d}W/\mathrm{d}t.$$

With dissipation of tidal energy at a rate $\mathrm{d}E/\mathrm{d}t$, the energy equation becomes

$$\frac{\mathrm{d}}{\mathrm{d}t}(E_k + E_p) + \frac{\mathrm{d}}{\mathrm{d}t}\nabla_s \cdot (\text{energy flux}) = \frac{\mathrm{d}W}{\mathrm{d}t} + \frac{\mathrm{d}E}{\mathrm{d}t}. \qquad (10.4.6)$$

When this equation is integrated over the ocean surface, the energy flux term vanishes as there is no transport of water across the ocean–continent boundaries. The kinetic and potential energy terms also vanish if the equation is integrated over one tidal period $P$ and

$$-\overline{\left\langle \frac{\mathrm{d}E}{\mathrm{d}t} \right\rangle} = \int_{\text{Ocean}} \left\langle \frac{\mathrm{d}W}{\mathrm{d}t} \right\rangle \mathrm{d}S, \qquad (10.4.7a)$$

with

$$\left\langle \frac{\mathrm{d}W}{\mathrm{d}t} \right\rangle = \frac{1}{P} \int_{T=T_0}^{T_0+P} \frac{\mathrm{d}W}{\mathrm{d}t}\, \mathrm{d}t. \qquad (10.4.7b)$$

If, as is commonly believed, either for want of contradictory information or for reasons of convenience and tradition, bottom friction in shallow seas is the dominant dissipating mechanism, then

$$\mathrm{d}E/\mathrm{d}t = \alpha \mathbf{u} \cdot \mathbf{F},$$

where $\mathbf{F}$ is the bottom stress vector and can be written as $\rho_w |\mathbf{u}| \mathbf{u}$. Thus dissipation is proportional to the cube of the tidal current velocity; $\alpha$ is the coefficient of friction. G. I. Taylor in 1919 adopted $\alpha = 0.002$;† Brettschneider (1967) proposed $\alpha = 0.003$. In the open seas, the solution of the Laplace tidal equations gives tidal velocities that are typically of the order of $1\ \mathrm{cm\ s^{-1}}$ and with $\alpha \simeq 0.002$, $\alpha \rho_w u^3 \simeq 0.002\ \mathrm{erg\ s^{-1}\ cm^2}$. Integrating over the world's oceans gives

† This value is based on the observed friction coefficient for wind over the grassland of Salisbury Plain. Curiously enough it appears that friction along the sea floor is dynamically similar to wind friction over grass (Taylor 1919).

$-\langle\overline{dE/dt}\rangle \simeq 10^{16}$ erg s$^{-1}$, compared with the required amount of about $4\times10^{19}$ erg s$^{-1}$ if the astronomically observed tidal accelerations are a consequence of dissipation in the oceans. This is the usual argument to show that the dissipation is limited to very shallow seas where the observed tidal currents are much larger than those in the open seas. Even the coastal shelves, where the water depth is of the order of 200 m, are usually thought to provide an inadequate energy sink. Thus

$$-\left\langle\overline{\frac{dE}{dt}}\right\rangle = \int_{S'}\int_{T_0}^{T_0+P} \rho_w \alpha |\mathbf{u}|\mathbf{u}\cdot\mathbf{u}\, dT\, dS', \qquad (10.4.8)$$

where the surface integral is carried out over the shallow seas of area $S'$.

If dissipation is limited to shallow seas then $\langle\overline{dE/dt}\rangle$ can be evaluated from the energy balance across the entrances to these waters. That is, with the divergence theorem, the rate of energy flux across the entrance during one cycle of the tide, *plus* the rate of energy added by the work done on the enclosed sea by the Moon, equals the rate at which energy is dissipated in the enclosed body of water. The total energy flux across the boundary $L$ includes the rate at which work is done by the water entering across the boundary (the first term on the right-hand side of equation 10.4.5) and the kinetic and potential energies carried along by the tidal current, although Jeffreys (1929) concluded that only the first is important.† That is

$$-\left\langle\overline{\frac{dE}{dt}}\right\rangle = g\rho_w \int_L \int_T D\mathbf{u}\cdot\boldsymbol{\xi}\, dT\, dL + \int_{S'} \left\langle\frac{dW}{dt}\right\rangle dS'. \qquad (10.4.9a)$$

Furthermore, the areas of the shallow seas are generally quite small and the work done by the Moon and Sun on them is usually considered to be negligible. For example, in Taylor's calculation for the Irish Sea, the rate of change of energy flux is $6.4\times10^{17}$ erg s$^{-1}$ while the rate at which work is done by the Moon on the Irish Sea is

† Garrett (1975) observed that the sum of the first two work terms on the right-hand side of (10.4.5),

$$\nabla_s\cdot(D\mathbf{u}\mathcal{U}) + \mathcal{U}\, d\xi/dt,$$

is often much less than either term and that it is not valid to ignore the first as has usually been the case. Omitting the first, but retaining the second, leads to an overestimation of the importance of direct forcing in the energy equation.

only $0.4 \times 10^{17}$ erg $s^{-1}$. The rate at which energy is dissipated therefore reduces to

$$-\left\langle \overline{\frac{\mathrm{d}E}{\mathrm{d}t}} \right\rangle = g\rho_{\mathrm{w}} \int_L \int_T D\mathbf{u}\cdot\boldsymbol{\xi}\,\mathrm{d}T\,\mathrm{d}L. \qquad (10.4.9b)$$

However, from Zahel's (1977) maps of the global distributions of energy dissipated by bottom friction and of the rate at which work is done on the ocean surface, it is not evident that the rate of work done on the shallow sea can always be ignored. For several regions $|\mathrm{d}W/\mathrm{d}t|$ is considerably larger than $\mathrm{d}E/\mathrm{d}t$ and $\mathrm{d}W/\mathrm{d}t$ may be positive or negative.

The integrals (10.4.7), (10.4.8) and (10.4.9) summarize the standard discussion of dissipation (see, for example, Munk & MacDonald 1960) which has changed very little over the last 50 yr. Application of all three methods has been beset by numerous difficulties. On reading the original accounts of these attempts one is immediately struck by the paucity of relevant observations and by the hypotheses and extrapolations that have been made in order to arrive at a global estimate. It is surprising that there is any agreement at all between the various estimates.

The bottom friction method (integral 10.4.8) was first used by G. I. Taylor in 1919 in his discussion of dissipation in the Irish Sea and was extended to the world's oceans by Jeffreys (1920) and by Heiskanen in 1921. Miller (1964, 1966) applied the energy flux method (integral 10.4.9$b$). This method also assumes that dissipation is restricted to shallow seas, but has the advantage that the actual mechanisms need not be specified. As the energy flux is proportional to the first power of the tidal current velocity, it is generally considered to be more precise than the bottom friction method, although it does require a knowledge of the tide amplitude and its phase relation with the current velocity along the boundary $L$. Miller ignored the $\mathrm{d}W/\mathrm{d}t$ term in (10.4.9$a$).

The first integral (10.4.7) – the evaluation of the rate at which work is done by the Sun and Moon on the ocean surface – requires the global ocean tide to be known everywhere and it has often been assumed that the present models are inadequate for this. The advantages of the method are that it neither requires an assumption about the nature of the energy sink in the oceans nor does it depend upon a knowledge of the tidal currents. The method was first used by Heiskanen in 1921 and later by Groves & Munk (1958). The most recent attempts at directly evaluating the integral (10.4.7) have been made by Pariyskiy, Kuznetsov & Kuznetsova (1972), Kuznetsov (1972) and Hendershott (1972), although the form of the ocean tide potential (6.2.2) and its influence on the lunar orbit (equations 10.2.5) suggest that this calculation can be considerably simplified.

### 10.4.2 *The work integral*

The rate at which work is done on the ocean consists of two parts: (i) the rate at which body forces work on the ocean

$$\rho_w \mathcal{U}\, d\xi/dt + \rho_w \nabla_s \cdot [\mathbf{u}(\xi + D)\mathcal{U}], \qquad (10.4.10a)$$

and (ii) the rate at which the sea floor, moving due to the solid Earth tide and the variable ocean load, works on the ocean

$$\rho_w g(\xi + D)\, d\xi^*/dt. \qquad (10.4.10b)$$

With (10.4.2) and the tide expansion (6.2.1), the total potential $\mathcal{U}_\beta$ is

$$\mathcal{U}_\beta = (1+k_2)U_{\beta,lm} + \sum_s \sum_t \sum_{+}^{-} \alpha_s (1+k_s')D^{\pm}_{\beta,st} \times \cos[2\pi f_\beta T \pm t\lambda - \varepsilon^{\pm}_{\beta,st}] \cdot P_{st}(\sin\phi), \qquad (10.4.11)$$

where

$$\alpha_s = 4\pi G R \rho_w/(2s+1).$$

The total solid tide of the Earth is given by (10.4.3) and (6.2.1)

$$\xi^*_\beta = \frac{h_2}{g} U_{\beta,lm} + \sum_s \sum_t \sum_{+}^{-} \alpha_s \frac{h_s'}{g} D^{\pm}_{\beta,st} \times \cos[2\pi f_\beta T \pm t\lambda - \varepsilon^{\pm}_{\beta,st}] P_{st}(\sin\phi) \qquad (10.4.12)$$

From (10.4.10)

$$\langle \dot{W} \rangle = \rho_w \left\langle \mathcal{U}\frac{d\xi}{dt}\right\rangle + \rho_w g\left\langle (\xi + D)\frac{d\xi^*}{dt}\right\rangle + \rho_w \langle \nabla_s \cdot [\mathbf{u}(\xi + D)\mathcal{U}]\rangle.$$

For elastic yielding $\langle d\xi^*/dt\rangle = 0$ and in the third term $\xi \ll D$. Hence

$$\langle \dot{W} \rangle = \rho_w \left\langle \mathcal{U}\frac{d\xi}{dt}\right\rangle + \rho_w g\left\langle \xi\frac{d\xi^*}{dt}\right\rangle + \rho_w \langle \nabla_s \cdot [\mathbf{u}D\mathcal{U}]\rangle.$$

Upon integrating over the ocean surface the last term vanishes if the boundaries are impermeable. Then

$$\langle \dot{W} \rangle = \rho_w \int_S \left\langle \mathcal{U} \frac{d\xi}{dt} \right\rangle dS + \rho_w g \int_S \left\langle \xi \frac{d\xi^*}{dt} \right\rangle dS$$

$$= \rho_w (1+k_2) \int_S \left\langle U_{lm} \frac{d\xi}{dt} \right\rangle dS + \rho_w h_2 \int_S \left\langle \xi \frac{dU_{lm}}{dt} \right\rangle dS, \quad (10.4.13)$$

(Lambeck 1977). Thus, apart from modifying the actual tide, the Earth's elastic deformation by the tidal load does not contribute to the mean rate at which work is done on the ocean layer.

Writing the potential $U_{lm}$ as (equation 6.1.7)

$$U_{lm} = U^0_{lmpq} P_{lm}(\sin\phi) \begin{bmatrix} \cos \\ \sin \end{bmatrix}_{l-m \text{ odd}}^{l-m \text{ even}} (v^* - m\lambda), \quad (10.4.14)$$

and with the tide defined by (6.2.1),

$$\int_S U_{lm} \frac{d\xi}{dt} dS = -2\pi f_\beta U^0_{lm} D^{\pm}_{\beta,lm} N^2_{lm} \begin{pmatrix} \sin \\ \cos \end{pmatrix}_{l-m \text{ odd}}^{l-m \text{ even}} \times (2\pi f_\beta T - \varepsilon^{\pm}_{\beta,m} \pm v_{lmpq}).$$

Furthermore, with (6.1.9)

$$\left\langle \int_S U_{lm} \frac{d\xi}{dt} dS \right\rangle = 2\pi f_\beta U^0_{lm} D^{\pm}_{\beta,lm} N^2_{lm} \begin{pmatrix} \sin \\ \cos \end{pmatrix}_{l-m \text{ odd}}^{l-m \text{ even}} \times \left[ \varepsilon^{\pm}_{\beta,lm} - \pi\left(\frac{r_\beta}{2} + m\right) \right].$$

Similarly,

$$\left\langle \int_S \xi \frac{dU_{lm}}{dt} dS \right\rangle = 2\pi f_\beta U^0_{lm} D^{\pm}_{\beta,lm} N^2_{lm} \begin{pmatrix} \sin \\ \cos \end{pmatrix}_{l-m \text{ odd}}^{l-m \text{ even}} \times \left[ \varepsilon^{\pm}_{\beta,lm} - \pi\left(\frac{r_\beta}{2} + m\right) \right].$$

Thus

$$\langle \dot{W} \rangle = 2\pi f_\beta \rho_w (1 + k_2 - h_2) U^0_{lm} D^{+}_{\beta,lm} R^2 N^2_{lm} \begin{pmatrix} \sin \\ \cos \end{pmatrix}_{l-m \text{ odd}}^{l-m \text{ even}} \times \left[ \varepsilon^{+}_{\beta,lm} - \pi\left(\frac{r_\beta}{2} + m\right) \right].$$

$N_{lm}$ is defined by (2.4.4). With the definition of $U^0_{lm}$ given by (10.4.14) and (6.1.7)

$$\langle \dot{W} \rangle = 2\pi f_\beta(1+k_2-h_2)\frac{4\pi GR^2 m_☾ \rho_w}{a}\left(\frac{R}{a}\right)^l \frac{D^+_{\beta,lm}}{2l+1}$$
$$\times F_{lmp}(I)G_{lpq}(e)\begin{bmatrix}\sin\\ \cos\end{bmatrix}^{l-m \text{ even}}_{l-m \text{ odd}}\left[\varepsilon^+_{\beta,lm}-\left(\frac{r_\beta}{2}+m\right)\right]. \quad (10.4.15)$$

The work method, integral (10.4.7), requires only the second-degree harmonics in the ocean tide. More specifically, only the harmonics with the same degree and order as the potential of the forcing function intervene in the mean rate at which energy is dissipated in the global oceans, the rate at which the work is done by the other harmonics being zero when averaged over one period and over the world's oceans. The consequence of this is that the rate of energy dissipation can be computed with relatively good precision since the second-degree terms of the various $M_2$ models are in quite good agreement, despite the variation in detail (Lambeck 1977). A further consequence is that, with (10.4.13), on an elastically yielding Earth the dissipation is still defined by the coefficient $D^+_{\beta,lm} \sin \varepsilon^+_{\beta,lm}$. This means that $\langle \dot{W} \rangle$ can be estimated from empirical tide models without requiring any information on the manner in which the tide loads the Earth.

The result (10.4.15) for the rate at which work is done on the ocean surface can also be estimated directly from the energy balance in the Earth–Moon system. From (10.2.7), with the tidal acceleration given by (10.2.4*b*) and the secular rates in the orbital elements given by (10.2.5), ignoring terms in $e^2$ and with $Mm_☾/(M+m_☾) \simeq m_☾$, this method gives

$$\left\langle \overline{\frac{\mathrm{d}E}{\mathrm{d}t}} \right\rangle = [n(l-2p+q)+m\dot{\Phi}]\frac{4\pi GR^2 m_☾ \rho_w}{a}\frac{(1+k'_l)}{2l+1}\left(\frac{R}{a}\right)^l$$
$$\times F_{lmp}(I)G_{lpq}(e)D^+_{\beta,lm}\begin{bmatrix}\sin\\ \cos\end{bmatrix}^{l-m \text{ even}}_{l-m \text{ odd}} \varepsilon^+_{\beta,lm}. \quad (10.4.16)$$

The frequency of the tidal wave is

$$-2\pi f_\beta \simeq (l-2p)\dot{\omega}+(l-2p+q)n_☾+m(\dot{\Phi}-\Omega),$$

but as $\dot{\omega}$ and $\dot{\Phi}$ are both small compared with $n_☾$ or $\Omega$

$$-2\pi f_\beta \simeq (l-2p+q)n_☾+m\dot{\Phi}.$$

Table 10.8. *Estimates of dissipation in the $M_2$ ocean tide using the energy integrals (10.4.7), (10.4.8) and (10.4.9)*

| Author | Method | $\langle\overline{dE/dt}\rangle$ ($10^{19}$ erg $s^{-1}$) | $\langle\overline{dE/dt}\rangle$ (corrected) ($10^{19}$ erg $s^{-1}$) |
|---|---|---|---|
| Jeffreys (1920) | Bottom friction | 1.1 | |
| Heiskanen | Bottom friction | 1.9[a] | |
| Groves & Munk (1958) | Torque | 3.2 | |
| Miller (1966) | Energy flux | 1.7 | |
| Pekeris & Accad (1969) | Bottom friction | 6.0 | 4.20[b] |
| Hendershott (1972) | Torque | 3.0 | |
| Kuznetsov (1972) | Torque (Zahel model) | 7.28 | 3.57[c] |
| | (Pekeris & Accad model) | 6.68 | 3.29[c] |
| Pariyskiy *et al.* (1972) | (Bogdanov & Magarik model) | 5.24 | 3.67[b] |
| Zahel (1976) | Bottom friction and turbulence | 3.8 | |

[a] Value given by Munk & MacDonald (1960)
[b] Reduced by a factor $(1+k-h)$ (see text)
[c] Reduced by a factor $(1+k-h)^2$ (see text)

The only difference between the expressions (10.4.15) and (10.4.16) now is the factor $(1+k_2-h_2)$ appearing in the former and $(1+k_2')$ in the latter, but the two are equivalent (equation 2.1.8).

### 10.4.3 *Estimates of global dissipation rates*

Table 10.8 summarizes the estimates of energy dissipation obtained by Jeffreys and Heiskanen. For the latter, the value, corrected by W. D. Lambert, given by Munk & MacDonald (1960) is used. Both these values are now of historical interest only. Miller's (1966) estimate is also given. The emphasis of tidal dissipation studies of recent years has been to estimate the rates directly from numerical models of the global ocean tide (table 10.8). The first attempt at this was by Pekeris & Accad (1969) who introduced into their tide model a friction force proportional to velocity (section 6.2.2). This, plus their definition of the ocean–continent boundary by the 1000-fathom depth contour, results in a dissipation that is much more uniformly distributed than in the calculations of Jeffreys and Miller. Pekeris & Accad stress that their estimate of $dE/dt = -6.0\times10^{19}$ erg s$^{-1}$ is provisional. It is nearly three times larger than previous estimates. Hendershott integrates the work done on the ocean by both the Moon and the elastic solid body tide. The latter reduces the dissipation from what it would be for the same tide on a rigid Earth by a factor $(1+k_2-h_2)$ or by about 70% (e.g. equation 10.4.15). Pariyskiy *et al.* (1972) numerically integrated the model of Bogdanov & Magarik (1967) for the work done on the ocean by the Moon. Their value should be reduced by the above factor to allow for the work done by the solid tide on the ocean. Kuznetsov (1972) has also integrated the models of Pekeris & Accad (1969) and of Zahel (1970). These results should also be reduced by 0.70. Both models are for tides on a solid Earth without having imposed upon them the boundary conditions that observed tides agree with computed tides, although Pekeris & Accad do adjust for their friction coefficient so as to obtain an approximate global agreement. To allow for the modification of the ocean tide by the elastic tide, both estimates may be further reduced by a factor of $(1+k_2-h_2)$, or a further 70%. The agreement between the *corrected* results is now quite satisfactory.

Table 10.9 summarizes the estimates for $dE/dt$ using (10.4.15) and the $M_2$ tide coefficients given in table 6.8. The agreement between these estimates, based on the various solutions of the Laplace tidal equations, is quite good, despite the quite different assumptions made about the way dissipation is introduced in the models. Both Hendershott and Bogdanov & Magarik exclude some of the important shallow seas from their solutions; Pekeris & Accad define the continent–ocean margin by the 1000-m depth contour and assume linear friction; Zahel allows for dissipation by turbulence. Yet all yield quite similar results for $dE/dt$. This is perhaps as it should be. The torques exerted on the Earth by the Moon are

Table 10.9. *Estimates of dissipation in the $M_2$ ocean tides using (10.4.15)*

| Model | | $\langle \overline{dE/dt} \rangle$ ($10^{19}$ erg $s^{-1}$) |
|---|---|---|
| Pekeris & Accad | $M_2$ | 2.98 |
| Zahel (1970) | $M_2$ | 3.28 |
| Hendershott | $M_2$ | 3.46 |
| Bogdanov & Magarik | $M_2$ | 3.49 |
| Zahel (1976) | $M_2$ | 4.35 |

described by the second-degree harmonics but the energy is dissipated by components at the other end of the wavelength spectrum. To estimate the torques, we are only interested in these second-degree terms, particularly in the phase lag, and what happens to the energy once it passes into higher modes need concern us no further. The efficacy of this breakup into the higher harmonics is apparently dominated by global ocean characteristics since the above calculations yield essentially the same results for $dE/dt$, despite the differences in methods. Geometry of the ocean–continent configuration, continental margins and sea floor topography would appear to be more important than what happens to the high-frequency part of the spatial spectrum in, say, a few localized shallow seas. To evaluate the acceleration of the Moon by estimating dissipation in shallow seas is equivalent to trying to re-establish the second-degree harmonic of the tide from very localized measurements, and clearly this is a difficult and uncertain exercise at best.

Using the results of table 6.5 for the ocean tide coefficients of frequencies other than $M_2$, the total rate of dissipation of tidal energy in the oceans can be established. The result (table 10.10) is in good agreement with the incomplete estimate of $4 \times 10^{19}$ erg $s^{-1}$ based on the astronomical observations, and stresses that the oceans provide the predominant sink for the tidal energy: dissipation in the solid Earth and Moon can at most be about 10% of the total.

Table 10.10. *Estimates of dissipation using (10.4.15) and the ocean tide coefficients of table 6.9*

| Tide | $\overline{\langle dE/dt \rangle}$ ($10^{19}$ erg $s^{-1}$) |
|---|---|
| $M_2$ | −3.35 |
| $N_2$ | −0.10 |
| $O_1$ | −0.09 |
| $S_2$ (ocean) | −0.57 |
| $T_2$ | −0.02 |
| $K_1$ | −0.12 |
| $P_1$ | −0.02 |
| $S_2$ (atmosphere) | +0.05 |
| Total | 4.22 |

### 10.4.4 *Energy dissipation mechanisms*

The calculations of Jeffreys and Miller indicate rather localized dissipation. Thus in Miller's (1966) calculation, about 14% of the total energy is dissipated in the Bering Sea and 12% in the Okhotz Sea. The Timor Sea, Patagonia Shelf and Hudson Strait account for another 24% and 10 smaller seas contribute a further 30%. In the earlier studies the energy sink in the Bering Sea was even more important, 70% in the case of Jeffreys' (1920) study and 25% in the case of Heiskanen's (1921) study. Dissipation in the Bering Sea dominates all these discussions. However, the tidal currents across the shelf seem to be less important than the values used by Jeffreys, Heiskanen and Miller. Maximum tidal currents around the Pribilof and St Mathew Islands on the edge of the Bering Shelf and elsewhere, have an average value of less than 50 cm $s^{-1}$ (National Ocean Survey 1975) and the open sea currents are likely to be less than 30 cm $s^{-1}$. Tidal amplitudes are of the order of 20 cm. The average depth of the shelf margin is about 60 m and its length is about 1500 km. The energy flux method (integral 10.4.9*b*) then gives

$$-\overline{\langle dE/dt \rangle} \simeq 5 \times 10^{17} \text{ erg s}^{-1}.$$

This is a maximum value and includes tides other than $M_2$. It is five times smaller than the value of $24\times10^{17}$ erg s$^{-1}$ found by Miller (1966). For $\alpha = 0.002$ and a shelf area of $1.1\times10^6$ km$^2$, the bottom friction method (integral 10.4.8) gives the same value for $\langle\overline{\mathrm{d}E/\mathrm{d}t}\rangle$. It appears most unlikely that the Bering Sea can play the dominant role that is suggested by the calculations of Jeffreys, Heiskanen and Miller, and, if from the above results we can extrapolate to other seas, Miller's total of $1.7\times10^{19}$ erg s$^{-1}$ represents very much an upper limit to the amount of energy that can be dissipated by bottom friction in shallow seas. A further hint that the bottom friction calculations may not be in order is already given by Hendershott (1973). The lower limit to the $Q$ of the global ocean, as estimated by Garrett & Munk (1971), is of the order of 25 (but see Webb 1973). Hendershott (1972) estimates a $Q$ of 34. But the analysis by Wunsch (1972) of the North Atlantic tide suggests a lower limit to $Q$ of about 5, much smaller than the global estimate, and unexpected from Miller's (1966) calculations which indicate that the North Atlantic is relatively dissipationless.

There is no shortage of ideas on alternative dissipation mechanisms, yet there is little quantitative evidence. Dissipation over the coastal shelves may be more important than is generally supposed. Defant (1961) suggests that the average tidal currents over these sea shelves is of the order of 30 cm s$^{-1}$, leading to a dissipation rate of 50 erg s$^{-1}$ cm$^{-2}$. The total shelf area is about $30\times10^6$ km$^2$ resulting in a total rate of dissipation of $1.5\times10^{19}$ erg s$^{-1}$, nearly one-half of the astronomically required value for the $M_2$ tide. Munk (1968) suggests that, due to an interaction with internal tides, the tidal currents at the bottom of the deep oceans may also be larger than generally thought so that the deep sea may contribute further to dissipation. His provisional estimate is $10^{18}$ erg s$^{-1}$. Jeffreys has suggested that dissipation along the open coastlines may be important since ordinary waves breaking on the coast are almost totally dissipated, there being a general absence of strong reflected waves along the shore. Proudman (1941) and Jeffreys (1968*b*) discuss this possibility in some detail and the latter concludes that dissipation by the breaking of the waves is more important than by bottom friction. Applying the mechanism to tidal waves, he concludes that it may be an important source for the loss of tidal energy.

Zahel (1970, 1977) has described dissipation by turbulence as well as by bottom friction. Tidal currents can inject energy into horizontal eddies by lateral stresses set up along the coast or continental shelves, by bottom topography or by adjacent tidal currents. This suggests that the dissipation by turbulence may be most important along the continental margins. Turbulent motion is usually introduced qualitatively into the tidal equations by an effective viscosity or eddy viscous force. This has proved useful in providing simple dissipative mechanisms in a number of oceanic circulation problems. Estimates of the lateral eddy viscosity coefficient $K_h$ vary over a wide range. Munk (1950) requires $K_h \simeq 5 \times 10^7 \text{ cm}^2 \text{ s}^{-1}$ if the energy acquired by the ocean circulation from the winds is dissipated by lateral stresses while values up to $10^9 \text{ cm}^2 \text{ s}^{-1}$ are required to account for features of the western boundary currents (Bowden 1962). For the Antarctic Circumpolar current Hidaka & Tsuchiya (1953) estimate $K_h \simeq 10^{10} \text{ cm}^2 \text{ s}^{-1}$. Zahel (1970, 1973) adopts a constant value for the world's oceans of $10^{11} \text{ cm}^2 \text{ s}^{-1}$ but reduces this to $5 \times 10^9 \text{ cm}^2 \text{ s}^{-1}$ in his most recent model in which he finds

$$-\overline{\langle \mathrm{d}E/\mathrm{d}t \rangle}_{\text{bottom friction}} = 0.7 \times 10^{19} \text{ erg s}^{-1},$$
$$-\overline{\langle \mathrm{d}E/\mathrm{d}t \rangle}_{\text{turbulent friction}} = 3.0 \times 10^{19} \text{ erg s}^{-1}.$$

The dominance of the turbulent friction may be a consequence in part of Zahel's use of impermeable coastlines defined by the 50-m depth contour excluding, thereby, a major part of the shallow seas. Gordeyev, Kagan & Rivkind (1974) adopt $K_h \simeq 10^7 \text{ cm}^2 \text{ s}^{-1}$ and conclude that dissipation by turbulence is not important. Later, Gordeyev *et al.* (1977) used $K_h \simeq 10^{11} \text{ cm}^2 \text{ s}^{-1}$. Clearly, more precise information on a representative value of the eddy viscosity, applicable to tidal problems, is required before the importance of this mechanism can be established quantitatively.

Other proposed mechanisms for tidal dissipation are no better at quantifying the actual rate at which energy is dissipated. Munk (1968) concludes that a significant fraction of the dissipation may take place by way of scattering into internal modes. Once the energy is in these modes it is lost to the surface or barotropic tides and it effectively represents the dissipation. Thus there are two approaches to estimating the amount of tidal energy that may be dissipated by the internal or baroclinic tides: by estimating the rate of conversion of energy from the barotropic tide to the baroclinic tide, or by investigating the actual mechanisms of dissipation of the energy in the baroclinic tides. Conversion of energy is believed to occur by two main mechanisms: (i) along large-scale topographic features such as the continental shelves or ocean ridges, and (ii) by interactions of the barotropic tide with small-scale topographic features whose scales are less than the wavelength of the internal waves (see, for example, Bell, 1975). Wunsch (1975*a*) concludes that the large-scale topographic features are unimportant, that they may convert only some $10^{16}$ erg $s^{-1}$. But Sandstrom's (1976) study, while not attempting an estimate of the global conversion, suggests that the continental edges and ocean ridges may be important in controlling the conversion of energy from surface to internal tides. Cox & Sandstrom's (1962) theory of the scattering into internal modes by an irregular bottom topography fares a little better; according to Munk (1968) a conversion of some $5\times10^{18}$ erg $s^{-1}$ may occur in this way while Sandstrom (1976) suggests that his earlier calculation with Cox leads to an underestimation. Wunsch (1975*c*), in summarizing the limited observational evidence, suggests that from 10% to 50% of the total tidal energy may be converted into the internal modes (see also Baines 1973; Thorpe 1975; and Schott 1977). The equilibrium $M_2$ tide contains about $5\times10^{23}$ erg (Munk & MacDonald 1960) while Hendershott's (1972) model yields a value of some $7\times10^{24}$ erg. The amount of energy stored in the internal tide, $E_i$, may therefore range from $5\times10^{22}$ erg to $3.5\times10^{24}$ erg. Dissipation of this energy may occur by several mechanisms including (i) viscous dissipation, (ii) by the breaking of internal waves, and (iii) by a non-linear interaction between the internal tides and the rest of the internal wave spectrum. Viscous dissipation has been investigated by LeBlond (1966) who finds that the first mode of the internal tide is dissipated with a decay time of about 20 tidal periods, assuming a vertical eddy coefficient of some $10^3$ $cm^2$ $s^{-1}$. Then

$$dE/dt \simeq (E_i/20\times 12\times60\times60)\ \text{erg s}^{-1}.$$

For the above values of $E_i$, $dE/dt$ ranges from $5.8\times10^{16}$ erg $s^{-1}$ to $4.1\times10^{18}$ erg $s^{-1}$. This is very much an upper limit since LeBlond's value for the vertical eddy coefficient is higher than usually assumed. Garrett & Munk (1972) have investigated the dissipation through the breaking of internal waves and conclude that this may amount to about $7\times10^{18}$ erg s$^{-1}$. Wunsch (1975*c*) argues that this value represents an upper limit. Wunsch also concludes that dissipation by resonant interactions is inadequate. Webb (1976) suggests that such resonant absorption may occur on some continental shelves but, if this is important, it would already be included in the energy flux calculations across the entrances to shallow shelves.

Whichever of these mechanisms is responsible for the dissipation of energy, there is evidence for the oceans to be close to resonance at the semi-diurnal frequency. If the phase lags of the $M_2$ and $S_2$ frequencies were the same then the age of the tide would be zero instead of the observed 1 d. Satellite results also indicate a different lag $\varepsilon_{22}^{+}$ and hence a different $Q$ for these two tides. Numerical models of the $M_2$ tide show some sensitivity to small changes in the model and Pekeris & Accad (1969) suspect that this tide is close to a resonant frequency. Calculations by Longuet-Higgins & Pond (1970), and in particular by Platzman (1975), show that the oceans possess several free modes whose frequencies are close to semi-diurnal. Platzman also finds a free mode with a frequency near diurnal. Thus changes in the ocean–continent geometry may have had important consequences on dissipation in the past. In particular, if ocean geometries existed in the past that result in free modes with frequencies distinctly different from the forcing frequency, the rate of dissipation could have been significantly less than its present value. As both the frequencies of the free modes and of the forcing function will vary slowly with time due to the secular tidal acceleration of the Earth, even if all other factors have remained constant, it does not appear feasible to predict if the present near-resonance conditions have existed over long time intervals during the past, without solving the free-oscillation problem for each case.

## 10.5 Discussion

### 10.5.1 *Tidal accelerations*

*Lunar acceleration.* From the equations (10.2.5) and the ocean tide parameters summarized in table 6.5, the secular changes in the

Table 10.11. *Estimates of the secular changes in $a_☾$, $e_☾$ and $I_☾$ due to the ocean tides compared with satellite estimates*

| Tide | $da_☾/dt$ ($10^{-7}$ cm $s^{-1}$) | $\dot{n}_☾$ ($10^{-23}$ rad $s^{-2}$) | $de_☾/dt$ ($10^{-19}$ $s^{-1}$) | $dI_☾/dt$ ($10^{-19}$ $s^{-1}$) | Error estimate (%) | Satellite solution $\dot{n}_☾$ ($10^{-23}$ rad $s^{-2}$) |
|---|---|---|---|---|---|---|
| $M_2$ | 1.29 | −1.34 | −0.45 | −3.46 | 10 | −1.19 |
| $N_2$ | 0.08 | −0.08 | 5.82 | −0.14 | 30 | −0.07 |
| $K_2$ | — | — | — | −0.02 | 30 | |
| $L_2$ | | | −0.10 | | 30 | |
| $2N_2$ | | | 0.21 | | 30 | |
| $K_1$ | — | | — | −1.38 | 20 | |
| $O_1$ | 0.07 | −0.07 | −0.02 | 0.80 | 20 | −0.07 |
| $Q_1$ | | | 0.27 | 0.83 | 30 | |
| Total | 1.44 ±0.15 | −1.49 ±0.15 | 5.73 ±1.75 | −4.17 ±0.47 | | −1.33 ±0.25 |

Moon's orbital elements, $da_☾/dt$, $de_☾/dt$, $dI_☾/dt$ and $\dot{n}_☾$ can be evaluated (table 10.11). Of these elements, the last can be directly compared with the astronomical estimate for the Moon's acceleration in longitude. The principal contribution to $\dot{n}_☾$ comes from the $M_2$ tide with smaller contributions coming from $N_2$ and $O_1$. All relevant tidal frequencies give a total acceleration in longitude of $(1.44 \pm 0.15)10^{-23}$ rad $s^{-2}$. The satellite estimate of the $M_2$ tide parameter tends to be smaller than the ocean model result and, if we scale the other contributions by a similar ratio, the satellite-based estimate for the lunar acceleration is $(1.33 \pm 0.26)10^{-23}$ rad $s^{-2}$. Both values are consistent with Muller's (1976) *best estimate* of $1.40 \times 10^{-23}$ rad $s^{-2}$ based on several astronomical sources. The good agreement between the astronomical and the satellite result is better than we have the right to expect in view of the latter's rather large error estimates. But this agreement does indicate that we have a powerful new method of estimating the tidal accelerations and improved results can be expected when long series of observations of satellites, such as GEOS 3 and STARLETTE, become available. While the agreement between the astronomical and oceanic estimate of $\dot{n}_☾$ is such that the primary role of the oceans in dissipating the tidal energy is established beyond any doubt, the uncertainties of both estimates are still uncomfortably large. In particular, we cannot draw any firm conclusion about the possible role of dissipation in the solid parts of the Earth and Moon other than conclude that it must be small.

*Eccentricity and inclination.* The present tidal variations in the eccentricity and inclination of the lunar orbit follow from (10.2.5), and the rates are small (table 10.11). The former is of the order of $5\times10^{-19}\,s^{-1}$, very much smaller than the value $(1.5\pm0.6)10^{16}\,s^{-1}$ deduced by Martin & Van Flandern (1970) from the lunar observations. Tides raised on the Moon are also quite inadequate to explain this difference, and the explanation for the observed value must be sought elsewhere; it cannot be caused by tidal dissipation as these authors suggest. If Martin & Van Flandern's results are confirmed, this would suggest remaining long-period discrepancies in the lunar and solar theories, which may also explain the different values for $\dot{n}_☾$ based on the telescope observations since the seventeenth century, modern observations with respect to the atomic time scale, and the eclipse solutions. The present tidal change of the inclination of the lunar orbit on the equatorial plane is insignificantly small as will be the inclination of the equator on the ecliptic (equation 67 of Kaula 1964). Martin & Van Flandern's analysis of the lunar motion does not indicate a significant variation in these elements.

*The Earth's secular acceleration.* The tidal acceleration of the Earth follows from (10.2.2) where $da_☾/dt$, $de_☾/dt$ and $dI_☾/dt$ follow from (10.2.5). Both lunar and solar tides must be considered. For the latter, the changes in the Earth's orbit are negligible, but the effect of the solar torque on the spin is not, due to the $a_\odot^2$ term entering into the solar equivalent of (10.2.2). The total tidal acceleration is

$$\begin{aligned}\dot{\Omega}_T &= (\dot{\Omega}_T|_a + \dot{\Omega}_T|_e + \dot{\Omega}_T|_I)_☾ + (\dot{\Omega}_T|_a + \dot{\Omega}_T|_e + \dot{\Omega}_T|_I)_\odot \\ &\equiv (\dot{\Omega}_T|_a)_☾ + \delta\dot{\Omega}_T, \qquad (10.5.1)\end{aligned}$$

where $\Omega_T|_{\kappa_i}$ denotes the contribution to the total acceleration due to the secular change in the element $\kappa_i$. Of the various contributions to (10.5.1) (table 10.12), the dominant part, some 80%, comes from

Table 10.12. *Estimates of the secular tidal acceleration $\dot{\Omega}_T$ from ocean models and from satellite observations. $\dot{\Omega}_T|_{\kappa_i}$ is the tidal acceleration due to a secular change in the lunar orbit elements $\kappa_i$*

| Tide | $\dot{\Omega}\|_a$ ($10^{-22}$ rad s$^{-2}$) | $\dot{\Omega}\|_e$ ($10^{-22}$ rad s$^{-2}$) | $\dot{\Omega}\|_I$ ($10^{-22}$ rad s$^{-2}$) | Total $\dot{\Omega}_T$ from satellite solutions ($10^{-22}$ rad s$^{-2}$) |
|---|---|---|---|---|
| $M_2$ | −5.44 | −0.01 | −0.49 | −5.29 |
| $N_2$ | −0.33 | 0.10 | −0.02 | (−0.25) |
| $K_2$ | | 0.01 | 0.01 | |
| $2N_2$ | | 0.01 | | |
| $S_2$ | −0.90 | | −0.08 | −1.54 |
| $T_2$ | −0.06 | 0.02 | | (−0.04) |
| $K_1$ | | | −0.20 | (−0.20) |
| $O_1$ | −0.28 | | 0.11 | (−0.17) |
| $Q_1$ | | 0.01 | | |
| $P_1$ | −0.05 | | 0.02 | (−0.03) |
| $S_2$ (atmos) | 0.08 | | 0.01 | 0.24 |
| Total | −6.98 | 0.11 | −0.65 | −7.28±1.50 |
| | | −7.52±0.75 | | |

$(\dot{\Omega}_T|_a)_☾$ and this quantity can either be estimated from the tidal theory or deduced from the astronomically observed $\dot{n}_☾$. From the former (table 10.11), $(\dot{\Omega}_T|_a)_☾ = -6.05 \times 10^{-22}$ rad s$^{-2}$ while the astronomical data give $-5.48 \times 10^{-22}$ rad s$^{-2}$. The total oceanic estimate of $\dot{\Omega}_T$ is $(-7.5 \pm 0.8)10^{-22}$ rad s$^{-2}$. The satellite solution (for tides other than $M_2$ and $S_2$, the ocean models have been used) gives $(-6.9 \pm 1.5)10^{-22}$ rad s$^{-2}$ and the astronomical estimate – $(\dot{\Omega}_T|_a)_☾$ from the observed $\dot{n}_☾$ plus ocean estimates for $\dot{\Omega}_T|_{e_☾}$, $\dot{\Omega}_T|_{I_☾}$ and solar ocean and atmospheric tides – gives $(-7.0 \pm 0.7)10^{-22}$ rad s$^{-2}$. The mean of these three estimates of $\dot{\Omega}_T$ is $(-7.1 \pm 0.3)10^{-22}$ rad s$^{-2}$.

With this mean value for $\dot{\Omega}_T$ and the mean value for $\dot{n}_☾$ of $1.4 \times 10^{-23}$ rad s$^{-2}$,

$$\dot{\Omega}_T \simeq (51 \pm 4)\dot{n}_☾, \qquad (10.5.2)$$

where the mean values of both $\dot{\Omega}_T$ and $\dot{n}_☾$ are assumed to have an accuracy of 5%.

The astronomical evidence for the observed acceleration $\dot{\Omega}_o$ of the Earth has been summarized in table 10.7. The nontidal acceleration of the Earth is $\dot{\Omega}_{NT} = \dot{\Omega}_o - \dot{\Omega}_T$. With the above mean value and the astronomical values in table 10.7,

$$\dot{\Omega}_{NT} = (1.6 \pm 0.6)10^{-22} \text{ rad s}^{-2} \qquad (10.5.3a)$$

if $\dot{G} = 0$, and

$$\dot{\Omega}_{NT} = (0.5 \pm 0.6)10^{-22} \text{ rad s}^{-2} \qquad (10.5.3b)$$

if $\dot{G} \neq 0$ (Lambeck, 1979*a*).

The non-tidal acceleration is the most unsatisfactory quantity due to its being the difference between two quantities, both of limited accuracy. If the gravitational constant does vary with time, $\dot{\Omega}_{NT}$ is insignificant. We are rapidly approaching the embarrassing situation of a phenomenon, for which there has never been a shortage of geophysical explanations, that now appears to be vanishing. It is reminiscent of the earlier discussion of tidal dissipation, and stresses once again the need for further improvements in both the observed and the theoretical accelerations. Now we cannot seek comfort in new methods such as lunar laser ranging or satellite orbit analyses since $\dot{\Omega}_o$ is the sum of the secular part and long-period irregularities. Only ancient astronomical observations can contribute and this emphasizes the need for a systematic search for records going further back into time than the presently available data.

*Constancy of tidal accelerations.* R. R. Newton (1970) suggested that the lunar acceleration may have undergone important changes over the last 3000 yr. He concluded this from his results based on satellite observations of the tidal perturbations and on his analysis of ancient and medieval eclipse records. His satellite results, corresponding to a present-day value for the dissipation, for $\dot{n}_☾$ are close to the Spencer Jones determination of $1.1 \times 10^{-23}$ rad s$^{-2}$, but this agreement must be considered as fortuitous rather than real, and his results (R. R. Newton 1968) must be discarded for the following reasons:

(1) Newton does not allow for the fourth-degree harmonics in the ocean tide. For the satellites used these terms are as important as the second-degree harmonics.

(2) He does not allow for the frequency dependence of the tide coefficients. This is particularly important as the $M_2$ and $O_1$ tidal perturbations cannot be separated from his data.
(3) His treatment of the loading of the Earth by the atmospheric tide is incorrect.
(4) The dispersion of individual results for the lunar and solar tides obtained from the perturbations in inclination and ascending node of four satellites is far greater than can be explained by the above effects and is indicative of further unmodelled perturbations in the orbital theory.

Newton's values for $\dot{n}_{\leftmoon}$ at epochs 200 BC and AD 1000 are not significantly different. R. R. Newton's (1972) value of $(3.92 \pm 0.79)10^{-23}$ rad s$^{-2}$ centred at epoch AD 1000 is quite different, but Muller (1975) argues that this value is in error due to Newton's use of partial eclipse records. Muller & Stephenson (1975) and Muller (1975) find no evidence for a change in the lunar acceleration. This makes good geophysical sense since a variation in $\dot{n}_{\leftmoon}$ by a factor of about 2 as suggested by R. R. Newton (1970) requires a comparable change in the coefficients $D^{+}_{22} \sin \varepsilon^{+}_{22}$. Newton suggests that important dissipation may occur by friction between the ocean and shelf ice, implying that the shelf ice controls the tidal bulge. Whatever the merits of this mechanism, there is no evidence that significant changes have occurred in the extent of shelf ice since sea level has not changed by more than a few metres during the last 3000 yr. R. R. Newton (1972) argues that there was a sudden change in the properties of tides around the seventh or eighth century and suggests that an apparent change of the Normandy coastline early in the eighth century may be evidence for such a change. In view of the evidence that localized tidal friction may not be very important, such speculations appear inappropriate. The fact that (i) sea level has not varied greatly over the last few thousand years (Fairbridge 1961, Mörner 1971), that (ii) there has not been any significant change in the sea floor topography or in the ocean–continent distribution, and that (iii) dissipation is apparently not controlled by phenomena in a few localized regions, rules out any significant change in the lunar tidal acceleration over this time interval.

Changes in the secular rate of the Earth's spin can be readily accepted as due to long-period variations associated with angular momentum and inertia changes of the Earth and with torques acting on the mantle. Muller & Stephenson (1975) and Muller (1976) discuss the astronomical evidence for such changes and conclude that both $\dot{n}_{\mathbb{C}}$ and $\dot{\Omega}$ have remained constant over the last 3000 yr.

### 10.5.2 *Dissipation in the solid Earth*

*Energy dissipation.* The amount of tidal energy dissipated in the Earth–Moon system follows from (10.2.7) or (10.4.15), and is summarized in table 10.13. The astronomical and satellite results agree to within 10%. The total rate of energy dissipation in the oceans is a comparable $4.3\times10^{19}$ erg s$^{-1}$, stressing once again that a very major part of the tidal energy is dissipated in the oceans and that the solid Earth does not possess an important energy sink. Energy dissipated in the $M_2$ tide is $3.06\times10^{19}$ erg s$^{-1}$ (astronomical estimate) or $3.35\times10^{19}$ erg s$^{-1}$ (tidal model estimate). In view of the uncertainties of these three estimates of $dE/dt$, the differences are not significant, and can only be used to provide very approximate estimates of the amount of energy dissipated within the solid Earth. The small difference between the astronomical and satellite values for $dE/dt$ leads to an estimate of the rate of dissipation in the Moon. Present results are inadequate for this, apart from confirming that it must be small.

*Limits on mantle Q.* In view of the uncertainties in the estimates of $dE/dt$, the fact that the tidal estimate is somewhat greater than the astronomical estimate is not significant, in particular as the satellite results suggest that the tidal estimates may be too high. If we take the difference between the upper limit, $-1.45\times10^{-23}$ rad s$^{-2}$, of the astronomical estimate for $\dot{n}_{\mathbb{C}}$ and the lower limit, $-1.35\times 10^{-23}$ rad s$^{-1}$, estimated from the tide models, we obtain what can be considered as an estimate of the maximum specific dissipation of the Earth. From (10.2.4*b*)

$$\delta\dot{n}_{\mathbb{C}} = -\frac{3n}{2a}\sum_{lmpq} 2K_{lm}[F_{lmp}(I)G_{lpq}(e)]^2(l-2p+q)\sin\varepsilon_{lmpq}, \tag{10.5.4}$$

where the contribution of Moon tides to $\delta\dot{n}_{\mathbb{C}}$ is neglected. The three

Table 10.13. *Summary of tidal accelerations estimated from astronomical, tidal and satellite data. $\delta\dot{\Omega}_T$ is the contribution to $\dot{\Omega}$ from terms other than those related directly to the lunar acceleration (10.5.1)*

| | Astronomical estimate | Tidal estimate | Satellite estimate |
|---|---|---|---|
| $\dot{n}_☾$ ($10^{-23}$ rad s$^{-2}$) | $-1.35\pm0.10$ | $-1.49\pm0.5$ | $-1.33\pm0.25$ |
| $\dot{\Omega}_T\|_{n_☾}$ ($10^{-22}$ rad s$^{-2}$) | $-5.48$ | $-6.05$ | $-5.40$ |
| $\delta\dot{\Omega}_T$ ($10^{-22}$ rad s$^{-2}$) | $-1.47$ | $-1.47$ | $-1.47$ |
| $\dot{\Omega}_T$ ($10^{-22}$ rad s$^{-2}$) | $-6.95$ | $-7.52$ | $-6.87$ |
| $dE/dt$ ($10^{19}$ erg s$^{-1}$) | $-3.94\pm0.30$ | $-4.26\pm0.45$ | $-3.90\pm0.70$ |

principal contributions to (10.5.4) come from the $M_2$, $N_2$ and $O_1$ tides and we assume that the phase lags $\varepsilon_{lmpq}$ are constant for these three frequencies. For large $Q$

$$\sin\varepsilon_{lmpq} = Q^{-1},$$

and $Q^{-1} \simeq 1/130$. More precise upper limits for the specific dissipation can only be established if both the astronomical data and the tide models are improved.

The mantle $Q$ can also be estimated, in principle at least, from a comparison of the satellite and numerical results for the coefficients $D^+_{22}\sin\varepsilon^+_{22}$ (or the $D^+_{21}\cos\varepsilon^+_{21}$) since the satellite result is a measure of the total response of the Earth to the tidal potential. From the results for $M_2$ summarized in table 6.5, the difference between the upper limit of the satellite solution and the lower limit of the model solution results in an equivalent residual phase lag of 0°.2, yielding a solid earth $Q^{-1}$ of 1/300 or less. For the $S_2$ solution, the satellite coefficients are somewhat larger than the ocean model coefficients and the difference could be interpreted as a measure of dissipation in the mantle. From table 6.9 the result is $Q^{-1} \simeq 1/250$ with limits of 1/160 and 1/480. The mean of the above three estimates for $Q^{-1}$ leads to a lower limit to the mantle $Q$ at the tidal frequency of about 200. Improved values for $Q^{-1}$ require (i) better satellite results, (ii) improved ocean models, and (iii) a correct treatment of the ocean–atmosphere interaction for the $S_2$ tide. Once these improvements have been realized, the specific function at diurnal frequencies can also be estimated.

# ELEVEN
# PALEOROTATION

## 11.1 Polar wander

The concept of polar wander, the large-scale wandering of the Earth's axis of rotation throughout geological time, goes back more than 100 yr and has its origin in observations of fossil plant and animal distributions, and in the scars, tillites and moraines of past glaciations. For example, the discovery that a subtropical climate existed in Spitsbergen at a time when Central and Southern Europe were subject to a tropical climate and when extensive glaciations occurred in Southern Africa, led to the conclusion that, in the western hemisphere, the Carboniferous equator must have lain far to the north of the present equator. A further degree of freedom to the interpretation of the paleontological and paleoclimatic data was introduced by Wegener and by F. B. Taylor with their concept of continental drift, in which the continents are postulated to have moved relative to each other over large distances throughout geologic time. For lack of convincing observational evidence and for an absence of compelling theoretical arguments, the notions of polar wander and of continental drift both remained at the periphery of scientific responsibility until rescued from this limbo by two important paleomagnetic discoveries: that large changes have occurred in the mean direction of the geomagnetic field and that this field has periodically reversed itself. Evidence for both changes is found throughout the Phanerozoic and Proterozoic, permitting some conclusions to be drawn about the reality of drift and wander for much of the geologic record. The separation of drift and wander has, however, remained problematical and this has plagued many of the subsequent discussions. Presumably if drift occurs, so will wander, the rotation axis moving along with the principal axis of maximum inertia as a redistribution of mass occurs, and the two will have a common explanation. Yet the available magnetic data do not permit a clear separation of the two. Additional geophysical hypotheses are required to effect a separation.

### 11.1.1 *Rotation of a quasi-rigid body*

Kelvin in 1863 apparently could see little objection to the polar wander hypothesis, the formation of the world's major mountains being sufficient to shift the Earth's inertia axis by some tens of degrees. G. H. Darwin concluded that if the Earth's rheology resembles that of a plastic body, polar wander may well occur on an extensive scale but that it would not exceed a few degrees if the Earth's behaviour is that of an essentially rigid body. Gold (1955) came to a similar conclusion: that large-scale polar wander is inevitable if the long-term mantle rheology is inelastic. The problem has been discussed further by Inglis (1957), Munk & MacDonald (1960) and Goldreich & Toomre (1969).

The problem is to determine the motion of the rotation axis of a body that evolves slowly with time. Goldreich & Toomre refer to such a body as a *quasi-rigid* body and they illustrate it with an extension of Gold's beetle model in which a colony of $N$ beetles roams a rigid rotating sphere (figure 11.1). The solution of the equations of motion is such that the rotation axis, once it lies close to the principal axis of either maximum or minimum inertia, continues to follow this axis. As the beetles crawl along random paths, at rates slow compared with the free-wobble frequency of the body, the position of the principal axis of maximum inertia of the combined sphere and beetle population evolves with time and the rotation axis follows it. This is for the motion viewed from the sphere itself. In an absolute frame it is the angular momentum axis that remains fixed in space and for all practical purposes this can be identified with the rotation axis if there are no torques acting on the body. Viewed from space, the rotation and principal axes appear fixed and the crust moves relative to it; a set of reference points fixed to the sphere will appear to oscillate with respect to the rotation axis. An important aspect of Goldreich & Toomre's model is that they find that the speed of the motion of the rotation axis exceeds the average speed of the beetles by a factor of the order $N^{1/2}$. Thus, if $N$ is large, even modest displacements on the sphere lead to large-scale polar wandering.

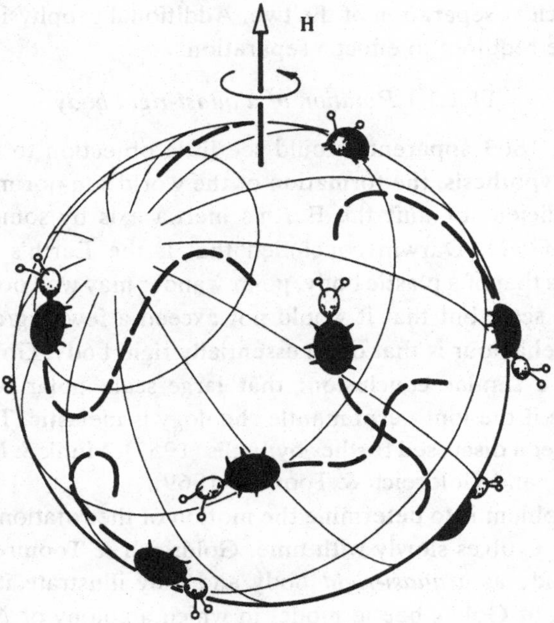


Figure 11.1. Goldreich & Toomre's (1969) quasi-rigid model.

Not surprisingly the actual Earth differs from this idealized model in several respects. Most important is that the Earth possesses a rotational bulge. If the Earth were in a truly hydrostatic equilibrium state this would not significantly modify the above model since the bulge re-orients itself relative to the rotational axis, should the latter shift in position. Should the bulge not adjust, any polar wander would be small indeed. A question raised by Munk and others is to what extent the bulge follows this hydrostatic equilibrium behaviour. The importance of this is illustrated by Gold's calculation in which he estimates that if South America were suddenly raised by 30 m the change in the inertia tensor would be such as to change the position of the pole by 0°.01. But if the rotational bulge follows the rotation axis the total displacement would be some 10°.

The Earth's bulge is described by a second-degree zonal coefficient in the geopotential. The difference $\Delta\bar{C}_{20}$ (section 2.4) between the observed and theoretical estimates, $-4.63\times10^{-6}$ for

Jeffreys' (1963) value and $-4.01\times10^{-6}$ for Caputo's (1965) value, has been thought sufficiently important, compared with other non-hydrostatic terms in the geopotential, to warrant special interpretation (Munk & MacDonald 1960; MacDonald 1963, 1966; McKenzie 1966). Compared with the other important second-degree terms

$$\Delta\bar{C}_{20}/(\bar{C}_{22}^2+\bar{S}_{22}^2)^{1/2}=1.7\text{–}1.5,$$

and there is approximately two-to-three times more potential energy contained in the non-equilibrium bulge than there is in the two coefficients $\bar{C}_{22}$, $\bar{S}_{22}$ describing the equatorial bulge of the Earth. MacDonald and McKenzie concluded that this difference was a consequence of a delay in the Earth's response to a change in the speed of rotation. As the Earth slows down under the influence of tidal forces, it takes time for the rotational bulge to adjust to the new angular velocity and the observed bulge is interpreted as a fossil bulge corresponding to the equilibrium state at some time in the past, when the Earth turned faster. The relaxation of the oblateness involves a change in the overall mass distribution of the mantle and it should be controlled by the flow properties of the mantle as a whole. McKenzie concludes that this fossil bulge hypothesis leads to a lower-mantle viscosity of the order of $10^{26}$ P compared with values of $10^{21}$–$10^{22}$ P for the upper mantle deduced from post-glacial uplift studies. Such high values in the lower mantle reduce the rate of convection in the lower mantle relative to that in the upper mantle (Stacey 1977).

However, Goldreich & Toomre show that neither this interpretation nor the conclusion is justified and that the discrepancy in the bulge is not anomalous but a consequence of the choice of the coordinate system adopted (see also O'Connell 1971). Thus if we consider a new set of axes $x_i'$, such that $x_3'$ is aligned with the equatorial axis of smallest moment of inertia and the $x_1'$ axis is aligned with the present pole, the potential coefficient $\Delta\bar{C}_{20}'$ becomes

$$(\Delta\bar{C}_{20}')^2=\tfrac{3}{4}[\bar{C}_{22}^2+\bar{S}_{22}^2]+\tfrac{1}{4}\Delta\bar{C}_{20}^2-\tfrac{\sqrt{3}}{2}\Delta\bar{C}_{20}[\bar{C}_{22}^2+\bar{S}_{22}^2]^{1/2}. \tag{11.1.1}$$

Also the degree variances $\sum_m(\bar{C}_{lm}^2+\bar{S}_{lm}^2)$ are invariant to a rotation of the axes, so that once $\Delta\bar{C}_{20}'$ is determined from (11.1.1), $(\bar{C}_{22}'^2+$

$\bar{S}'^2_{22}$) is also known, and one can compute

$$\Delta\bar{C}'_{20}/[(\bar{C}'_{22})^2+(\bar{S}'_{22})^2]^{1/2} \simeq 1.8-1.4,$$

a ratio similar to that obtained before, but now the $\Delta\bar{C}'_{20}$ is associated with the ellipticity of the equator while the $\bar{C}'_{22}$, $\bar{S}'_{22}$ are associated with the meridional ellipticity: the non-hydrostatic bulge is no more anomalous than the equatorial bulge, and if a special explanation is to be given to the one, it must also be given to the other. But as Goldreich & Toomre remark, no one has yet proposed that the large negative geoid anomaly on the equator south of India has anything to do with some past rotation about the corresponding equatorial axis.

The main point of Goldreich & Toomre's study is to emphasize that for a quasi-rigid body, large-scale displacements of the pole can be readily explained by relatively small horizontal displacements of material in the mantle. Secondly, they conclude that the non-hydrostatic part of the rotational bulge is not anomalously large, and that it is not necessary to suppose that the Earth's response to long-wavelength deformations is much different from its response to shorter-wavelength deformations: that there is no need to invoke steep gradients in mantle viscosity. Thus the Earth can, in fact, be considered as a quasi-rigid body whose rotational bulge flows in conformity with the changing position of the rotation axis. If mantle convection causes the axes of inertia to shift, so will the rotation axis wander; both drift and wander can be expected to occur but, in their model, the rate of wander will be significantly larger than the rate of drift.

### 11.1.2 *Paleomagnetic evidence*

The history of the Earth's magnetic field beyond the last few centuries lies in the record of the fossil magnetization of rocks. The observation of a primary magnetization of rocks permits the position of the magnetic pole to be estimated if, amongst other factors, the field can be assumed to correspond to that of an axial geocentric dipole (see McElhinny & Merrill 1975; Evans 1976; Merrill & McElhinny 1977).

A recent review of the paleomagnetic data and of the polar wander paths is given by McElhinny (1973) for all continents, from the Precambrian to the Recent. A more limited review is given by

Van der Voo & French (1974) for the continents bordering the Atlantic, from the Late Carboniferous to the Eocene. The resulting pole paths determined in these two studies for the North American and European data are in general agreement and confirm the early observation by Runcorn (1956) that the North American path was systematically to the west of the European path, one of the first paleomagnetic observations pointing to continental drift.

The paleomagnetic evidence, and more so the marine magnetic anomalies about the mid-ocean ridges, have led to the plate tectonics hypothesis, the modern counterpart of Wegener's continental drift model. The plate tectonics model is thoroughly reviewed by Le Pichon, Francheteau & Bonnin (1973). Briefly, the Earth's surface is divided into a number of lithospheric plates whose thicknesses vary from perhaps 70 km to 150 km and which overlie a less-rigid asthenosphere. Some of the plates carry mainly oceanic crust, others mainly continental crust or both. The margins of the plates are of three types: (i) zones of extension where new lithosphere is formed and which correspond mostly to the ocean ridges, (ii) zones of compression where old lithosphere is subducted back into the mantle and which correspond mainly to the regions where deep ocean trenches occur, and (iii) interconnecting transform faults. Together they complete a continuous network of active plate margins. One important aspect of the model is that, viewed on a geological time scale, the plates are essentially rigid away from the margins, and the relative geometry of points on the same plate remains undistorted. Another important property of the model is that the plate margins themselves are mobile and it is this that permits plates such as the Antarctic and African to be surrounded by spreading margins. Relative motion between two rigid plates $A$ and $B$ can be described by a rotation ${}_A\Omega'_B$ of one plate about some axis that is fixed with respect to both plates (figure 11.2). Two parameters fix the position of the pole of this axis, and the rate of rotation completes the vector. Marine magnetic anomalies, geometry of transform faults and earthquake slip vectors have all been used to determine the positions and rates of the poles of rotation for each of the plates (Minster *et al.* 1974).

The paleomagnetic pole places the plate in latitude and in azimuth but the longitude remains indeterminate. Thus the motion of

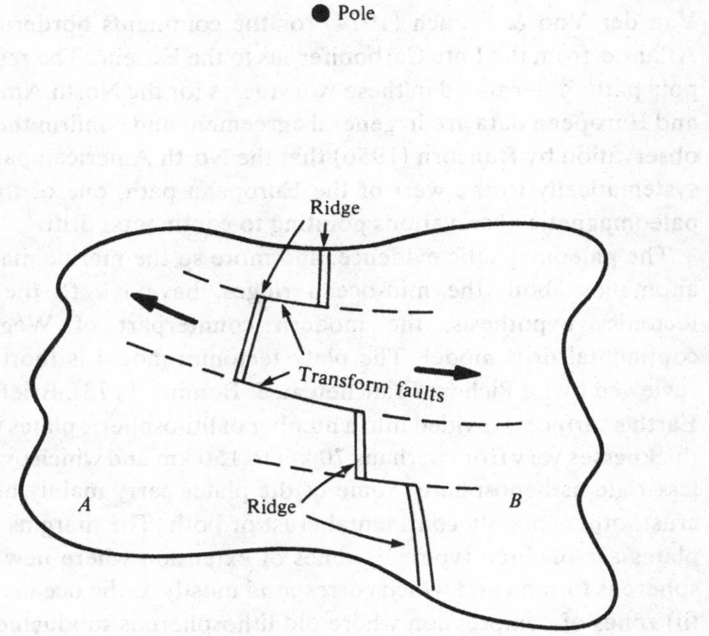


Figure 11.2. On a sphere the motion of plate $A$ relative to $B$ can be represented by a rotation ${}_A\Omega'_B$ about some pole that is fixed relative to both plates. The transform faults form small circles centred at the pole (from Morgan 1968).

the plate $A$ can always be described relative to the geographic axis $\hat{\mathbf{x}}_3$ by an angular velocity vector ${}_{\mathbf{x}_3}\mathbf{\Omega}'_A$ that lies in the geographic equator. If the relative velocity vector ${}_A\mathbf{\Omega}'_B$ between $A$ and the second plate $B$ is known, the apparent pole path of the latter is given by

$${}_{\mathbf{x}_3}\mathbf{\Omega}'_B = {}_{\mathbf{x}_3}\mathbf{\Omega}'_A + {}_A\mathbf{\Omega}'_B - ({}_A\mathbf{\Omega}'_B\hat{\mathbf{x}}_3)\hat{\mathbf{x}}_3,$$

which also lies in the equator of $\mathbf{x}_3$. Thus, while the paleomagnetic observations are independent measures of motion, they do not give complementary estimates of relative motions. Drift and wander cannot be separated and the latter becomes an unnecessary concept in paleomagnetism and plate tectonics (Munk & MacDonald 1960; McKenzie 1972). This can also be illustrated in the following manner. To fix each plate requires three parameters. If $N$ continents are drifting and the pole wanders, there will be $(3N-$

1) + 2 unknowns – if one longitude is arbitrarily fixed – for any given geological epoch. As independent observations we have the position of the plate $A$ and the relative motions between the plates, a total of $2+3(N-1)$. Thus the number of unknowns, $3N+1$, always exceeds the number of observations, $3N-1$, for the $N$ plates. A unique solution is possible only if polar wander is not introduced. Other conditions must be imposed if polar wander is to be separated from the plate tectonics motions. Recent attempts at establishing an *absolute* reference for the plate motions have been based on the concept that mantle *hot spots* form a nearly fixed frame (Morgan 1971; Minster *et al.* 1974), or they have been based on calculations in which the plate motions have been constrained to plausible physical properties (Minster *et al.* 1974; Kaula 1975; Solomon & Sleep 1974).

McKenzie (1972) suggests that, while polar wander is not necessary within the present plate tectonics framework, it may nevertheless be a useful one if the motion of the pole relative to any one plate is very much faster than the motion between the plates themselves. This is illustrated in figure 11.3. The linear velocity of the pole relative to plate $A$ is

$$_A\mathbf{v}_{\mathbf{x}_3} = R({}_A\mathbf{\Omega}'_{\mathbf{x}_3} \wedge \hat{\mathbf{x}}_3).$$

If for the $N$ plates

$$\sum_{n=1}^{N} {}_n\mathbf{v}_{\mathbf{x}_3} \gg \sum_{n=1}^{N} |{}_n\mathbf{v}_{\mathbf{x}_3} - \mathbf{v}_m|,$$

where

$$\mathbf{v}_{\mathrm{m}} = \frac{1}{N}\sum_{n=1}^{N} {}_n\mathbf{v}_{\mathbf{x}_3},$$

then it may be useful to consider the part $\mathbf{v}_{\mathrm{m}}$ as polar wander and the relative velocities ${}_n\mathbf{v}_{\mathbf{x}_3} - \mathbf{v}_{\mathrm{m}}$ as drift. Thus, in the case illustrated in figure 11.3($a$), polar wander is not a useful concept but in the case illustrated in figure 11.3($b$) it may be. To reduce the influence of small fast-moving plates, McKenzie proposes that the velocities be weighted in proportion to their areas. This procedure has been followed by McElhinny (1973) using paleomagnetic data with the result shown in figure 11.3($a$). These relative plate velocities,

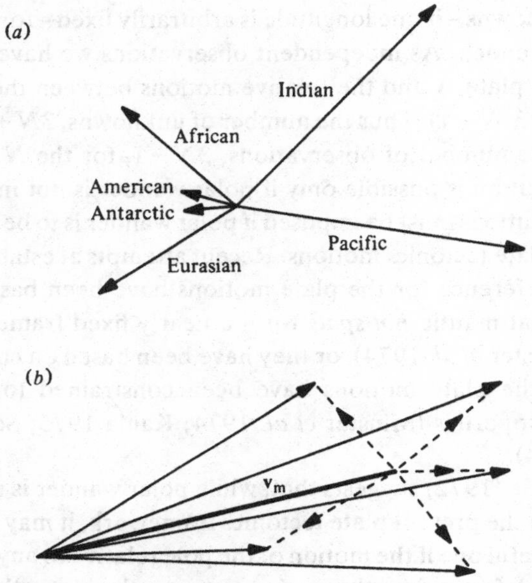


Figure 11.3. (*a*) Vectors of the linear velocity, weighted by areas of the plate, of the six major plates for the past $50 \times 10^6$ yr (from McElhinny 1973). The mean of these vectors is much smaller than most of the individual values. (*b*) Vectors for a hypothetical case when the mean $\mathbf{v}_m$ of the area-weighted velocity vectors is of the same magnitude as the individual vectors $\mathbf{v}_i$. In this case $\mathbf{v}_m$ can be considered as polar wander and $\mathbf{v}_i - \mathbf{v}_m$ (broken lines) as drift of the plates. In case (*a*) polar wander is not a useful concept but in case (*b*) it may be (after McKenzie 1972).

weighted by their areas, represent the situation for much of the Tertiary; there is no evidence that large-scale wander has occurred during the last $50 \times 10^6$ yr. Jurdy & Van der Voo (1974) developed a similar approach using the paleomagnetic pole for one plate and relative plate motions for the other plates. Their method is further discussed by McKenzie (1975) and Jurdy & Van der Voo (1975*b*) but their conclusion is the same as that reached by McElhinny. In a later paper they extended the analysis back to the Early Cretaceous and conclude that there is no compelling evidence for large-scale polar wander during the last $10^8$ yr (Jurdy & Van der Voo 1975*a*).

Further back into time the evidence for both polar wander and plate motions becomes much less certain, particularly as marine magnetic anomalies are not found further back in time than the

Mesozoic. The paleomagnetic data do suggest that there may have been a period of either very rapid drift or wander during the Cambrian (McElhinny & Briden 1971), but the ensemble of the Paleozoic data suggests that continental drift rather than polar wander was predominant.

### 11.1.3 *Discussion*

If Goldreich & Toomre's (1969) polar wander model is valid, the paleo- and marine-magnetic data should point to a distribution of relative linear velocity vectors for the plates that is similar to that illustrated in figure 11.3(*b*) since polar wander will exceed drift by an average of $N^{1/2}$. But the observations of the past $110\times10^6$ yr point to the situation, illustrated in figure 11.3(*a*), where there is no indication of a dominant polar wander. It appears unlikely that the polar wander model discussed by Goldreich & Toomre is immediately applicable. One possible reason for this is that the lower mantle is, after all, too viscous to permit the rotational bulge to adjust itself to changes in the position of the rotation axis. But there is little evidence for this. Cathles's (1975) work in the post-glacial rebound problem suggests that there is no need for a rapid increase in viscosity with depth and that the sea level data may, in fact, exclude such a model. This conclusion is confirmed by Peltier & Andrews (1976).† O'Connell (1977) reviews the flow properties of mantle materials and also believes that there is no compelling argument for adopting a high viscosity for the lower mantle.‡ An alternative explanation to this apparent paradox is that the mass redistribution associated with the drift occurs mainly in a vertical direction and that it is confined to the upper mantle. Such an essentially isostatic response will not contribute much to changes in the inertia tensor and the polar wander will be reduced from that predicted by the uncompensated beetles on the rigid sphere model. In a similar way the contribution of convection in the lower mantle to changes in the inertia tensor may only be of the second order if some form of internal equilibrium is approached. Lateral seismic

† See also Nakiboglu & Lambeck (1979).

‡ Kaula (personal communication) has pointed out that the melting point gradient in the mantle exceeds the adiabatic temperature gradient in the lower mantle and since viscosity is proportional to exp (melting point temperature/temperature) the mantle viscosity can be expected to increase with depth.

anomalies for the mantle are often a magnitude larger than predicted from the gravity field and equations of state, and one way of reconciling this difference is by assuming some form of compensation whereby positive density anomalies are overlain by negative anomalies and vice versa (Lambeck 1976; McQueen & Stacey 1976; Kaula, 1977*a*). Thus, while changes in the density distribution associated with convection may have a characteristic time scale of the order of $10^8$ yr, a first-order compensation of these anomalies occurs with a much shorter time scale. Only the residual effects will modify the inertia tensor. In the case of the Goldreich & Toomre model, gravity will be strongly correlated with the positions of the beetles, but such a correlation between gravity and topography is not evident for the Earth (Lambeck 1976).

## 11.2 Tidal accelerations in the geological past

The torque created by the Moon's attraction on the tidal bulge transfers energy and angular momentum from the Earth's spin to the lunar orbital motion and, with the present configuration of $\Omega > n_☾$, the Moon moves away from the Earth. The rate of change of the semi-major axis of the lunar orbit is given by (10.2.4*a*). The present rate is small, about 5 cm yr$^{-1}$, but viewed on a geological time scale this change becomes important. Writing (10.2.4*a*) as

$$\dot{a}_☾ = \mathscr{A} a_☾^{-11/2},$$

and integrating from the present time $t_0$ to some time $t$ in the past

$$a_☾^{13/2} = (a_☾|_{t_0})^{13/2} + \tfrac{13}{2}\mathscr{A}(t - t_0), \qquad (11.2.1.)$$

with

$$\mathscr{A} = \tfrac{3}{16}(1 + \cos I_☾)^4 \{G m_☾ R^5 k_2 / [G(M + m_☾)]^{1/2}\} \sin \varepsilon_2.$$

If we consider only the dominant $M_2$ tide and assume a constant equivalent phase lag of 6°, the Moon would have been within 10 earth-radii about $1.5 \times 10^9$ yr ago. For an average lag of 3° this would have occurred some $3.0 \times 10^9$ yr ago.

Conservation of angular momentum requires that the Earth's spin at time $t$ is (equation 10.2.2)

$$\Omega(t)=\frac{1}{C(t)}\Big[(H_{☾}\cos I_{☾})|_{t_0}+(C\Omega)|_{t_0} -\frac{Mm_{☾}}{M+m_{☾}}\,a_{☾}^2(t)n_{☾}(t)\cos I_{☾}(t)\Big], \qquad (11.2.2)$$

assuming that the solar tidal torque has remained constant and that there are no other torques operating on the Earth. For $a_{☾}=10$ earth-radii, $\Omega \simeq 3.5\times 10^{-4}\ \mathrm{s}^{-1}$ and the l.o.d. is 5 h. Because of the $(1/a_{☾})^{-11/2}$ term in $\mathrm{d}a_{☾}/\mathrm{d}t$, both the lunar orbit and the Earth's spin evolve rapidly prior to this position and a condition of some interest occurs when $\Omega = n_{☾}$ – when the lunar month equals the terrestrial day – for at this time the phase lag of the $M_2$ tide vanishes and tidal evolution ceases. Thus

$$\frac{(GM)^{1/2}}{a_{☾}^{3/2}}\Big[1+\frac{1}{C}\,\frac{Mm_{☾}}{M+m_{☾}}\,a_{☾}^2\cos I_{☾}\Big]C(t)=[(H_{☾}\cos I_{☾})|_{t_0}+(C\Omega)|_{t_0}].$$

This occurs at either 2.4 or 86.4 earth-radii. The first case corresponds to $n_{☾}=\Omega=1$ rev/4.8 h. The second corresponds to $n_{☾}=\Omega=1$ rev/47 d and, in the absence of other forces or tidal frequencies, the Moon's orbit will not expand beyond 86.4 earth-radii. With (11.2.1) and $\varepsilon_{2200}\simeq 6°$ this will occur in about $14\times 10^9$ yr.

If the Moon is initially in an eccentric orbit, the transfer of angular momentum and energy is greater at perigee than at apogee, the satellite is accelerated more in the former position than in the latter, and the eccentricity increases with time, always supposing that $n_{☾}<\Omega$. This orbital evolution is described by the $N_2$ tide, the so-called radial tide. With time, the lunar orbit spirals outwards from the Earth in an increasingly eccentric orbit. At $n_{☾}=\frac{2}{3}\Omega$ the frequency of the $N_2$ tide goes to zero and the subsequent evolution is determined by the lesser terms in $\mathrm{d}e/\mathrm{d}t$ and by the solar tide. This occurs near 3 earth-radii and 80 earth-radii. As the present eccentricity of the lunar orbit is small, much of the past evolution can be studied by assuming zero eccentricity. The exception will be during an eventual close approach phase for, when the Moon is within 2.4 earth-radii, $n>\Omega$ and at this point the evolution will be extremely rapid and unpredictable because other factors will dominate the process.† In particular, the Moon will be within the Roche limit of about 3.0 earth-radii; at this point the self-attraction of the satellite is exceeded by the tidal attraction of the planet, and the satellite becomes gravitationally unstable and breaks up (see, for example, Stacey 1977). Even if the satellite did not break up, the tides would be so large that the tidal response would no longer be linear and the present orbital theories would fail.

The third element of interest in studying the orbital evolution is the inclination of the lunar orbit. At present, the plane of the orbit is inclined at some 5° to the ecliptic and, since the solar attraction on the Moon exceeds the attraction on the Moon by the Earth's permanent equatorial bulge, the orbit rotates in space with nearly constant inclination to the ecliptic, while the inclination with respect to the equator varies periodically with time. If, in the past, the Moon was much closer to the Earth, the attraction by the equatorial bulge would have dominated and the inclination would have been constant with respect to the equator but variable with respect to the ecliptic. This leads to the concept of a critical distance $a_c$ such that when $a \gg a_c$ the satellite will maintain a constant inclination to the ecliptic while for $a \ll a_c$ the satellite maintains a constant inclination to the equator. For the Earth–Moon system $a_c \simeq 10$ earth-radii (Goldreich 1966) and for $a > 20$ earth-radii the inclination of the Moon's orbit can be considered constant with respect to the ecliptic. In an inclined orbit, the displacement of the tidal bulge due to the delay $\Delta t$ in the response is a function of the position of the Moon according to $\cos \phi \; \Omega \, \Delta t$ and lies in a plane nearly parallel to the equator. At the maximum latitude, $\phi = I$, the displacement is less than at the equator and the torque is a minimum. When the Moon is over the equator the tidal torque is a maximum. Integrating this torque over one orbital period gives a non-zero average torque whose vector lies in a plane normal to the line of nodes with a component directed towards the point of maximum orbital elevation. This torque tends to bring together the angular momentum vector and the rotation axis. Thus, in the presence of tidal torques,

† The Martian satellites, Phobos and Deimos lie close to the planet and the orbital eccentricities undergo significant changes (Lambeck 1979*c*).

the orbital inclination varies slowly with time: for $a > a_c$ the orbit moves towards the ecliptic, for $a < a_c$ this plane moves towards the equator.

The exchange of energy and angular momenta in the Earth–Moon system is expressed geometrically by the three equations (10.2.4) for the secular rates of change in $a$, $e$, and $I$. As already indicated, the integration into the past of only these three equations is not valid since they ignore the interaction between the Moon's motion and the gravitational attraction due to the Sun and the Earth's oblateness. But as long as $a > a_c$, and we now consider $I$ as the inclination on the ecliptic, this does not seem to be very important. The solar tide should also be introduced as should be the tide raised on the Moon. More important is the assumption that must be made about the phase lags, for dissipation occurs mainly in the oceans and it is unlikely that it has remained constant throughout geologic time. In consequence, the time scale of the evolution remains uncertain and it is this that we wish to control with the paleontological evidence. Independently of this, the evolution of the eccentricity and inclination as a function of semi-major axis is valid since no matter how d' sipation occurs, the lagged ellipsoid remains a most adequate description of the tidal bulge. An exception is when the Moon approaches close to the Earth, for then the fourth-degree terms in the ocean tide expansion may become important. But this requires an approach close to, or even within, the Roche limit, and the question is largely academic (Lambeck 1975*a*).

The integration of the equations of motion for the Moon was first carried out by G. H. Darwin. More recent solutions have been attempted by Gerstenkorn (1955), Slichter (1963), MacDonald (1964), Sorokin (1965), Goldreich (1966) and Singer (1968). These studies all agree in that (i) there has been a minimum Earth–Moon distance at some time in the past, (ii) the inclination of the lunar orbit was substantially greater in the past than it is now, and (iii) the eccentricity of the lunar orbit increases as the distance increases. Figure 11.4 summarizes Goldreich's results for $\varepsilon_{2mpq}$ = constant. The two branches of the curves determine the limits between which the inclinations and obliquity oscillate during each precession period. The most significant result, also noted by Gerstenkorn

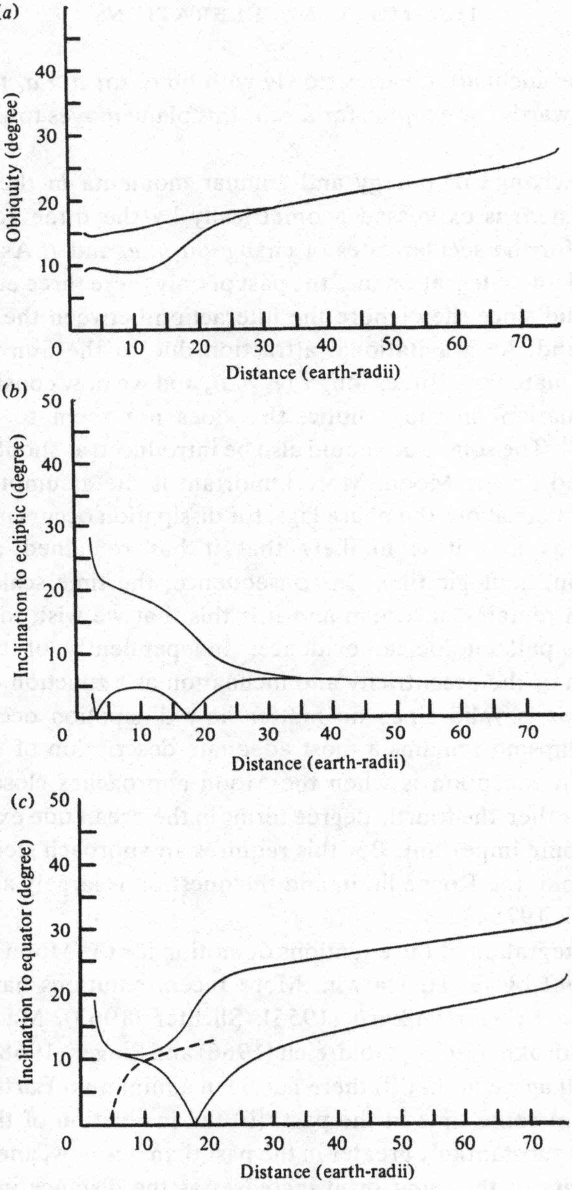


Figure 11.4. Variations in (*a*) obliquity of the Earth's equator to the ecliptic, (*b*) inclination of Moon's orbit to the ecliptic, and (*c*) inclination of the Moon's orbit to the Earth's equator, for $\varepsilon_{2mpq}$ = constant (from Goldreich 1966). Rubincam's results are indicated by the broken line (section 11.4).

(1955), is that the inclination of the lunar orbit on the equator must have been non-zero when the Moon was near 10 earth-radii and that the Moon could not have moved in an equatorial orbit. Theories of the lunar origin should be able to explain this condition.

Differences between the above-mentioned integrations do occur, and they have been discussed by Gerstenkorn (1967). The most complete integration into the past is the work by Gerstenkorn (1955) and Goldreich; both assumed that the orbit remained circular throughout and that dissipation in the Moon can be neglected. Gerstenkorn considers a phase lag that is proportional to frequency, while Goldreich's results are for a constant $\varepsilon_{2mpq}$, although he states that the evolution scenario for $a > a_c$ is not significantly modified by a frequency-dependent lag. A frequency-dependent phase lag will modify the evolution very significantly when $a < a_c$. A constant lag angle would appear reasonable if the dissipation occurs within the solid Earth, by mechanisms similar to that by which seismic energy is dissipated, if the apparently broad, frequency-independent, seismic absorption band encompasses the tidal frequencies. However, as discussed in chapter 10, most of the energy is dissipated in the oceans. Gerstenkorn's phase lag model corresponds to a Kelvin–Voigt rheology (see page 21), and while this is probably unrealistic at the tidal periods, if dissipation occurs within the solid Earth, the equivalent phase lags do have a similar frequency dependence (see table 10.2). Goldreich also considers the case treated by Slichter and MacDonald for constant geometric phase angles $\delta = (\Omega - n)\,\Delta t$. This implies that the time lag varies with the position of the Moon in its orbit and results in a complex dependence of the energy dissipation on frequency. This is of little consequence in the present tidal evolution but its consequences may be severe during the close approach stage, when orbital inclinations and eccentricities are high (Gerstenkorn 1967). The most detailed study of the lunar motion during this close approach phase is that by Gerstenkorn (1969) who considers both the eccentricity of the lunar orbit and the dissipation within the Moon.

The mathematical developments of the dynamical orbital evolution can now be considered in a satisfactory state. Much more problematical is the lack of a sound geophysical basis for the extrapolation into the past. Can we, from the presently inadequate understanding of the dissipation mechanisms, assume that conditions in the past, particularly during close approach, have been constant? Gerstenkorn, Ruskol (1966) and others still believe that the tidal dissipation occurs in the solid Earth, making the extrapolation into the past believable once the initial assumption is accepted. Dissipation in the oceans is important today, and presumably has been important for as long as there have been oceans. The problem is that even if the average rate of dissipation has been one-half of the present rate, the Moon is brought uncomfortably close to the Earth about $3\times10^9$ yr ago. Neither the terrestrial nor the lunar geology bears clear evidence for such a catastrophic event (section 11.4.2). Hence there is considerable interest in attempting to define further the time scale of the evolutionary process. Growth rhythms in coral, bivalve and stromatolite fossils have been interpreted in terms of such astronomical cycles. The coral and bivalve records extend back to the Ordovician, some $4.5\times10^8$ yr ago, and while these data do not provide stringent controls on evolutionary theories, they may provide a constraint on the time scale: an average rate of dissipation deduced from these data can possibly be considered as representative of dissipation for all times when oceans were present, that is, well back into the Precambrian. Prior to the Paleozoic the only source of information on past astronomical cycles lies in the stromatolite formations that extend back to the Cryptozoic. Possibly we are asking too much of these organic structures if we expect them to lead us to the origin of the Moon; and if they were only aware of what geophysicists are trying to read into them today, surely they would have adopted quite different living habits. But perhaps they may provide rough constraints on orbital evolution models during the otherwise data-sparse Precambrian.

Fossil coral and bivalve records are believed to provide measures of three astronomical cycles: (i) the number $N_1$ of solar days per year; (ii) the number $N_2$ of solar days per synodic month; and (iii) the number $N_3$ of synodic months per year.

Observations of $N_1$ give a direct estimate of the rotation of the Earth since

$$N_1 = \Omega(t)/n_{\odot} - 1, \qquad (11.2.3a)$$

and

$$dN_1/dt = \dot{\Omega}(t)/n_\odot. \qquad (11.2.3b)$$

The length-of-year is assumed to have remained constant over geologic time since (i) the tidal evolution of the Earth's orbital motion is very small (page 296) and (ii) the mass of the Earth is unlikely to have changed significantly since the Archean.

The frequency of the synodic month is given by $n_☾(t) - n_\odot$, if the precession rates $\dot{\omega}_☾$ and $\dot{\Phi}_☾$ of the lunar orbit remain small compared with $n_☾$. The number of solar days per synodic month is therefore

$$N_2 = [\Omega(t) - n_\odot]/[n_☾(t) - n_\odot], \qquad (11.2.4a)$$

and

$$dN_2/dt = \dot{\Omega}/(n_☾ - n_\odot) - \dot{n}_☾(\Omega - n_\odot)/(n_☾ - n_\odot)^2. \qquad (11.2.4b)$$

The number of synodic months per year is

$$N_3 = [n_☾(t) - n_\odot]/n_\odot, \qquad (11.2.5a)$$

with

$$dN_3/dt = \dot{n}_☾(t)/n_\odot. \qquad (11.2.5b)$$

If tidal friction is the only phenomenon responsible for $\dot{\Omega}$ and $\dot{n}_☾$, the integration of the equations (10.2.2$b$) gives the change in the quantities $N_1$, $N_2$, $N_3$ with time. In these integrations we have assumed that the contribution to $\dot{\Omega}$ from the solar tide has remained constant except for a change in the polar moment of inertia $C$, since $C$ is a function of $\Omega$. That is (equation 2.4.15),

$$C(t) = I[1 + (2k_s R^5/9GI)\Omega^2(t)].$$

We assume that there has been no change in the dissipating mechanisms. At present there is no reasonable alternative to this assumption and we consider this as a hypothesis against which the *paleontological clock* results eventually have to be tested. The integration is carried out with $\varepsilon_{22pq} = 2\varepsilon_{21pq}$ as is suggested by the present situation. Figure 11.5 gives the results. If the various assumptions made are valid and the interpretation of the paleontological data is correct and precise, the observed $N_1$, $N_2$, $N_3$ should fall somewhere on these curves and determine the average equivalent phase lag and accelerations.

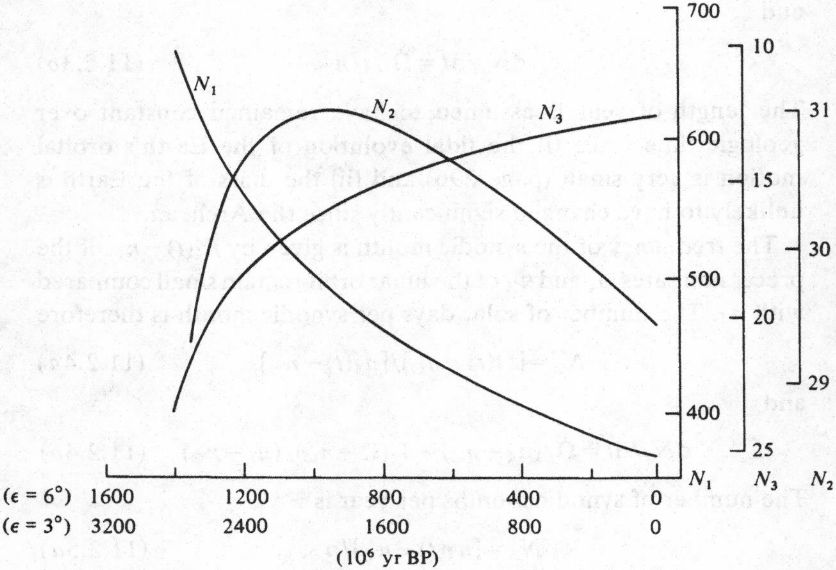


Figure 11.5. Number of solar days per year as a function of age ($N_1$), number of solar days per synodic month ($N_2$), and number of synodic months per year ($N_3$). The time scales are based on different values for the equivalent phase lag of the semi-diurnal tides.

## 11.3 Paleontological clocks†

Several groups of organisms leave records in the skeletal parts of their accreting tissue in the form of sequential and repetitive layers. These layers are interpreted as growth increments, and the sequence of layers appears to be a consequence of modulation of growth by internal rhythms inherent to the animal and of environmental conditions. Variations in the environmental conditions result in a variable growth rate that may be reflected in the skeletal parts of the organism itself. If these conditions can be identified, the comparative study of fossil organisms with their modern living counterparts may enable some conclusions to be drawn about fluctuations in environmental conditions throughout the geological

† Recently, Scrutton (1978) also reviewed the published paleontological evidence for the paleorotation with quite similar conclusions as to the reliability of the data.

past. Of particular interest are growth patterns that are controlled by astronomical phenomena. A comparison of the frequencies found in living and in fossil specimens may then yield indications of the constancy or otherwise of certain astronomical periodicities; in particular, they may give an indication of the Earth's paleorotation. Runcorn (1966) refers to such organisms as paleontological clocks, although this term should not be confused with biological clocks.

Three groups of organisms have been discussed for this geophysical application: corals, molluscs and stromatolites. Daily increments appear to be controlled by the successive alternation of daylight and darkness; in some organisms growth occurs, for example, mainly in daylight, in response to the available light, while at night growth may be either reduced or involve different biological and chemical factors. The daily pattern may be recognized by changes in colouration, in texture or in biological, physical and chemical properties of the layers. Annual variations in growth rate may be caused by changes in the length of daylight, in water temperature or in food supply, resulting in a seasonal modulation of the thickness of the daily pattern. Lesser periodicities may also occur due to lunar influences. Variable moonlight directly affects growth in some organisms by triggering spawning, or by modifying the available food supply. Periodic tidal influences could be important; animals living in an intertidal environment have their growth strongly modified by the tidal control of food supply, by contact with the atmosphere or by variable sedimentation rates. Superimposed on these more-or-less regular variations will be random or erratic events caused, for example, by the interruption of the cycle by storms, by interference of other marine animals or by extremes in water temperature.

The two principal rhythms that have been investigated in the recent literature are (i) the number of what are believed to be daily growth increments between successive seasonal growth marks and (ii) the number of daily increments between lesser growth marks that are believed to indicate successive synodic months due to some lunar influence on growth. These two quantities permit a separation of the accelerations of the Earth and Moon and, if the fossil record extends far back into geological time, they may provide important constraints on the dynamical evolution of the lunar orbit during the geological past. This simple argument is, however, strewn with difficulties. The actual growth pattern is often complex, due to it being the consequence of several interacting environmental factors and, while there is a growing body of empirical evidence concerning these factors, there is still little understanding of the actual physical controlling processes. This would not be a very serious handicap if it could be established that the controlling factors have not changed in the past, but, as both the environmental conditions and the species themselves have evolved with time, we cannot be sure of this. Equally important is the correct identification of the growth rhythm and this has not always been so. In particular, the bivalve evidence has received several different interpretations in the few years that they have been studied for their potential geochronological value. A further difficulty is the presence of random disturbances in the growth rhythm. Such growth lines will be characteristic of the sample and their effect on the accelerations would be minimized if many statistically representative samples were available for study. But relatively few well-preserved fossil records are available and the quality of the results will be degraded by these random events.

It is hardly surprising, then, that the paleontologists stress that the evidence for changes in the astronomical cycles is circumstantial, that the mechanisms governing the growth cycles remain unknown and that the hypothesis rests in large part on the observation that the observed periodicities in both modern and fossil corals correspond to what we would expect, if the present accelerations of the Earth and the Moon have remained constant throughout the past. The problem is akin to the early discussions of the ancient astronomical observations where the eclipse records were dated from the astronomical evidence itself. Such a foundation leaves any conclusions about the actual accelerations during the geological past on an uncertain footing and subject to the impression that perhaps one sees in the paleontological record what one expects or hopes to see. How many of the isolated published figures, unsupported by detailed descriptions of the sample and of the methods for identifying growth increments, and with no indication of the number of counts made nor of the spread of individual values, are a consequence of this? As Pannella (1975) rightly wonders: what would be the state of the field today if J. W. Wells had published in 1963 a much different number for the number of days in the Devonian year? Yet the more substantiated studies do, together, indicate that the paleontological record shows certain trends for the accelerations and that it may ultimately provide more than just gross constraints on the evolution of the Earth–Moon system.

### 11.3.1 *Corals*

*The evidence.* Seasonal fluctuations in the rate of coral growth were first reported by R. P. Whitefield in 1898 who described undulations on some surfaces of living corals and suggested that these represented annual growth increments associated with seasonal water temperature changes. The detailed mechanisms by which growth occurs and the factors controlling the rate of growth are still inadequately understood but there is general agreement that the growth rate varies annually, due to water temperature fluctuations, variations in the nutrient supply or by reproductive activity (Dodge & Vaisnys 1976). Apart from being a possible indicator of environmental conditions, the seasonal annulations or varves would be of little interest except that the corals also exhibit a much finer growth increment structure between the annual lines. The skeletal part of corals consists of several elements, one of which, the epitheca, reveals a fine structure of ridges that are parallel to the growing edge (figure 11.6). These ridges are interpreted as growth increments and they suggest a periodic fluctuation in the rate of calcium carbonate secretion. The rhythm of deposition of these growth increments in modern reef-forming corals is believed to be daily; that is, the calcium carbonate intake decreases during night-time or darkness and increases during day-time (Goreau 1959; Barnes 1972). Modern reef-forming, or hermatypic, corals possess symbiotic algae in their tissues whose photosynthetic activity appears to aid calcification during hours of daylight. An indirect indication that the growth ridges are daily is that modern corals typically add about 360 such increments per year, suggesting that the solar day controls the frequency of deposition (Wells 1963; Barnes 1972). Barnes concludes that the fine growth increments are solar daily but that the reliability of the corals in depositing one layer per day is limited.

Figure 11.6. Middle Devonian coral epitheca from Michigan, U.S.A., illustrating 13 well-developed bands, each with an average of 30.8 ridges (supplied by C. T. Scrutton).

Whether Paleozoic fossil coral epitheca displayed the same habits as their modern counterparts is problematical, and requires much further study before the ancient fossils can provide an unambiguous record of any changes in the number of days per year. While the daily rhythm of the hermatypic corals appears to be reasonably well established, ahermatypic corals, which do not contain the algae symbionts and which flourish in effective darkness, also exhibit regular growth rhythms, although apparently at much slower cycles than the reef corals. Factors other than variable daylight may be important here in modulating the growth rates. No relevant studies on living specimens of such corals appear to have been made and this will make any extrapolation, from present reef-forming corals to ahermatypic fossil corals, a questionable process. Indirect evidence again suggests that the solar day has remained the dominant periodicity in the corals studied; the Devonian corals studied by Wells (1963) show about 400 daily growth increments between successive seasonal annulations, in keeping with the *expected* value, if the present tidal acceleration of the Earth has remained roughly constant over the last $3–4 \times 10^8$ yr.

Not all corals show the clear seasonal annulations of the specimens studied by Wells and this is generally attributed to geographic variations in the environmental stimuli. Corals grown near the equator, for example, may exhibit little seasonal banding since water temperatures and food supply have remained nearly constant. Thus the Lower Carboniferous corals of Northern England studied by Johnson & Nudds (1975) show little or no seasonal annulation, having grown near the Carboniferous equator. Such corals may provide information on paleolatitudes but they contain no information on the number of days per year. Wells in 1937 did suggest that lesser annulations observed in some corals might reflect monthly growth fluctuations and, because he observed that their periodicities were similar to those of the synodic month, he suggested that these annulations in the growth are regulated by the Moon. Such a lunar control can be exercised in several ways. Some living corals are known to have breeding cycles that are controlled by the Moon, presumably by moonlight. While the corals are preoccupied with reproduction, less calcium carbonate is deposited and the daily growth increments are more closely grouped together than during the remainder of the cycle. Moonlight may also increase the time available for photosynthesis and hence increase the thickness of the growth-band increments. Alternatively, moonlight controls the coral's supply of nutrients by causing a vertical migration of phytoplankton concentrations. Tidal effects may also play a role in that near-surface corals may be exposed during certain periods. For unequal semi-diurnal tides, for which the phase lags of the solar and lunar tides are not equal, the tides will be a minimum at every neap tide, otherwise the periodicity will be near 14 d. Controlled experiments by Buddemeier & Kinzie (1975) on living corals suggests that the cycle is monthly but the actual mechanism remains obscure. In all cases, except for a possible 14-d tidal influence, the lunar control of the growth rhythms will result in a periodicity that is the synodic month, the time between successive new Moons and twice the interval between successive spring or neap tides. The most detailed studies of the lesser annulations in the fossil record are those by Scrutton (1965) of Devonian specimens. The one specimen studied that possessed both seasonal and lesser annulations indicated 13 of the latter per year; the other specimens showed from 27 to 34 daily increments per lesser annulation. Both counts are in general agreement with the hypothesis that the lesser annulations are indeed monthly.

In interpreting the fossil record, several factors must be kept in mind. (i) Few well-preserved corals are available for the geological past. The daily growth increments on the surface of the epitheca are usually less than a few tens of microns thick and their preservation, even in modern epitheca, is relatively rare; the epitheca do not appear to contain important internal markers of their daily growth. Thus most published studies are based on only a few specimens and it will be difficult to obtain global statistical averages of the counts for any one epoch. (ii) The seasonal annulations are annual only in the sense of occurring once a year, not necessarily repeating themselves every 365 d. Growth dependence on water temperature, for example, may result in intervals between successive seasonal varves or in *climatic years* that are longer or shorter than 365 d, when the seasonal fluctuations in the water temperature are not strictly annual. Thus corals with a large number of successive and well-preserved seasonal annulations are desirable, so that the average values of the number of days per year can be established. This will also be true for the monthly bands which may not exhibit a strictly synodic period. Statistically the monthly counts will be more precise since the number of successive monthly bands in a specimen will in general be larger than the number of successive seasonal annulations. On the other hand, the change in the number of monthly counts with time is not very large (figure 11.5). (iii) The diurnal pattern is sometimes disrupted, either by a halt in growth or by a reduction in the thickness of the daily increments, to such an extent that successive layers can no longer be separated. This could occur, for example, when the water temperature drops below a certain critical level or if water turbidity reduces the photosynthetic activity of the algal symbiont. Counts of some modern corals by Wells (1963) for example, give about 360 d per seasonal annulation, but whether this low value is due to the coral *forgetting* to deposit its daily layer or due to the seasonal annulation not being strictly of 365-d periodicity, is not clear.

Counts of the number of fine growth increments per seasonal annulation obtained by Wells (1970) are summarized in table 11.1 for several geological epochs. These values include earlier values published by Wells (1963, 1966). He provides no information on the number of seasonal annulations counted on each specimen. The Middle Devonian results appear to be the most precise as they are based on fossil epitheca of four different species found in three different localities. The scatter of the results for the individual specimens (Wells 1970, Figure 5) suggests a standard deviation of a single specimen of about 6 d. The standard deviation of the mean value is of the order of 2 d but biases in the actual growth rhythms, in the interpretation of imperfect growth ridges and in the identification of the seasonal bands, probably result in an accuracy of the mean value that is worse than this. We assume standard deviations of $6/\sqrt{(\text{number of samples})}$ d. Scrutton's (1965) one specimen for the Middle Devonian that exhibited seasonal annulations resulted in a value of 401 increments per seasonal annulation, in good agreement with Wells's average value of about 395. The only other published values for the number of increments per seasonal annulation is from the work by Mazzullo (1971) who studied Silurian and Devonian corals and Brachiopods. The latter samples apparently displayed marked annulations with finer banding suggestive of diurnal increments. No information is available on the number of seasonal annulations counted per specimen. Mazzullo's published values are maximum counts rather than

Table 11.1. *Summary of the number of growth increments per seasonal annulation in fossil corals*

| Epoch | Geologic time ($10^6$ yr) | No. of counts | No. of samples | Range of counts | Adopted s.d. |
|---|---|---|---|---|---|
| Wells 1970 | | | | | |
| Upper Carboniferous | −300 | 385 | 2 | 380–90 | 4.2 |
| Lower Carboniferous | −320 | 398 | 1 | | 6 |
| Middle Devonian | −370 | 398 | 12 | 385–405 | 1.7 |
| Middle Silurian | −420 | 400 | 1 | | 6 |
| Upper Ordovician | −440 | 412 | 1 | | 6 |
| Modern | 0 | 360 | Several | ? | 4 |
| Scrutton (1970) | | | | | |
| Middle Devonian | −370 | 401 | 1 | | |
| Mazzullo (1971) | | | | | |
| Middle Devonian | −370 | 405 | 17 | 393–410 | |
| Silurian | −420 | 419 | 3 | 416–21 | |

averaged counts and he argues that these are more precise due to the tendency for growth increments to be absent or very small at certain times. Whether or not this is so is debatable (Barnes 1971), but in any case the interval between successive seasonal annulations will not be strictly constant and the use of a maximum count will bias the results to values for those years when the climatic year was longer than the actual year. The use of the maximum-count method leads to values for the Devonian that are about 15 increments greater than the values found by Wells. S. J. Mazzullo (personal communication 1977) has supplied partial information on the number of increments counted per annulation: 17 fossil records of the Middle Devonian give an average count of 405 and three fossils of the Silurian give 419 increments per annulation (table 11.1).

Scrutton (1965, 1970) studied 10 Middle Devonian specimens representing several different genera found in three locations. These coral epitheca showed the clear groupings of the basic fine growth increments that have tentatively been attributed to a lunar influence and as representing the synodic month cycle. A total of

Table 11.2. *Summary of the number of growth increments per lesser annulation in fossil corals*

| Epoch | Geologic time ($10^6$ yr) | No. of counts | No. of samples | Range of counts | Adopted s.d. |
|---|---|---|---|---|---|
| Scrutton (1970) | | | | | |
| Middle Devonian | −370 | 30.66 | 10 | 27–35 | 0.5 |
| Johnson & Nudds (1975) | | | | | |
| Lower Carboniferous | −330 | 30.2 | At least 6 | ? | 0.4 |

113 such bands have been counted with up to 16 consecutive bands in a single specimen. Counts between successive monthly bands varied from 27 to 35 while the average count for a single specimen showed a much smaller range, from 29.0 to 31.5 with an average of 30.6 for all 10 specimens (Scrutton 1965) (table 11.2). The standard deviation of a single monthly band is 1.1 and the standard deviation of the mean is of the order of 0.1. Biases in growth habits, identification and counting of growth increments undoubtedly are larger than this but, because of the large number of bands counted, they are probably smaller than those associated with the annual counts. Scrutton (1970) suggests that, in view of a certain skew shown by his data, the mode may be a more meaningful estimate (see also Dolman 1975). Because of likely biases in the results, such refinements are probably not yet warranted. We adopt for the standard deviations $1.1/\sqrt{(\text{number of samples})}$ d. Johnson & Nudds (1975) have provided monthly counts for the Lower Carboniferous. Apparently six different genera were studied but the authors give no information on the number of specimens, on the number of monthly bands counted nor on the range of the individual counts. Mazzullo (1971) also gives values for the number of growth increments per synodic month, but whether these values were actually counted or simply computed from the annual counts is unclear: his ratio of the number of increments per year to the number of increments per month yields exactly 13 for all specimens given in his table 1.

*The coral accelerations.* In no single instance are counts of $N_1$, $N_2$ and $N_3$ available for the same date. Wells and Scrutton's results give $N_1$ and $N_2$ for the Devonian, and $\dot{\Omega}$ and $\dot{n}_{☾}$ can be estimated for this epoch. Likewise, by combining the results of Wells with those of Johnson & Nudds, $\dot{n}_{☾}$ and $\dot{\Omega}$ can be estimated for the Lower Carboniferous. For the other epochs the data of Wells enable only $\dot{\Omega}$ to be estimated as a function of time. Perhaps the simplest solution is to solve simultaneously for the accelerations $\dot{\Omega}$ and $\dot{n}_{☾}$ from all of these data by assuming that $\Omega(t)$ and $n_{☾}(t)$ vary linearly with time and that the accelerations are constant. A preliminary solution of the data suggests that the average accelerations over the last $400 \times 10^6$ yr were about half of their present values so that the Earth–Moon distance during the Devonian was only a few per cent smaller than the present value. The nature of the data probably does not warrant a more sophisticated treatment. Writing

$$\left.\begin{aligned}\Omega(t) &= \Omega(t_0) + \Delta\Omega + \dot{\Omega}t \\ n_{☾}(t) &= n_{☾}(t_0) + \Delta n_{☾} + \dot{n}_{☾}t,\end{aligned}\right\} \tag{11.3.1.}$$

the condition equations corresponding to the two types of observations $N_1$ and $N_2$ follow from (11.2.3*a*) and (11.2.4*a*) as

$$\frac{1}{n_\odot}\begin{pmatrix}1 & 0 & t & 0 \\ 1 & -\beta & t & -\beta t\end{pmatrix}\begin{pmatrix}\Delta\Omega \\ \Delta n_{☾} \\ \dot{\Omega} \\ \dot{n}_{☾}\end{pmatrix} = \begin{pmatrix}N_1 + 1 - \Omega(t_0)/n_\odot \\ \{[n_{☾}(t_0) - n_\odot]/n_\odot\}N_2 + 1 - \Omega(t_0)/n_\odot\end{pmatrix} \tag{11.3.2.}$$

with

$$\beta = [\Omega(t_0) - n_\odot]/[n_{☾}(t_0) - n_\odot].$$

These equations are solved using the data by Wells, Scrutton and Johnson & Nudds, assuming that the systematic biases in the growth ring counts are the same for all samples, and using as weights $1/v^2$ where $v$ is the standard deviation given in the last column of tables 11.1 and 11.2. A meaningful separation of $\Delta n_{☾}$ and $\dot{n}_{☾}$ is not possible with $N_2$ estimates for only two nearby epochs and we impose the condition that for the present epoch

$$N_2 = 29.5 \pm 0.05. \qquad (11.3.3.)$$

The solution of the coral data gives

$$\Omega(T) = [366.2 - (5.13 \pm 3.93) - (0.100 \pm 0.012)t]n_{\odot}, \qquad (11.3.4a)$$
$$n_{\text{☾}} = [13.3 - (0.16 \pm 0.23) - (0.0024 \pm 0.0007)t]n_{\odot},$$

where the epoch $t$ is in units of $10^6$ yr and negative when it refers to the past. The accelerations are

$$\left.\begin{aligned} \dot{\Omega} &= (-6.3 \pm 0.7)10^{-22}\ \text{rad s}^{-2}, \\ \dot{n}_{\text{☾}} &= (-1.5 \pm 0.4)10^{-23}\ \text{rad s}^{-2}, \end{aligned}\right\} \qquad (11.3.4b)$$

with a correlation coefficient of 0.53. The ratio $\dot{\Omega}/\dot{n}_{\text{☾}}$ is $42.0 \pm 11.1$ compared with 51.4 for the present value (equation 10.5.2). This suggests that there has been no significant non-tidal acceleration in the Earth's spin on a time scale of $10^8$ yr.

### 11.3.2 *Bivalves*

*The evidence.* The use of molluscs for studying periodicities in the environmental factors governing daily and annual growth was first suggested by Barker (1964). Since then, bivalves have received particular attention, due in part to their relatively easy culture, permitting the growth behaviour of living specimens to be readily studied. Yet, as for corals, there is little understanding of the physical processes controlling growth rhythms. Much of the evidence is again circumstantial in that the periodicities found appear to agree with both present astronomical periodicities and *predicted* periodicities for past epochs. Bivalves possess some other advantages over corals. The shells possess internal growth lines and these are more readily preserved intact than the external growth increments on the coral epitheca. Secondly, the bivalves – both now and in the past – have a wider distribution and a more extensive fossil record exists. On the other hand the growth rhythms of bivalves seem to be more complex than that of corals due to interactions of diurnal and tidal stimuli but this may merely reflect that modern bivalves have been more studied than living corals.

Figure 11.7 Bivalve *Clinocardium nuttalli* showing the external growth ridges (from Evans 1975).

Daily growth lines, due to a deposition of calcium carbonate layers alternating with thin organic layers, are visible in most shells, both superficially and internally, in cross-sections (figure 11.7). The calcium carbonate deposition occurs during periods of shell opening and an interruption in the process occurs during prolonged closing. During this interval the thin organic layer is deposited. Thus any periodic stimulus that forces the bivalve to close during certain intervals leaves periodic markers in the shell record of animal growth. Seasonal variations in the thickness of the daily increments are readily visible but, as for corals, these events are not strictly periodic. Cold water bivalves, in particular, may lead to misleading values for the length-of-year since, during cold winter months, growth slows down very significantly or may even halt temporarily. Breeding events are also recorded in the shells by reduced growth rates at times of spawning. These breaks appear to occur at regular intervals – once or twice per year in some species, more frequently in others – often with a multiplicity of 29 daily increments, suggesting lunar control.

Pannella & MacClintock (1968) found that specimens in intertidal waters show an increase in growth at a rate of one increment per solar day, and concluded that the solar day is the basic unit and that the synodic month is expressed by a variation in the thickness of the increments. Clark (1968) and House & Farrow (1968) also concluded that the growth rhythms formed daily. Later, Pannella (1972) suggested that the deposition of the growth lines may be affected by tidal oscillations; that is, growth is interrupted during low tide when the shellfish is uncovered. The larger the tidal range between high and low tide, the thicker will be the growth increment, and the specimens will show a grouping together of thin increments into clusters separated by the interval between successive neap tides. For unequal semi-diurnal tides, the interval may be twice as long. This tidal influence is clearly illustrated by the experiments of Evans (1972) who finds that internal growth lines of some bivalves in areas of unequal semi-diurnal tides possess complex patterns, with increments forming sometimes at daily intervals and sometimes at semi-diurnal intervals (figure 11.8). Pannella & MacClintock's (1968) bivalve *Mercenaria*, living in intertidal waters where the semi-diurnal tides are nearly equal, do not show such complex patterns and do not appear to add two growth lines per lunar day. Other mechanisms appear to govern their growth. Clark (1974) suggests that combinations of lunar and solar periodicities, such as the number of exposures during daylight, may control growth. Pannella (1975) concludes that calcification in the shellfish occurs at night and follows a solar cycle due to diurnal variations in light, food supply or water temperature. During periods of prolonged closing, such as occurs when the animal is exposed during low tide, this process is interrupted and a thin organic layer is formed. If the shell is exposed for a major part of night-time, such as would occur at neap tide for an animal living near the mean water level, no calcite deposition occurs at all and the animal misses a growth ring. Alternatively, if the shell is exposed briefly near the

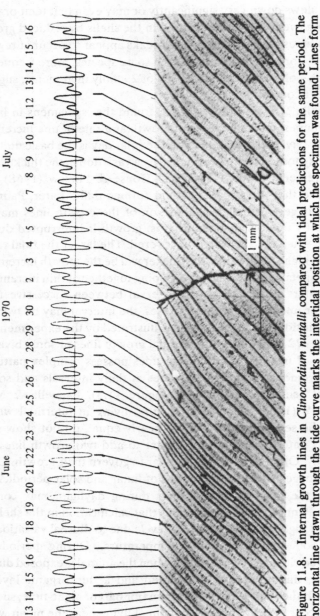

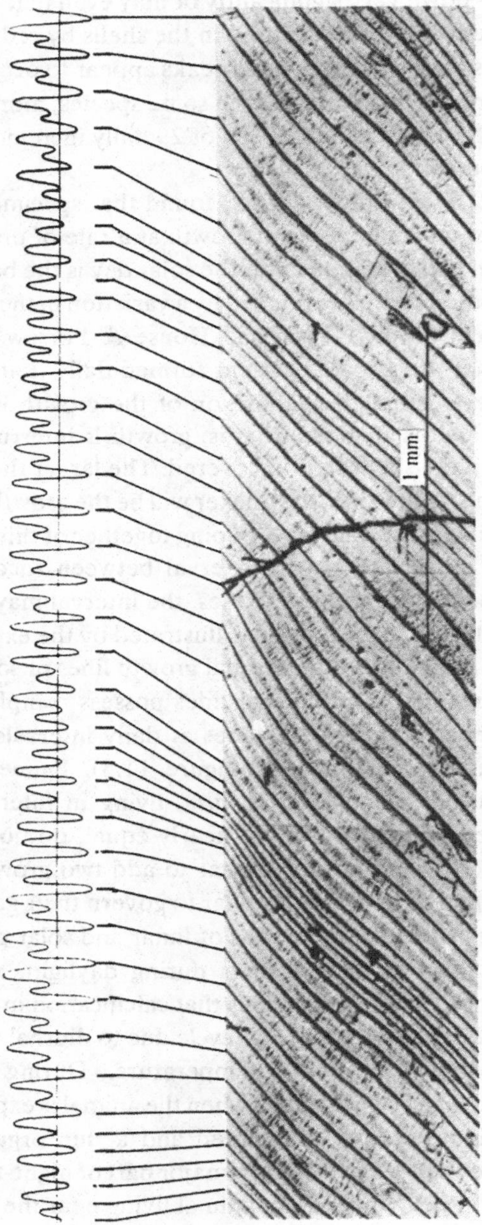


Figure 11.8. Internal growth lines in *Clinocardium nuttalli* compared with tidal predictions for the same period. The horizontal line drawn through the tide curve marks the intertidal position at which the specimen was found. Lines form when the tide drops below this position (from Evans 1972).

middle of the night, calcification is only temporarily interrupted and resumed when the animal is again covered. Now two calcium carbonate layers are deposited, separated by the thin organic layer formed during exposure. Complex growth patterns will develop in this manner and will depend on (i) the position of the animal with respect to mean sea level, (ii) the type and range of tides, (iii) the time of day of occurrence of spring and neap tides, and (iv) the time of year, as this controls the duration of daylight. The consequences are clear; growth patterns of animals living in intertidal waters will vary considerably from one specimen to the next depending on their position, even when found in the same area. Pannella (1975) argues that these patterns can be recognized and taken into consideration in establishing the frequencies of the astronomical cycles.

Pannella (1972) has studied numerous bivalve fossils from the Upper Cretaceous to the Middle Devonian. These results (table 11.3) all show a gradual increase in the number of increments per annulation with increasing age. For comparison, counts on recent specimens give an average of 359.3 increments per annulation. As for Wells's coral results, the estimates are a little below the expected values if the growth cycle corresponds to solar days per year. The results for the Devonian are in general agreement with the coral data. For precision estimates we adopt the same rule as for the corals, namely $6/\sqrt{(\text{number of samples})}$ d. Some caution in interpreting these results is again necessary since some of the complexities in the growth rhythms discussed above, were not fully recognized at the time of Pannella's study. Pannella gives additional results for the number of increments per lesser annulation which he interprets as a measure of the number of days per synodic month (table 11.4). These results also indicate a gradual increase in the length of the synodic month with age, similar to the coral results. The recent results are underestimated by about 1.2% compared with the duration of the actual synodic month. We adopt as standard deviation $v$, Pannella's estimate of the $v$ of a single count divided by the square root of the number of samples studied for a given epoch.

Berry & Barker (1968) find comparable results for the number of increments per lesser annulation that are suggestive of a tidal frequency, there being on an average 25.0 such cycles between major seasonal annulations. Berry & Barker (1975) have further

Table 11.3. *Summary of results from Pannella (1972) for the number of increments per annulation in fossil bivalves. The number of annulations is the total for all specimens studied for the given period.*

| Period | Geologic time ($10^6$ yr) | Increments per annulation | No. of specimens | No. of annulation | Range of counts | Adopted s.d. |
|---|---|---|---|---|---|---|
| Recent | 0 | 359 | 9 | 32 | 353–66 | 2.1 |
| Upper Cretaceous | −70 | 375 | 3 | 16 | 371–9 | 3.5 |
| Middle Triassic | −220 | 372 | 3 | 7 | 365–75 | 3.5 |
| Upper Carboniferous | −290 | 383 | 3 | 11 | 380–9 | 3.5 |
| Lower Carboniferous | −340 | 398 | 2 | 9 | 397–9 | 4.2 |
| Middle Devonian | −360 | 406 | 1 | 6 | | 6.0 |

Table 11.4. *Summary of results from Pannella* (1972) *for the number of increments per lesser annulation observed in fossil bivalves*

| Period | Geologic time ($10^6$ yr) | Increments per lesser annulation | No. of specimens | No. of lesser annulation | Range of counts | Adopted s.d. |
|---|---|---|---|---|---|---|
| Recent | 0 | 29.2 | 7 | 186 | 29.0–29.6 | 0.4 |
| Upper Tertiary | −14 | 29.4 | 5 | 197 | 29.2–29.8 | 0.5 |
| Middle Tertiary | −38 | 29.8 | 2 | 40 | 29.6–29.9 | 0.8 |
| Lower Tertiary | −54 | 29.6 | 3 | 141 | 29.4–30.0 | 0.6 |
| Upper Cretaceous | −70 | 29.9 | 7 | 159 | 29.6–30.2 | 0.4 |
| Middle Triassic | −220 | 29.7 | 3 | 77 | 29.4–30.0 | 0.6 |
| Upper Carboniferous | −300 | 30.2 | 4 | 59 | 29.9–30.7 | 0.6 |
| Upper Devonian | −350 | 30.4 | 4 | 168 | 30.2–30.5 | 0.6 |
| Upper Ordovician | −445 | 30.3 | 2 | 84 | 29.8–30.7 | 0.8 |

investigated these lesser annulations in bivalves from the present to the Devonian. Their specimens show pronounced clustering of growth lines in groups of about 15. Accepting their interpretation that these rhythms are indicative of tidal influence, and Berry & Barker stress that the underlying assumption is not proven, this gives a measure of $\frac{1}{2}N_2$ (table 11.5). Their results are similar to those of Pannella. We adopt as measure of the standard deviation of $N_2$, $2\times(v$ of single sample$)/\sqrt{(\text{number of samples})}$.

The number of lesser annulations per seasonal band has also been counted by Pannella (1972) for Recent, Cretaceous and Triassic fossils with results that are in close agreement with those that can be calculated from the independent results in tables 11.3 and 11.4. Pannella suggests that this may indicate the number of synodic months per year. Table 11.6 gives Pannella's results. The adopted standard deviations are estimated from those given in tables 11.3 and 11.4 for $N_1$ and $N_2$ according to

$$v_{N_3}^2 = v_{N_1}^2/N_1^2 + v_{N_2}^2/N_2^2.$$

Berry & Barker (1968) give a comparable result for the Cretaceous if their lesser annulations do indeed indicate a fortnightly tidal phenomenon. Mazzullo (1971) finds that for the Silurian and Devonian there were 13 lesser annulations per seasonal growth ring. Neither of these studies provides details on the number of annulations counted nor on the range of results. Mazzullo's result is presumably a maximum count. Some of Scrutton's Devonian corals also indicate 13 bands per year.

*The bivalve accelerations.* For three epochs, Recent, Upper Cretaceous and Middle Triassic, Pannella (1972) has estimated all three quantities $N_1$, $N_2$ and $N_3$. With (11.2.3*a*), (11.2.4*a*), (11.2.5*a*) and the assumption (11.3.1) the observation equations are of the form

$$\frac{1}{n_\odot}\begin{pmatrix} 1 & 0 & t & 0 \\ 1 & -\beta & t & -\beta t \\ 0 & 1 & 0 & t \end{pmatrix}\begin{pmatrix} \Delta\Omega \\ \Delta n_☾ \\ \dot{\Omega} \\ \dot{n}_☾ \end{pmatrix} = \begin{pmatrix} N_1 + 1 - \Omega(t_0)/n_\odot \\ \{[n_☾(t_0) - n_\odot]/n_\odot\}N_2 + 1 - \dfrac{\Omega(t_0)}{n_\odot} \\ N_3 + 1 - n_☾(t_0)/n_\odot \end{pmatrix}. \tag{11.3.5}$$

Table 11.5. *Summary of results by Berry & Barker* (1975) *for the number of growth lines per lesser annulation in fossil bivalves*

| Epoch | Age ($10^6$ yr) | No. of species | No. of samples | No. of clusters | Average count | S.d. of one sample | Range of counts | Adopted s.d. of mean |
|---|---|---|---|---|---|---|---|---|
| Pleistocene | 1 | 3 | 19 | 115 | 14.74 | 0.30 | 14.2–15.0 | 0.14 |
| Pliocene | 4 | 2 | 9 | 40 | 14.83 | 0.24 | 14.3–15.0 | 0.16 |
| Oligocene | 30 | 2 | 10 | 49 | 14.82 | 0.15 | 14.7–15.0 | 0.09 |
| Eocene | 48 | 1 | 3 | 14 | 14.87 | 0.23 | 14.6–15.0 | 0.27 |
| Palaeocene | 60 | 1 | 6 | 20 | 14.82 | 0.21 | 14.7–15.0 | 0.17 |
| Cretaceous | 100 | 4 | 17 | 1189 | 14.88 | 0.25 | 14.7–15.0 | 0.12 |
| Jurassic | 160 | 1 | 10 | 12 | 14.90 | 0.28 | 14.7–15.3 | 0.18 |
| Triassic | 230 | 1 | 10 | 17 | 14.91 | 0.11 | 14.8–15.0 | 0.07 |
| Carboniferous | 310 | 3 | 17 | 29 | 15.09 | 0.35 | 14.7–15.5 | 0.17 |
| Devonian | 370 | 1 | 1 | 8 | 15.25 | 0.30 | — | 0.60 |

Table 11.6. *Summary of results for the number of lesser annulations per seasonal annulation*

| Period | Geologic time ($10^6$ yr) | Lesser annulations per annulation | No. of specimens | Range of counts | Adopted s.d. |
|---|---|---|---|---|---|
| Pannella | | | | | |
| Recent | 0 | 12.3 | 7 | 12.0–12.7 | 0.17 |
| Cretaceous | −70 | 12.6 | 3 | 12.5–12.8 | 0.26 |
| Middle Triassic | −220 | 12.6 | 3 | 12.5–12.7 | 0.26 |
| Berry & Barker | −70 | 12.5 | | | |

For other epochs, estimates of only one or two of the $N_i$ are available. As for the coral data, the most direct approach is to solve for the accelerations using all data simultaneously. With only Pannella's data, using the adopted standard deviations given in tables 11.3–11.5, the solution is

$$\Omega(T) = [366.2 - (4.8 \pm 1.8) - (0.095 \pm 0.009)t]n_{\odot}, \tag{11.3.5a}$$
$$n_{☾}(T) = [13.3 - (0.04 \pm 0.09) - (0.0021 \pm 0.0005)t]n_{\odot}.$$

The constraint (11.3.3) on $N_2$ need not be imposed now as there are observed values for $N_2$ near epoch $t = 0$. The accelerations are

$$\left.\begin{aligned} \dot{\Omega} &= -(5.9 \pm 0.6)10^{-22} \text{ rad s}^{-2}, \\ \dot{n}_{☾} &= -(1.3 \pm 0.3)10^{-23} \text{ rad s}^{-2}, \end{aligned}\right\} \tag{11.3.5b}$$

with a correlation of 0.65. Then

$$\dot{\Omega}/\dot{n}_{☾} \simeq 45.9 \pm 8.5. \tag{11.3.6}$$

The solution is very compatible with that obtained from the coral data. Berry & Barker's counts for $N_2$ give an unsatisfactory separation of $\dot{\Omega}$ and $\dot{n}_{☾}$ but the rate of change of $N_2$ with time appears reasonably well established. These data suggest (i) accelerations that are somewhat less than those from Pannella's solution (figure 11.9), and (ii) systematic errors in the counts of opposite sign to those from Pannella.

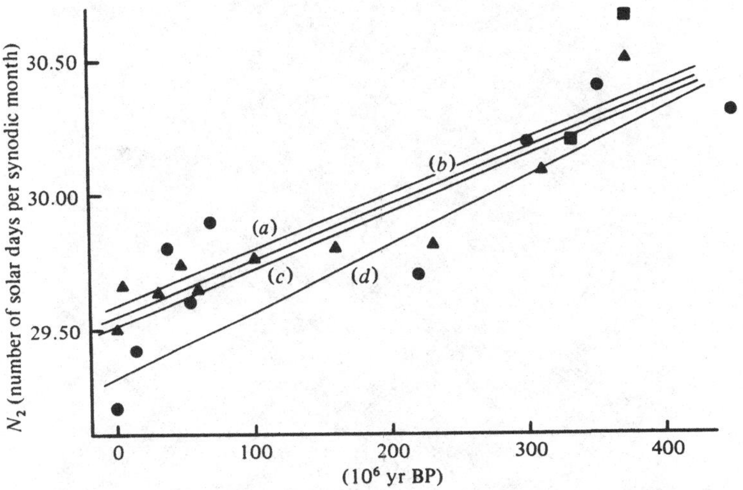


Figure 11.9. Estimates for the number of days per synodic month ($N_2$) according to Pannella (●), Berry & Barker (▲) and the coral data (■). The four lines represent $N_2(t)$ based on (*a*) solution (11.3.4*a*) of coral data, (*b*) an unweighted best fit to Berry & Barker's observed data, (*c*) an unweighted best fit to Pannella's observed $N_2$ data, and (*d*) solution (11.3.5*a*) from all of Pannella's bivalve data.

### 11.3.3 *Stromatolites*

Stromatolites are organo-sedimentary structures produced by an interaction between the growth and metabolic activity of micro-organisms and sedimentation processes. The micro-organisms are mainly blue-green algae or cyanophytes. They form more-or-less cohesive layers, or algal mats, that spread over loose sediments and trap or bind particles that are deposited on them. Responding actively to light, the algae grow upwards and, under continuing sedimentation, they continually colonize newly deposited particles as the entire colony is gradually displaced upwards. It is the fossil record of this cumulative growth-sedimentation record that makes up the stromatolitic paleontological clock (figure 11.10).

Both growth and sedimentation rates may be strongly modified by environmental stimuli, and the stromatolites show a propensity for more-or-less regular banding. Various factors may contribute. During daylight, the algal growth tends to be rapid and directed upwards towards the source of energy. At night the growth rate is

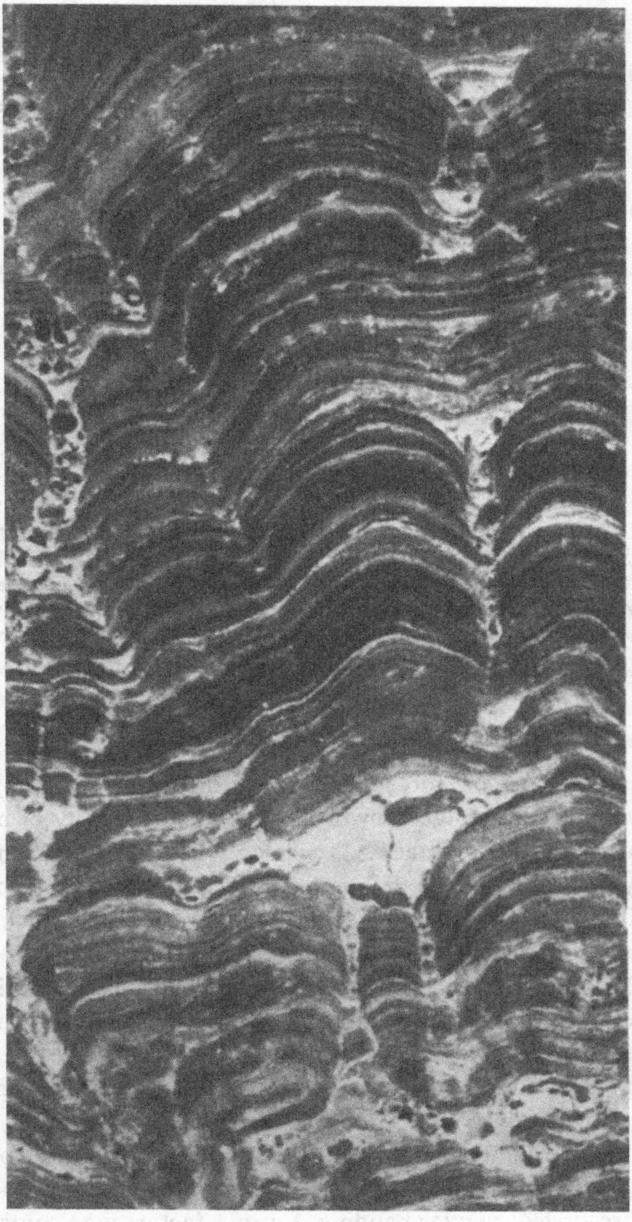

Figure 11.10. Precambrian stromatolite from Montana (from Runcorn 1966).

much reduced and prostrate. Not only is there a textural difference of the laminae, but the day-time layer is organic rich while the night-time layer is sediment rich. Sedimentation rates may vary periodically, in particular if the stromatolites prosper in peritidal environments; growth may be much modified during periods of exposure to air at low tides or tides may result in a variation in the sediment supply. Seasonal variations in temperature, salinity or other factors may cause seasonal patterns in growth rate. Or, the variety of organisms cohabiting the algal mat may bloom at different times resulting in further laminations in the fabric texture. Even in perfectly preserved fossil records, the interpretation of the layering in terms of astronomical periodicities involves all the difficulties and ambiguities that have already been discussed for the coral and bivalve records. Superimposed on these more-or-less periodic cycles are erratic events. Growth may cease for limited periods, storms may destroy part of the record, invasions by other animals may disrupt the process. The success of the stromatolites as paleontological clocks therefore depends on whether these various stimuli can be identified and separated, and whether it is possible to extrapolate from observations of living forms to fossil forms going back more than $3 \times 10^9$ yr in time. The collection of papers edited by Walter (1976) summarizes the present state of our knowledge of the stromatolites.

Morphological forms of stromatolites are varied and include stratified, domed, columnar and nodular forms. They will form under many conditions and have occurred throughout the geological record since the Archean. They can form in almost any free-standing body of water, marine, lake or hot spring pool, provided that certain conditions are met: that the conditions for growth of the micro-organisms are appropriate, that the growth rate of these organisms exceeds their rate of destruction and decomposition, and that the sedimentation rates are sufficient to produce a laminar structure, yet not too high to prevent the micro-organic colonization of the new layers. Furthermore, an absence of disturbances created by the burrowing, grazing and hunting activities of other forms of life is essential if a minimally disturbed record is sought. It is probably this condition more than any other that explains the widespread presence of stromatolites in the Proterozoic. Fossil remains of more recent formations are widespread and living forms exist in some regions today but nowhere do these formations appear to be as extensive as they were in the Precambrian. Coral and bivalve fossils do not extend beyond the Phanerozoic and they can provide information on the Earth's paleorotation for barely 10% of the Earth's lifetime. Thus the Proterozoic stromatolites, apart from their fundamental geological and biological implications, are of considerable geophysical interest. The oldest known stromatolites are those in the Rhodesian Bulawayan formation whose age probably exceeds $3 \times 10^9$ yr, and if these can provide useful data, the time span over which the paleo-tide and paleorotation record can be established will be increased sixfold over the present record based upon invertebrate fossils.

A number of experiments on living stromatolites indicate that the laminae form with a daily periodicity due to the predominant role played by light stimulating the algae growth (see, for example, Monty 1965). Other observations show that the rates of sedimentation may be controlled by the tides, either by exposing the growth or by variable tidal currents, resulting in a tidal record in the sequence of laminae (Gebelein 1969). In intertidal zones, Gebelein & Hoffman (1968) find that two layers per day are formed, each consisting of an organic-rich layer formed at low tide and a sediment-rich layer formed at high tide. Presumably such patterns will interact with the light–dark variations to provide layers of organic material that show different textural characteristics. Hardie & Ginsberg (1971) find that the layering of stromatolites formed in tidal flats is mainly due to periodic storms.

The most useful stromatolites for establishing periodicities in layering appear to be digital columnar forms (Pannella 1976), as these are thought to have been formed in subtidal waters (see, for example, Awrawik 1976; Playford & Cockbain 1976), where their physical disruption is less likely than in the intertidal zone and where their growth and sedimentation processes may also be the most regular and controlled by diurnal cycles. Such formations are found living in some areas, such as Shark Bay in Western Australia, and their general morphology is similar to forms found in the important Gunflint formation of the Middle and Lower Proterozoic. These latter are also believed to be formed subtidally

(Awrawik 1976) although the microorganisms are quite different from those inhabiting the Shark Bay specimens. McGugan (1967) recognized three possible periodicities in Upper Cambrian stromatolites: layers of alternating organic and calcite material, bands of 9–12 laminae each, and further bands of 400–20 laminae. He tentatively interpreted these as being either daily, monthly and annual bands or subannual, annual and longer-period bands. Pannella, MacClintock & Thompson (1968) studied Upper Cambrian specimens and interpreted the banding of the fine laminae as tidal and synodical patterns. The reliability of the stromatolites as fossil clocks remains, however, to be established. Hoffman (1973) considers them to be very unreliable, while Pannella (1972) concludes that they are less reliable than other fossil clocks but that they may nevertheless provide approximate constraints on the evolution of the Earth–Moon system during the Precambrian.

Pannella (1972) has investigated the growth patterns in fossil stromatolites from the Recent to the Archean but only small digitate forms, from the Biwabik–Gunflint formation exposed on the northern shores of Lake Superior and dated at about $2 \times 10^9$ yr, provide what are apparently significant results. Two specimens counted by Pannella give 446–8 laminae per seasonal line. Lesser bands were found to possess an average of 33.4 laminae. Because of the facility with which growth layers may be destroyed by environmental factors, Pannella recommends that for stromatolites maximum counts are used as a measure of the periodicities of astronomical cycles. This leads to 39 laminae per lesser band. Pannella interprets these bands as representing the synodic month interval and that the fine laminations formed daily. Mohr (1975) has also studied the Biwabik stromatolites and finds a quite different result. From 17 samples and a total of 341 bands, he estimates an average of 12.8 laminae per band. The method used by Mohr for counting the laminae appears to be quite objective and the result should be free of any preconceived idea of what the numbers should be. The interpretation of the lesser bands appears difficult as there are no clear seasonal bands in the samples studied. Mohr suggests that the laminae are indicative of daily growth cycles and that they are modulated by a variable sediment supply that follows a fortnightly pattern. Thus his result indicates an average of 26 d per synodic month. The discrepancy between this result and that obtained by Pannella is disturbing and needs to be resolved before one can have some confidence in the results, particularly as it appears improbable that Pannella's value could ever have been attained during the orbital history of the Moon. As illustrated in figure 11.5, the number of solar days per synodic month attains a maximum of about 32 d and then decreases. Possibly the only way to attain much larger values would be by supposing that large non-tidal accelerations of the Earth's spin occurred. Mohr's study indicates that the counts of the laminae per band has a bimodal distribution, probably in consequence of a non-identification of some bands or of periods when only every second band formed clearly, due to a particular pattern of the spring and neap tide sequence. Pannella's results are also suggestive of a bimodal distribution (figure 11.11) but this would point to a 20-d synodic month for the Middle Proterozoic. Mohr does not give values for the maximum counts obtained for each sample but his histogram suggests that this may be as high as 30.

The Bulawayan stromatolites show some banding that is indicative of growth modulated by periodic environmental events, but as no clear seasonal banding is evident the identification of these events is problematical. Pannella's results again indicate a bimodal distribution with maxima at 10 and 20 laminae per band. Using a linear extrapolation of the number of days per synodic month for the present and for the Gunflint stromatolites, he concludes that there were about 40 d per synodic month in the Late Archean. But this is one case where the preconceived ideas lead one astray. Assuming that the bimodal distribution is caused by factors similar to that for the Gunflint series, these counts rather suggest a synodic month of 20-d duration. Contrary to Pannella's conclusion, this value has a less drastic geophysical consequence than does his value of 40 d per month. It suggests that the Moon was very close to the Earth at about this time. Pannella's comment that maximum counts are probably better indicators than average counts means that the value of 20 d per month is a lower limit and that there could be as many as 24–5 d per synodic month (figure 11.11) at this time. Much further work is required before we can have much confidence in these values.

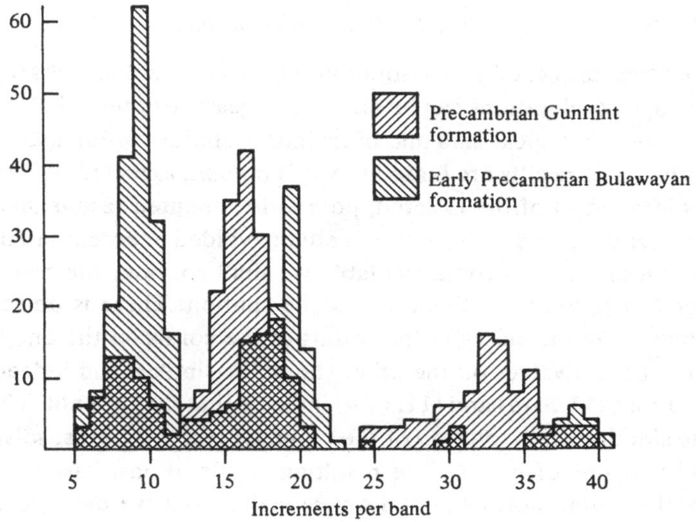


(*b*)

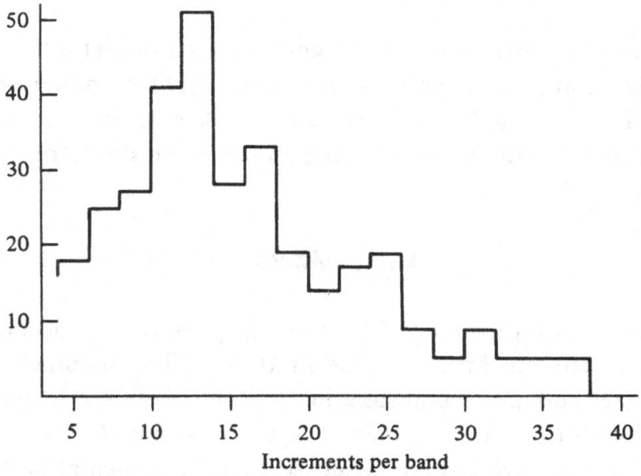


Figure 11.11. Histograms of counts of tidal growth patterns in stromatolites. (*a*) From Pannella (1972) for the Precambrian Gunflint formation and for the Early Precambrian Bulawayan formation. (*b*) From Mohr (1975) for the Precambrian Biwabik formation which is contemporary with the Gunflint stromatolites.

## 11.4 Paleorotation and the lunar orbit

### 11.4.1 *Paleoaccelerations*

That the coral and bivalve solutions give comparable results for the average accelerations can be taken as a measure of the reliability of the paleontological data and of its interpretation, or it may be that both sets of results are biased towards *a priori*, *expected* results. By avoiding most of the isolated, poorly documented results given in the literature, the latter option is either avoided or accentuated, but until more data become available we shall consider the results as significant, yet treat them with some caution. There is no reason why systematic biases in the counts of the corals on the one hand and of the bivalves on the other should be similar, and indeed the solutions (11.3.4*b*) and (11.3.5*b*) indicate that they are not. Thus in the simultaneous solution of the two data types, we must solve for the two sets of biases. The resulting accelerations, based on the coral solution (11.3.4*a*) and on Pannella's bivalve data (solution 11.3.5*a*) are

$$\dot{\Omega} = -(5.2 \pm 0.2)\, 10^{-22}\ \mathrm{rad\ s^{-2}},$$
$$\dot{n}_{☾} = -(1.2 \pm 0.2)\, 10^{-23}\ \mathrm{rad\ s^{-2}}.$$

The introduction of Berry & Barker's data, allowing for the fact that the biases appear to be different, gives slightly smaller accelerations. The above solution is also compatible with the absence of any significant non-tidal torque acting on the Earth since, for the above solution,

$$\dot{\Omega}/\dot{n}_{☾} \simeq 42.9 \pm 5.2,$$

compared with the value of 51.4 for the present tidal accelerations of the Earth and Moon (equation 10.5.2). This assumes that the solar tides and the secular rates in the eccentricity and inclination of the Moon's orbit have remained constant during the last $4 \times 10^8$ yr or so. As we have already assumed that the principal parts of the total accelerations have remained constant, this additional assumption can hardly be important. Thus the combined data set can also be solved with the condition that angular momentum is conserved;

that is, by effectively imposing the condition (10.5.2), the solution is

$$\left.\begin{aligned}\dot{\Omega} &= -(5.2 \pm 0.2)\ 10^{-22}\ \text{rad s}^{-2}\\ \dot{n}_{☾} &= -(1.0 \pm 0.1)\ 10^{-23}\ \text{rad s}^{-2}.\end{aligned}\right\} \tag{11.4.1}$$

The precision estimates are only as good as the validity of the underlying assumptions made about the growth rhythms. These values are in surprisingly good agreement with the modern astronomical values (equations 10.3.13 and 10.3.14)

$$\begin{aligned}\dot{\Omega} &= -(5.5 \pm 0.5)\ 10^{-22}\ \text{rad s}^{-2},\\ \dot{n}_{☾} &= -(1.35 \pm 0.10)\ 10^{-23}\ \text{rad s}^{-2}.\end{aligned}$$

After $5 \times 10^8$ yr the dominant contribution to $\dot{n}_{☾}$ still comes from the $M_2$ tide and the average equivalent lag angles of the tides are about 75% of their present values. Apparently the present rate of tidal dissipation is about 1.4 times greater than the average value since the Ordovician. Such a change can be attributed to changes in the ocean configuration and to changes in the ocean volume. The period of the ocean free oscillations is given very approximately by $2L/\sqrt{(gh)}$, where $L$ is the typical horizontal length scale of the ocean basin and $h$ the typical ocean depth. The length scale of the ocean basins may well have been much greater prior to the breakup of Pangea about $2 \times 10^8$ yr ago. This would increase the periods of free oscillations and, with higher frequencies in the past for the main tidal forcing functions, the dissipation may well have been much less throughout the geological past. Similarly, substantial reduction in sea level during epochs of severe glaciation may have led to reduced dissipation. If the average lag angle over the last $4 \times 10^8$ yr is less than the present value, this could indicate either (i) a fluctuating lag angle or (ii) a slow secular increase in the lag angle. The former may occur due to periodic changes in sea level associated with the glacial epochs and in this case we can extrapolate further back into time using the average value. This brings the Moon close to the Earth about $2.0 \times 10^9$ yr ago. The latter situation may occur during the gradual separation of the continents from their Pangea configuration and in this case the dissipation $4 \times 10^8$ yr ago would actually have been less than the average value found. Now we can easily avoid bringing the Moon too close to the Earth during the age of the

Earth–Moon system. For example, if prior to the Devonian the dissipation was only one-half of the average value deduced from the paleontological evidence, the Moon would not have approached the Earth to within 10 earth-radii during the last $3.8 \times 10^9$ yr.

What further constraints can be imposed in this guessing game? A detailed comparison of the individual counts $N_i$ with those computed from the mean accelerations (11.4.1) indicates considerable scatter and no real trend (Lambeck 1978*a*). This implies either a fluctuating phase lag or, more probably, noise in the data. Extrapolating back into time with the observed average lag, results in nearly 700 d $yr^{-1}$ for the epoch $2 \times 10^9$ yr, very much higher than the two stromatolite counts of 446–8 d $yr^{-1}$ by Pannella (1972). If these stromatolite values are significant, the average lag angle from the Devonian to Proterozoic must have been about 70% of the average value from the present to the Devonian, or about 3°. The interpretation of Mohr's (1975) result, as implying 26 d per synodic month requires, however, that the average phase lag over the last $2 \times 10^9$ yr has been of the order of 5° and that the Moon was very close to the Earth at this time. On other grounds (section 11.4.2) this is most improbable. Pannella's (1972) results for the Middle Proterozoic are equally ambiguous. His results, but not his interpretation of them, for the Lower Archean indicate that the close approach did not occur until at least this time. The rate of change of the number of days per synodic month is so rapid during the close approach phase of the orbit evolution, that even a maximum count of about 25 d per synodic month does not change the estimated time of close approach by more than a fraction of an aeon. If the close approach occurred before about $4.0 \times 10^9$ yr ago, the expected count would still be about 30 d per synodic month.

A further constraint on the time-scale problem has been suggested by Brosche & Sündermann (1977), who have attempted to model the $M_2$ tide for the Pangea configuration at about 230 million years ago. Such a computation requires that the dissipation mechanism and the ocean–continent geometry are known. As they assume that dissipation is by bottom friction, such a calculation is dependent on the depths and extent of the shallow seas; but past ocean depths are particularly poorly known. Their initial results indicate that the dissipation in the Pangea oceans is only about one-third of that in the present seas. Such a reduction is what we would expect from the simple argument above concerning the free-oscillation periods of the oceans and is also in general agreement with the paleontological evidence but what more can be said about its significance?

### 11.4.2 *Geological evidence*

During a close-approach stage of the evolution of the lunar orbit, tides raised on both the Earth and the Moon will have been very much larger than they are today and the rate of tidal energy dissipation would be substantial. In Gerstenkorn's model, close approach occurred close to, or within, the Roche limit, and the geological consequences are severe indeed. Even if the Moon survives this event without disintegrating, the energy dissipated in the two bodies would have been enough to lead to the wholesale melting of the mantle and of the Moon. Thus the Gerstenkorn event should be firmly imprinted on the geological record of both the Moon and the Earth. The early part of the geological record is, however, uncertain and subject to much controversy, but there does seem to be general agreement on the following sequence of events. From about $4.0 \times 10^9$ yr BP to $3.5 \times 10^9$ yr BP the Earth's crust was mainly granitic and only later, from about $3.5 \times 10^9$ yr BP to $2.7 \times 10^9$ yr BP does the crust contain larger quantities of sedimentary and volcanic rocks. In the early stages the crust was thin and mobile, with the consequence that no recognizable continental blocks developed. These were apparently formed only much later, about $2.7–2.0 \times 10^9$ yr BP, during a gradual thickening of the crust and the lithosphere. Thus the period prior to about $2.7 \times 10^9$ yr BP can be considered as one characterized by great tectonic mobility, a general absence of rigid features such as cratons or platforms, high heat flow and a thin lithosphere. Over the next $0.7 \times 10^9$ yr, a gradual transition occurred from this to the present style of global plate tectonics. Only in this interval does a clear distinction between stable blocks – the present continental cratons – and mobile belts become apparent. The period from $2.2 \times 10^9$ yr BP to the present is considered to be one of episodes of continental drift or plate tectonics similar to the well-documented drift that has occurred since the Early Mesozoic. Sutton (1973), Burke & Dewey (1973),

Anhaeusser (1975) and Shaw (1976) review some of the evidence for the crustal evolution during the Archean.

Oceans were present on the Earth by about $3.9 \times 10^9$ yr BP but the basins formed later, presumably during the process of continental plate thickening. Initially small, the subsequent sea floor spreading resulted in the larger basins that are characteristic throughout the Phanerozoic. The rate of transfer of the angular momentum from the Earth to the Moon depends on the magnitude and phase of the second-degreee harmonics in the ocean tides. Large shallow seas, frequently interrupted by reefs and small, mobile continents presumably result in a reduced amplitude of the tidal bulge, and dissipation may also be much reduced compared with the rates after the formation of extensive ocean basins. Dissipation prior to $2 \times 10^9$ yr BP and, in particular prior to $2.7$–$2.9 \times 10^9$ yr BP, would have been less than occurred later. Higher mantle temperatures and thinner lithosphere may, on the other hand, have led to increased dissipation within the solid Earth.

The Moon, formed at about the same time as the Earth, first underwent a period of large-scale differentiation that lasted for perhaps $0.3 \times 10^9$ yr during which its mantle and crust were formed. The intense bombardment of the lunar surface, to which the Earth was also subject, continued until about $3.9 \times 10^9$ yr BP and this was followed by a period of eruption of mare basalts, mostly on the near side of the Moon, which continued until about $3.1 \times 10^9$ yr BP. The following three aeons were times of relative quiescence (Wasserburg *et al.* 1977).

As to when, during these times of upheaval, the Gerstenkorn event may have occurred has been the subject of much speculation. Concerning the terrestrial record, about the only agreement is that (i) it has not occurred during the last $2 \times 10^9$ yr and (ii) speculation on its possible occurrence prior to about $4.0 \times 10^9$ yr BP is futile as the geological record is entirely absent for the first $0.5 \times 10^9$ yr of the Earth's life time. Extensive studies of Precambrian rock ages (see, for example, Moorbath 1976) indicate groupings of radiometric dates at several epochs, around approximately $3.8$–$3.5 \times 10^9$ yr BP, $2.8$–$2.5 \times 10^9$ yr BP and $1.9$–$1.6 \times 10^9$ yr BP. Other events occurred later in the Precambrian. These ages are indicative of major events in the crustal evolution during which various mantle isotopes were homogenized. Several attempts have been made to relate one or another of these epochs to the Gerstenkorn event: Cooper, Richards & Stacey (1967) suggested a date of $2.5 \times 10^9$ yr BP; Cloud (1968) suggested a date around $3.8 \times 10^9$ yr BP. There is no compelling case to associate either date, or any other epoch of major upheavals in the crustal evolution, with the close approach of the Moon. Cloud does make the suggestion that during close approach the primordial hydrosphere would have been largely lost and, due to melting of mantle materials and the release of volatiles, the present hydrosphere would be formed. Evidence for a primitive hydrosphere is not very compelling, however, and the subsequent release of volatiles is readily associated with the crustal evolution process and does not require a close approach of the Moon. Turcotte, Cisne & Nordmann (1977) suggest a date of $2.85 \pm 0.25 \times 10^9$ yr BP for the Gerstenkorn event. This conclusion is based on the one very questionable stromatolite data point obtained by Pannella from the Bulawayan formation. They stress that the data of $2.85 \times 10^9$ yr BP coincide with (i) the end of the mobile period of Archean tectonics and the beginning of crustal thickening and stable continent formation on the Earth, and (ii) the end of the mare volcanism on the Moon. They argue that significant tidal interactions between the Earth and Moon occurred over an extended period between about $3.8 \times 10^9$ yr BP and $2.8 \times 10^9$ yr BP, and that the tidal energy dissipated in both the Earth and in the Moon was responsible for both the Archean and mare volcanism. The obvious problem with this argument is to explain how the Moon could have been stored close to the Earth for such an extended period of time. No orbital evolution scenario allows for this and recourse has to be made to vague resonance configurations (Alfvén & Arrhenius 1969; see also section 11.4.3). Once this act of faith is accepted, Turcotte *et al.*'s next conjecture, that this period of close approach also initiated rapid biological evolution, needs little further imagination.

The relatively quiescent nature of the evolution of the lunar surface during the last three aeons is good evidence that close approach must have occurred earlier than about $3.0 \times 10^9$ yr BP if it occurred at all. The flooding of the basins between $3.9 \times 10^9$ yr BP and $3.1 \times 10^9$ yr BP could possibly be a consequence of large-scale partial melting during the close approach but this explanation is not without its own difficultiies. In particular, the large-scale relief of the lunar highlands has been preserved, and the crystallization ages of many rocks are much older than the ages of the mare basalts. Melting could not have been wholesale. Furthermore, the filling of the basins has not been synchronous, but extended over some $0.8 \times 10^9$ yr. Thus, if the tidal dissipation does provide the energy source for large-scale partial melting, the Moon must have remained resonably close to the Earth for an extended period but not too close to cause extensive melting. Possibly the formation of the mare basalts as one stage in the lunar crustal evolution provides a more satisfactory answer. In conclusion, there is no compelling geological evidence to suggest that the Gerstenkorn event occurred during the last $4.0 \times 10^9$ yr and most likely the Moon was already well outside the Roche limit at this time.

### 11.4.3 *Constraints on theories of lunar evolution*

In the absence of any clear landmarks to fix the lunar orbit at any time in the geologically distant past, a variety of conjectures has been proposed to explain the origin and evolution of the Moon. Traditionally these hypotheses fall into three main categories: (i) capture of the Moon by the Earth, (ii) fission of the Moon from the Earth, (iii) simultaneous, but separate, formation of the Earth and Moon in an orbit about each other including the coagulation or accretion of numerous small moons in orbit around the Earth. Various combinations of these hypotheses have also been proposed. Once the two bodies are formed in more-or-less their present forms, the tidal evolution of the Moon's orbit must be such that at some point it connects to the evolutionary curves of the type given in figure 11.4. The number of proposed or implied connecting points is great and the curves do not appear to provide very stringent constraints, particularly while the time scale remains uncertain.

A complete review of lunar origin theses is not attempted here and, in particular, the chemical and petrological evidence is not considered. The reader is referred to the recent reviews by Kaula (1977*b*) and J. V. Smith (1976), to papers in the proceedings of the

Lunar Science Conferences, to the Royal Society Discussion (1977) and to Ringwood (1979). Only some aspects of the formation and the subsequent orbital evolution, to which evidence from the paleorotation data and the tidal evolution calculations may be relevant, can be discussed here.

The capture model assumes that the Moon has been formed independently of the Earth and later captured into an Earth-orbiting trajectory. Gerstenkorn (1955) was the first to study this possibility in detail and he concluded that the Moon was captured in an almost hyperbolic and retrograde orbit, which subsequently evolved under tidal forces to its present orbit. Subsequent calculations by Gerstenkorn (1969), allowing for dissipation of tidal energy in the Moon as well as within the Earth, indicated that capture could have occurred from a polar or even a prograde orbit. Alfvén (1963), Urey (1963), Singer (1968, 1970), Anderson (1972), Clark, Turcotte & Nordmann (1975) and others have all considered various aspects of the Moon's formation and its subsequent capture in an Earth orbit. Many possibilities have been considered: a formation time for the Moon that is earlier or later than that of the Earth, a formation close to the Earth, in the asteroid belt, near Venus, inside the orbit of Mercury, or in the Earth–Sun Lagrangian point, the choice being apparently guided by the various authors' interpretation of the physical and chemical evidence rather than by the dynamical consequences.

Most proponents of the capture hypothesis in its simplest form assume that capture occurred while the Moon was in a retrograde or high-inclination orbit, when the perigee distance was very small. In fact, all of Gerstenkorn's (1969) extrapolations show that the Moon must have been very close to, or within, the Roche limit during the close-approach phase. If the Moon spent any length of time within this limit, it would be fractured into pieces, some of which would be captured and others of which would escape. If the captured pieces later accreted into the present Moon (see, for example, J. V. Smith 1974) this would mean that the close approach stage occurred very early in the history of the Earth–Moon system and this cannot solve the time-scale problem. Gerstenkorn (1969) and others favour a more recent close approach and argue that the interval spent by the Moon inside the Roche limit is so short that there was not enough time for the Moon to disintegrate. Non-disintegrative capture appears, however, improbable if the approach velocity is high. Disintegrative capture has of course the drawback that the backwards extrapolation beyond the time of close approach is invalid and there is no way of reconstituting the general form of the orbit prior to this event.

The main argument against the capture hypothesis is, according to Kaula (1971), its improbability not its impossibility. For a body to be captured, its velocity must be reduced by some form of dissipation, and hypotheses in which the Moon is originally far from the Earth will require large velocity decreases during the close approach. This renders capture improbable, since situations where capture is most likely are those where the dissipation is least. Thus the Moon is plausibly captured only from a very slow approach velocity and this would occur only if the initial orbit was close to that of the Earth (Kaula & Harris 1973; see also Opik 1972; and Harris & Kaula 1975).

In the fission hypothesis, the Moon represents a mass which broke away from a fast-spinning Earth. The hypothesis originated with Darwin in 1898 and is an obvious consequence of his backward integration of the lunar equations of motion bringing the Moon to the Earth's surface. In one version of this hypothesis, the Earth in a liquid state is broken up by the solar tides. The free-oscillation periods of the liquid Earth are of the order of 2 h and, if the solar tide has a comparable frequency, resonance occurs; the tidal amplitude grows in size until instability is reached. For 2-h solar tides the period of the Earth's rotation must be about 4 h and this is comparable to the period of rotation of a primitive body whose angular momentum is the same as that of the present Earth–Moon system. This is the resonance hypothesis. The obvious limitation is that tidal energy will be dissipated and this limits the growth of the tidal bulge (Jeffreys 1930). In another version, rotational instability is created by a very high angular velocity of the body. For a liquid its period of rotation must be of the order of 2–3 h, depending on its density gradient. The problem with this model is to explain the loss of the angular momentum from the initially high-spinning state to that in the present system. The present angular momentum is only some 27% and the present energy is only some 6% of the corresponding values for the original body. Tidal deformations could account for part of this energy loss but not for all of it. Theories of the instability of rotating radially inhomogeneous bodies are discussed by Roberts (1963) and James (1964).

Ringwood (1960) suggested that the rotational instability may have been associated with the formation of the core. Wise (1963) also suggested such a mechanism. In this model the Earth is originally homogeneous and rotating rapidly. Gravitational settling during the formation of the metallic core leads to a reduction in the moment of inertia and to a spin-up until the rotational instability is attained. The model is discussed in some detail by Wise (1969) and by O'Keefe (1969, 1970). The problem of the angular momentum and energy budget remains: in O'Keefe's model the final state represents only about 60% of the angular momentum and 30% of the rotational energy of the original parent body. Ways around this are to postulate the loss of a heavy atmosphere after separation (Ringwood 1960, 1972).

In the original binary system argument, the Earth and Moon grew at the same time, as independent bodies, by a gradual accumulation from diffuse and cold gases and dust. The Moon and Earth are assumed to have formed close to each other so that there was a binary planet system from the start. An alternative version of this argument is the coagulation or many–Moon hypothesis in which the Moon was formed from a swarm of smaller bodies trapped in orbit around the Earth during the final stages of accretion of the latter body. Many small moons of variable size are assumed to have orbited around the Earth in a ring, spiralling outwards at different rates under the influence of tidal dissipation, and accreting to form the present Moon (Ruskol 1972, 1973). Harris & Kaula (1975) present a model in which joint accretion of the two bodies occurs at a time when the Earth was about one-tenth of its present size while the Moon remained at about 10 earth-radii from the Earth. In Ringwood's (1970) model, the planetesimals are the remains of a primitive and massive atmosphere about the Earth.

Difficulties in explaining the chemical evidence by any of the models discussed above have led to a number of combined hypotheses which have found greater favour amongst planetary scientists, both on chemical and on astronomical grounds. In these models the

Moon is fromed partly from material fissioned from the Earth and partly from captured material. By an appropriate choice of the relative importance of the two mechanisms, some of the dynamical problems introduced above can also be disposed of: if the Moon was predominantly fissioned, the capture of additional material could resolve the excess angular momentum problem, while, if the Moon was predominantly captured, collision with material created by fission some time earlier could make an effective energy sink.

Attempts to solve the time-scale problem, or a desire to retain the Moon close to the Earth for an extended time, have given rise to suggestions that resonance configurations may have existed for part of the lunar orbit's history. Alfvén & Arrhenius (1969), who very much favour the capture theory, propose that interactions between the Moon's orbit and the longitude-dependent part of the Earth's gravity field can retard the orbit evolution during the early stages. Such resonances are similar to those met in orbit theories for close Earth satellites. Most of the likely resonances occur very close to the Earth. Kaula & Harris (1973) conclude that the most likely resonance is a 2:1 resonance, in which the Earth rotates twice during the lunar month. This occurs at 3.8 earth-radii from the Earth but it is quite unstable. At this distance the eccentricity of the lunar orbit will evolve sufficiently rapidly for the Moon to escape quickly from the resonance configuration. A second type of resonance, proposed by R. G. Hipkin in 1972, assumes a resonant coupling between Venus (or another planet) and the Moon, in which angular momentum is transferred from the Earth–Moon system to Venus by a non-zero average gravitational torque. Yoder (1973) has demonstrated, however, that this hypothesis is hopelessly inadequate: that the gravitational torque of Venus on the Moon is much too small to maintain the resonance against either the tidal torque or against the periodic perturbations in the orbits of the Earth and Venus.

Apart from the time-scale problem there is also the inclination problem discussed in section 11.2. The present inclination of the lunar orbit upon the ecliptic is some 5° and this rules out an origin by fission or accretion, unless some mechanism wrenched the Moon out of its equatorial orbit in the early stages of its orbit evolution or, subsequently, out of an ecliptic orbit once the Moon has passed beyond the critical radius. Kaula (1971) has suggested that the removal of the primordial atmosphere could cause the orbit to be inclined to the equator. Others have suggested that an infall of substantial planetesimals during the final stages of accretion tipped the Earth's rotation axis. The cratering records of the Moon, Mars and Mercury show that large accreting fragments were common for the first few $10^8$ yr after the formation of the bulk of the planets. Yet other authors have argued that any non-uniformity of tidal dissipation may significantly modify the orbital evolution and that extrapolations such as figure 11.4 will be invalid (see, for example, Munk 1968; Alfvén & Arrhenius 1969). But this is incorrect (chapter 10); any enhanced localized dissipation can only modify the coefficient $D^{+}_{22} \sin\varepsilon^{+}_{22}$. It can modify the time scale but not the relations $\mathrm{d}I/\mathrm{d}a$ and $\mathrm{d}e/\mathrm{d}a$. One possible way of modifying the inclination is by postulating past situations in which the relative importance of the tide components is different from what it is at present. For example, the main contributions to $\mathrm{d}I/\mathrm{d}t$ come from the $M_2$, $K_1$ and $O_1$ frequencies, with the first two bringing the Moon into the equator and the last moving the Moon out of the equator with time. Situations may have existed in the past where the free-oscillation periods of some of the oceans resonated with the $O_1$ tide but not with the others. Such a situation is unlikely today since $K_1$ and $O_1$ have quite similar frequencies, but this has not always been so in the past when $n_☾$ was of the same magnitude as $\Omega$. A solution like this is, however, unsatisfactory as it is unlikely to have been more than a transient phenomenon without any long-term consequence on the lunar orbit. Possibly a more satisfactory explanation is to introduce frequency-dependent-$Q$ models, such as a Maxwell model (section 2.2). This has been done by Rubincam (1975) who, following Darwin and O'Keefe (1972), considers the evolution of the lunar orbit under the assumption that dissipation occurs in the mantle by linear viscous damping. In such a model, the tangent of the lag angle is proportional to the frequency of the tide component and to the viscosity. For viscosities greater than about $10^{16}$ P the lag angles go to 90° when the tidal frequencies are significantly non-zero, and the amplitude of the tidal bulge becomes vanishingly small. When $n_☾ = 2\Omega$ the $M_2$ and $K_1$ frequencies are non-zero but the $O_1$ frequency is close to zero, and this tide may dominate the evolution. Pursuing Darwin's hypothesis, Rubincam finds that if there is an initial perturbation of some 3° in the inclination of the lunar orbit on the equator at the time that $\Omega \simeq 2n_{☾}$ – at 3.8 earth-radii from the Earth – the present 5° inclination to the ecliptic can be explained, provided that the Earth's viscosity at that time exceeded $10^{18}$ P. In this model, the evolution of the orbit is quite different from that proposed by Gerstenkorn and Goldreich (figure 11.4) and stresses the rheology dependence of the evolution once the Moon approaches the Earth. The principal objection against this hypothesis is that it presupposes a linear rheology, but when the tidal stresses reach several hundred bars, as they do when the Moon approaches the planet, the rheology will become non-linear.

# BIBLIOGRAPHY


Abe, K. (1970). Determination of seismic moment and energy from the Earth's free oscillation. *Phys. Earth Planet. Inter.*, **4**, 49–61.

Accad, Y. & C. L. Pekeris (1978). Solution of the tidal equations for the $M_2$ and $S_2$ tides in the world oceans from a knowledge of the tidal potential alone. *Phil. Trans. R. Soc. Lond.*, **A290**, 235–66.

Acheson, D. J. & R. Hide (1973). Hydromagnetics of rotating fluids. *Rep. Prog. Phys.*, **36**, 159–221.

Agnew, D. C. & W. E. Farrell (1978). Self-consistent equilibrium ocean tides. *Geophys. J.*, **55**, 171–81.

Aki, K. (1967). Scaling law of seismic spectrum. *J. Geophys. Res.*, **72**, 1217–31.

Aki, K. (1972). Earthquake mechanisms. *Tectonophysics*, **13**, 423–46.

Alfvén, H. (1963). The early history of the Moon and the Earth. *Icarus*, **1**, 357–63.

Alfvén, H. & G. Arrhenius (1969). Two alternatives for the history of the Moon. *Science*, **165**, 11–17.

Alldredge, L. R. (1975). A hypothesis for the source of impulses in geomagnetic secular variations. *J. Geophys. Res.*, **80**, 1571–8.

Alldredge, L. R. (1977). Deep mantle conductivity. *J. Geophys. Res.*, **82**, 5427–31.

Alterman, Z., H. Jarosch & C. L. Pekeris (1959). Oscillations of the Earth. *Proc. R. Soc. Lond.*, **A252**, 80–95.

Anderle, R. J. (1973). Determination of polar motion from satellite observations. *Geophys. Surv.*, **1**, 147–61.

Anderle, R. J. (1976). Comparison of Doppler and optical pole position over twelve years. *Naval Surface Weapons Center, Dalgren Laboratory Technical Report 3464*, 10 pp.

Anderle, R. J. & L. K. Beuglass (1970). Doppler satellite observations of polar motion. *Bull. Géod.*, **96**, 125–41.

Anderson, D. L. (1967). The anelasticity of the Earth. *Geophys. J.*, **14**, 135–63.

Anderson, D. L. (1972). The origin of the Moon. *Nature*, **239**, 263–365.

Anderson, D. L. (1974). Earthquakes and the rotation of the Earth. *Science*, **182**, 49–50.

Anderson, D. L. (1975). Earthquakes, volcanoes, climate and the rotation of the Earth (abstract). *EOS, Trans. Am. Geophys. Union*, **56**, 346.

Anderson, D. L. (1977). Composition of the mantle and core. *Annu. Rev. Earth Planet. Sci.*, **5**, 179–202.

Anderson, D. L. & R. S. Hart (1976). An Earth model based on free oscillations and body wave data. *J. Geophys. Res.*, **81**, 1461–75.

Anderson, D. L. & R. S. Hart (1978*a*). Attenuation models of the Earth. *Phys. Earth Planet. Inter.*, **16**, 289–306.

Anderson, D. L. & R. S. Hart (1978*b*). The *Q* of the Earth. *J. Geophys. Res.*, **83**, 5869–72.

Anderson, D. L. & J. B. Minster (1979). The frequency dependence of *Q* in the Earth and implications for mantle rheology and Chandler wobble. *Geophys. J.*, **58**, 431–40.

Anderson, D. L., H. Kanamori, R. S. Hart & H. P. Liu (1977). The Earth as a seismic absorption band. *Science*, **196**, 1104–6.

Andrews, J. T. (ed.) (1974). *Glacial Isostasy*. Dowden, Hutchinson & Ross, Stroudsburg, Pennsylvania, 491 pp.

Angell, J. K. & Korshover, J. (1970). Quasi-biennial, annual and semi-annual zonal wind and temperature harmonic amplitudes and phases in the stratospheres and low mesosphere of the Northern Hemisphere. *J. Geophys. Res.*, **75**, 543–50.

Anhaeusser, C. R. (1975). Precambrian tectonic environments. *Annu. Rev. Earth Planet. Sci.*, **3**, 31–53.

Arakawa, A. (1975). General circulation of the atmosphere. *Rev. Geophys. Space Phys.*, **13**, 668–80.

Arur, M. G. & I. I. Mueller (1971). Latitude observations and the detection of continental drift. *J. Geophys. Res.*, **76**, 2071–6.

Awrawik, S. M. (1976). Gunflint stromatolites: microfossil distribution in relation to stromatolite morphology. In Walter (1976), pp. 311–20.

Backus, G. E. (1967). Converting vector and tensor equations to scalar equations in spherical coordinates. *Geophys. J.*, **13**, 71–101.

Baines, P. G. (1973). The generation of internal tides by flat-bump topography. *Deep-Sea Res.*, **20**, 179–205.

Baker, T. F. & G. W. Lennon (1976). The investigation of marine loading by gravity variation profiles in the United Kingdom. In *Proceedings of the Seventh International Symposium on Earth Tides* (ed. G. Szádeczky-Kardoss) pp. 463–78. Akademiai Kiado, Budapest.

Baker, D. J., W. D. Nowlin, R. D. Pillsbury & H. L. Bryden (1977). Antarctic circumpolar current: space and time fluctuations in the Drake Passage. *Nature*, **268**, 696–9.

Ball, R. H., A. B. Kahle & E. H. Vestine (1969). Determination of surface motions of the Earth's core. *J. Geophys. Res.*, **74**, 3659–80.

Balmino, G. (1978). On the product of Legendre functions as encountered in geodynamics. *Stud. Geophys. Geod.*, **22**, 107–18.

Balmino, G., K. Lambeck & W. M. Kaula (1973). A spherical harmonic analysis of the Earth's topography. *J. Geophys. Res.*, **78**, 478–81.

Banks, R. J. (1969). Geomagnetic variations and the electrical conductivity of the upper mantle. *Geophys. J.*, **17**, 457–87.

Banks, R. J. (1972). The overall conductivity distribution of the Earth. *J. Geomagn. Geoelectr.*, **24**, 337–351.

Banks, R. J. & E. C. Bullard (1966). The annual and 27 day magnetic variations. *Earth Planet. Sci. Lett.*, **1**, 118–20.

Barker, R. M. (1964). Microtextural variations in pelecypod shells. *Malacologia*, **2**, 69–86.

Barnes, D. J. (1971). *A Study of Growth, Structure and Form in Modern Coral Skeletons.* Thesis, University of Newcastle upon Tyne, 180pp.

Barnes, D. J. (1972). The structure and formation of growth-ridges in scleractinian coral skeletons. *Proc. R. Soc. Lond.*, **B182**, 331–50.

Barraclough, D. R. (1974). Spherical harmonic analyses of the geomagnetic field for eight epochs between 1600 and 1910. *Geophys. J.*, **36**, 497–513.

Barraclough, D. R. (1976). Spherical harmonic analysis of the geomagnetic secular variation – a review of methods. *Phys. Earth Planet. Inter.*, **12**, 365–82.

Barraclough, D. R., J. H. Harwood, B. R. Leaton & S. R. C. Malin (1975). 17 models of the geomagnetic field at Epoch 1975. *Geophys. J.*, **43**, 645–59.

Bauer, A. (1966). Le bilan de masse de l'inlandsis du Groenland n'est pas positif. *Bull. Int. Ass. Surf. Hydr.*, **11**, 8–12.

Bell, T. H. (1975). Topographically generated internal waves in the open ocean. *J. Geophys. Res.*, **80**, 320–27.

Belmont, A. D. & D. G. Dartt (1970). The variability of tropical stratospheric winds. *J. Geophys. Res.*, **75**, 3133–45.

Bender, P. L., D. G. Currie, R. H. Dicke *et al.* (1973). The lunar laser ranging experiment. *Science*, **182**, 229–38.

Ben Menahem, A. & D. Harkrider (1964). Radiation patterns of seismic surface waves from buried dipolar point sources in a flat stratified Earth. *J. Geophys. Res.*, **69**, 2605–20.

Ben Menahem, A. & M. Israel (1970). Effects of major seismic events on the rotation of the Earth. *Geophys. J.*, **19**, 367–93.

Berdichevsky, M. N., E. B. Fainberg, N. M. Rotanova, J. B. Smirnov & L. L. Vanjan (1977). Deep electromagnetic investigations. Submitted to *Ann. Géophys.*

Berry, A. (1961). *A Short History of Astronomy*. Dover, New York, 440 pp.

Berry, W. B. N. & R. M. Barker (1968). Fossil bivalve shells indicate longer month and year in Cretaceous than present. *Nature*, **217**, 938–9.

Berry, W. B. N. & R. M. Barker (1975). Growth increments in fossil and modern bivalves. In Rosenberg & Runcorn (1975), pp. 9–24.

BIH. (1965). *Bulletin Horaire du Bureau International de l'Heure*, sér. J, numero 7.

Bogdanov, K. T. & V. A. Magarik (1967). Numerical solutions for the world's semi-diurnal ($M_2$ and $S_2$) tides. *Dokl. Akad. Nauk. SSSR*, **172**, 1315–17.

Bogdanov, K. T. & V. A. Magarik (1969). A numerical solution of the problem of tidal wave propagation in the world ocean. *Bull. Acad. Sci. USSR, Atmos. Ocean. Phys.*, **5**, 757–61.

Bondi, H. & T. Gold (1955). On the damping of the free nutation of the Earth. *Mon. Not. R. Astron. Soc.*, **115**, 41–6.

Bondi, H. & R. A. Lyttleton (1948). On the dynamical theory of the rotation of the Earth. *Proc. Camb. Phil. Soc.*, **44**, 345–59.

Bowden, K. F. (1962). General properties of turbulence. In *The Sea*, vol. 1 (ed. M. N. Hill), pp. 802–25. John Wiley, New York.

Braginski, S. I. & V. V. Nikolaichik (1973). Estimation of electrical conductivity of the lower mantle of the earth by electromagnetic signal delay. *Phys. Solid Earth*, 601–2.

Brennan, B. J. & F. D. Stacey (1977). Frequency dependence of elasticity of rock – a test of seismic velocity dispersion. *Nature*, **268**, 220–2.

Brettschneider, G. (1967). Anwendung des hydrodynamisch-numerischen Verfahrens zur ermittlung der $M_2$ Mitschwingungsgezeit der Nordsee. *Mitt. Inst. Meeresk. d. Univ. Hamburg*, no. 7.

Brillinger, D. R. (1973). An empirical investigation of the Chandler wobble and two proposed excitation processes. *Bull. Int. Stat. Inst.*, **45**, 413–34.

Britten, J. P. (1967). *On the Quality of Solar and Lunar Observations and Parameters in Ptolemy's Almagest.* Thesis, Yale University, 260 pp.

Brosche, P. & J. Sündermann (1977). Effect of oceanic tides on the rotation of the Earth. In *Scientific Applications of Lunar Laser Ranging* (ed. J. D. Mullholland), pp. 133–41. D. Reidel, Dordrecht.

Brouwer, D. (1952). A study of the change on the rate of rotation of the Earth. *Astron. J.*, **57**, 125–46.

Brouwer, D. & G. M. Clemence (1961). *Methods of Celestial Mechanics.* Academic Press, London & New York, 598 pp.

Brown, E. (1926). The evidence for changes in the rate of rotation of the Earth and their geophysical consequences. *Trans. Astron. Obs. Yale Univ.*, **3**, 207–35.

Brune, J. N. (1968). Seismic moment, seismicity, and rate of slip along major fault zones. *J. Geophys. Res.*, **73**, 777–84.

Brune, J. N. & G. R. Engen (1969). Excitation of mantle Love waves and definition of mantle wave magnitude. *Bull. Seism. Soc. Am.*, **59**, 923–33.

Brune, J. N. & C. Y. King (1967). Excitation of mantle Rayleigh waves of period 100 seconds as a function of magnitude. *Bull. Seism. Soc. Am.*, **57**, 1355–65.

Buddemeier, R. W. & R. A. Kinzie (1975). The chronometric reliability of contemporary corals. In Rosenberg & Runcorn (1975), pp. 135–46.

Budyko, M. I. (1974). *Climate and Life.* Academic Press, London & New York, 508 pp.

Bukowinski, M. S. & L. Knopoff (1976). Electronic structure of iron and the Earth's core. *Geophys. Res. Lett.*, **3**, 45–8.

Bullard, E. C. (1948). The secular change in the Earth's magnetic field. *Mon. Not. R. Astron. Soc., Geophys. Suppl.*, **5**, 248–57.

Bullard, E. C., C. Freedman, H. Gellman & J. Nixon (1950). The westward drift of the Earth's magnetic field. *Phil. Trans. R. Soc.*, **A243**, 67–92.

Bullen, K. E. (1963). *An Introduction to the Theory of Seismology* (third edition). Cambridge University Press, 381 pp.

Burke, K. & J. Dewey (1973). An outline of Precambrian plate tectonics. In *Implications of Continental Drift to the Earth Sciences*, vol. 2 (ed. D. H. Tarling & S. K. Runcorn), pp. 1035–45, Academic Press, London.

Busse, F. H. (1968). Steady fluid flow in a precessing spheroidal shell. *J. Fluid Mech.*, **33**, 739–51.

Busse, F. H. (1975). A model of the geodynamo. *Geophys. J.*, **42**, 437–59.

Byzova, N. L. (1947). Influence of seasonal air mass shifts on the motion of the Earth's axis. *Dokl. Akad. Nauk. SSSR*, **58**, no. 3.

Cannon, W. H. (1974). The Chandler annual resonance and its possible geophysical significance. *Phys. Earth Planet. Inter.*, **9**, 83–90.

Capitaine, N. (1975). Total effect of any nearly-diurnal wobble of the Earth's axis of rotation in latitude and time observations, *Geophys. J.*, **43**, 573–88.

Caputo, M. (1965). The minimum strength of the Earth. *J. Geophys. Res.*, **70**, 955–63.

Cartwright, D. E. (1968). A unified analysis of tides and surges round north and east Britain. *Phil. Trans. R. Soc. Lond.*, **A263**, 1–55.

Cartwright, D. E. (1977). Oceanic Tides. *Rep. Prog. Phys.*, **40**, 665–708.

Cartwright, D. E. & A. C. Edden (1973). Corrected tables of tidal harmonics. *Geophys. J.*, **33**, 253–64.

Cartwright, D. E. & A. C. Edden (1977). Spectroscopy of the tide generating potentials and their relationship to observed features in the ocean tide. *Ann. Geophys.*, **33**, 179–82.

Cathles, L. M. (1975). *The Viscosity of the Earth's Mantle*. Princeton University Press, 286 pp.

Cazenave, A. (1975). *Interactions Entre les Irregularites de la Vitesse de Rotation de la Terre et les Phenomenes Meteorologiques et Climatiques.* Thesis, Université Paul Sabatier, Toulouse, France, 258 pp.

Cazenave, A. & S. Daillet (1977). Determination de la maree oceanique $M_2$ avec STARLETTE. Internal Report of the Groupe de Recherches de Géodesie Spatiale. Centre Nationale d'Etudes Spatiales, Toulouse, 39 pp.

Cazenave, A., S. Daillet & K. Lambeck (1977). Tidal studies from the perturbations in satellite orbits. *Trans. R. Soc. Lond.*, **A284**, 595–606.

Challinor, R. D. (1971). Variations in the rate of rotation of the Earth. *Science*, **172**, 1022–5.

Chapman, S. & R. S. Lindzen (1970). *Atmospheric Tides*, D. Reidel, Dordrecht, 200 pp.

Chen, W. P. & P. Molnar (1977). Seismic moments of major earthquakes and the average rate of slip in Central Asia. *J. Geophys. Res.*, **82**, 2945–69.

Chinnery, M. A. (1965). The vertical displacements associated with transcurrent faulting. *J. Geophys. Res.*, **70**, 4627–32.

Chinnery, M. A. (1975). The static deformation of an Earth with a fluid core: A physical approach. *Geophys. J.*, **42**, 461–75.

Chinnery, M. A. & R. G. North (1975). The frequency of large earthquakes. *Science*, **190**, 1197–8.

Claerbout, J. F. (1969). Frequency mixing in Chandler Wobble data. *Trans. Am. Geophys. Union*, **50**, 119.

Clark, G. R. (1968). Mollusk Shell: Daily growth lines. *Science*, **161**, 800–2.

Clark, G. R. (1974). Growth lines in invertebrate skeletons. *Annu. Rev. Earth Planet. Sci.*, **2**, 77–99.

Clark, S. P., D. L. Turcotte & J. C. Nordmann (1975). Accretional capture of the Moon. *Nature*, **258**, 219–20.

Cleary, J. R. & R. S. Anderssen (1979). Seismology and the internal structure of the Earth. In *The Earth: Its Origin, Structure and Evolution* (ed. M. W. McElhinny), pp. 137–75. Academic Press, London & New York.

Clemence, G. M. (1943). The motion of Mercury 1765–1937, *Astron. Pap. Am. Ephem. Naut. Alm.*, **XI**, 9–224.

Clemence, G. M. (1948). On the system of astronomical constants. *Astron. J.*, **53**, 169–79.

Cloud, P. E. (1968). Atmospheric and hydrospheric evolution on the primitive Earth. *Science*, **160**, 729–36.

Coleman, P. J. (1971). Solar wind torque on the geomagnetic cavity. *J. Geophys. Res.*, **76**, 3800–5.

Colombo, G. & I. Shapiro (1968). Theoretical model for the Chandler Wobble. *Nature*, **217**, 156–7.

Cooper, J. A., J. R. Richards & F. D. Stacey (1967). Possible new evidence bearing on the lunar capture hypothesis. *Nature*, **215**, 1256.

Counselman, C. C. (1976). Radio Astrometry. *Annu. Rev. Astron. Astrophys.*, **14**, 197–214.

Courtillot, V. and J.-L. Le Mouël (1976). On the long period variations of the Earth's magnetic field from 2 months to 20 years. *J. Geophys. Res.*, **81**, 2941–50.

Cox, C. S. & H. Sandstrom (1962). Coupling of internal and surface waves in water of variable depth. *J. Ocean. Soc. Jap.*, **20**, 499–513.

Crepon, M. (1973). Sealevel, atmospheric pressure and geostrophic adjustment. *Woods Hole Oceanographic Institute Contribution 3208*. Woods Hole Oceanographic Institute, Mass., 30 pp.

Crossley, D. J. (1975). Core undertones with rotation. *Geophys. J.*, **42**, 477–88.

Crossley, D. J. & D. Gubbins (1975). Static deformation of the Earth's liquid core. *Geophys. Res. Lett.*, **2**, 1–4.

Curott, D. R. (1966). Earth deceleration from ancient solar eclipses. *Astron. J.*, **71**, 264–9.

Currie, R. G. (1968). Geomagnetic spectrum of internal origin and lower mantle conductivity. *J. Geophys. Res.*, **73**, 2779–86.

Currie, R. G. (1974). Period and $Q$ of the Chandler wobble. *Geophys. J.*, **38**, 179–85.

Currie, R. G. (1975). Period $Q_p$ and amplitude of the pole tide. *Geophys. J.*, **43**, 73–86.

Currie, R. G. (1976). The spectrum of sea level from 4 to 40 years. *Geophys. J.*, **46**, 513–20.

Dahlen, F. A. (1971*a*). Comments on paper by D. E. Smylie & L. Mansinha. *Geophys. J.*, **23**, 355–8.

Dahlen, F. A. (1971*b*). The excitation of the Chandler wobble by earthquakes. *Geophys. J.*, **25**, 157–206.

Dahlen, F. A. (1973). A correction to the excitation of the Chandler wobble by earthquakes. *Geophys. J.*, **32**, 203–17.

Dahlen, F. A. (1974). On the static deformation of an Earth model with a fluid core. *Geophys. J.*, **36**, 461–85.

Dahlen, F. A. (1976). The passive influence of the oceans upon the rotation of the Earth. *Geophys. J.*, **46**, 363–406.

Dahlen, F. A. (1979). The period of the Chandler wobble. In Federov *et al.* (1979).

Dahlen, F. A. & S. B. Fels (1978). A physical explanation of the static core paradox. *Geophys. J. R. Astron. Soc.*, **55**, 317–32.

Daillet, S. (1977). *Etudes des Marees Oceaniques par les Perturbations Orbitales des Satellites Artificiels.* Thesis, Université Paul Sabatier, Toulouse, France, 177 pp.

Danjon, A. (1958). The contribution of the impersonal astrolabe to fundamental astronomy. *Mon. Not. R. Astron. Soc.*, **118**, 411–31.

Danjon, A. (1960*a*). The impersonal astrolabe. *Telescopes* (ed. G. P. Kuiper & B. M. Middlehurst), pp. 115–37. Chicago University Press.

Danjon, A. (1960*b*). Sur un changement du regime de la rotation de la terre survenu au mois de juillet 1959. *C. R. Acad. Sci.*, **B250**, 1399–402.

Dartt, D. G. & A. D. Belmont (1964). Periodic features of the 50-millibar zonal winds in the tropics. *J. Geophys. Res.*, **69**, 2887–93.

Dartt, D. G. & A. D. Belmont (1970). A global analysis of the variability of the quasi-biennial oscillation. *Q. J. R. Met. Soc.*, **96**, 186–94.

Davies, G. F. & J. N. Brune (1971). Regional and global fault slip rates from seismicity. *Nature*, **229**, 101–7.

Débarbat, S. & B. Guinot (1970). *La Methode des Hauteurs Egales en Astronomie.* Gordon & Breach, Paris, 138 pp.

Defant, A. (1961). *Physical Oceanography*, vol. II. Pergamon Press, Oxford & New York, 598 pp.

De Prit, A., J. Henrard & A. Rom (1971). Analytical lunar ephemeris. *Astron. Astrophys.*, **10**, 257–69.

Dicke, R. H. (1966). The secular acceleration of the Earth's rotation. In Marsden & Cameron (1966), pp. 98–164.

Dicke, R. H. (1969). Average acceleration of the earth rotation and the viscosity of the deep mantle. *J. Geophys. Res.*, **74**, 5895–902.

Dickman, S. R. (1977). Secular trend of the Earth's rotation pole: consideration of motions of the latitude observatories. *Geophys. J.*, **51**, 229–44.

Dickman, S. R. (1979). Consequences of an enhanced pole tide. *J. Geophys. Res.*, **84**, 5447–56.

Dietrich, G. (1944). Die Schwingungssysteme der halb- und eintagigen Tiden in den Ozeanen. *Veroeff. Meereskd. Univ. Berl.*, **A41**, 7–68.

Djurovic, D. (1976). Determination du nombre de Love *k* et du facteur Λ affectant les observations du temps universal. *Astron. Astrophys.* **47**, 325–32.

Djurovic, D. & P. Melchior (1972). Recherche des termes de marée dans les variations de la vitesse de rotation de la Terre. *Acad. R. Belg., Ser.* **5**, **73**, 1248–57.

Dodge, R. E. & J. Vaisnys (1976). Annual banding in corals: Climatological implications. *Geol. Soc. Am. Prog. Annu. Meet.*, 838.

Dolman, J. (1975). A technique for the extraction of environmental and geophysical information from growth records in invertebrates and stromatolites. In Rosenburg & Runcorn (1975), pp. 191–221.

Doodson, A. T. (1921). The harmonic development of tide generating potential. *Proc. R. Soc. Lond.*, **100**, 305–29.

Doodson, A. T. & H. D. Warburg (1941). *Admiralty Manual of Tides.* H.M. Stationery Office, London, 270 pp.

Duda, S. J. (1965). Secular seismic energy release in the CircumPacific belt. *Tectonophysics*, **2**, 409–52.

Duncombe, R. L. (1958). Motion of Venus 1750–1949. *Astron. Pap. Am. Ephem. Naut. Alm.*, **XVI**, 1–258.

Dziewonski, A. M. & F. Gilbert (1974). Temporal variation of the seismic moment tensor and the evidence of precursive compression for two deep earthquake. *Nature*, **247**, 185–8.

Eckert, W. J. (1965). On the motions of the perigee and node and the distribution of mass in the Moon. *Astron. J.*, **70**, 787–92.

Elsasser, W. (1950). The Earth's interior and geomagnetism. *Rev. Mod. Phys.*, **22**, 1–35.

Elsasser, W. M. and H. Takeuchi (1955). Non uniform rotation of the Earth and geomagnetic drift. *Trans. Am. Geophys. Union*, **36**, 584–90.

Evans, J. W. (1972). Tidal increments in the cockle *Clinocardium nuttalli. Science*, **176**, 416–17.

Evans, J. W. (1975). Growth and micromorphology of two bivalves exhibiting non daily growth lines. In Rosenberg & Runcorn (1975), pp. 119–33.

Evans, M. E. (1976). Test of the dipolar nature of the geomagnetic field throughout Phanerozoic time. *Nature*, **262**, 676–7.

Fairbridge, R. W. (1961). Eustatic changes in sea level. *Phys. Chem. Earth*, **4**, 99–185.

Fairbridge, R. W. & O. A. Krebs, (1962). Sea level and the southern oscillation. *Geophys. J.*, **6**, 532–45.

Farrell, W. E. (1970). *Gravity Tides.* Thesis, University of California, San Diego.

Farrell, W. E. (1972). Deformation of the Earth by surface loads. *Rev. Geophys. Space Phys.*, **10**, 761–97.

Farrell, W. E. & J. A. Clark (1976). On postglacial sea level. *Geophys. J.*, **46**, 648–67.

Federer, C. A. (1975). Evapotranspiration. *Rev. Geophys. Space Phys.*, **13**, 442–5.

Federov, Y. P. (1963). *Nutation and Forced Motion of the Earth's Pole.* Pergamon Press, Oxford & New York, 152 pp.

Federov, Y. P., A. A. Rorsun, S. P. Major, N. T. Panchenko, V. K. Taraby & Ya.A. Yatskiv (1972). New determination of the polar motion from 1890–1969. In Melchior & Yumi (1972), pp. 12–13.

Federov, E. P., M. L. Smith & P. L. Bender (eds.) (1979). *Nutation and the Earth's rotation.* D. Reidel, Dordrecht (in press).

Felsentreger, T. L., J. G. Marsh and R. G. Williamson (1978). Tidal perturbations on the satellite 1967–92A. *J. Geophys. Res.*, **83**, 1837–42.

Felsentreger, T. L., J. G. Marsh, R. G. Williamson & K. Lambeck (1979). Ocean tide parameters and the acceleration of the Moon's mean longitude from satellite orbit analysis. Symposium 19, Int. Ass. Geod. Geophys., Canberra.

Feynman, R. P., R. B. Leighton & M. Sands (1963). *The Feynman Lectures on Physics*, vol. 1. Addison-Wesley, Reading, Mass.

Filloux, J. H. (1971). Deep sea tide observations from the north-eastern Pacific. *Deep-Sea Res.*, **18**, 275–84.

Finsterwalder, R. (1954). Photogrammetry and glacier research. *J. Glac.*, **1**, 306–15.

Fliegel, H. F. & T. P. Hawkins (1967). Analysis of variations in the rotation of the Earth. *Astron. J.*, **72**, 544–50.

Fotheringham, J. K. (1915). The secular acceleration of the Moon's mean motion as determined from the occultations in the Almagest. *Mon. Not. R. Astron. Soc.*, **75**, 377–94.

Fotheringham, J. K. (1918). The secular acceleration of the Sun as determined from Hipparchus' equinox observations. *Mon. Not. R. Astron. Soc.*, **78**, 406–23.

Fotheringham, J. K. (1935). Two Babylonian eclipses. *Mon. Not. R. Astron. Soc.*, **95**, 719–23.

Frostman, T. O., D. W. Martin & W. Schwerdtfeger (1967). Annual and semiannual variations in the length of day, related to geophysical effects. *J. Geophys. Res.*, **72**, 5065–73.

Fukao, Y. & M. Furumoto (1975). Mechanism of large earthquakes along the eastern margin of the Japan Sea. *Tectonophysics*, **24**, 247–66.

Fung, Y. C. (1965). *Foundations of Solid Mechanics*. Prentice-Hall, New Jersey, 525 pp.

Gans, R. F. (1972). Viscosity of the Earth's core. *J. Geophys. Res.*, **77**, 360–6.

Gaposchkin, E. M. (1972). Analysis of pole position from 1846–1970. In Melchior & Yumi (1972), pp. 19–32.

Gaposchkin, E. M. (1973). Smithsonian Standard Earth (III). *Smithson. Astrophys. Obs. Spec. Rep.*, **353**, 388 pp.

Gaposchkin, E. M. (1974). Earth's gravity field to the eighteenth degree. *J. Geophys. Res.*, **79**, 5377–411.

Gaposchkin, E. M. & K. Lambeck (1971). Earth's gravity field to the sixteenth degree and station coordinates from satellite and terrestrial data. *J. Geophys. Res.*, **76**, 4855–83.

Garland, G. D. (1957). The figure of the Earth's core and the non-dipole field. *J. Geophys. Res.*, **62**, 486–7.

Garrett, C. (1975). Tides in gulfs. *Deep-Sea Res.*, **22**, 23–35.

Garrett, C. & W. Munk (1972). Ocean mixing by breaking internal waves. *Deep-Sea Res.*, **19**, 823–32.

Garrett, G. J. R. & W. H. Munk (1971). The age of the tide and the 'Q' of the oceans. *Deep-Sea Res.*, **18**, 493–503.

Garthwaite, K., D. B. Holdridge & J. D. Mulholland (1970). A preliminary special perturbation theory for the lunar motion. *Astron. J.*, **75**, 1133–9.

Gates, W. L. & J. Imbrie (1975). Climatic change. *Rev. Geophys. Space Phys.*, **13**, 726–31.

Gebelein, C. D. (1969). Distribution, morphology and accretion rate of Recent subtidal algal stromatolites, Bermuda. *J. Sediment. Petrol.*, **39**, 49–69.

Gebelein, C. D. & P. Hoffman (1968). Inertidal stromatolites from Cape Sable Florida (abstract). *Geol. Soc. Am. Spec. Pap.*, **121**, 109.

Gerstenkorn, H. (1955). Über gezeiten Reibung beim zwei Körberproblem. *Z. Astrophys.*, **36**, 245–74.

Gerstenkorn, H. (1967). On the controversy over the effect of tidal friction upon the history of the Earth–Moon system, *Icarus*, **7**, 60–167.

Gerstenkorn, H. (1969). The earliest past of the Earth–Moon system. *Icarus*, **11**, 189–207.

Gilbert, F. (1971). Excitation of the normal modes of the Earth by earthquake sources. *Geophys. J.*, **22**, 223–6.

Gilbert, F. (1972). Inverse problems for the Earth's normal modes. *The Nature of the Solid Earth* (ed. E. C. Robertson), pp. 125–46, McGraw-Hill, New York.

Gilbert, F. & A. M. Dziewonski (1975). An application of normal mode theory to the retrieval of structural parameters and source mechanisms from seismic spectra. *Phil. Trans. R. Soc. Lond.*, **A278**, 187–269.

Gill, A. E. & P. O. Niiler (1973). The theory of the seasonal variability in the ocean. *Deep-Sea Res.*, **20**, 141–77.

Goad, C. C. & B. C. Douglas (1978). Lunar tidal acceleration obtained from satellite-derived ocean tide parameters. *J. Geophys. Res.*, **83**, 2306–10.

Gold, T. (1955). Instability of the Earth's axis of rotation. *Nature*, **175**, 526–9.

Gold, T. & S. Soter (1969). Atmospheric tides and the resonant rotation of Venus. *Icarus*, **11**, 356–66.

Goldreich, P. (1966). History of the lunar orbit. *Rev. Geophys. Space Phys.*, **4**, 411–39.

Goldreich, P. & S. J. Peale (1968). The dynamics of planetary rotations. *Annu. Rev. Astron. Astrophys.*, **6**, 287–320.

Goldreich, P. & A. Toomre (1969). Some remarks on polar wandering. *J. Geophys. Res.*, **74**, 2555–67.

Goldstein, H. (1970). *Classical Mechanics.* Addison-Wesley, Reading, Mass., 399 pp.

Golovkov, V. P. & G. I. Kolomiytseva (1971). Morphology of 60-year geomagnetic field variations in Europe. *Geomagn. Aeron.*, **11**, 571–4.

Golovkov, V. P., G. I. Kolomiytseva, N. N. Berdichevskiy & N. M. Rotanova (1971). On Earth's conductivity determination by the data on secular geomagnetic field variations. *Geomagn. Aeron.*, **11**, 954–6.

Gordeyev, R. G., B. A. Kagan & E. V. Polyakov (1977). The effects of loading and self attraction on global ocean tides. *J. Phys. Ocean.*, **7**, 161–70.

Gordeyev, R. G., B. A. Kagan & V. Y. Rivkind (1973). A numerical solution of the tidal dynamics equations in the world ocean. *Dokl. Akad. Nauk. SSSR*, **209**, 340–3.

Gordeyev, R. G., B. A. Kagan & V. Ya. Rivkind (1974). Estimation of the rate of dissipation of tidal energy in the open ocean. *Oceanol. Acad. Sci. USSR*, **14**, 177–9.

Gordon, R. B. & C. W. Nelson (1966). Anelastic properties of the Earth. *Rev. Geophys.*, **4**, 457–74.

Goreau, T. F. (1959). The physiology of skeleton formation in corals. *Biol. Bull.*, **116**, 59–75.

Graber, M. A. (1976). Polar motion spectra based upon doppler, IPMS and BIH data. *Geophys. J.*, **46**, 75–85.

Greenspan, H. P. (1968). *The Theory of Rotating Fluids.* Cambridge University Press, 328 pp.

Groves, G. & W. Munk (1958). A note on tidal friction. *J. Mar. Res.*, **17**, 199–214.

Gubbins, D. (1976). Observational constraints on the generation process of the Earth's magnetic field. *Geophys. J.*, **47**, 19–39.

Guinot, B. (1958). L'astrolabe Impersonnel Danjon, variation de la latitude. *Bull. Astron.*, **22**, 1–71.

Guinot, B. (1970*a*). Short period terms in universal time. *Astron. Astrophys.*, **8**, 26–8.

Guinot, B. (1970*b*). Work of the Bureau International d l'Heure on the rotation of the Earth. In Mansinha, Smylie & Beck (1970), pp. 54–62.

Guinot, B. (1972). The Chandlerian Wobble from 1900 to 1970. *Astron. Astrophys.*, **19**, 207–14.

Guinot, B. (1974). A determination of the Love number *k* from the periodic waves of UT1. *Astron. Astrophys.*, **36**, 1–4.

Guinot, B. & M. Feisel (1969). *Annual Report for 1968.* Bureau International de l'Heure, Paris.

Guinot, B., M. Feissel & N. Granveaud (1971). Annual Report for 1970, Bureau International de l'Heure, Paris.

Gutenberg, B. (1941). Changes in sea level post glacial uplight and mobility of the Earth's interior. *Bull. Geol. Soc. Am.*, **52**, 721–72.

Hales, S. (1750). Some considerations on the causes of Earthquakes. *Phil. Trans. R. Soc.*, **A46**, 669–81.

Hardie, L. A. & R. N. Ginsberg (1971). The sedimentary record of a tidal flat lamination. *Geol. Soc. Am. Abstr. Prog.*, **3**, 591.

Harris, A. W. and W. M. Kaula (1975). Co-accretion model of satellite formation. *Icarus*, **24**, 516–24.

Harris, A. W. & J. G. Williams (1977). Earth rotation study using lunar laser ranging data. In *Scientific Applications of Lunar Laser Ranging* (ed. J. D. Mulholland), pp. 179–89. D. Reidel, Dordrecht.

Harrison, J. C., N. F. Ness, I. M. Longman, R. F. S. Forbes, E. A. Kraut & L. B. Slichter (1963). Earth-tide observations made during the year, *J. Geophys. Res.*, **68**, 1497–516.

Hassan, S. M. (1961). Fluctuations in the atmospheric inertia: 1873–1950, *Meteorological Monographs*, vol. 4, no. 24, pp. 1–48. American Meteorological Society, Boston, Mass.

Haubrich, R. A. (1970). An examination of the data relating pole motion to earthquakes. In Mansinha, Smylie & Beck (1970), pp. 149–57.

Haubrich, R. A. & W. Munk (1959). The Pole Tide. *J. Geophys. Res.*, **64**, 2373–88.

Heiskanen, W. (1921). Über den Einfluss der Gezeiten auf die säkuläre Acceleration des Mondes. *Ann. Acad. Sci. Fennicae.*, **A18**, 1–84.

Hellerman, S. (1967). An updated estimate of the wind stress on the world ocean, *Mon. Weather Rev.*, **95**, 607–626 (see also (1968) **96**, 62–74).

Hendershott, M. C. (1972). The effects of solid earth deformations on global ocean tides, *Geophys. J.*, **29**, 389–402.

Hendershott, M. C. (1973). Ocean tides. *EOS, Trans. Am. Geophys. Union*, **54**, 76–86.

Hendershott, M. C. & W. Munk (1970). Tides. *Annu. Rev. Fluid Mech.*, **2**, 205–24.

Hicks, S. D. (1973). Trends and variability of the yearly mean sea level 1893–1971. *National Oceanic and Atmospheric Administration Technical Memorandum, National Ocean Survey* no. 12. US Department of Commerce, Rockville, Md, 13 pp.

Hidaka, K. & M. Tsuchiya (1953). On the Antarctic circumpolar current. *J. Mar. Res.*, **12**, 214–22.

Hide, R. (1966). Free hydromagnetic oscillations of the Earth's core and the theory of the geomagnetic secular variation. *Phil. Trans. R. Soc. Lond.* **A259**, 615–50.

Hide, R. (1969). Interaction between the earth's liquid core and solid mantle. *Nature*, **222**, 1055–6.

Hide, R. (1970). On the Earth's core–mantle interface. *Q. J. R. Met. Soc.*, **96**, 579–90.

Hide, R. (1977). Towards a theory of irregular variations in the length of day and core–mantle coupling. *Phil. Trans. R. Soc. Lond.*, **A284**, 547–54.

Hide, R. (1978). How to locate the electrically conducting fluid core of a planet from external magnetic observations. *Nature*, **271**, 640–1.

Hide, R. and K. I. Horai (1968). On the topography of the core–mantle interface. *Phys. Earth Planet. Inter.*, **1**, 305–8.

Higgins, G. H. & G. C. Kennedy (1971). The adiabatic gradient and the melting point gradient in the core of the Earth. *J. Geophys. Res.*, **76**, 1870–8.

Hipkin, R. G. (1975). Tides and the rotation of the Earth. In Rosenberg & Runcorn (1975), pp. 319–35.

Ho, P.-Y. (1966). *The Astronomical Chapters of the Chin Shu.* Mouton, Paris, 271 pp.

Hoffman, H. J. (1973). Stromatolites: characteristics and utility. *Earth Sci. Rev.*, **9**, 339–73.

Holland, G. L. & T. S. Murty (1970). On the pole tide and related Chandler oscillations. In *Report on the Symposium on Coastal Geodesy*, (ed. R. Sigl), pp. 368–89. Institute for Astronomical and Physical Geodesy, Technical University, Munich.

Holmberg, R. R. (1952). A suggested explanation of the present value of the velocity of rotation of the earth. *Mon. Not. R. Astron. Soc. Geophys. Suppl.*, **6**, 325–30.

Hosoyama, K., I. Naito & N. Sato (1976). Tidal admittance of the pole tide. *J. Phys. Earth Jap.*, **24**, 43–50.

House, M. R. & G. E. Farrow (1968). Daily growth banding in the shell of the cockle *Cardium edule. Nature* **218**, 1384–6.

Iijima, S. & S. Okazaki (1966). On the biennial component in the rate of rotation of the Earth. *J. Geod. Soc. Jap.*, **12**, 91.

Inglis, D. (1957). Shifting of the Earth's axis of rotation. *Rev. Mod. Phys.*, **29**, 9–19.

Ishii, H. & I. Naito (1974). Note on the secular variations of $Z$ term. *Publ. Int. Lat. Obs. Mizusawa*, **9**, 235–40.

Israel, M., A. Ben Menahem & S. J. Singh (1973). Residual deformation of real Earth models with application to the Chandler wobble. *Geophys. J.*, **32**, 219–47.

Izsak, I. G. (1964). Tesseral harmonics of the geopotential and corrections to station coordinates. *J. Geophys. Res.*, **69**, 2621–30.

Jackson, D. D. and D. L. Anderson (1970). Physical mechanisms of seismic-wave attenuation. *Rev. Geophys. Space Phys.*, **8**, 1–63.

Jacobs, J. A. (1975). *The Earth's Core*. Academic Press, London & New York, 253 pp.

James, R. A. (1964). The structure and stability of rotating gas masses. *Astrophys. J.*, **140**, 552–82.

Jeffreys, B. (1965). Transformation of tesseral harmonics under rotation. *Geophys. J. R. Astron. Soc.*, **10**, 141–5.

Jeffreys, H. (1920). Tidal friction in shallow seas. *Phil. Trans. R. Soc. Lond.*, **A221**, 239–64.

Jeffreys, H. (1929). *The Earth* (second edition). Cambridge University Press, 346 pp.

Jeffreys, H. (1930). The resonance theory of the origin of the Moon. *Mon. Not. R. Astron. Soc.*, **91**, 169–173.

Jeffreys, H. (1940). The variation of latitude. *Mon. Not. R. Astron. Soc.*, **100**, 139–55.

Jeffreys, H. (1963). On the hydrostatic theory of the figure of the Earth. *Geophys. J.*, **8**, 196–202.

Jeffreys, H. (1968*a*). The variation of latitude. *Mon. Not. R. Astron. Soc.*, **141**, 255–68.

Jeffreys, H. (1968*b*). Waves and tides near the shore. *Geophys. J.*, **16**, 253–7.

Jeffreys, H. (1970). *The Earth* (fifth edition). Cambridge University Press, 525 pp.

Jeffreys, H. (1973). Tidal friction. *Nature*, **246**, 346.

Jeffreys, H. & B. Jeffreys (1962). *Methods of Mathematical Physics*. Cambridge University Press, 718 pp.

Jeffreys, H. and R. O. Vicente (1957*a*). The theory of nutation and the variation of latitude. *Mon. Not. R. Astron. Soc.*, **117**, 556–75.

Jeffreys, H. & R. O. Vicente (1957*b*). The theory of nutation and the variation of latitude: The Roche model core. *Mon. Not. R. Astron. Soc.*, **117**, 162–73.

Jeffreys, H. & R. O. Vicente (1966). Comparison of forms of the elastic equations for the Earth. *Mem. Acad. R. Belg.*, **37**, 5–31.

Jessen, A. (1964). Chandler's period in the mean sea level. *Tellus*, **16**, 513–16.

Jobert, G. (1973*a*). Deformation d'une sphère élastique gravitante. In *Traité de Geophysique Interne*, vol. 1 (ed. J. Coulomb & G. Jobert), pp. 171–80. Masson, Paris.

Jobert, G. (1973*b*). Marees terrestres. In *Traite de Geophysique Interne*, vol. 1 (ed. J. Coulomb & G. Jobert), pp. 507–27. Masson Paris.

Jochmann, H. (1976). Der einfluss von luftmassenbewegungen in der Atmosphäre auf die polbewegung. *Veroeff Zentralinst. Phys. Erde, Akad. Wiss. DDR*, **35**, 1–38.

Johnson, G. A. L. & J. R. Nudds (1975). Carboniferous coral geochronometers. In Rosenberg & Runcorn (1975), pp. 27–41.

Johnson, I. M. & D. E. Smylie (1977). A variational approach to whole-Earth Dynamics. *Geophys. J.*, **50**, 35–54.

Jurdy, D. M. & R. Van der Voo (1974). A method for the separation of true polar wander and continental drift, including results for the last 55 m.y. *J. Geophys. Res.*, **79**, 2945–52.

Jurdy, D. M. & R. Van Der Voo (1975*a*). True polar wander since the Early Cretaceous. *Science*, **197**, 1193–6.

Jurdy, D. M. & R. Van der Voo (1975*b*). Reply. *J. Geophys. Res.*, **80**, 3373–4.

Kagan, B. A. (1977). *Global Interactions of Ocean and Earth Tides.* Gidrometeoizdat, Leningrad, 45 pp.

Kakuta, C. (1961). The magnetic torque on the impulsive change of the rotation of the Earth. *Publ. Astron. Soc. Jap.*, **13**, 361–8.

Kakuta, C. (1965). Note on the relaxation time of magnetic coupling in the Earth's rotation. *Publ. Astron. Soc. Jap.*, **17**, 337–8.

Kanamori, H. (1970). The Alaska earthquake of 1964: radiation of long period surface waves and source mechanism. *J. Geophys. Res.*, **75**, 5029–40.

Kanamori, H. (1972). Mechanism of Tsunami earthquakes. *Phys. Earth Planet. Inter.*, **6**, 346–59.

Kanamori, H. (1976*a*). Are earthquakes a major cause of the Chandler wobble? *Nature*, **262**, 254–5.

Kanamori, H. (1976*b*). Re-examination of the Earth's free oscillations excited by the Kamchatka earthquake of November 4, 1952. *Phys. Earth Planet. Inter.*, **11**, 216–26.

Kanamori, H. (1977*a*). The energy release in great earthquakes. *J. Geophys. Res.*, **82**, 2981–7.

Kanamori, H. (1977*b*). Seismic and aseismic slip along subduction zones and their tectonic implications. In *Island Arcs, Deep Sea Trenches, and Back-Arc Basins*, vol. 1 (ed. M. Talwani & W. C. Pitman), pp. 163–74. American Geophysical Union, Washington.

Kanamori, H. & D. L. Anderson (1975*a*). Amplitude of the Earth's free oscillations and long-period characteristics of the earthquake source. *J. Geophys. Res.*, **80**, 1075–8.

Kanamori, H. & D. L. Anderson (1975*b*). Theoretical basis of some empirical relations in seismology. *Bull. Seism. Soc. Am.*, **65**, 1073–95.

Kanamori, H. & D. L. Anderson (1977). Importance of physical dispersion in surface wave and free oscillation problems. *Rev. Geophys. Space Phys.*, **15**, 105–12.

Kanamori, H. & J. J. Cipar (1974). Focal process of the great Chilean earthquake May 22, 1960. *Phys. Earth Planet. Inter.* **9**, 128–36.

Kaula, W. M. (1963). Elastic models of the mantle corresponding to variations in the external gravity field. *J. Geophys. Res.*, **68**, 4967–78.

Kaula, W. M. (1964). Tidal dissipation by solid friction and the resulting orbital evolution. *Rev. Geophys. Space Phys.*, **2**, 661–85.

Kaula, W. M. (1966). *Introduction to Satellite Geodesy.* Blaisdell, Waltham, Mass., 124 pp.

Kaula, W. M. (1969). Tidal friction with latitude dependent amplitude and phase angle. *Astron. J.*, **74**., 1108–14.

Kaula, W. M. (1971). Dynamical aspects of lunar origin. *Rev. Geophys. Space Phys.*, **9**, 217–38.

Kaula, W. M. (1975). Absolute plate motions by boundary velocity minimizations. *J. Geophys. Res.*, **80**, 244–8.

Kaula, W. M. (1977*a*). Geophysical inferences from statistical analysis of the gravity field. In *The Changing World of Geodetic Science*, (ed. U. A. Uotila). *Ohio State University, Department of Geodetic Science Report 250*, pp. 119–41.

Kaula, W. M. (1977*b*). On the origin of the Moon, with emphasis on bulk composition. *Proceedings of the 8th Lunar Science Conference*, pp. 321–31.

Kaula, W. M. (1979). Problems in understanding vertical movements and Earth Rheology. In *Earth Rheology, Isostasy and Eustasy* (ed. N. A. Mörner). John Wiley, New York (in press).

Kaula, W. M. & A. W. Harris (1973). Dynamically plausible hypotheses of lunar origin. *Nature*, **245**, 367–9.

Kaula, W. M. and A. W. Harris (1975). Dynamics of lunar origin and orbital evolution. *Rev. Geophys. Space Phys.*, **13**, 363–371.

Kaula, W. M., K. Lambeck, W. Markowitz, *et al.* (1973). The rotation of the Earth and polar motion. *EOS, Trans. Am. Geophys. Union*, **54**, 792–8.

Kaula, W. M., G. Schubert, J. T. Wasson, T. J. Ahrens, D. L. Anderson & D. O. Muhleman (1980) *Solid Earth and Planets.* John Wiley, New York (in press).

Kikuchi, N. (1971). Annual and yearly variations of the atmospheric pressure and the Earth's rotation. *Proc. Int. Lat. Obs. Mizusawa*, no. 11.

King, R. W., C. C. Counselman & I. I. Shapiro (1978). Universal time: results from lunar laser ranging. *J. Geophys. Res.*, **83**, 3377–81.

Kinoshita, H. (1977). Theory of the rotation of the rigid Earth. *Celestial Mechanics.*

Kolomiytseva, G. I. (1972). Distribution of electric conductivity in the mantle of the Earth, according to data on secular geomagnetic field variations. *Geomagn. Aeron.*, **12**, 938–41.

Kolsky, H. (1956). The propagation of stress pulses in visco-elastic solids. *Phil. Mag.*, **1**, 693–710.

Korsun, A. A., S. P. Major, L. V. Ryklova & Ya. S. Yatskiv (1974). On the annual component of the polar motion. *Veroeff Zentralinst. Phys. Erde, Akad. Wiss. DDR*, **30**, 153–9.

Kort, V. G. (1962). The Antarctic Ocean. *Sci. Am.*, **207**, 103–12.

Kovach, R. L. & D. L. Anderson (1967). Study of the energy of the free oscillations of the Earth. *J. Geophys. Res.*, **72**, 2155–68.

Kovalevsky, J. (1977). Lunar orbital theory. *Phil. Trans. R. Soc. Lond.*, **A284**, 565–71.

Kozai, Y. (1968). Love's number of the Earth derived from satellites observations. *Publ. Astron. Soc. Jap.*, **29**, 24–6.

Kung, E. C. (1968). On the momentum exchange between the atmosphere and Earth over the Northern Hemisphere. *Mon. Weather Rev.*, **96**, 337–41.

Kuo, J. T. & R. C. Jachens (1977). Indirect mapping of ocean tides by solving the inverse problem for the tidal gravity observations. *Ann. Geophys.*, **33**, 73–82.

Kuo, J. T., R. C. Jachens, M. Ewing & G. White (1970). Transcontinental tidal gravity profile across the the United States. *Science*, **168**, 968–71.

Kuznetsov, M. V. (1972). Calculation of the secular retardation of the earth's rotation from up to date cotidal charts. *Izv. Acad. Sci. USSR Phys. Solid Earth*, **2**, 3–11.

Lamb, H. (1932). *Hydrodynamics* (sixth edition). Cambridge University Press, 738 pp.

Lamb, H. H. (1970). Volcanic dust in the atmosphere. *Phil. Trans. R. Soc. Lond.*, **A266**, 425–533.

Lamb, H. H. (1972). *Climate, Present, Past and Future*, Methuen, London, 613 pp.

Lamb, H. H. & A. I. Johnson (1966). Secular variations of the atmosphere circulation since 1750. *Geophys. Mem. Lond.*, no. 110.

Lambeck, K. (1971). Determination of the Earth's pole of rotation from laser range observations to satellite. *Bull. Géod.* **101**, 263–81.

Lambeck, K. (1973). Temporal variations of rotational origin in the absolute value of gravity. *Studia Geophys. Geod.*, **17**, 269–71.

Lambeck, K. (1975*a*). Effects of tidal dissipation in the oceans on the Moon's orbit and the Earth's rotation. *J. Geophys. Res.*, **80**, 2917–25.

Lambeck, K. (1975*b*). The Chandler annual resonance. *Phys. Earth Planet. Inter.*, **11**, 166–8.

Lambeck, K. (1976). Lateral density anomalies in the mantle. *J. Geophys. Res.*, **81**, 6333–40.

Lambeck, K. (1977). Tidal dissipation in the oceans: astronomical, geophysical and oceanographic consequences. *Phil. Trans. R. Soc. Lond.* **A287**, 545–94.

Lambeck, K. (1978*a*). The Earth's paleorotation. *Tidal Friction and the Rotation of the Earth* (ed. P. Brosche & J. Sunderman), pp. 145–53. Springer-Verlag, Berlin, Heidelberg & New York.

Lambeck, K. (1978*b*). Progress in geophysical aspects of the rotation of the Earth. In *Ninth Research Conference on Application of Geodesy to Geodynamics, Ohio State University, Department of Geodetic Science Report 280*, pp. 1–11.

Lambeck, K. (1979*a*). The history of the Earth's rotation. In *The Earth: Its Origin, Structure and Evolution* (ed. M. W. McElhinny), pp. 59–81. Academic Press, London & New York.

Lambeck, K. (1979*b*). Methods and geophysical applications of satellite geodesy. *Rep. Prog. Phys.*, **42**, 547–628.

Lambeck, K. (1979*c*). On the orbital evolution of the Martian satellites. *J. Geophys. Res.* **84**, 5651–58.

Lambeck, K. & A. Cazenave (1973). The Earth's rotation and atmospheric circulation. I, Seasonal variations. *Geophys. J.*, **32**, 79–93.

Lambeck, K. & A. Cazenave (1974). The Earth's rotation and atmospheric circulation. II, The Continuum, *Geophys. J.* **38**, 49–61.

Lambeck, K. & A. Cazenave (1976). Long term variations in length of day and climatic change. *Geophys. J.*, **46**, 555–73.

Lambeck, K. & A. Cazenave (1977). The Earth's variable rate of rotation: a discussion of some meteorological and oceanic causes and consequences. *Phil. Trans. R. Soc. Lond.* **A284**, 495–506.

Lambeck, K., A. Cazenave & G. Balmino (1974). Solid Earth and Ocean tides estimated from satellite orbit analyses. *Rev. Geophys. Space Phys.*, **12**, 421–34.

Lambeck, K. & P. Hopgood (1979). New results for the angular momentum exchange between the solid Earth and atmosphere, 1963–1973. Symposium 6, Int. Ass. Geod. Geophys., Canberra.

Lambert, A. (1970). The response of the earth to loading by the ocean tides around Nova Scotia. *Geophys. J.*, **19**, 449–77.

Lambert, W., F. Schlesinger & E. Brown (1931). The variation of latitude. *Bull. Nat. Res. Counc.*, **16**, 245–77.

LeBlond, P. (1966). On the damping of internal gravity waves in a continuously stratified ocean. *J. Fluid Mech.*, **25**, 121–42.

Le Mouël, J. L. (1976*a*). L'induction dans le globe. In *Traité de Géophysique Interne, 2. Magnétisme et Géodynamique* (ed. J. Coulomb & G. Jobert) pp. 129–60. Masson, Paris.

Le Mouël, J. L. (1976*b*). L'origine du champ magnetique terrestre. In *Traité de Géophysique Interne, 2. Magnétisme et Géodynamique* (ed. J. Coulomb & G. Jobert) pp. 161–200. Masson, Paris.

Le Pichon, X. (1968). Sea floor spreading and continental drift. *J. Geophys. Res.*, **73**, 3661–97.

Le Pichon, X., J. Francheteau & J. Bonnin (1973). *Plate Tectonics*. Elsevier, Amsterdam, 300 pp.

Le Roy Ladurie, E. (1973). *Times of Feast, Times of Famine. A History of Climate Since the Year 1000*. George Allen & Unwin, London, 428 pp.

Lisitzin, E. (1974). *Sea Level Changes*. Elsevier, Amsterdam, 286 pp.

Lisitzin, E. & J. G. Pattullo (1961). The principal factors influencing the seasonal oscillation of sea level. *J. Geophys. Res.*, **66**, 845–52.

Liu, H.-P., D. L. Anderson & H. Kanamori (1976). Velocity dispersion due to anelasticity; implications for seismology and mantle composition. *Geophys. J.*, **47**, 41–58.

Lliboutry, L. (1965). *Traité de Glaciologie*, vol. II. Masson, Paris, 610 pp.

Loewe, F. (1964). Das grönländische inlandeis nach neven feststellungen. *Erdkunde*, **18**, 189–202.

Loewe, F. (1967). The water budget of Antarctica. In *Proceedings of the Symposium on Pacific Antarctic Sciences* (ed. Y. Nagata), Tokyo.

Longman, I. M. (1962). A Green's function for determining the deformation of the Earth under surface-mass loads. 1. Theory. *J. Geophys. Res.*, **67**, 845–50.

Longman, I. M. (1963). A Green's function for determining the deformation of the Earth under surface mass loads. 2. Computations and numerical results. *J. Geophys. Res.*, **68**, 485–96.

Longuet–Higgins, M. S. & G. S. Pond (1970). The free oscillations of fluid on a hemisphere bounded by meridians of longitude. *Phil. Trans. R. Soc.*, **A266**, 193–223.

Lowes, F. J. (1974). Spatial power spectrum of the main geomagnetic field and extrapolation to the core. *Geophys. J.*, **36**, 717–37.

Luther, D. S. & C. Wunsch (1975). Tidal charts of the Central Pacific Ocean. *J. Phys. Ocean.*, **5**, 222–30.

McCarthy, D. M. (1972). Secular and non polar variation of Washington latitude. In Melchior & Yumi (1972) pp. 86–96.

McCord, T. B. (1968). The loss of retrograde satellites in the solar system. *J. Geophys. Res.*, **73**, 1497–1500.

MacDonald, G. J. F. (1963). The deep structure of the oceans and the continents. *Rev. Geophys.*, **1**, 587–655.

MacDonald, G. J. F. (1964). Tidal friction. *Rev. Geophys.*, **2**, 467–541.

MacDonald, G. J. F. (1966). The figure and long-term mechanical properties of the Earth. In *Advances in Earth Sciences* (ed. P. M. Hurley), pp. 199–245. MIT Press, Cambridge, Mass.

McDonald, K. L. (1957). Penetration of the geomagnetic secular field through a mantle with variable conductivity. *J. Geophys. Res.*, **62**, 117–141.

McDonald, K. L. & R. H. Gunst (1967). An analysis of the Earth's magnetic field from 1835 to 1965. *ESSA Technical Report, Institutes for Environmental Research IER 46-IES 1*, 87 pp.
McElhinny, M. W. (1973). *Palaeomagnetism and Plate Tectonics.* Cambridge University Press, 358 pp.
McElhinny, M. W. & J. C. Briden (1971). Continental drift during the Palaeozoic. *Earth Planet Sci. Lett.*, **10**, 407–16.
McElhinny, M. W. & R. T. Merrill (1975). Geomagnetic secular variations over the past 5 m.y. *Rev. Geophys. Space Phys.*, **13**, 687–708.
McGarr, A. (1976). Upper limit to Earthquake size. *Nature*, **262**, 378–9.
McGugan, A. (1967). Possible use of algal stromatolite rhythms in geochronology (abstract). *Geol. Soc. Am. Prog. Annu. Meet.*, 145.
McKee, W. D. (1971). A note on the sea level oscillations in the neighbourhood of the Drake Passage. *Deep-Sea Res.*, **18**, 547–9.
McKenzie, D. P. (1966). The viscosity of the lower mantle. *Geophys. J.*, **14**, 297–305.
McKenzie, D. P. (1972). Plate tectonics. *The Nature of the Solid Earth* (ed. E. Robertson), pp. 323–60. McGraw-Hill, New York.
McKenzie, D. P. (1975). Comments on paper by D. M. Jurdy and R. Van der Voo. *J. Geophys. Res.*, **80**, 3371–2.
McKenzie, D. P. & R. L. Parker (1967). The North Pacific: an example of tectonics on a sphere. *Nature*, **216**, 1276–80.
McQueen, H. W. S. & F. D. Stacey (1976). Interpretation of low degree components of gravitational potential in terms of undulations of mantle phase boundaries. *Tectonophysics*, **34**, T1–T8.
Madden, R. A. & P. R. Julian (1971). Detection of a 40–50 day oscillation in the zonal wind in the tropical Pacific. *J. Atmos. Sci.*, **28**, 702–8.
Madden, R. & J. Stokes (1975). Evidence of global-scale 5-day waves in a 73-year pressure record. *J. Atmos. Sci.*, **32**, 831–6.
Maksimov, I. V. & V. P. Karklin (1965). The pole tide in the Baltic Sea. *Dokl. Akad. Nauk. SSSR*, **161**, 580–2.
Malin, S. R. C. (1969). Geomagnetic secular variation and its changes, 1942.5 to 1962.5. *Geophys. J.*, **17**, 415–41.
Malin, S. R. C. & A. D. Clark (1974). Geomagnetic secular variation, 1962.5 to 1967.5 *Geophys. J.*, **36**, 11–20.
Malvern, L. E. (1969). *Introduction to the Mechanics of a Continuous Medium.* Prentice-Hall, New Jersey, 713 pp.
Manabe, S., D. G. Hahn & J. L. Holloway (1974). The seasonal variation of the tropical circulation as simulated by a global model of the atmosphere. *J. Atmos. Sci.*, **31**, 43–83.
Manabe, S. & T. B. Terpstra (1974). The effect of mountains on the general circulation of the atmosphere as identified by numerical experiments. *J. Atmos. Sci.*, **31**, 3–42.
Mandelbrot, B. B. & K. McCamy (1970). On the secular pole motion and the Chandler wobble. *Geophys. J.*, **21**, 217–32.

Manley, G. (1954). Discussion. *J. Glac.*, **1**, 312–13.

Mansinha, L. & D. E. Smylie (1967). Effects of Earthquakes on the Chandler wobble and the secular pole shift. *J. Geophys. Res.*, **72**, 4731–43.

Mansinha, L. & D. E. Smylie (1970). Seismic excitation of the Chandler wobble. In Mansinha, Smylie & Beck (1970), pp. 122–34.

Mansinha, L., D. E. Smylie & A. E. Beck (eds.) (1970). *Earthquake Displacement Fields and the Rotation of the Earth.* D. Reidel, Dordrecht, 308 pp.

Mansinha, L., D. E. Smylie & C. H. Chapman (1979). Seismic excitation of the Chandler wobble revisited. *Geophys. J.*, **59**, 1–17.

Markowitz, W. (1960). The photo zenith telescope and the dual rate Moon position camera. In *Telescopes* (ed. G. P. Kuiper & B. Middlehurst) pp. 88–114. University of Chicago Press.

Markowitz, W. (1968). Concurrent astronomical observations for studying continental drift, polar motion, and the rotation of the Earth. In Markowitz & Guinot (1968), pp. 25–32.

Markowitz, W. (1970). Sudden changes in rotational acceleration of the Earth and secular motion of the pole. In Mansinha, Smylie & Beck (1970), pp. 69–81.

Marsden, B. G. & A. G. W. Cameron (eds.) (1966). *The Earth–Moon System.* Plenum Press, New York, 288 pp.

Martin, C. F. (1969). *A Study of the Rate of Rotation of the Earth from Occultations of Stars by the Moon*, 1927–1960. Thesis, Yale University, 83 pp.

Martin, C. F. and T. C. Van Flandern (1970). Secular changes in the lunar elements. *Science*, **168**, 246–7.

Mazullo, S. J. (1971). Length of the year during the Silurian and Devonian periods: New Values. *Geol. Soc. Am. Bull.*, **82**, 1085–6.

Melchior, P. (1957). Latitude variation. *Phys. Chem. Earth*, **2**, 212–43.

Melchior, P. (1972). Past and future of research methods in problems of the Earth's rotation. In Melchior & Yumi (1972), XI–XXIII.

Melchior, P. (1978). *The Tides of the Planet Earth.* Pergamon Press, Oxford, 609 pp.

Melchior, P., J. T. Kuo & B. Ducarme (1976). Earth tide gravity maps for Western Europe. *Phys. Earth Planet. Inter.*, **13**, 184–96.

Melchior, P. & S. Yumi (eds.) (1972). *Rotation of the Earth.* D. Reidel, Dordrecht, 244 pp.

Merriam, J. B. (1976). *The dissipation of tidal energy in the solid Earth.* Thesis, York University, Toronto, 97 pp.

Merriam, J. B. (1979). Zonal tides and changes in the length of day. Submitted to *Geophys. J.*

Merriam, J. B. & K. Lambeck (1979). Comments on the Chandler wobble *Q. Geophys. J.*, **59**, 281–6

Merrill, R. T. & M. W. McElhinny (1977). Anomalies in the time-averaged paleomagnetic field and their implications for the lower mantle. *Rev. Geophys. Space Phys.*, **15**, 309–23.

Miles, J. W. (1974). On Laplace's tidal equations. *J. Fluid Mech.*, **66**, 241–60.

Miller, A. J. (1974). Periodic variation of atmospheric circulation at 14–16 days. *J. Atmos. Sci.*, **31**, 720–6.

Miller, G. R. (1964). *Tsunamis and Tides.* Thesis, University of California, San Diego, 120 pp.

Miller, G. R. (1966). The flux of tidal energy out of the deep oceans. *J. Geophys. Res.*, **71**, 2485–9.

Miller, S. P. (1973). *Observations and Interpretation of the Pole Tide.* Thesis, Massachusetts Institute of Technology, 97 pp.

Miller, S. P. & C. Wunsch (1973). The Pole Tide. *Nature*, **246**, 97–102.

Minster, J. B., T. H. Jordan, P. Molnar & E. Haines (1974). Numerical modelling of instantaneous plate tectonics. *Geophys. J.*, **36**, 541–76.

Mitchell, J. (1976). An overview of climatic variability and its causal mechanisms. *Quaternary Res.*, **6**, 481–91.

Moens, M. (1976). Solid Earth tide and Arctic oceanic loading tide at Longyearbyen (Spitsbergen). *Phys. Earth Planet. Inter.*, **13**, 197–211.

Moffatt, H. K. (1978). Topographic coupling at the core–mantle interface, *Geophys. Astrophys. Fluid Dyn.*, **9**, 279–88.

Mohr, R. E. (1975). Measured periodicities of the Biwabike stromatolites and their geophysical significance. In Rosenberg & Runcorn (1975), pp. 43–55.

Molodensky, M. S. (1961). The theory of nutation and diurnal Earth tides. *Comm. Obs. R. Belg.*, **288**, 25–56.

Molodensky, S. M. (1977). On the relation between the Love numbers and the load coefficients. *Fis. Zemli*, no. 3, 3–7.

Monty, C. L. V. (1965). Recent algal stromatolites in the Windward lagoon, Andros Island, Bahamas. *Ann. Soc. Geol. Belg.*, **88**, 296–76.

Moorbath, S. (1976). Age and isotope constraints for the evolution of the archaean crust. In *The Early History of the Earth* (ed. B. F. Windley), pp. 351–60, John Wiley, New York.

Morgan, W. J. (1968). Rises, trenches, great faults and crustal blocks. *J. Geophys. Res.*, **73**, 1959–82.

Morgan, W. J. (1971). Convection plume in the lower mantle. *Nature*, **230**, 42–3.

Mörner, N. (1971). Eustatic changes during the last 20,000 years and a method of separating the isostatic and eustatic factors in an uplifted area. *Palaeogeogr. Palaeoclimat. Palaeocol.*, **9**, 153–81.

Morrison, L. V. (1972). The secular accelerations of the Moon's orbital motion and the Earth's rotation. *Moon*, **5**, 253–64.

Morrison, L. V. (1973). Rotation of the Earth and the constancy of *G*. *Nature*, **241**, 519–20.

Morrison, L. V. (1978). Tidal deceleration of the Earth's rotation deduced from astronomical observations in the period AD 1600 to the present. In *Tidal Friction and the Earth's Rotation* (ed. P. Brosche & J. Sundermann), pp. 22–7. Springer-Verlag, Berlin, Heidelberg & New York.

Morrison, L. V. (1979). Redetermination of the decade fluctuations in the rotation of the Earth in the period 1861–1978. *Geophys. J.*, 58, 349–60.

Morrison, L. V. and C. G. Ward (1975). The analysis of the transits of Mercury. *Mon. Not. R. Astron. Soc.*, **173**, 183–206.

Muller, P. M. (1975). *An Analysis of the Ancient Astronomical Observations with the Implications for Geophysics and Cosmology.* Thesis, University of Newcastle.

Muller, P. M. (1976). Determination of the cosmological rate of change of *G* and the tidal accelerations of Earth and Moon from ancient and modern astronomical data. *Report SP 43-36.* Jet Propulsion Laboratory, Pasadena, California, 24 pp.

Muller, P. M. & F. R. Stephenson (1975). The acceleration of the Earth and Moon from early astronomical observations. In Rosenberg & Runcorn (1975), pp. 459–534.

Munk, W. (1950). On the wind driven ocean circulation. *J. Met.*, **7**, 79–93.

Munk, W. H. (1966). Discussion. In Marsden & Cameron (1966), p. 276.

Munk, W. H. (1968). Once again – tidal friction. *Q. J. R. Astron. Soc.*, **9**, 352–75.

Munk, W. H. & E. M. Hassan (1961). Atmospheric excitation of the earth's wobble. *Geophys. J.*, **4**, 339–58.

Munk, W. & R. Haubrich (1958). The annual pole tide. *Nature*, **182**, 42.

Munk, W. H. & G. J. F. MacDonald (1960). *The Rotation of the Earth.* Cambridge University Press, 323 pp.

Munk, W. & R. Revelle (1952). On the geophysical interpretation of irregularities in the rotation of the Earth. *Mon. Not. R. Astron. Soc., Geophys. Suppl.*, **6**, 331–47.

Munk, W. H., F. Snodgrass & M. Wimbush (1970). Tides offshore: transition from California coastal to deep sea waters. *Fluid Dyn.*, **1**, 161–235.

Murray, C. A. (1978). On the precession and nutation of the Earth's axis of figure. *Mon. Not. R. Astron. Soc.*, **183**, 677–85.

Myerson, R. J. (1970). Evidence for association of earthquakes with the Chandler wobble. In Mansinha, Smylie & Beck (1970), pp. 159–68.

Nakiboglu, S. M. & K. Lambeck (1979). Deglaciation effects on the rotation of the Earth. *Geophys. J.* (in press).

Naito, I. (1974). Mean Pole, *Z* Term and Kuroshio. *J. Ocean. Soc. Jap.*, **30**, 168–78.

Naito I. & H. Ishii (1974). Secular variations and spectral structure of *Z* term in latitude variations. *Publ. Astron. Soc. Jap.*, **26**, 485–94.

National Ocean Survey (1975). *Tidal Current Tables (1976), Pacific Coast of North America and Asia*; and *Tide Tables (1976), West Coast of North and South America*. National Oceanic and Atmospheric Administration, US Department of Commerce.

Neumann, G. (1968). *Ocean Currents*. Elsevier, Amsterdam, 352 pp.

Newell, R. E., J. W. Kidson, D. G. Vincent & G. J. Boer (1974). *The General Circulation of the Tropical Atmosphere*, vol. 2, MIT Press, Cambridge, Mass., 371 pp.

Newton, C. W. (1971*a*). Mountain torques in the global angular momentum balance. *J. Atmos. Sci.*, **28**, 623–8.

Newton, C. W. (1971*b*). Global angular momentum balance; Earth torques and atmospheric fluxes. *J. Atmos. Sci.*, **28**, 1329–41.

Newton, R. R. (1968). A satellite determination of tidal parameters and Earth deceleration. *Geophys. J.*, **14**, 505–39.

Newton, R. R. (1970). *Ancient Astronomical Observations and the Accelerations of the Earth and Moon*. Johns Hopkins University Press, Baltimore, Md, 309 pp.

Newton, R. R. (1972). *Medieval Chronicles and the Rotation of the Earth*. Johns Hopkins University Press, Baltimore, Md, 825 pp.

Newton, R. R. (1976). *Ancient Planetary Observations and the Validity of Ephemeris Time*. Johns Hopkins University Press, Baltimore, Md, 749 pp.

O'Connell, R. J. (1971). Pleistocene glaciation and the velocity of the lower mantle. *Geophys. J.*, **23**, 299–327.

O'Connell, R. J. (1977). On the scale of mantle convection. *Tectonophysics*, **38**, 119–36.

O'Connell, R. J. & B. Budiansky (1977). Viscoelastic properties of fluid-saturated cracked solids. *J. Geophys. Res.*, **82**, 5719–35.

O'Connell, R. J. & B. Budiansky (1978). Measures of attenuation in dissipative media. *Geophys. Res. Lett.*, **5**, 5–8.

O'Connell, R. J. & A. M. Dziewonski (1976). Excitation of the Chandler wobble by large earthquakes. *Nature*, **262**, 259–62.

Oesterwinter, C. & C. J. Cohen (1972). New orbital elements for Moon and planets. *Celestial Mech.*, **5**, 317–95.

Okal, E. A. (1976). A surface wave investigation of the rupture mechanism of the Gobi–Altai (December 4, 1957) earthquake. *Phys. Earth Planet. Inter.*, **12**, 319–28.

Okal, E. A. (1977). The July 9 and 23, 1905 Mongolian earthquakes: A surface wave investigation. *Earth Planet. Sci. Lett.*, **34**, 326–31.

Okazaki, S. (1975). On the amplitude changes of seasonal components in the rate of rotation of the Earth. *Publ. Astron. Soc. Jap.*, **27**, 367–78.

Okazaki, S. (1977). On a relation between irregular variations of the Earth's rotation and anomalous changes of the atmospheric circulation. *Publ. Astron. Soc. Jap.*, **29**, 619–29.

O'Keefe, J. A. (1969). Origin of the Moon. *J. Geophys. Res.*, **74**, 2758–67.

O'Keefe, J. A. (1970). The origin of the Moon. *J. Geophys. Res.*, **75**, 6565–74.

O'Keefe, J. A. (1972). Inclination of the Moon's orbit: The early history. *Irish Astron. J.*, **10**, 241–50.

Ooe, M. (1978). An optimal complex ARMA model of the Chandler wobble. *Geophys. J.*, **53**, 445–57.

Oort, A. H. & H. D. Bowman (1974). A study of the mountain torque and its interannual variations in the Northern Hemisphere. *J. Atmos. Sci.*, **31**, 1974–82.

Opik, E. J. (1972). Comments on lunar origin. *Irish Astron. J.*, **10**, 190–238.

Ostrovsky, A. Ye. (1976). Results of observations of tidal tilts of the Earth's surface on the territory of the USSR for the period of 1957–1972. In *Proceedings of the Seventh International Symposium on Earth Tides* (ed. G. Szádeczky-Kardoss), pp. 121–6. Akademiai Kiado, Budapest.

Pannella, G. (1972). Paleontological evidence on the Earth's rotational history since Early Precambrian. *Astrophys. Space Sci.*, **16**, 212–37.

Pannella, G. (1975). Paleontological clocks and the history of the Earth's rotation. In Rosenberg & Runcorn (1975), pp. 253–84.

Pannella, G. (1976). Geophysical inferences from stromatolite lamination. In Walter (1976), pp. 673–85.

Pannella, G. & C. MacClintock (1968). Biological and environmental rhythms reflected in molluscan shell growth. *J. Palaeontol. Mem.*, **42**, 64–80.

Pannella, G., C. MacClintock & M. N. Thompson (1968). Paleontological evidence of variations in length of synodic month since late Cambrian. *Science*, **162**, 792–6.

Pariyskiy, N. N., M. V. Kuznetsov & L. V. Kuznetsova (1972). The effect of oceanic tides on the secular deceleration of the earth's rotation. *Izv. Acad. Sci. USSR Phys. Solid Earth* (English translation), no. 2, 3–12.

Parke, M. E. (1978). *Global numerical models of the open ocean tides M2, S2, K1 on an elastic Earth.* Thesis, University of California, San Diego, 156 pp.

Pattullo, J. G. (1963). Seasonal changes in sea level. In *The Sea*, vol. 2 (ed. M. N. Hill), pp. 485–96. John Wiley, New York.

Peale, S. J. (1973). Rotation of solid bodies in the solar system. *Rev. Geophys. Space Phys.*, **11**, 767–93.

Pedersen, G. P. H. & M. G. Rochester (1972). Spectral analysis of the Chandler wobble. In Melchior & Yumi (1972), pp. 33–8.

Pekeris, C. L. & Y. Accad (1969). Solution of Laplace's equations for $M_2$ tide in the world oceans. *Phil. Trans. R. Soc. Lond.*, **A265**, 413–36.

Pekeris, C. L. & Y. Accad (1972). Dynamics of the liquid core of the Earth. *Phil. Trans. R. Soc. Lond.*, **A273**, 237–60.

Peltier, W. R. (1974). The impulse response of a Maxwell Earth. *Rev. Geophys. Space Phys.*, **12**, 649–69.

Peltier, W. R. (1976). Glacial isostatic adjustment – II. The inverse problem. *Geophys. J.*, **46**, 669–706.

Peltier, W. R. & J. T. Andrews (1976). Glacial-isostatic adjustment – I. The forward problem. *Geophys. J.*, **46**, 605–46.

Pertsev, B. P. (1969). The effect of ocean tides upon Earth tide observations. *Comm. A-9 Ser. Geophys. Obs. R. Belg.*, **96**, 113–15.

Pertsev, B. P. (1977). $M_2$ ocean tide corrections to tidal gravity observations in Western Europe. *Ann. Geophys.*, **33**, 63–6.

Pilnik, G. P. (1970). A correlation analysis of the Earth tides and nutation. *Sov. Astron.*, **14**, 1044–56.

Plafker, G. (1969). Tectonics of the March 27, 1964 Alaska earthquake. *Geol. Surv. Prof. Pap.*, *543-I*, 74 pp.

Platzman, G. W. (1971). Ocean Tides. In *Lectures in Applied Mathematics*, vol. 14, part 2, pp. 239–92. American Mathematical Society, Providence, RI.

Platzman, G. W. (1975). Normal modes of the Atlantic and Indian Oceans. *J. Phys. Ocean.*, **5**, 201–21.

Playford, P. E. & A. E. Cockbain (1976). Modern algal stromatolites at Hamelin Pool. In Walter (1976), pp. 389–411.

Press, F. (1965). Displacements, strains and tilts at teleseismic distances. *J. Geophys. Res.*, **70**, 2395–412.

Press, F. & P. Briggs (1975). Earthquakes, Chandler wobble, rotation and geomagnetic changes: a pattern recognition approach. *Nature*, **256**, 270–3.

Price, A. T. (1970). The electrical conductivity of the Earth. *Q. J. R. Astron. Soc.*, **11**, 23–42.

Proudman, J. (1941). The effect of coastal friction on the tides. *Mon. Not. R. Astron. Soc., Geophys. Suppl.*, **5**, 23–6.

Proudman, J. (1960). The condition that a long period tide shall follow the equilibrium law. *Geophys. J.*, **3**, 244–9.

Proverbio, E., F. Carta & F. Mazzoleni (1972). In Melchior & Yumi (1972), pp. 43–5.

Proverbio, E. & V. Quesada (1973). Analysis of secular polar motion and continental drift. *Bull. Géod.*, **109**, 281–91.

Reid, J. L. & A. W. Mantyla (1976). The effect of the geostrophic flow upon coastal sea elevations in the northern North Pacific Ocean. *J. Geophys. Res.*, **81**, 3100–3110.

Rikitake, T. (1966). *Electromagnetism and the Earth's Interior.* Elsevier, Amsterdam, 308 pp.

Ringwood, A. E. (1960). Some aspects of the thermal evolution of the Earth. *Geochim. Cosmochim. Acta*, **20**, 241–9.

Ringwood, A. E. (1970). Origin of the Moon: The precipitation hypothesis. *Earth Planet. Sci. Lett.*, **8**, 131–40.

Ringwood, A. E. (1972). Some comparative aspects of lunar origin. *Phys. Earth Planet. Inter.*, **6**, 366–76.

Ringwood, A. E. (1975). *Composition and Petrology of the Earth's Mantle.* McGraw-Hill, New York, 618 pp.

Ringwood, A. E. (1979). *The Origin of the Earth and Moon.* Springer-Verlag, Berlin, Heidelberg & New York, 295 pp.

Ritsema, A. R. (1972). Discussion. *Tectonophysics,* **13**, 579–80.

Roberts, P. H. (1963). On highly rotating polytropes, 1. *Astrophys. J.*, **137**, 1129–41.

Roberts, P. H. (1971). Dynamo theory. *Lectures in Applied Mathematics,* vol. 14, part 2, pp. 129–206. American Mathematical Society, Providence, RI.

Roberts, P. H. (1972). Electromagnetic core–mantle coupling. *J. Geomagn. Geoelectr.*, **24**, 231–59.

Roberts, P. H. & K. Stewartson (1965). On the motion of a liquid in a spheroidal cavity of a precessing rigid body. *Proc. Camb. Phil. Soc.*, **61**, 279–88.

Robertson, D. S., W. E. Carter, B. E. Corey *et al.* (1979). Recent results of radio interferometric determinations of polar motion and Earth rotation. *IAU Symposium, 82,* Cadiz, Spain.

Rochester, M. G. (1960). Geomagnetic westward drift and irregularities in the Earth's rotation. *Phil. Trans. R. Soc. Lond.*, **A252**, 531–55.

Rochester, M. G. (1962). Geomagnetic core–mantle coupling. *J. Geophys. Res.*, **67**, 4833–6.

Rochester, M. G. (1968). Perturbations in the Earth's rotation and geomagnetic core–mantle coupling. *J. Geomagn. Geoelectr.*, **20**, 387–402.

Rochester, M. G. (1970). Core–mantle interactions: Geophysical and astronomical consequences. In Mansinha, Smylie & Beck (1970), pp. 136–48.

Rochester, M. G. (1973). The Earth's rotation, *EOS, Trans. Am. Geophys. Union*, **54**, 769–80.

Rochester, M. G. (1974). The effect of the core on the Earth's rotation. *Veroeff. Zentralinst. Phys. Erde, Akad. Wiss. DDR*, **30**, 77–89.

Rochester, M. G. (1976). The secular decrease of obliquity due to dissipative core–mantle coupling. *Geophys. J.*, **46**, 109–26.

Rochester, M. G., O. G. Jensen & D. E. Smylie (1974). A search for the Earth's nearly diurnal free wobble. *Geophys. J.*, **38**, 349–63.

Rochester, M. G. & D. E. Smylie (1965). Geomagnetic core–mantle coupling and the Chandler wobble. *Geophys. J.*, **10**, 289–315.

Rochester, M. G. & D. E. Smylie (1974). On changes in the Earth's inertia tensor. *J. Geophys. Res.*, **79**, 4948–51.

Roden, R. B. (1963). Electromagnetic core–mantle coupling. *Geophys. J.*, **7**, 361–74.

Romanowicz, B. & K. Lambeck (1977). The mass and moment of inertia of the Earth. *Phys. Earth Planet. Inter.*, **15**, P1–P4.

Rosen, R. D., M.-F. Wu & J. P. Peixoto (1976). Observational study of the interannual variability in certain features of the general circulation. *J. Geophys. Res.*, **81**, 6383–9.

Rosenberg, G. D. & S. K. Runcorn (eds.) (1975). *Growth Rhythms and the History of the Earth's Rotation.* John Wiley, New York, 559 pp.

Rossiter, J. R. (1962). Long-term variations in sea-level. In *The Sea*, vol. 1 (ed. M. N. Hill), pp. 590–610. John Wiley, New York.

Rossiter, J. R. (1967). An analysis of annual sea level in European waters. *Geophys. J.*, **13**, 259–99.

Routh, E. J. (1905). *Advanced Dynamics of a System of Rigid Bodies* (sixth edition). Macmillan, London, 484 pp. (reprinted (1955) by Dover, New York).

Royal Society Discussion (1977). *Phil. Trans. R. Soc. Lond.*, **A285**, 606 pp.

Rubincam, D. P. (1975). Tidal friction and the early history of the Moon's orbit. *J. Geophys. Res.*, **80**, 1537–48.

Runcorn, S. K. (1955). The electrical conductivity of the Earth's mantle. *Trans. Am. Geophys. Union*, **36**, 191–8.

Runcorn, S. K. (1956). Paleomagnetic comparisons between Europe and North America. *Proc. Geol. Assoc. Can.*, **8**, 77–85.

Runcorn, S. K. (1966). Corals as paleontological clocks. *Sci. Am.*, **215**, 26–33.

Runcorn, S. K. (1968). Polar wandering and continental drift. In *Continental Drift, Secular Motion of the Pole and Rotation of the Earth* (ed. W. Markowitz & B. Guinot), pp. 80–5. Reidel, Dordrecht.

Runcorn, S. K. (1970). A possible cause of the correlation between earthquakes and polar motions. In Mansinha, Smylie & Beck (1970), pp. 181–7.

Ruskol, E. L. (1966). On the past history of the Earth–Moon system. *Icarus*, **5**, 221–7.

Ruskol, E. L. (1972). The origin of the Moon – 3. Some aspects of the dynamics of the circumterrestrial swarm. *Sov. Astron.*, **15**, 646–54.

Ruskol, E. L. (1973). On the model of the accumulation of the Moon compatible with the data on the composition and the age of lunar rocks. *Moon*, **6**, 190–201.

Sailor, R. V. & A. M. Dziewonski (1978). Measurements and interpretation of normal mode attenuation. *Geophys. J.*, **53**, 559–81.

Saito, M. (1974). Some problems of static deformation of the Earth. *J. Phys. Earth Jap.*, **22**, 123–40.

Sandstrom, H. (1976). On topographic generation and coupling of internal waves. *Geophys. Fluid Dyn.*, **7**, 231–70.

Sasao, T., I. Okamoto & S. Sakai (1977). Dissipative core–mantle coupling and nutational motion of the Earth. *Publ. Astron. Soc. Jap.*, **20**, 83–105.

Sasao, T., S. Okubo & M. Saito (1979). A simple model on dynamical effects of stratified fluid core upon nutational motion of the Earth. In Federov, Smith & Bender (1979).

Sato, R. & A. F. Espinosa (1967). Dissipation factor of the torsional mode ${}_0T_2$ for a homogeneous-mantle Earth with a soft-solid or a viscous-liquid core. *J. Geophys. Res.*, **72**, 1761–7.

Sawyer, J. F. A. & F. R. Stephenson (1970). Literary and astronomical evidence for a total eclipse of the Sun observed in Ancient Ugarit on 3 May 1375 BC. *Bull. School Orient. Afr. Stud.*, **33**, 467–89.

Schatzman, E. (1966). Interplanetary torques. In Marsden & Cameron (1966), 12–25.

Schott, F. (1977). On the energetics of baroclinic tides in the North Atlantic. *Ann. Geophys.*, **33**, 41–62.

Schutz, C. & W. Gates (1971). Global climatic data for surface, 800 mb, 400 mb: January. *ARPA Report R-195-ARPA.* Rand Corporation, Santa Monica, Calif.

Schutz, C. & W. Gates (1972). Global climatic data for surface, 800 mb, 400 mb: July. *ARPA Report R-1029-ARPA.* Rand Corporation, Santa Monica, Calif.

Schutz, C. & W. Gates (1973). Global climatic data for surface, 800 mb, 400 mb: April. *ARPA Report R-1317-ARPA.* Rand Corporation, Santa Monica, Calif.

Schutz, C. & W. Gates (1974). Global climatic data for surface, 800 mb, 400 mb: October. *ARPA Report R-1425-ARPA.* Rand Corporation, Santa Monica, Calif.

Scrutton, C. T. (1965). Periodicity in Devonian coral growth. *Palaeontology*, **7**, 552–8.

Scrutton, C. T. (1970). Evidence for a monthly periodicity in the growth of some corals. In *Palaeogeophysics* (ed. S. K. Runcorn), pp. 11–16. Academic Press, London & New York.

Scrutton, C. T. (1978). Periodic growth features in fossil organisms and the length of the day and month. In *Tidal Friction and the Earth's Rotation* (ed. P. Broche & J. Sunderman), pp. 154–96. Springer-Verlag, Berlin, Heidelberg & New York.

Shankland, T. J. (1975). Electrical conduction in rocks and minerals: parameters for interpretation. *Phys. Earth Planet. Inter.*, **10**, 209–19.

Shapiro, I. I. & C. A. Knight (1970). Geophysical applications of long-baseline radio interferometry. In Mansinha, Smylie & Beck (1970), pp. 284–301.

Shapiro, I. I., D. S. Robertson, C. A. Knight *et al.* (1974). Transcontinental baselines and the rotation of the Earth measured by radio interferometry. *Science*, **186**, 920–2.

Shaw, D. M. (1976). Development of the early continental crust. Part 2. Precambrian, Protoarchean and later eras. *The Early History of the Earth* (ed. B. F. Windley), pp. 33–53. John Wiley, New York.

Siderenkov, N. S. (1968). Methods for estimating the influence of atmospheric circulation on the Earth's rotational velocity. *Sov. Astron.*, **12**, 303–8.

Siderenkov, N. S. (1969). The influence of atmospheric circulations on the Earth's rotational velocity. *Sov. Astron.*, **12**, 706–14.

Siderenkov, N. S. (1973). The inertia tensor of the atmosphere, the annual variations of its components and the variations of the Earth's rotation. *Izv. Atmos. Ocean. Phys.*, **9**, 339–51.

Siderenkov, N. S. & D. I. Stekhnovskiy (1971). The total mass of the atmosphere and its seasonal redistribution. *Izv. Atmos. Ocean. Phys.*, **7**, 979–89.

Singer, S. F. (1968). The origin of the Moon and geophysical consequences. *Geophys. J.*, **15**, 205–26.

Singer, S. F. (1970). Origin of the Moon by capture and its consequences. *Trans. Am. Geophys. Union*, **51**, 637–41.

Slichter, L. B. (1963). Secular effects of tidal friction. *J. Geophys. Res.*, **68**, 4281–8.

Slichter, L. B. (1972). Earth Tides. In *The Nature of the Solid Earth* (ed. E. C. Robertson), pp. 285–320. McGraw-Hill, New York.

Smagorinsky, J. (1969). Numerical circulation of the global atmosphere. In *The Global Circulation of the Atmosphere* (ed. G. A. Corby), pp. 24–41. Royal Meteorological Society, London.

Smart, W. M. (1962). *Spherical Astronomy.* Cambridge University Press, 430 pp.

Smith, D. E., R. Kolenkiewicz, P. J. Dunn, H. H. Plotkin & T. S. Johnson (1972). Polar motion from laser tracking of artificial satellites. *Science*, **178**, 405–6.

Smith, D. E., F. J. Lerch, J. G. Marsh, C. A. Wagner, R. Kolenkiewicz & M. A. Khan (1976). Contributions to the national geodetic satellite program. *J. Geophys. Res.*, **81**, 1006–26.

Smith, J. V. (1974). Origin of Moon by disintegrative capture with chemical differentiation followed by sequential accretion. *Lunar Sci.*, **V**, 718–20.

Smith, J. V. (1976). Development of the Earth–Moon system with implications for the geology of the early Earth. *The Early History of the Earth* (ed. B. F. Windley), pp. 3–19. John Wiley, New York.

Smith, M. L. (1977). Wobble and nutation of the Earth. *Geophys. J.*, **50**, 103–40.

Smith, S. W. (1972). The anelasticity of the mantle. *Tectonophysics*, **13**, 601–22.

Smylie, D. E. (1965). Magnetic diffusion in a spherically-symmetric conducting mantle. *Geophys. J.*, **9**, 169–84.

Smylie, D. E. (1974). Dynamics of the outer core. *Veroeff. Zentralinst. Phys. Erde., Acad. Wiss. DDR*, **30**, 91–104.

Smylie, D. E., G. K. C. Clarke & L. Mansinha (1970). Deconvolution of the pole path. In Mansinha, Smylie & Beck (1970), pp. 99–112.

Smylie, D. E., G. K. C. Clarke & J. J. Ulrych (1973). Analysis of irregularities in the Earth's rotation. *Methods Comput. Phys.*, **13**, 391–430.

Smylie, D. E. & L. Mansinha (1968). Earthquakes and the observed motion of the rotation pole. *J. Geophys. Res.*, **73**, 7661–73.

Smylie, D. E. & L. Mansinha (1971). The elasticity theory of dislocations in real Earth models and changes in the rotation of the Earth. *Geophys. J.*, **23**, 329–54.

Solomon, S. C. & N. H. Sleep (1974). Some simple physical models for absolute plate motions. *J. Geophys. Res.*, **79**, 2557–67.

Sorokin, N. A. (1965). The relative orientation of the Earth's equator and the Moon's orbit. *Sov. Astron.*, **9**, 826–9.

Spencer Jones, H. (1932). *Ann. Cape Obs.*, **13**, part 3.

Spencer Jones, H. (1939). The rotation of the Earth and the secular accelerations of the Sun, Moon and planets. *Mon. Not. R. Astron. Soc.*, **99**, 541–58.

Stacey, F. D. (1970). A re-examination of core–mantle coupling as the cause of the wobble. In Mansinha, Smylie & Beck (1970), pp. 176–80.

Stacey, F. D. (1977). *Physics of the Earth* (second edition). John Wiley, New York, 414 pp.

Stacey, F. D., D. Conley & H. S. McQueen (1979). Spectral character of the geomagnetic field and the electrical conductivity of the lower mantle. *J. Geophys. Res.* (in press).

Stephenson, F. R. (1972). *Some Geophysical, Astrophysical and Chronological Deductions from Early Astronomical Records.* Thesis, University of Newcastle, 204 pp.

Stewartson, K. & P. H. Roberts (1963). On the motion of a liquid in a spheroidal cavity of a precessing rigid body. *J. Fluid Mech.*, **17**, 1–20.

Stolz, A. (1976*a*). Changes in the position of the geocentre due to seasonal variations in air mass and ground water. *Geophys. J.*, **44**, 19–26.

Stolz, A. (1976*b*). Changes in the position of the geocentre due to variations in sea-level. *Bull. Géod.*, **50**, 159–68.

Stolz, A. & D. Larden (1979). Seasonal displacement of an Earth model by the atmosphere. *J. Geophys. Res.*, **84**, 6185–94.

Stoyko, A. (1968). Mouvement séculaire du pôle et la variation des latitudes des stations du SIL. In Markowitz & Guinot (1968), pp. 52–6.

Stoyko, A. (1969). Temps des ephemerides, temps atomique, temps rotationnel et leur comparaison. *Bull. Acad. R. Belg.*, ser. 5, **55**.

Stoyko, A. (1973). Le mouvement du pôle instantané. *Vistas Astron.*, **13**, 51–134.

Stoyko, A. & N. Stoyko (1956). Les fluctuations périodiques de la rotation de la Terre pendant les années 1933–1940. *Bull. Acad. R. Belg.*, ser 5, **42**, 693–702.

Stoyko, A. & N. Stoyko (1969). Rotation de la terre, phénomènes géophysiques et activité du soleil. *Bull. Acad. R. Belg.*, ser. 5, **55**, 279–285.

Stuart, W. D. & M. J. S. Johnston (1974). Tectonic implications of anomalous tilt before central California earthquakes (abstract). *Trans. Am. Geophys. Union.*, **58**, 1196 pp.

Sutton, J. (1973). Some changes in continental structure since early Precambrian time. In *Implications of Continental Drift to the Earth Sciences*, vol. 2 (ed. D. H. Tarling & S. K. Runcorn), pp. 1071–81. Academic Press, London & New York.

Suzuki, Y. & R. Sato (1970). Viscosity determination in the Earth's outer core from ScS and SKS phases. *J. Phys. Earth Jap.*, **18**, 157–70.

Takeuchi, H. (1950). On the Earth tide of the compressible Earth of variable density and elasticity. *Trans. Am. Geophys. Union*, **31**, 651–89.

Takeuchi, H. (1967). *Theory of the Earth's Interior.* Blaisdell, Waltham, Mass., 131 pp.

Takeuchi, H. & W. M. Elsasser (1954). Fluid motions near the core and the irregular variations in the Earth's rotation. *J. Phys. Earth Jap.*, **2**, 39–44.

Takeuchi, H., M. Saito & N. Kobayashi (1962). Statical deformations and free oscillations of a model Earth. *J. Geophys. Res.*, **67**, 1141–54.

Taylor, G. I. (1919). Tidal friction in the Irish Sea. *Phil. Trans. R. Soc.*, **A220**, 1–93.

Terrien, J. (1976). Standards of length and time. *Rep. Prog. Phys.*, **39**, 1067–108.

Thatcher, W. (1974). Strain release mechanism of the 1906 San Francisco earthquake. *Science*, **184**, 1283–5.

Thorpe, S. A. (1975). The excitation, dissipation, and interaction of internal waves in the deep oceans. *J. Geophys. Res.*, **80**, 328–38.

Tiron, K. D., Y. N. Sergeev & A. N. Michurin (1967). Tidal charts for the Pacific, Atlantic and Indian Oceans. *Vest. Leningrad Univ., Ser. Geol. Geogr.*, **24**, 123–35.

Toksöz, M. N., A. H. Dainty, S. C. Solomon & K. R. Anderson (1974). Structure of the moon. *Rev. Geophys. Space Phys.*, **12**, 539–67.

Toomre, A. (1966). On the coupling of the Earth's core and mantle during the 26 000 year precession. In Marsden & Cameron (1966), pp. 33–45.

Toomre, A. (1974). On the nearly diurnal wobble of the Earth. *Geophys. J.*, **38**, 335–48.

Tozer, D. C. (1959). The electrical properties of the Earth's interior. *Phys. Chem. Earth*, **3**, 414–36.

Treshnikov, A. F., I. V. Maksimov & B. V. Gindysh (1966). The great eastward drift of the Antarctic Ocean. *Probl. Antarkt., Leningrad*, **22**, 18–34.

Tuller, S. E. (1968). World distribution of mean monthly and annual precipitable water. *Mon. Weather Rev.*, **96**, 785–97.

Turcotte, D. L., J. L. Cisne & J. C. Nordmann (1977). On the evolution of the lunar orbit. *Icarus*, **30**, 251–66.

Urey, H. C. (1963). The origin and evolution of the solar system. In *Space Science* (ed. D. P. LeGalley), pp. 123–68. John Wiley, New York.

Van der Voo, R. & R. B. French (1974). Apparent polar wandering for the Atlantic bordering continents: Late Carboniferous to Eocene. *Earth Sci. Rev.*, **10**, 99–119.

Van Flandern, T. C. (1970). The secular acceleration of the moon. *Astron. J.*, **75**, 657–8.

Van Flandern, T. C. (1975). A determination of the rate of change of *G*. *Mon. Not. R. Astron. Soc.*, **170**, 333–42.

Van Hylckama, T. E. A. (1970). Water balance and Earth unbalance. In *International Association of Scientific Hydrology, Proceedings of the Reading Symposium World Water Balance, Publication 92*, pp. 434–553. AIHS–UNESCO.

Van Loon, H. (1971). A half-yearly variation of the circumpolar surface drift in the Southern Hemisphere. *Tellus*, **23**, 511–16.

Van Loon, H. (1972). Half-yearly oscillations in the Drake Passage. *Deep-Sea Res.*, **19**, 525–7.

Van Loon, H. & K. Labitzke (1973). Comments on the half yearly wave in the stratosphere. *J. Geophys. Res.*, **79**, 470–1.

Vestine, E. H. (1953). On variations of the geomagnetic field, fluid motions and the rate of the Earth's rotation. *J. Geophys. Res.*, **58**, 127–245.

Vestine, E. H. & A. B. Kahle (1968). The westward drift and geomagnetic secular change. *Geophys. J.*, **15**, 29–37.

Vestine, E. H., L. La Porte, I. Lange, C. Cooper & W. C. Hendric (1947). Description of the Earth's main magnetic field and its secular change, 1905–1945. *Carnegie Institution, Washington Publication*, 578 pp.

Vicente, R. & S. Yumi (1969). Co-ordinates of the Pole (1899–1968) referred to the Conventional Internal Origin, *Publ. Int. Lat. Obs., Mizusawa*, **7**, 41.

Vicente, R. & S. Yumi (1970). Revised values (1941–1962) of the coordinates of the Pole referred to the CIO. *Publ. Int. Lat. Obs., Mizusawa*, **7**, 109.

Wahl, E. W. & R. A. Bryson (1975). Recent changes in Atlantic surface temperatures. *Nature*, **254**, 45–6.

Walcott, R. I. (1970). Isostatic response to loading of the crust in Canada. *Can. J. Earth Sci.*, **7**, 716–27.

Walker, A. & A. Young (1955). The analysis of the observations of the variation of latitude. *Mon. Not. R. Astron. Soc.*, **115**, 443–59.

Wallace, J. M. (1966). Long period wind fluctuations in the tropical stratosphere. *Report 19, Planetary Circulation Project.* Department of Meteorology, MIT, Cambridge, Mass.

Wallace, J. M. (1973). General circulation of the tropical lower stratosphere. *Rev. Geophys. Space Phys.*, **11**, 191–222.

Wallace, J. M. & C. P. Chang (1969). Spectrum analysis of large scale wave disturbances in the tropical lower troposphere. *J. Atmos. Sci.*, **26**, 1010–25.

Wallace, J. M. & R. E. Newell (1966). Eddy fluxes and the biennial stratospheric oscillation. *Q. J. R. Met. Soc.*, **92**, 481–9.

Walter, M. R. (ed.) (1976). *Stromatolites.* Elsevier, Amsterdam, 790 pp.

Warburton, R. J., C. Beaumont & J. M. Goodkind (1975). The effect of ocean tide loading on tides of the solid Earth observed with the superconducting gravimeter. *Geophys. J.*, **43**, 707–20.

Wasserburg, G. J., D. A. Papanastassiou, F. Tera & J. C. Huneke (1977). Outline of a lunar chronology. *Phil. Trans. R. Soc. Lond.*, **A285**, 7–22.

Watanabe, H. & T. Yukutake (1975). Electromagnetic core–mantle coupling associated with changes in the geomagnetic dipole field. *J. Geomagn. Geoelectr.*, **27**, 153–75.

Webb, D. J. (1973). On the age of the semi-diurnal tide. *Deep-Sea Res.*, **20**, 847–52.

Webb, D. J. (1976). A model of continental shelf resonances. *Deep-Sea Res.*, **23**, 1–15.

Webster, A. G. (1959). *The Dynamics of Particles.* Dover, New York, 588 pp.

Weertman, J. (1970). The creep strength of the Earth's mantle. *Rev. Geophys. Space Phys.*, **8**, 145–68.

Wells, F. J. (1972). *On the Separation of the Annual and Chandler Spectral Components in Astronomic Latitude and Solar Motion Data.* Thesis, Brown University, 228 pp.

Wells, F. J. & M. A. Chinnery (1973). On the separation of the spectral components of polar motion. *Geophys. J.*, **34**, 179–92.

Wells, J. W. (1963). Coral growth and geochronometry. *Nature*, **197**, 448–950.

Wells, J. W. (1966). Paleontological evidence of the rate of the Earth's rotation. In Marsden & Cameron (1966), pp. 70–81.

Wells, J. W. (1970). Problems of annual and daily growth rings in corals. Photogrammetry. US Coast and Geodetic Survey, Washington.

Whitten, C. A. (1971). Preliminary investigation of the correlation of polar motion and major earthquakes. *Internal Report, Office of Geodesy and Photogrammetry*. US Coast and Geodetic Survey, Washington.

Williams, J. G. (1977). Present scientific achievements from lunar laser ranging. *Scientific Applications of Lunar Laser Ranging* (ed. J. D. Mulholland), pp. 37–48. D. Reidel, Dordrecht.

Williams, J. G., W. S. Sinclair & C. F. Yoder (1979). Tidal acceleration of the Moon. *Science* (in press).

Wilson, C. R. (1975). *Meteorological Excitation of the Earth's Wobble.* Thesis, University of California, San Diego, 107 pp.

Wilson, C. R. & R. A. Haubrich (1976*a*). Meteorological excitation of the Earth's wobble. *Geophys. J.*, **46**, 707–44.

Wilson, C. R. & R. A. Haubrich (1976*b*). Atmospheric contributions to the excitation of the Earth's wobble, 1901–1970. *Geophys. J.*, **46**, 745–60.

Wilson, C. R. & R. A. Haubrich (1977). Earthquakes, weather and wobble. *Geophys. Res. Lett.*, **4**, 283–4.

Wise, D. V. (1963). An origin of the Moon by rotational fusion during formation of the Earth's core. *J. Geophys. Res.*, **68**, 1547–54.

Wise, D. V. (1969). Origin of the Moon from the Earth: Some new mechanisms and comparisons. *J. Geophys. Res.*, **74**, 6034–45.

Wittman, A. (1974). Numerical simulation of the Mercury transit black drop phenomenon. *Astron. Astrophys.*, **31**, 239–43.

Woolard, E. W. (1953). Theory of the rotation of the Earth around its center of mass. *Astron. Pap. Am. Ephem. Naut. Alm.*, **15**, 3–165.

Woolard, E. W. & G. M. Clemence (1966). *Spherical Astronomy.* Academic Press, London & New York, 453 pp.

Wunsch, C. (1967). The long-period tides. *Rev. Geophys.*, **5**, 447–75.

Wunsch, C. (1972). Bermuda sea level in relation to tides, weather and baroclinic fluctuations. *Rev. Geophys. Space Phys.*, **10**, 1–49.

Wunsch, C. (1974*a*). Dynamics of the pole tide and the damping of the Chandler wobble. *Geophys. J.*, **39**, 539–50.

Wunsch, C. (1974*b*). Simple models of the deformation of an Earth with a fluid core – I. *Geophys. J.*, **39**, 413–19.

Wunsch, C. (1975*a*). Errata. *Geophys. J.*, **40**, 311.

Wunsch, C. (1975*b*). Simple models of the deformation of an Earth with a fluid core – II. Dissipation and magnetohydrodynamic effects. *Geophys. J.*, **41**, 165–84.

Wunsch, C. (1975*c*). Internal tides in the ocean. *Rev. Geophys. Space Phys.*, **13**, 167–82.

Yanai, M. & M. Murakami (1970). A further study of tropical wave disturbances by use of spectrum analysis. *J. Met. Soc. Jap.*, **48**, 185–97.

Yatskiv, Ya. S. (1972). On the comparison of diurnal nutation derived from separate series of latitude and time observations. In Melchior & Yumi (1972), pp. 200–5.

Yoder, C. F. (1973). *On the establishment and evolution of orbit–orbit resonances.* Thesis, University of California, Santa Barbara, 303 pp.

Yoder, C. F. (1979). Effects of the spin–spin interaction and the inelastic tidal deformation on the lunar physical librations. In *Natural and Artificial Satellite Motion* (ed. P. Nacozy and S. Ferrez-Mello), pp. 211–21. University of Texas Press, Austin.

Yukutake, T. (1965). The solar cycle contribution to the secular change in the geomagnetic field. *J. Geomagn. Geoelectr.*, **17**, 287–309.

Yukutake, T. (1972). The effect of change in the geomagnetic dipole moment on the rate of the Earth's rotation. *J. Geomagn. Geoelectr.*, **24**, 19–47.

Yukutake, T. (1973). Fluctuations in the Earth's rate of rotation related to changes in the geomagnetic dipole field. *J. Geomagn. Geoelectr.*, **25**, 195–212.

Yumi, S. (1970). Stability of station and the Earth's figure. In Mansinha, Smylie & Beck (1970), pp. 45–52.

Yumi, S. & Y. Wako (1970). Secular motion of the Pole. In Mansinha, Smylie & Beck (1970), pp. 82–7.

Zahel, W. (1970). Die Reproduktion gezeiten Bewegungsvorgange in weltozean Mittels des hydrodynamisch-numerischen Verfahrens. *Mitt. Inst. Meeresk. d. Univ. Hamburg*, **17**.

Zahel, W. (1973). The diurnal K tide in the world ocean – A numerical investigation. *Pure Appl. Geophys.*, **109**, 1819–25.

Zahel, W. (1977). A global hydrodynamic-numerical 1° model of the ocean tide. *Ann. Geophys.*, **33**, 31–40.

Zahel, W. (1978). *Tidal Friction and the Rotation of the Earth* (ed. P. Brosche & J. Sundermann). Springer-Verlag, Berlin, Heidelberg & New York, pp. 98–124.

Zetler, B. D. (1971). Radiational ocean tides along the coasts of the United States. *J. Phys. Ocean.*, **1**, 34–8.

# AUTHOR INDEX


Abe, K., 235
Accad, Y., 9, 11, 22–3, 125, 127–9, 201, 223, 327–8, 334
Acheson, D. J., 259
Adams, J. C., 69
Agnew, D. C., 145
Aki, K., 231–2
Alfven, H., 289, 393, 395, 398–9
Alldredge, L. R., 255–6
Alterman, Z., 8
Anderle, R. J., 104–5
Anderson, D. L., 11–2, 14, 16–22, 203, 206, 221, 231, 235–6, 280–1, 296, 395
Anderson, K. R., 296
Anderssen, R. S., 11
Andrews, J. T., 20, 270, 351
Angell, J. K., 177, 190
Anhaeusser, C. R., 392
Arakawa, A. 170
Arrhenius, G. 289, 393, 398–9
Arur, M. G., 92
Awrawik, S. M., 394, 385

Backus, G. E., 6
Bader, H., 272
Baines, P. G., 333
Baker, D. J., 165
Baker, T. F., 111–2
Ball, R. H., 259
Balmino, G., 52–3, 121, 127, 133, 136, 293
Banks, R. J., 255
Barker, R. M., 371, 375, 378–81
Barnes, D. J., 363, 368
Barraclough, D. R., 261–2
Bauer, A., 272
Beaumont, C., 112
Beck, A. E., ix
Bell, T. H., 333
Belmont, A. D., 174, 176–7, 179, 190
Bender, P. L., 1, 105
Ben Menahem, A., 223–4, 226–7, 230
Berdichevsky, M. N., 256
Berry, A., 300
Berry, W. B. M., 69, 375, 378–81
Beuglass, L. K., 104
Boer, G. J., 173–4, 176–7, 179–80
Bogdanov, K. T., 125–9, 131, 328
Bondi, H. 206, 210, 247
Bonnin, J., 347
Bowden, K. F., 332
Bowman, H. D., 169–70, 172
Braginski, S. I., 257
Braha, T., 66
Brennan, B. J., 17
Brettschneider, G., 321
Briden, J. C., 351
Briggs, P., 281
Brillinger, D. R., 100, 102–3
Britten, J. P. 304
Brosche, P. 390
Brouwer, D., 73–5, 78–9, 112–13, 133, 247, 303
Brown, E., 31, 78–9, 85, 300
Brune, J. N., 230–1
Bryden, H. L., 165
Bryson, R. A., 270
Buddemeier, R. W., 366
Budiansky, B., 15–6
Budyko, M. I., 160
Bukowinski, M. S., 24–5
Bullard, E. C., 247–8, 250, 252, 254–5, 258
Bullen, K. E., 11–2, 197
Bürgi, J., 66
Burke, K., 391

Busse, F. H., 1, 208, 210, 259
Byzova, N. L., 147

Cameron, A. G. W., ix
Cannon, W. H. 45
Capitaine, N., 39
Caputo, M., 27, 345
Carta, F., 220
Carter, W. E., 106
Cartwright, D. E., 116, 124, 131
Cathles, L. M., 20, 270, 351
Cazenave, A., 83–4, 127, 133, 136–8, 142, 167, 169–70, 173–4, 179, 182, 185, 188, 191–2, 194, 240, 268, 271, 275–9, 293, 298
Cecchine, G., 65, 220
Celoria, G., 313
Challinor, R. D., 191
Chandler, S. C., 86, 220
Chang, C. P., 190
Chapman, C. H., 223–4
Chapman, S., 130
Chen, W. P., 236, 238
Chinnery, M. A., 8, 100, 222, 223, 232–6, 238, 281
Cipar, J. J., 225, 235–6
Cisne, J. L., 393
Claerbout, J. F., 103
Clark, A. D., 261
Clark, G. R., 373
Clark, J. A., 269
Clark, S. P., 395
Clarke, G. K. C., 61, 103
Cleary, J. R., 11
Clemence, G. M., 62, 73, 112–13, 133, 301–3
Cloud, P. E., 393
Cockbain, A. E., 384
Coleman, P. J., 284
Cohen, C. J., 74, 303
Colombo, G., 99
Conley, D., 255, 258
Cooper, C., 258, 261
Cooper, J. A., 393
Corey, B. E., 106
Counselman, C. C., 106
Cowell, P., 69
Cox, C. S., 333
Crepon, M., 149
Crossley, D. J., 24, 223
Curott, D. R., 311, 315–7
Currie, D. G., 105
Currie, R. G., 102–3, 105, 123, 213, 255

Dahlen, F. A., 122, 144, 198–201, 208, 216, 223–9, 235, 238, 282
Daillet, S., 137–8
Dainty, A. H., 296
Danjon, A., 66, 191
Dartt, D. G., 174, 176–7, 179, 190
Darwin, G. H., ix, 37, 123, 195, 198, 287, 343
Davies, G. F., 231
Débarbat, S., 66
Defant, A., 331
Delaunay, C., 69, 300
De Pritt, A., 300
De Sitter, W., 78, 311, 315–16
Dewey, J., 391
Dicke, R. H., 105, 259, 262, 289, 312, 315–16
Dickman, S. R., 94, 216
Dietrich, G., 127–9
Djurovic, D., 145
Dodge, R. E., 363
Dolgouchine, L. D., 273
Dolman, J., 369
Doodson, A. T., 116
Douglas, B. C., 138
Ducarme, B., 111
Duda, S. J., 236
Duncombe, R. L., 70
Dunn, P., 104
Dziewonski, A. M., 11–12, 18–19, 22, 223–4, 229, 232, 234, 236, 238, 243–4

Eckert, W. J., 317
Edden, A. C., 116, 131
Elsasser, W., 247–8, 267
Engen, G. R., 231–2
Espionosa, A. F., 24
Euler, 31–2, 69, 85, 300
Evans, J. W., 372–4
Evans, M. E., 346
Ewing, M., 111

Fainberg, E. B., 256
Fairbridge, R. W., 271–2, 275, 339

Farrell, W. E., 9, 11–14, 111, 119, 145, 269
Farrow, G. E., 373
Federer, C. A., 160
Federov, Y. P., 1, 90
Feissel, M., 68
Fels, S. B., 223
Felsentreger, T. L., 132, 138
Ferrel, W., 69
Feynman, R. P., 16
Filloux, J. H., 124
Finsterwalder, R., 274
Fletcher, J., 282
Fliegel, H. F., 182, 188
Forbes, R. F. S., 111
Fotheringham, J. K., 306–8, 311, 316
Francheteau, J., 347
Freedman, C., 248, 250, 258
French, R. B., 347
Frostman, T. O., 173, 182
Fukao, Y., 237
Fung, Y. C., 15, 21
Furumoto, M., 237

Gans, R. F., 24
Gaposchkin, E. M., 26, 88, 96, 99, 134
Garland, G. D., 267
Garrett, C., 322, 333
Garrett, G. J. R., 331
Garthwaite, K., 300
Gates, W. L., 153, 170
Gebelein, C. D., 384
Gellman, H., 248, 250, 258
Gerstenkorn, H., 287, 355, 357–8, 391, 395, 400
Gilbert, F., 8, 11–12, 22, 199, 229, 237
Gill, A. E., 163, 165
Gindysh, B. V., 165
Ginsberg, R. N., 384
Ginzel, F. K., 313
Goad, C. C., 138
Gold, T., 206, 210, 289, 343–4
Goldreich, P., 27, 131, 287, 289, 343–6, 351–7, 400
Goldstein, H., 30
Golovkov, V. P., 256, 263–4
Goodkind, J. M., 112
Gordeyev, R. G., 125, 127, 332
Gordon, R. B., 14
Goreau, T. F., 363
Graber, M. A., 205
Granveaud, N., 68
Greenspan, H. P., 208
Groves, G., 324, 327
Gubbins, D., 24–5, 223
Guinot, B., 66, 68, 81–2, 84, 88, 100–2, 108, 145, 191
Gunst, R. H., 261
Gutenberg, B., 11–12, 271

Hahn, D. G., 170
Haines, E., 92, 347, 349
Harkrider, D., 230
Hardie, L. A., 384
Harris, A. W., 105, 288, 396–8
Hales, S., 281
Halley, E., 69, 300
Hansen, P. A., 69
Harrison, J. C., 111
Hart, R. S., 11, 16, 18–19, 22
Harwood, J. H., 261
Hassan, E. S. M., 148, 152, 156, 158–9, 239–40
Haubrich, R. A., 102–3, 142, 148, 153, 171, 179, 198, 212–13, 222, 234, 238, 240–1, 243–4
Hawkins, T. P., 182, 188
Heiskanen, W., 298, 323–4, 327, 330
Hellerman, S., 169
Hendershott, M. C., 111, 120, 124–8, 324, 327–9, 331, 333
Henrard, J., 300
Hendric, W. C., 258, 261
Herglotz, G., 11
Hicks, S. D., 271
Hidaka, K., 332
Hide, R., 209, 246–7, 253, 256, 259, 267
Higgins, G. H., 25
Hipkin, R. G., 289, 398
Ho, P-Y., 306
Hoffman, H. J., 385
Hoffman, P., 384
Holdridge, D. B., 300
Holland, G. L., 212, 214
Holloway, J. L., 170
Holmberg, R. R., 130
Hopgood, P., 167

Hopkins, W., 38, 246
Horai, K. I., 268
Hosoyama, K., 213–14
Hough, S., 11, 38, 123, 197, 246
House, M. R., 373
Huneke, J. C., 392
Huygens, C., 66

Iijima, S., 177
Imbrie, J., 170
Inglis, D., 343
Ishii, H., 87
Israel, M., 223–4, 226–7
Izsak, I.,113

Jachens, R. C., 111–12
Jackson, D. D., 14
Jacobs, J. A., 22, 25
James, R. A., 397
Jarosch, H., 8
Jeffreys, B., 111–12
Jeffreys, H., ix, 23–4, 26, 88, 96–7, 102–3, 112, 146–7, 197, 201–2, 205–6, 219, 238, 286, 288, 290, 298, 322–3, 327, 330, 331, 345, 396
Jensen, O. G., 39, 196
Jessen, A., 214
Jobert, G., 6, 111
Jochmann, H., 148, 157–8, 179
Johnson, A. I., 271
Johnson, G. A. L., 365, 369
Johnson, I. M., 24
Johnson, T. S., 104
Johnston, M. I. S., 238
Jordan, T. H., 92, 347, 349
Jarosch, H., 8
Julian, P. R., 190
Jurdy, D. M., 350

Kagan, B. A., 125, 127, 332
Kahle, A. B., 259–62
Kakuta, C., 248
Kanamori, H., 16–7, 220–1, 223, 225, 231–2, 235–9, 283
Kant, I., 69
Karklin, V. P., 212–13
Kaula, W. M., 6, 9, 20, 52–3, 113–14, 133–4, 203, 223, 287–8, 291, 293, 295, 336, 349, 351–2, 394, 396–9
Kelvin, Lord, ix, 10, 38, 86, 123, 195, 201, 220, 246, 343
Kennedy, G. C., 25
Khan, M. A., 26
Kidson, J. W., 173–4, 176–7, 179, 180
Kikuchi, N., 148, 157–8
King, C. Y., 231–2
King, R. W., 106
Kinoshita, H., 1, 33
Kinzie, R. A., 366
Knight, C. A., 106
Knopoff, L., 24–5
Kobayashi, N., 11
Kolenkiewicz, R., 26, 104
Kolomiytseva, G. I., 256, 263–4
Kolsky, H., 17
Korshover, J., 177, 190
Korsun, A. A., 100
Kort, V. G., 165
Kovach, R. L., 203
Kovalevsky, J., 300
Kozai, Y., 136
Kraut, E. A., 111
Krebs, O. A., 271–2, 275
Kung, E. C., 169
Kuo, J. T., 111–12
Kuznetsov, M. V., 324, 327–8
Kuznetsova, L. V., 324, 327–8

Labitzke, K., 175
Lagrange, J. L., 69, 300
Lamb, H. 38, 124, 319
Lamb, H. H., 148, 160, 165, 239, 271, 279–82
Lambert, A., 111
Lambert, W. D., 31, 85, 327
Lange, I., 258, 261
Laplace, P. S., 69, 123, 300
Larden, D., 152, 156
Larmor, J., 195, 197–8, 220
Leaton, B. R., 261
La Porte, L., 258, 261
LeBlond, P., 333
Leighton, R. B., 16
Le Mouël, J-L., 250, 255–6
Lange, I., 258, 261
Lennon, G. W., 111–12
Le Pichon, X., 92, 347
Lerch, F. J., 26

Le Roy Ladurie, E., 274
Lindzen, R. S., 130
Lisitzin, E., 150, 271
Liu, H-P.,
Lliboutry, L., 269, 272–4
Loewe, F., 272–3
Lomnitz, C., 206
Longman, I. M., 9, 11, 109, 111
Longuet-Higgins, M. S., 334
Love, A. E. H., 10–11, 71
Lowes, F. J., 258, 261
Luther, D. S., 124, 127
Lyttleton, R. A., 247

MacClintock, C., 373, 385
McCamy, K., 92, 102–3
McCarthy, D. M., 91
McCord, T. B., 288
MacDonald, G. J. F., ix, 27, 34, 44, 51, 69, 79, 80, 87–8, 103, 108, 149, 151, 156, 164–7, 173, 191, 195–6, 205–6, 211, 216, 221–2, 238, 268, 284, 287–8, 291, 300, 303, 311, 315, 323, 327, 333, 343, 345, 348, 355, 357
McDonald, K. L., 255, 261
McElhinny, M. W., 346, 349, 350–1
McGarr, A., 235
McGugan, A., 385
McKee, W. D., 166
McKenzie, D. P., 232, 345, 348–50
McQueen, H. W. S., 255, 258, 352
Madden, R. A., 190–1
Magarik, V. A., 125–9, 131, 328
Major, S. P., 90, 100
Maksimov, I. V., 165, 212–13, 239
Malin, S. R. C., 261–2
Malvern, L. E., 15
Manabe, S., 168, 170, 172
Mandelbrot, B. B., 92, 102–3
Manley, G., 274
Mansinha, L., ix, 61, 222–4, 226
Mantyla, A. W., 150, 164
Markowitz, W., 65, 78, 80, 91, 94, 223, 238, 282
Marsden, B. G., ix
Marsh, J. G., 26, 132, 138
Martin, C. F., 74, 317, 336
Martin, D. W., 173, 182
Mazzoleni, F., 220
Mazzullo, S. J., 367–9, 378
Melchior, P., 65, 88, 94, 111, 145
Merriam, J. B., 145, 203–4
Merrill, R. T., 346
Michurin, A. N., 125–6
Miles, J. W., 125
Miller, A. J., 190, 192
Miller, G. R., 323, 327, 330–1
Miller, R., 166
Miller, S. P., 212–14
Milne, J., 220
Minster, J. B., 19–20, 92, 206, 296, 347, 349
Mintz, Y., 166, 170
Mitchell, J., 191
Moens, M., 119
Moffatt, H. K., 267
Mohr, R. E., 384–7, 390
Molnar, P., 92, 236, 238, 347, 349
Molodensky, M. S., 197–201
Molodensky, S. M., 10
Monty, C. L. V., 384
Moorbath, S., 392
Moore, A. F., 254
Morgan, W. J., 348–9
Mörner, N., 339
Morrison, L. V., 70, 73–5, 500–3, 308, 317–18
Mueller, I. I., 92
Muhleman, D. O., 6
Mulholland, J. D., 300
Muller, P. M., 289, 300–3, 306–7, 311–14, 316–18, 335, 339–40
Munk, W. M., 27, 37, 44, 51, 69, 79, 80, 87–8, 103, 108, 111, 120, 124, 142, 148–9, 151–2, 156, 158–9, 164–7, 173, 191, 195–6, 198, 205–6, 211–13, 216, 221–2, 239–42, 248, 268, 284, 286, 288–9, 300, 303, 311, 315, 323–4, 327, 331–3, 343–5, 348, 399
Murakami, M., 190
Murray, C. A., 33
Murty, T. S., 212, 214
Myerson, R. J., 220

Naito, I., 87, 213, 214
Nakiboglu, S. M., 285, 351
Nelson, C. W., 14
Ness, N. F., 111

Neumann, G., 125
Newcomb, S., 86, 195–6, 307
Newell, R. E., 173–174, 176–7, 179–80, 189–90
Newhall, X., 317
Newton, C. W., 168–70, 172
Newton, R. R., 70, 136, 304–7, 309–11, 313–17, 338–9
Niiler, P. O., 163, 165
Nikoloaichik, V. V., 257
Nixon, J., 248, 250, 258
Nordmann, J. C., 393, 395
North, R. G., 232–6
Nowlin, W. D., 165
Nudds, J. R., 365, 369

O'Connell, R. J., 15–16, 223–4, 229, 232–4, 238, 289, 243–4, 345, 351
Oesterwinter, C., 74, 302–3
Okal, E. A., 236
Okamoto, I., 202
Okazaki, S., 82, 177, 185, 187, 191, 194
O'Keefe, J. A., 397, 399
Okubo, S., 202, 266
Ooe, M., 102–3
Oort, A. H., 169–70, 172
Opik, E. J., 396
Ostrovsky, A. Ye., 111

Panchenko, N. T., 90
Pannella, G., 362, 373, 376–8, 384–6, 390, 393
Papanastassiou, D. A., 392
Pariyskiy, N. N., 324, 327–8
Parke, M. E., 127
Parker, R. L., 232
Pattullo, J. G., 150
Peale, S. J., 133, 289
Pedersen, G. P. H., 99, 100
Peixoto, J. P., 194
Pekeris, C. L., 8, 9, 11, 22–3, 125, 127–9, 201, 223, 327–8, 334
Peltier, W. R., 9, 20–1, 109, 351
Penman, H. L., 160
Pertsev, B. P., 111
Pillsbury, R. D., 165
Pilnik, G. P., 145
Plafker, G., 232
Platzman, G. W., 124, 217, 293, 334
Playford, P. E., 384
Plotkin, H. H., 104
Poincaré, H., 38, 201
Polyakov, E. V., 125, 127, 332
Pond, G. S., 334
Press, F., 221, 281
Price, A. T., 254
Proudman, J., 123, 331
Proverbio, E., 88, 92, 220

Quesada, V., 92

Rayleigh, Lord, 123
Reid, J. L., 150, 164
Revelle, R., 248
Richard, J., 66
Richards, J. R., 393
Rikitake, T., 250, 252
Ringwood, A. E., 12, 395, 397
Ritsema, A. R., 206
Rivkind, V. Y., 127, 332
Roberts, P. H., 1, 208, 248, 253–4, 397
Robertson, D. S., 106
Rochester, M. G., 24, 28–9, 39, 99, 100, 147, 196, 206, 210, 247–50, 252–3, 255, 263, 265–6, 284
Roden, R. B., 248, 252
Rom, A., 300
Romanowicz, B., 26–7
Rosen, R. D., 194
Rosenhead, L., 10, 147
Rossiter, J. R., 123, 212, 271
Rotanova, M. M., 256
Routh, E. J., 31, 34, 37
Rubincam, D. P., 356, 399, 400
Rudnick, P., 239
Runcorn, S. K., 90, 255, 265, 347, 361, 383
Ruskol, E. L., 358, 397
Ryklova, L. V., 100

Sailor, R. V., 18–19
Saito, M., 11, 202, 223, 266
Sakai, S., 202
Sands, M., 16
Sandstrom, H., 333
Sasao, T., 202, 266
Sato, N., 213–14

Sato, R., 24
Sawyer, J. F. A., 331–2
Schatzman, E., 284
Schlesinger, F., 31, 85
Schott, F., 333
Schubert, G., 6
Schutz, C., 153
Schwerdtfeger, W., 172, 182
Schweydar, W., 123
Scrutton, C. T., 360, 366–70
Sergeev, Y. N., 125–6
Shankland, T. J., 257
Shapiro, I. I., 99, 106
Shaw, D. M., 392
Shida, T., 10
Siderenkov, N. S., 148, 150, 152, 156–8, 166–7, 169, 193–4, 242–3, 275
Sinclair, W. S., 286
Singer, S. F., 287, 355, 395
Singh, S. J., 223–4, 226–7
Sleep, N. H., 349
Slichter, L. B., 111, 287, 355, 357
Sludskii, F., 38
Smagorinsky, J., 168, 172
Smart, W. M., 62
Smirnov, J. B., 256
Smith, D. W., 26, 104
Smith, J. V., 394, 395
Smith, M. L., 1, 39, 198–202, 210, 215, 224, 229
Smith, S. W., 17
Smylie, D. E., ix, 24, 28–9, 39, 61, 103, 147, 196, 206, 220, 222–4, 227, 248, 255, 263, 265
Snodgrass, F., 124
Solomon, S. C., 296, 349
Sorokin, N. A., 355
Soter, S., 289
Spencer Jones, H., 70, 72-4, 301–2, 338
Spitaler, R., 147
Stacey, F. D., 6, 17, 21, 205, 254–5, 257–9, 266, 290, 345, 352, 354, 393
Stalin, J., 264
Stekhnovsky, D. I., 156
Stephenson, F. R., 308, 311–14, 316–17, 339–40
Stewartson, K., 1, 208
Stokes, J., 191
Stolz, A., 147, 152, 156
Stokyo, A., 74, 86, 88, 90–1, 181, 220, 280
Stokyo, N., 181, 220, 280
Stuart, W. D., 238
Sündermann, J., 390
Sutton, J., 391
Suzuki, Y., 24

Takano, K., 165
Takeuchi, H., 6, 9, 11, 197, 248, 267
Taraby, V. K., 90
Taylor, F. B., 343
Taylor, G. I., 288, 321, 323
Terpstra, T. B., 168, 172
Tera, F., 392
Terrien, J., 67
Thatcher, W., 237
Thompson, M. N., 385
Thornthwaite, C., 160
Thorpe, S. A., 333
Tiron, K. D., 125–6
Tisserand, F., 37
Toksöz, M. N., 296
Toomre, A., 1, 24, 27, 39, 196, 206, 211, 247, 343–6, 351–2
Tozer, D. C., 257
Treshnikov, A. F., 165
Tsuchiya, M., 332
Tuller, S. E., 151, 161
Turcotte, D. L., 393, 395

Ulrych, J. J., 103
Urey, H. C., 395

Vaisnys, J., 363
Van der Voo, R., 347, 350
Van Flandern, T. C., 301–3, 317, 336
Van Hylckama, T. E. A., 161–2
Vanjan, L. L., 256
Van Loon, H., 166, 175
Vestine, E. H., 248, 258–62
Vicente, R. O., 23, 88, 197, 201
Vincent, D. G., 173–4, 176–7, 197–80
Volterra, V., 195, 239

Wahl, E. W., 270
Wako, Y., 91
Wagner, C. A., 26
Walcott, R. K., 271

Walker, A., 88, 92, 103
Wallace, J. M., 176–8, 189–90
Walter, M. R., 383
Warburg, H. D., 116
Warburton, R. J., 112
Ward, C. G., 70, 73, 301–3, 317–18
Wasserburg, G. J., 392
Wasson, J. T., 6
Watanabe, H., 254
Webb, D. J., 331, 334
Webster, A. G., 32
Weertman, J., 14
Wegener, A., 343
Wells, F. J., 96, 100
Wells, J. W., 362–3, 365, 367–8, 370, 375
White, G., 111
Whitefield, R. P., 363
Whitten, C. A., 221
Williams, J. G., 105, 286, 317
Williams, K. P., 132, 301
Williamson, R. G., 132, 138
Wilson, C. R., 102–3, 148, 153, 156–9, 165, 170–1, 179, 234, 239–44
Wimbush, M., 124
Wise, D. V., 397
Wittman, A., 73
Woolard, E. W., 1, 32–3, 62, 300
Wu, M-F., 194
Wunsch, C., 123–4, 127, 149, 205, 211, 213–14, 218, 223, 331, 333–4

Yanai, M., 190
Yatskiv, Ya. S., 39, 90, 100
Yoder, C. F., 286, 296, 398
Young, A., 88, 92, 103
Yukutake, T., 248, 252–5, 260, 262–3, 289
Yumi, S., 65, 87–8, 90–1

Zahel, W., 125–9, 323, 327–9, 332
Zelter, B. D., 131

# SUBJECT INDEX

It is intended that this index serve as a reader's guide through the book rather than give a detailed classification of the contents. The principal subjects are listed under key-word headings. The further subdivision under these key-words will aid in locating specific aspects of the subject.


anelastic deformation, *see also Q*
  Chandler wobble period, 208
  complex modulus, 15
  dissipation mechanisms, 13–18
  frequency dependence, 15–17, 19–20, 198
astronomical observations, *see also* time
  accuracy, 75–7, 81–2, 88, 95
  ancient, 304–14
  errors due to catalogue, 82, 94, 187
  instrumentation, 64–6
  latitude, 63
lunar orbit evolution, 299–301
Moon, Sun and planets, 69–73
  telescope, 301–3
atmospheric circulation, *see* winds *and* zonal angular momentum
atmospheric excitation functions, *see also* atmospheric pressure
  Chandler wobble, 238–243
  long-period fluctuations in l.o.d., 193–4, 275–8
  seasonal changes in l.o.d., 173–9, 181–90
  seasonal changes in wobble, 150–9, 179–81
atmospheric pressure
  changes in inertia tensor, 152
  excitation of Chandler wobble, 238
  excitation functions, 150–9, 239, 276
  inverted barometer, 148–50
  long-period fluctuations, 279
  monsoon, 153
  ocean response 148–50
  seasonal fluctuations, 148
  water vapour pressure, 151
atmospheric tides
  ocean response, 131–2
pressure variations, 129
radiational tide, 131
celestial mechanics
  coordinate transformation, 113
  dynamics of lunar orbit, 290–9
  Lagrange equations, 133
  orbital elements, 113, 115
  tidal perturbations in satellite orbit, 134
  tidal perturbations in lunar orbit, 292, 294
centre of mass, seasonal shift, 87, 147
Chandler wobble, *see also* Eulerian precession
  atmospheric excitation, 238–43
  core effect on period, 197, 201–3
  discovery, 85–6, 195
  dissipation in core, 208–11
  dissipation in mantle, 206, 208
  dissipation in oceans (*see* pole tide)
  elastic mantle and period, 198–201
  energetics, 203–5
  Love number, 197, 219
  period, 196–8, 219
  *Q*, 19, 195, 205–6, 217
  resonance with annual wobble, 45–6
  seismic excitation, 220–3 (*see also* seismic excitation)
  spectrum, 97–104
  variation in amplitude and phase, 100, 219
core, *see also* magnetic field *and* core–mantle coupling
  Brunt–Vaisala frequency, 22
  density, 22
  dynamics, 23
  long-period oscillations, 23
  seismic structure, 21–2
  viscosity, 24, 209–10
core–mantle coupling,
  Chandler wobble, 201–2
  dissipative, 208

core–mantle coupling—*cont.*
electromagnetic and l.o.d., 59, 248–54
electromagnetic and wobble, 263–6
general formulation, 247
inertial or pressure, 38, 201–2, 246
observational evidence, 258–63
passive, 246
for seasonal oscillations, 180–1, 187, 192
topographic, 246, 266–8
viscous, 208–11, 247

decade fluctuations in l.o.d., *see also* core–mantle coupling
astronomical evidence, 173–80
atmospheric excitation, 193, 275–7
Brouwer's model, 78
Brown's model, 78
oceanic excitation, 276–8
relation with climate, 4, 278–81
seismic excitation, 281–3
solar wind, 284
spectral investigations, 79–80
diurnal wobble, 33, 39, 196
dynamics of rotation
angular momentum, 30, 33, 291, 353
Euler's equations, 30–1
Liouville equation, 34
non-rigid motion, 33–36
perturbation equations, 35
Poinsot representation, 33–4, 40
rigid body rotation, 30–33
separation of wobble and l.o.d., 37

Earth, *see also* core, mantle, gravity field
dimensions and shape, 27
early evolution, 391–3
free oscillation, 8, 18, 23
Kelvin model, 10
seismic model, 11
elastic deformations, *see also* Love numbers
boundary conditions, 8–9
equations of motion, 6
general solution, 7
load Love numbers, 10
energy dissipation in oceans, *see also* pole tide

energy dissipation in oceans—*cont.*
bottom friction, 218, 321–2, 220–1
breaking of waves, 331
energy equation, 320
energy flux, 322–3, 330
estimates of global dissipation, 327–29
internal tides, 331–4
*Q*, 331
tidal dissipation in past, 389–90
turbulent friction, 125, 129, 332
virtual internal friction, 125
work integral, 321, 324–6
Eulerian precession, *see also* Chandler wobble
deformable body, 42–4
dynamic equations, 31
frequency, 35
kinematic equations, 32
excitation functions, *see also under entries for specific mechanisms*
angular momentum versus torque, 48, 166–7
Cartesian and spherical coordinates, 48–9
comparison with astronomical data, 47, 60
definition, 35
modification of by mantle elasticity, 43
rotational deformations, 42
schematic, 52–60

gravity field
density anomalies, 9
departures from hydrostatic equilibrium, 27, 345–6
dynamical oblateness, 26, 27
moments of inertia, 25
spherical harmonics expansion, 25
Stokes potential coefficients, 25, 27
groundwater
balance, 160
conservation of water mass, 161
excitation functions, 162
variability, 242, 278

inertia tensor, 25–6, 30
polar moment, 27

inertia tensor—*cont.*
rotational deformation, 41
temporal variations, 29
tidal deformation, 107
international time and latitude services
Bureau International de l'Heure (BIH), 68, 80–2, 88
comparison of BIH and ILS, 88–90
International Latitude Service, (ILS), 36, 86–7
International Polar Motion Service (IPMS), 86, 90

Legendre polynomials
addition theorem, 113
definition, 25
normalisation, 26
products, 121
length-of-day (l.o.d.) *see also* tidal accelerations
accuracy, 75–7, 81
amplitude spectrum, 3, 77, 84
decade fluctuations, 73
definition, 63
high-frequency and irregular changes, 84–5
historical records, 2, 299–314
new observation techniques, 105
paleontological evidence, 388–9
seasonal fluctuations, 82, 166–70, 173–9
secular acceleration, 289, 337–9
tidal variations, 138–42
Love numbers
Chandler, 208, 219
complex, 21
definition, 10
fluid, 28–9
for elastic Earth, 12–13
frequency dependence, 18, 20
load, 10, 50
mantle structure, 12
Maxwell body, 21
observations, 111, 192
satellite observations, 135–6
secular, 29
tidal effective, 293
lunar orbit evolution
dynamics of evolution, 290–9

lunar orbit evolution—*cont.*
eccentricity, 336, 353
energetics, 297
evolution due to ocean tides, 293–4
future evolution, 287, 353
geological constraints, 391–4
inclination, 336, 354–7, 398–400
past evolution, 286, 352–7, 389–90
time-scale problem, 398

magnetic field
archeomagnetic, 254
at core–mantle boundary, 249, 251, 258
Eurasian anomaly, 256, 263
long-period fluctuations, 255
paleomagnetism, 346
reversals, 342
secular variation impulses, 255
solar cycle, 255–6
South African anomaly, 254
torroidal field, 249–59
mantle
anelasticity, 16–17
electrical conductivity, 254–8
seismic velocities, 12
*Q*, 18–19, 340–1
viscosity, 345, 351
meteorological excitation
atmospheric pressure, 150–9
centre of mass, 87, 147
comparison with l.o.d., 8, 181–90
comparison with wobble, 179–81
groundwater, 162
high frequency, 190–2
low frequency, 193–4
seasonal, variability of, 97, 185
Moon, *see also* lunar orbit
evolution, 392
formation, 394–8
origin, 394–98
*Q*, 296
tides, 295–6

new observation techniques
long-baseline interferometry, 106
lunar laser ranging, 105
satellite tracking, 104–5

nutation, *see also* Eulerian precession *and* diurnal wobble
constraint on core viscosity, 24
fluid core effects, 1, 37–9
motion of rotation axis in space, 32–3
observational evidence, 1

ocean–continent functions, 51–2, 121, 142, 164, 268
ocean current
Bering Sea, 330
Circum-Antarctic, 165
excitation function, 166
geostrophic, 164–5
tidal, 321–2, 332
ocean tide, *see also* pole tide
due to rotation change, 144
effect on rotation, 127–8, 140–2, 145, 185
equilibrium, 121–4, 138
frictional forces, 125
interaction with solid tide, 111
Laplace equations, 319
models, 124–129
near resonance, 129, 293, 334
*Q*, 331
satellite observations, 139
spherical harmonic expansion, 119–21

paleontological observations of l.o.d.
bivalve acceleration, 378–81
bivalve growth rhythms, 371–3
coral accelerations, 370–1
coral growth rhythms, 363–6
paleoaccelerations, 388–9
paleontological clocks, 358–62
stromatolites, 381–5
Precambrian stromatolites, 385–6
plate tectonics
effect on polar motion, 92–4
hypothesis, 347
relative motion, 347–50
separation from polar wander, 342
polar motion, *see also* Chandler wobble, meteorological excitation *and* polar wander
amplitude spectrum, 2, 90–2, 97–9
comparison of results, 89–90

polar motion—*cont.*
deconvolution, 61
early search, 85–6
errors in data, 95–6
impulse response functions, 60
Kimura term, 87, 147
low-frequency variations, 90–1, 94, 245, 273–5
observations, 86–9
paleomagnetic evidence, 346–7
secular, 2, 91, 94, 274–5, 282, 285
seasonal variations 3, 95–7
tectonic displacements of stations, 92–4
tidal variations, 138–142
polar wander
adjustment of equatorial bulge, 344–5
Gold's beetles, 343–4, 351
mantle viscosity, 345, 351
paleomagnetism, 346–51
Pleistocene deglaciation, 285
separation from continental drift, 342, 350–1
pole tide
atmospheric, 239
dissipation, 216–8
global representation, 142–144
oceanic, 212–4
seasonal, 108, 142
Chandler wobble period, 215
post-glacial rebound, 20, 285, 289
precession, *see also* Eulerian precession
constant, 27

*Q*, *see also* anelastic deformation
definition, 14
frequency dependence, 18–20, 206, 208, 219
linear damped system, 44
mantle, 340–1
Moon, 351
Maxwell and Kelvin–Voigt solids, 21
structure of Earth, 18–19

reference systems
inertial, 32
motion of stations, 92–4

reference systems—*cont.*
  relation between inertial and terrestrial, 32
  terrestrial, 36–7
rotational deformation
  adjustment of equatorial bulge, 344–4
  excitation functions, 42
  inertia tensor, 41
  potential of centrifugal force, 40
  radial deformation, 40

sea level, *see also* ocean tide *and* pole tide
  decade fluctuations, 274–5
  deduced from ice budget, 272–3, 278
  excitation function, 58, 163–4, 269–74
  exchange with groundwater, 161–2
  exchange with ice sheets, 268
  long-term variations, 270–2
  response to air pressure, 148–50
  sterric changes, 163–4
  water temperature, 278
seismic excitation
  aseismic contribution, 236–8, 242–3
  comparison with astronomical data, 233–4, 243
  cumulative, 223, 233–4
  earthquake parameters, 230, 235
  excitation function, 57
  l.o.d., 281–3
  moment-magnitude relation, 230–2, 235–8
  silent earthquakes, 237
  static displacement field, 229
  Tsunami earthquakes, 237–8
solar torque, 246, 284

tidal accelerations, *see also* tidal phase lag *and* energy dissipation in oceans
  astronomical evidence, 69–71, 301–3
  constancy of lunar acceleration, 338–340
  Earth's secular acceleration, 3, 69, 289–91, 314–8, 336–8
tidal accelerations—*cont.*
  historical evidence, 304–14
  lunar acceleration in longitude, 3, 17, 292, 300, 344–6
  non-tidal acceleration, 289
  paleontological evidence, 370–1, 378–81, 388–9
  satellite results, 335
tidal phase lag
  age of tide, 129
  correction for atmospheric tide, 132
  equivalent lags, 294–5
  gravity and tilt tide, 119
  satellite motion, 118, 137–8
tide, *see also* Love numbers, ocean tide *and* pole tide
  changes in rotation, 4, 67, 84, 107–8, 138–40, 145, 182
  deflections of vertical, 81, 108
  generating potential, 112, 116–18
  lag in response, 118–19
  loading of crust, 109
  observations, 111–12
  potential, 118–19
  relative importance of solid and ocean, 107
time
  atomic time, 68
  clocks, 66, 67
  ephemeris time, 68
  universal time, 62–3, 66, 73

wind, *see also* zonal angular momentum
  angular momentum versus torque, 166–7
  seasonal fluctuations in zonal wind, 172–3
  torques and l.o.d., 169–70
  torques and wobble, 170

zonal angular momentum
  annual, 173–4
  biennial, 175–8, 187–90
  semi-annual, 174
  variability of biennial, 179, 189–90

Printed in the United States
By Bookmasters